Genetic Data Analysis for Plant and Animal Breeding

Fikret Isik • James Holland • Christian Maltecca

Genetic Data Analysis for Plant and Animal Breeding

Fikret Isik
Department of Forestry and
   Environmental Resources
North Carolina State University
Raleigh, NC, USA

Christian Maltecca
Department of Animal Science
North Carolina State University
Raleigh, NC, USA

James Holland
United States Department of Agriculture – Agricultural
   Research Service
Department of Crop and Soil Sciences
North Carolina State University
Raleigh, NC, USA

ISBN 978-3-319-85586-8   ISBN 978-3-319-55177-7   (eBook)
DOI 10.1007/978-3-319-55177-7

This Springer imprint is published by Springer Nature
The registered company is Springer International Publishing AG
The registered company address is: Gewerbestrasse 11, 6330 Cham, Switzerland

To Onur, Paula, my parents Akbal and Sevket (Isik)
To Andrea, Pablo, Lucía, and my parents (Holland)
To my parents and family (Maltecca)

Plant and animal illustrations:

Professor Hüsnü Dokak
Director of Hacettepe University Art Museum
Hacettepe University, Faculty of Fine Arts
Ankara, Turkey

Preface

We wrote this book to fill the gap between textbooks of quantitative genetic theory and software manuals that provide details on analytical methods but little context or perspective on which methods may be most appropriate for particular applications. We do not cover the basics of quantitative genetics theory; we recommend readers be familiar with two of the classic introductory texts on the subject, *Introduction to Quantitative Genetics*, 4th Ed. by D.S. Falconer and Trudy Mackay, and *Genetics and Analysis of Quantitative Traits* by Michael Lynch and Bruce Walsh. We hope to apply the concepts of quantitative genetics to particular analytical solutions that will be useful to plant and animal breeders, focusing mainly on methods to predict breeding values. We attempt to demonstrate analyses in freely available software (such as R packages) where possible, but we also include considerable attention to the commercial software ASReml because it provides so much flexibility and utility to analysis of breeding program data. Free (but time-limited) trials of ASReml are available, and "Discovery" versions of the software are freely available to public institutions in many developing countries (http://www.vsni.co.uk/free-to-use/asreml-discovery). In addition, we include some information on SAS analyses for comparison, because SAS is widely used in the breeding community.

This book is composed of two major sections. The first section (Chaps. 1, 2, 3, 4, 5, 6, 7, and 8) covers the topic of classical phenotypic data analysis for prediction of breeding values in animal and plant breeding programs. In Chap. 1, we introduce ASReml software because it is one the more popular, and we believe one of the most powerful, softwares available for analyzing data in breeding programs using mixed models analyses. Chapter 2 includes a brief review of linear mixed models and compares them to ordinary least squares analyses of variance, with which some readers may be more familiar. This is followed by a general introduction to variance-covariance structures used in mixed models (Chap. 3). Chapters 4 and 5 cover prediction of breeding values using sire (or general combining ability) models and animal models. Chapter 6 is about multivariate models used when breeders want to analyze multiple traits simultaneously and estimate genetic correlations among traits. Chapter 7 introduces spatial analyses for field experimental designs used in tree and crop breeding to account for environmental heterogeneity *within* environments. Chapter 8 introduces genotype-by-environment (GE) interactions in multi-environmental trails and various variance-covariance structures to model GE and the heterogeneity of error variation *among* environments.

In the second section (Chaps. 9, 10, 11, and, 12), we provide the concepts and an overview of available tools for using DNA markers for predictions of genetic merit in breeding populations. With recent advancements in DNA sequencing technologies, genomic data, especially single nucleotide polymorphism (SNP) markers, have become widely available for animal and plant breeding programs in recent years. Analyses of DNA markers for prediction of genetic merit are a relatively new and very active research area, with new methods and improvements of older methods being proposed and tested constantly. The algorithms and software to implement these algorithms are changing as we speak. Therefore, Sect. 2 intends to be an introduction to the topic, touching on some of the more widely used methods and softwares currently available. Readers should be aware that the methods discussed here are likely to be modified and improved in the near future, and that new statistical packages will be introduced. We present this material, however, in the hopes of providing a solid grounding in the basics of handling large marker data sets and using them to predict breeding values. In Chap. 9, we describe characteristics of typical DNA marker data sets and introduce some software tools useful for exploratory analyses (visualization, summary, and data manipulation) of marker data. Chapter 10 focuses on imputation of missing genotypes. Chapter 11 covers the use of DNA markers to predict genomic relationships between individuals in breeding populations and the use of genomic best linear unbiased prediction (GBLUP) to predict breeding values even of individuals that have not been phenotyped. Chapter 12 reviews the statistical background of more advanced genomic selection methods with several examples.

This book is intended for students in plant or animal breeding courses and for professional breeders interested in using these tools and approaches in their breeding programs. We love to hear from users about suggestions for improvements and corrections to the text.

We tried our best to give credit to resources we used to write this book. Apologies if we missed something; please let us know so that we can include the source in a future edition. Many friends, colleagues, and graduate students helped with the writing, revising, and editing the original lecture notes and exercises, which turned out to be a huge task. We acknowledge the contributions of Greg Dutkowski, Salvador Gezan, Steve McKeand, Jérôme Bartholome, Trevor Walker, Jaime Zapata-Valenzuela, Funda Ogut, Kent Gray, YiJian Huang, Patrick Cumbie, Alfredo Farjat, Terrance Ye, Jeremy Howard, Francesco Tiezzi, Brian Cullis, Tori Batista Brooks, Austin Heine, April Meeks, Paula Barnes Cardinale, Onur Troy Isik, Amanda Lee, Mohammad Nasir Shalizi, Edwin Lauer, and Miroslav Zoric for reviewing drafts of chapters and providing feedback. Thiago Marino, Jason Brewer, Heather Manching, and Randy Wisser generously provided unpublished maize data to use as an example in Chap. 11. Christophe Plomion (INRA, France) provided maritime pine data to use in Chaps. 9, 10, 11, and 12. Tree Improvement Program at NC State University provided unpublished pine progeny test data to use in several chapters.

Our colleague Dr. Ross Whetten of NC State University edited Chaps. 1, 2, 3, 4, and 5 and Chap. 9. Ross helped developed the training workshops on which this book is based, he tested many scripts, and provided us with invaluable feedback. We are grateful to Ross.

We are also very grateful to Hüsnü Dokak, Professor of Arts at Hacettepe University, Ankara, Turkey, for sketches/drawings of animals and plants used in the chapters.

Raleigh, NC, USA   Fikret Isik
                                                                                                           James Holland
                                                                                                     Christian Maltecca

Software Requirements

Several programs were used in this book. If you do not already have the following programs installed on your computer, we recommend you download and install them.

ASReml: ASReml is a powerful tool for analysis of linear mixed models. Download from https://www.vsni.co.uk/downloads/asreml. Make sure you obtain a license from VSN, the company that distributes the program. See program website for details. It typically takes several days to a week from requesting to receiving a license. For starters, Luis Apiolaza's website about ASRem is an excellent source: http://uncronopio.org/ASReml/HomePage

ConTEXT: ConTEXT is a small, fast, and powerful freeware text editor for Windows, available at http://www.contexteditor.org/. We used it to write ASReml standalone command files and examine output.

R: Download R from http://cran.r-project.org/, choosing the Windows, Mac, or Linux version according to the OS on your computer. All R versions are free. For the exercises, you need to install several packages (and the other packages they depend on). After installing R, start the R program from the desktop shortcut and copy-paste the following R script into the R window to install the required packages.

```
is.installed <- function(mypkg) is.element(mypkg,
installed.packages()[,1])

source("http://bioconductor.org/biocLite.R")
packBIOC=list("GeneticsPed","chopsticks")

for(i in 1:length(packBIOC)){
if(!is.installed(packBIOC[[i]])){biocLite(packBIOC[[i]])}
cat(paste("----------",packBIOC[[i]],"----------",sep="\t"));cat("\n")
}
#-copy to here, paste, and let R finish before copying the rest -
packCRAN=list("MASS","pedigree","rrBLUP","BLR","multicore","plyr")

for(i in 1:length(packCRAN)){
mip=as.character(packCRAN[i])
if(!is.installed(packCRAN[[i]])){install.packages(mip,dependencies =
T)}

cat(paste("----------",packCRAN[[i]],"----------",sep="
\t"));cat("\n")
}
```

Example data sets and code scripts: All of the example codes shown in this book are available for download from this book website: https://faculty.cnr.ncsu.edu/fikretisik/breedingbook/. We recommend keeping all of the example data sets together in a common folder so the examples shown in this book can be run "as-is" except for changing the file paths to that folder.

Installing packages from local source: We used a set of scripts bundled in package for the genomic selection chapter. These are not loaded on CRAN and are made available to readers. The installation process is slightly different for Mac/Unix and Windows. Please do the installation of the package after you have run the small script above.

MAC/Unix: Place the package `GSa_1.0.tar.gz` on your desktop. Open a terminal and change directory to your desktop (in Mac this will be something like `cd/Users/"NAME"/ Desktop` while in Unix it will likely be `cd/home/"NAME"/Desktop`). Run the following command `R CMD INSTALL GSa_1.0.tar.gz`. Note that you can do the same from the GUI on a Mac but this is simpler.

Windows: Place the file `GSa_1.0.zip` on your desktop. If you have a 64bit machine R will install both versions. In R change directory to your desktop (you can use the buttons of the GUI to do so) then run the following line `install.packages("GSa_1.0.zip", repos=NULL)`.

Contents

About the Authors

Fikret Isik is a Professor of Quantitative Genetics and Breeding in the Department of Forestry and Environmental Resources and the Associate Director of the Tree Improvement Program at North Carolina State University.

James Holland is a research geneticist with the United States Department of Agriculture – Agriculture Research Service, and a Professor in the Department of Crop and Soil Sciences at North Carolina State University.

Christian Maltecca is an Associate Professor of Quantitative Genetics and Breeding in the Animal Science Department at North Carolina State University.

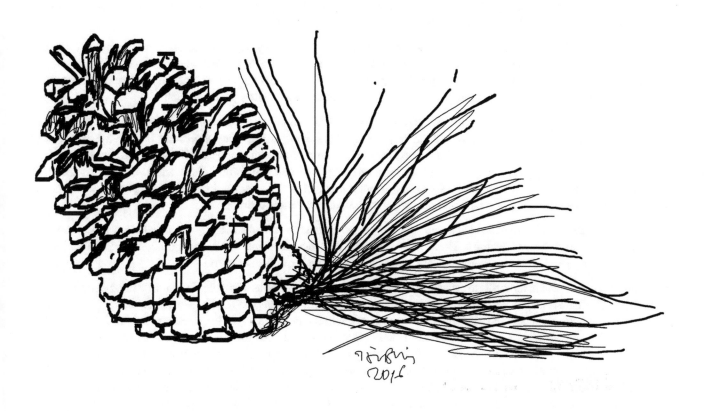

Electronic supplementary material: The online version of this chapter (doi:10.1007/978-3-319-55177-7_1) contains supplementary material, which is available to authorized users.

F. Isik et al., *Genetic Data Analysis for Plant and Animal Breeding*, DOI 10.1007/978-3-319-55177-7_1

Abstract

ASReml has been become a default software for analysis of linear mixed models. The Average Information Sparse Matrix algorithm of ASReml makes the software very fast to solve large number of mixed model equations. The software is flexible to fit complex variance structures in mixed models. We introduce ASReml stand alone and a brief introduction to ASReml-R in this chapter. Fitting more complex variance structures in mixed models using ASReml is given in Chaps. 2, 3, 4, 5, 6, 7, and 8.

Why ASReml?

Mixed models are commonly used to analyze many types of data produced by different disciplines. ASReml is a comprehensive software package developed for linear mixed model analysis. It uses Average Information and Sparse Matrix Algorithms to solve linear mixed model equations (Gilmour et al. 2014).

ASReml uses restricted maximum likelihood (REML) to estimate parameters. The method finds the parameter estimates **that are most likely** given the data, by maximizing the likelihood function $L(\boldsymbol{\beta}, \mathbf{V}|\mathbf{y})$. This function expresses the likelihood of a model (summarized by the components $\boldsymbol{\beta}$ (fixed effects) and $\mathbf{V}$ (variances and covariances of random effects) given a vector of observed data ($\mathbf{y}$). REML produces parameter estimates that are efficient and consistent.

ASReml was developed to solve large mixed model equations. ASReml is faster and computationally more efficient than SAS Proc Mixed (which relies on a Newton-Raphson algorithm) in solving mixed model equations. It is relatively easy to fit simple linear mixed models. It is also flexibly coded to analyze complex designs, such as diallels and multivariate models, and to handle pedigrees easily for quantitative genetics analyses. ASReml has certain disadvantages, primarily that it is specifically designed for mixed models analysis and is not a comprehensive statistical or data management tool. Also, it requires a good understanding of mixed models theory to fit complex variance structures.

ASReml software must be downloaded from: http://www.vsni.co.uk/downloads/asreml. The installation procedures are available from the download website. Users must also obtain a license from VSNI and install the license following instructions from VSNI. Free time-limited trial licenses are available for first-time users.

WinASReml is included in the ASReml installation for windows and can serve as a useful first start to writing command files (which have file names that end in '.AS') and manage ASReml projects (outputs). However, several text editor programs can also be used to write command files and view output files. We will focus on the use of the freeware ConTEXT editor in this book, as a nice text highlighter for ASReml command files has been developed for this software. Finally, ASReml models can be coded and fit, and output stored as objects in R using the ASReml-R package. We will introduce ASReml-R in this book, but readers should be aware that models are coded differently in ASReml-R than in the standalone ASReml. We believe it is simpler to focus on the standalone version, particularly as the models become more complex, as the coding for models tends to be more transparent in the standalone version. However, the ASReml-R package allows users to tightly integrate the mixed models analyses with data management, visualization, and other statistical analyses that can be performed in R, and this can provide a substantial benefit to users. Please note that the ASReml-R package must also be downloaded directly from VSNI, and it requires a separate installation from the standalone version, it is not available from the comprehensive R archive network (CRAN). Only a single valid license is required to run either (or both) standalone ASReml and ASReml-R for a given computer.

ASReml Release 4 introduced functional specification to define the variance structures for mixed models. In functional specification, the model random terms and residuals are wrapped with the variance functions. In other words, they are not defined after the model, as was the case in Release 3. The goal is to simplify the coding of variance structures and make the syntax less error-prone. This model specification is similar to ASReml- R.

ASReml Workflow

A typical standalone ASReml workflow is shown in Fig. 1.1. The user has a data file (typically in comma separated values, *csv*, format) and prepares a command file (with extension .as) in a text editor. The command file references the data file, instructs ASReml how to name and handle the columns of data in the data file, and includes a linear model that should be fit

Fig. 1.1 Typical standalone
ASReml workflow

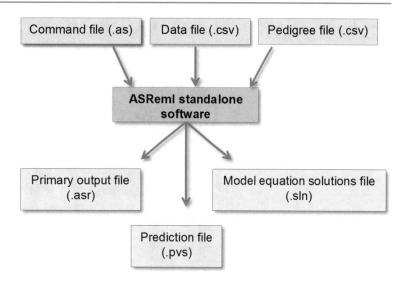

to the data. This command file is submitted to ASReml. ASReml produces a number of output files resulting from the analysis, including a results summary (*.asr* file), solutions to the mixed model equation effects (*.sln* file), and if requested, predictions for some factors in the model (.pvs file).

Setting Up ConTEXT Editor to Create and Execute ASReml Command Files

For Windows environments, we recommend using the ConTEXT text editor to write ASReml command files. It is free and supports multiple document interfaces; it can be downloaded from http://www.contexteditor.org/. A highlighter file that ConText uses to highlight ASReml command files to help scripting and editing is provided with the ASReml software itself. As of ASReml version 4, the highlighter file is named 'ASReml.chl' and is located in the 'Context' subfolder of the 'ASReml4' folder that is normally installed under 'Program Files' on Windows computers. Users can copy the 'ASReml.chl' file to the folder containing ConText highlighters (usually, 'C:\Program Files\ConTEXT\Highlighters').

Here is how you can set up ConTEXT text editor to run ASReml job files:

1. Open ConTEXT text editor.
2. Click on *Options* and then *Environment Options* to bring up a dialog box
3. Click on *Execute keys*
4. Click on *Add*
5. Click on *F9* key
6. Type file name extensions *as, asd, csv, txt, dat, tab* etc... in the *Extension edit* box (Fig. 1.2)

The above setup will create ASReml job templates when the data file is open in ConText (hit the F9 key or hit the icon 1 in ConText). Similarly, ASReml job file (.as) can be run when the job file is open. We can also setup ConText to run another type of file, called *.pin* file to calculate functions of variance component e.g. heritability. The *.pin* file is prepared by the user. We will see examples later in the chapter. Here is how we setup ConText text editor to run the *.pin* files (Figs. 1.3 and 1.4).

Starting with ASReml

Data should be in ASCII text format (.txt, .csv, .asd, .dat, etc...). MS Excel spreadsheets cannot be used directly for analysis. Instead, if you have data in an Excel file, save the file using the comma separated values (.csv) format.

Example 1.1 Pine provenance-progeny data
A provenance-progeny test of a pine species was established. Provenance refers to the different seed sources of a forest tree species adapted to different environments. They are like genetic groups (or possibly sub-populations of the same species).

Fig. 1.2 Setting environment options in ConTEXT to run ASReml

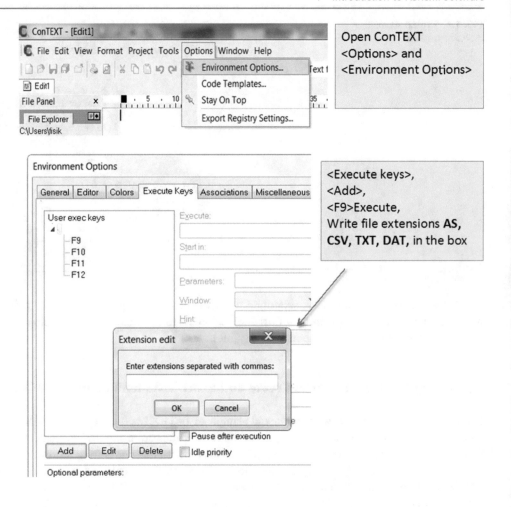

Fig. 1.3 Setting environment options in ConTEXT to run >PIN file in ASReml

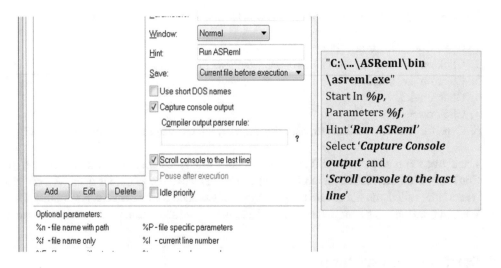

The maternal parent of each tree was known, but the seeds were derived from open-pollination (random outcrossing) of those female parents, so the male parents of each tree were not known. This represents a half-sib mating design within each provenance. There are four provenances and 36 half-sib families in total. The field experimental design was a randomized complete block design with five blocks and from two to six trees measured within each plot. Data are given for each tree. The first five lines of the data set (*Pine_provenance.csv*) are given below:

Fig. 1.4 Setting up environment options to run PIN files

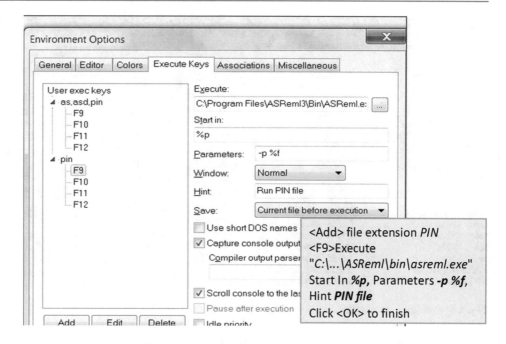

treeid	female	male	prov	block	plot	height	diameter	volume
191.1	191	0	10	1	1	10.7	15.0	0.072
191.2	191	0	10	1	1	11.5	22.0	0.167
191.3	191	0	10	1	1	12.1	23.8	0.206
191.4	191	0	10	1	1	12.0	22.7	0.186
191.5	191	0	10	1	1	12.2	21.5	0.169
191.6	191	0	10	1	1	12.5	22.8	0.195

treeid	A unique tree ID. There are 914 levels
female	Mother tree. There are 36 levels
male	Pollen parent or father tree (they are all unknown, so they are all coded as 0)
prov	Provenance number. There are four provenances
block	Block number. There are five blocks
plot	Plot (experimental unit) number
height	Covariate or measured trait
diameter	Covariate or measured trait
volume	Covariate or measured trait

Template job file

You may open a data file (*csv, txt, dat, tab* etc.) in ConTEXT and press F9 to generate a template job file (".as" suffix). If ConTEXT is setup correctly as described above, it can generate a template job file directly from the data files. The new file will take the name of the data file (for example, Pine_provenance.csv will create a file called Pine_provenance.as). Note that ASReml will only generate a template .as command file if there is not already a file in the directory that has the same .as file name that it would have generated (so if Pine_provenance.as already exists in the directory, then executing Pine_provenance.csv will not overwrite the existing command file, it will run the commands in that file). Alternatively, you may simply start from scratch by providing an appropriate header line and typing the name of fields in the ConTEXT editor *following the exact order* in the data file.

Let's look at the template command file for the *Pine_provenance.csv* data (Fig. 1.5).

Now open the job command template *.as* file in ConTEXT and inspect the beginning of the file. The command file has some header lines then 'data field definitions' that names the variables in the data file and their formats (e.g., numeric or alphanumeric), followed by a linear model. ASReml usually does not interpret all of the data formats correctly, and the template linear model is a useful placeholder but often a nonsensical model. In this example, the file created is named

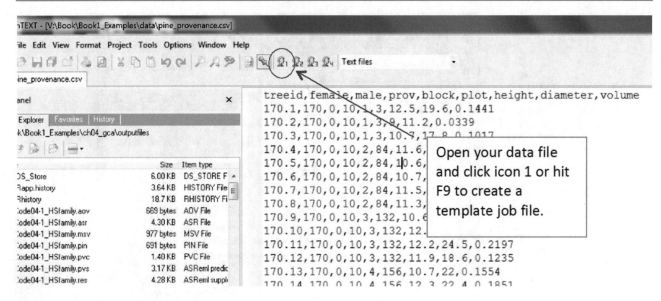

Fig. 1.5 Creating a job file (.AS) and correcting fields in the file

Pine_provenance.as. You will note that there are a few incomplete or wrong fields for the data set. You may also note some small variation in the format of.as command files generated in this way:

Uncorrected (template) ASReml job file example (*Pine_provenance.as*)

```
# !WORKSPACE 100 !RENAME !ARGS // !DOPART $1
Title: Pine_provenance.
#treeid,female,male,prov,block,plot,height,diameter,volume
#191.1,191,0,10,1,1,10.7,15,0.0722
#191.2,191,0,10,1,1,11.5,22,0.167
#191.3,191,0,10,1,1,12.1,23.8,0.2056
#191.4,191,0,10,1,1,12,22.7,0.1855
  treeid  !I      # 191_1_1_4
  female  !I      # 191
  male    *       # 0
  prov    !I      # 10
  block   *       # 1
  plot    *       # 1
  height  !I      # 12
  diameter        # 22.7
  volume          # 0.1855
# Check/correct these field definition
Pine_provenance.csv   !SKIP 1
height  ~ mu ,      # Specify fixed model
       !r           # Specify random model
residual units
```

The lines starting with # are comments that are ignored by ASReml.

Here are three fields that are not correct:

- The *treeid* field is alphanumeric. Qualifier **!I** (for integer variables) is not correct. It must be **!A** (for alphanumeric)
- *Height* is not an integer factor but a response variable. Remove **!I**
- The linear mixed model is not complete.

Here we show a corrected command file (provided in example scripts as *Pine_provenance.as*) for the *Pine_provenance.csv* data and the definitions of the fields.

Code example 1.1
Example of a corrected ASReml job file with simple linear model (Code 1-1_*Corrected Pine_provenance.as*)

```
# !WORKSPACE 100 !RENAME !ARGS // !DOPART $1
Title: Pine_provenance. Corrected job file
#treeid,female,male,prov,block,plot,height,diameter,volume
#191.1,191,0,10,1,1,10.7,15,0.0722
#191.2,191,0,10,1,1,11.5,22,0.167
#191.3,191,0,10,1,1,12.1,23.8,0.2056
#191.4,191,0,10,1,1,12,22.7,0.1855
 treeid    !A
 female    !I
 male      *
 prov      !I
 block     *
 plot      *
 height
 diameter
 volume
Pine_provenance.csv    !SKIP 1

height   ~ mu prov,     # Specify fixed model
         !r  female block block.female # Specify random model
residual units
```

The command (.as) file has specific formatting features that must be adhered to:

- There must be a title at the top of the file.
- For ASReml versions before 4, there must be a blank field (empty space) before each 'data field definition' line (the names of the data file columns, e.g., treeid, female, and male). This is no longer a requirement in version 4.
- Column names ('fields') can be followed by a qualifier (called a 'field_type') that indicates the type of data in that column. See the section "Data Field Definitions" that follows for details.
- Fields without a qualifier and those with decimal points in the data are taken as variates or covariates (e.g., height, diameter, volume).
- Names are case sensitive. *Height* and *height* are not the same.
- The data file name (**Pine_provenance.csv**) and model formula (**height~**...) that follow the data field definitions *should not* be preceded by spaces (no space).
- A linear model is required. The trait to be analyzed is given first in the model, followed by '~' and the model terms. An overall mean (intercept) is not fit in the model by default, so it should be given as 'mu' if desired (this is the opposite of default model specification in SAS Proc GLM, SAS Proc Mixed, or the lm() linear model fitting function in R). Fixed effect terms follow, then random model terms are given following **!r**. Although specification of the residual variance structure is not required, the template model specifies a form of the residual variance: "residual units" (the meaning of "units" will be introduced in Chap. 7). If the fixed and random parts of the linear model are written on multiple lines, commas must be used to indicate that the model specification continues on the following line. So, the template model has a comma after the specification of the fixed effects, and before the line with the random effects, but notice that there is no comma between the line with random effects and the line specifying the residual variance structure.
- In this simple example, we want to partition the variation into fixed sources due to field seed provenances and also into random sources due to female parents and experimental units block and the plot effect (block.female interaction). In this example, female parents are unique for each provenance, so they are 'nested in' provenance. However, since each female is given a unique identifier that is not repeated in any other provenance, we do not need to code the model term explicitly as nested.

```
height   ~ mu prov,       # Specify fixed model
          !r female.prov , # random effects
             block ,
             block.female.prov
residual units
```

- Our example fits the linear model: $Y_{ijkl} = \mu + B_i + P_j + F_{jk} + BF_{ijk} + \varepsilon_{ijkl}$, where Y is the measured tree height data, P_j is the fixed effect of the jth provenance, F_{jk} is the random effect of the kth female parent nested in the jth provenance, B_i is the random effect of the ith field block, BF_{ijk} is the random plot effect or female by block interaction and ε_{ijk} is the random residual error effect of the lth tree from female parent jk in the ith block.
- The data file name can be given as if it is in the same folder as the job command file (.AS), but if it is in a different directory, the path must be given. We can use the **!FOLDER** qualifier to specify the location of the data file or other input files (e.g. pedigree file) The **!FOLDER** qualifier is usually placed on a separate line BEFORE the data filename line or pedigree file name. Example;

```
   ...
   diameter
   volume

!FOLDER V:\Book\Book1_Examples\data       # data file location
Pine_provenance.csv  !SKIP 1

height   ~ mu prov,       # Specify fixed model
          !r  female block block.female   # Specify random model
residual units
```

Model building is an iterative process, so it is best to start by running a simple model like this example, reviewing its output (.asr), and then proceeding by adding more terms to the model one at a time.

Data Field Definitions

Following the title are the data field definitions, these correspond to the columns of the data file, and their order must correspond to the order of the variable columns in the data file.

General syntax: SPACE *label [field type]*

- You must have at least one whitespace before the data field, the whitespace is what distinguishes a data field definition from other parts of the ASReml job program. (ASReml version 4 has relaxed this requirement, but we recommend its use anyway to make the job scripts easier to read).
- *Label* identifies the field and must start with a letter. Note that ASReml is case sensitive (unlike SAS, for example). So, if you label a field as "Block" it must be spelled as "Block" in the model statements to follow; "block or BLOCK" would not be recognized as the same field.
- *Field type* defines how a variable is interpreted as it is read and whether it is regarded as a factor, covariate, or response variable in the linear model
- For the *factor* variables use *,!I, !A or the number of levels, depending on the field type.
- For *covariates* or *response* variables leave the field type blank or specify **1**
- You may put more than one data field definition on a single line. For example, the data field definitions for **height, diameter,** and **volume** can be on the same line in the example above.

Here are a few frequently used *field types*:

* or *n*	Are used when the data field has sequential values 1,...,*n* directly coding for the factor. For Example, if plots are coded as 1 through 12 in the data set, we would use **plot *** or **plot 12**, meaning that the levels of field plot range from 1 to 12.
!A	Means the field is a factor and coded as alphanumeric. For example, if we had 20 Sites coded as "NC11", "FL09", etc., we would use: **Site !A**. You may give the number of levels (20) after **!A** (e.g. !A 20) but this is not required if the levels are not greater than 1,000. If the levels are greater than 1,000 then an approximate number (greater than 1000) must be listed after !A.
!I	The field is a numeric factor but not sequential integer. For example, if we had blocks coded 11, 24, 49, 80, we would use **block ! I**. The qualifier **!I** must be followed by *n* if more than 1,000 levels are present. For example if there are 1,200 unique blocks, it must be written as **block !I 1200**.
!CSKIP *c*	is used to skip data fields (columns) when reading the data. For example, in the command file the *treeid* field is commented out by # and will not be read from data because **CSKIP 1** is used after the subsequent data field as shown below. This is useful if you want ASReml to skip the data fields that are not needed for analysis.

```
# !WORKSPACE 100 !RENAME !ARGS // !DOPART $1
Title: Pine_provenance. Corrected job file
#treeid,female,male,prov,block,plot,height,diameter,volume
# treeid    !A
   !CSKIP 1
   female   !I
   male *
   ...
```

!LL *c*	By default ASReml truncates field labels (level names) to 16 characters. If some labels are longer than 16 characters and it is important to have them in the output files, you may provide extra space for these fields using **!LL c**. In example below, let's say the **treeid** labels are longer than 16 characters. We can use **!LL 24** to accommodate the long labels. Even with the use of **!LL 24**, only up to 20 characters will be printed in the prediction output file (.pvs)

```
Title: Pine_provenance
#treeid,female,male,prov,block,plot,height,diameter,volume
   treeid !A  !LL 24 #treeid labels are longer than 16 characters
   female !I
   male   *
   ...
```

!SORT	Declared after **!A** or **!I** on a field definition line. It will cause ASReml to sort the levels so that labels occur in alphabetic/numeric order for the analysis. ASReml by default orders the factors by the order in which they appear in the data set. For example, if there are two locations with names 'South' and 'North' and South appears first, then ASReml keeps the order as 1 = South, 2 = North. If we use **!SORT** qualifier, ASReml will sort the field in alphabetical order (1 = North, 2 = South) and fits the effect in alphabetical order to form the design matrices in mixed models. Example: **location !A !SORT**
!PRUNE	If fewer levels are actually present in the factor than were declared, ASReml will reduce the factor size to the actual number of levels. Let's say there are 60 females in the data but we declared 100 levels after the **female** field. The program will prune the number of levels to 60 for *female* effect and the extra levels will not appear in the. SLN file. Example: **female !A 100 !PRUNE**

Transformation of Response Variables

Many data transformation functions are available directly in ASReml, and they are invoked as the fields are declared. See ASRreml manual for the complete list. Here are some examples of field definitions that transform a variable called height (Gilmour et al. 2014):

`height !M0`	Changes the height data points entered as 0 to missing values
`height !M<=0 !M>100`	Converts height data values of = <0 and >100 to missing
`height !^0`	Takes natural logarithms of the height values
`height !D<=0`	Deletes records which have 0, negative, or missing values in the field
`!+,!-, !*, !/`	Arithmetic operations. For example, `height !/10`: divides height by 10, `height !*5`: Multiplies height by 5
`Logheight !=height !^0`	Takes the natural log of height and creates a new variable, `Logheight`

In the following example, a new variable (named *Logheight*) is created by taking the logarithmic transformation of height. Also volume is multiplied by 10 to change the scale.

```
Title: Pine_provenance.
#treeid,female,male,prov,block,plot,height,diameter,volume
 treeid  !A
 female !I
 male *
 prov    !I    block *    plot  *
 height
 Logheight !=height !^0
 diameter
 volume   !*10

!FOLDER V:\Book\Book1_Examples\data    # data file location
Pine_provenance.csv   !SKIP 1

height  ~ mu prov ,    # Specify fixed model
         !r  female   block block.female # Specify random model
        residual units
```

It is usually more convenient or flexible to use a spreadsheet, R, or SAS to do transformations or data manipulations rather than using ASReml.

Data File and Job Control Qualifiers

Data file qualifiers

Following the field definition lines, we include a line indicating the name of the **data file**. ASReml recognizes the data file line because it is the first line following the field definition line, which is not indented. If the data file is in the same folder as the .AS program, we do not need to specify the data file path or put the file name in quotes. If not, however, we must provide the full path of the data file name in quotes, e.g., "**C:\users\document\Data.csv**".

On the same line, following the data file name we can include some qualifiers for how the data file is handled. Some important ones are:

`!SKIP n`	Causes the first *n* records of the data file to be ignored. Typically these lines contain headings (labels) for the data fields. For example, `Pine_provenance.csv` `!SKIP 1` skips the first line (column names) in the data.
`!FILTER v [ !SELECT n]` `[!EXCLUDE n]`	Enables a subset of the data to be analyzed; *v* is the number or the name of a data field. For example *block*.
`!FILTER block !SELECT 2`	Selects data only from block 2 to include in the analysis. If the block IDs are string of characters and numbers such as BL01, then you should put the selected block ID in double quotes as "BL01".

```
Title: Pine_provenance.
 treeid  !A    female !I    male  *    prov  !I    block  *    plot  *
 height
 diameter
 volume

!FOLDER V:\Book\Book1_Examples\data    # data file location
Pine_provenance.csv   !SKIP 1   !FILTER block !SELECT 2

height   ~ mu prov ,    # Specify fixed model
         !r   female block block.female  # Specify random model
```

!EXCLUDE n   Records of the level are ignored. In the following example records of block 2 are ignored (not included in the analysis). **!FILTER block !EXCLUDE 2**. NOTE: If **!FILTER** is specified but **!SELECT** and **!EXCLUDE** are omitted, records with zero in field *v* are excluded from analysis.

```
Title: Pine_provenance.
 treeid  !A    female !I    male  *    prov  !I    block  *    plot  *
 height
 diameter
 volume

!FOLDER V:\Book\Book1_Examples\data    # data file location
Pine_provenance.csv   !SKIP 1   !FILTER block !EXCLUDE 2

height   ~ mu prov ,    # Specify fixed model
         !r   female block block.female # Specify random model
```

!SUM   Provides a summary description of the variables in the data. It provides counts for levels of the factors, overall statistics (mean, standard deviation, minimum, maximum) of response variables and correlations between variables. The output is saved in a file with .ass extension. We strongly recommend users to check the file to make sure that the input data are correct. Here is an example of part of an .ass file

---

```
Title. Code example data file qualifiers     16 Jun 2015 13:47:20

------> treeid has     914 levels
Distribution of frequencies of cell counts
    914 level(s) with 1 observations

------> female has     36 levels
```

Level:	1	2	3	4	5	6	7	8	9	10
Count:	29	25	23	26	25	25	27	25	27	24
Level:	11	12	13	14	15	16	17	18	19	20
Count:	27	30	26	21	23	25	24	22	22	23
Level:	21	22	23	24	25	26	27	28	29	30
Count:	27	23	27	25	24	25	26	24	26	24
Level:	31	32	33	34	35	36				
Count:	27	27	29	27	28	26				

```
------> male has constant value  1.0000

------> prov has        4 levels
Level:       1       2       3       4
Count:      77     306     267     264

------> block has       5 levels
Level:       1       2       3       4       5
Count:     192     187     188     179     168

...

------> volume

Histogram of volume: MaxFreq 67, Range  0.30000E-02 0.42930
          *
         **
         **
        ***   *   *
        **** **  *
      **********
     ************ *
     ************ * *
    **************** **
    ********************
   *********************
   **********************
   ***********************
   **************************
   *******************************    *    **

Minimum  0.30000E-02 (ignoring    0 zeros)
Mean      0.11809     Standard Deviation 0.66110E-01
Maximum  0.42930              914 observations distributed
   16   19   29   40   43   42   51   67   64   50   45   53   47   45   51
   30   36   24   35   29   21   13   14    8    8    9    5    5    5    3
    1    3    0    0    0    0    0    1    0    0    0    1    1

Correlations and counts of volume with
 height         diameter
  0.84505   914   0.95625   914
```

Job qualifiers

Job control qualifiers and arguments that affect the whole program are listed after the data file. An example with some frequently used job control qualifiers is shown below:

```
Pine_provenance.csv    !DOPATH 1 !CONTINUE [f] !DDF 1
                       !DISPLAY 3 !MAXIT 15
                       !X volume !Y height
                       !X height (!Y volume) !G block
```

!DOPATH n or **DOPART n**	Allows one to write multiple models in the job file and run models selectively without editing the .as job file. **!DOPATH n** and **!DOPART n** are used interactively. Which part is run controlled by the **n** argument of the **!DOPART/!DOPATH**. The **n** can instructs ASReml to run the part. See the section "Processing Multiple Analyses with One Command File" later in this chapter for details.
!DDF i	Requests computation of the approximate denominator degrees of freedom (df) according to (Kenward and Roger 1997, 2009) for the testing of fixed effects terms. There are three options for i: $i = -1$ suppresses computation, $i = 1$ uses numerical methods, and $i = 2$ (the default) uses algebraic methods. If testing fixed effects is not an important goal, consider suppressing the computation of denominator df, as their computation may significantly increase computational demands.
!MAXIT n	Sets the maximum number of iterations. The default is 10. ASReml iterates for n iterations unless convergence is achieved first. Convergence is presumed when the REML log-likelihood changes less than 0.002* current iteration number and the individual variance parameter estimates change less than 1%. If convergence is not reached you may use the **!CONTINUE** qualifier to start from the final parameter estimates and run the job again.
!DISPLAY n	Used to determine which graphical displays are produced. $n = 1$ for a variogram for spatial analyses, $n = 2$ for a trait histogram, $n = 4$ shows row and column trends in spatial analysis $n = 8$ shows perspective plots of residuals. You may use n to represent the sums of codes for desired graphics. For example, $n = 3$ produces both a variogram and a histogram.
!X volume **!Y** height	Scatter plot of height and volume.
!X volume	Histogram of volume
!X volume **(!Y** height**)** **!G block**	Histogram of height and volume for each block.
!NODISPLAY	Used to suppress graphics display.

Code example 1.2
Example of using data file qualifiers (*Code 1-2_Data file qualifiers.as*).

```
Title: Pine_provenance. Data file qualifiers
  treeid  !A   female !I   male *   prov  !I   block *   plot *
  height
  diameter
  volume  !*10


!FOLDER V:\Book\Book1_Examples\data      # data file location

Pine_provenance.csv !SKIP 1   !SUM   !MAXIT 15   !DISPLAY 3   !X height
!CONTINUE !TSV   !DOPATH 2     !X height !Y volume

!PATH 1
height   ~ mu prov,      # Specify fixed model
         !r  female block block.female # Specify random model

!PATH 2
height   ~ mu prov,      # Specify fixed model
         !r female.prov ,
            block
            block.female.prov # Specify random model
```

!OUTFOLDER [*folder path*]	Writes most of the output files in a folder other than the working directory. It must be placed at top of the job file. Other arguments (!ARGS, !RENAME etc.) can be listed before the !OUTFOLDER qualifiers.

```
!OUTFOLDER V:\Book\Book1_Examples\ch01_asreml/outputfiles
Title: Pine_provenance. Data file qualifiers
 treeid  !A    female !I   male *   prov !I   block *   plot *
 height
 diameter
 volume  !*10

!FOLDER V:\Book\Book1_Examples\data        # data file location
Pine_provenance.csv !SKIP 1

height  ~ mu prov ,       # Specify fixed model
          !r  female  block block.female  # Specify random model
```

Specifying Terms in the Linear Model

Before you can write an appropriate model for analysis using ASReml (or any software!) you need to understand the mating design (how individuals are related), the treatment design (if external factors were applied to the experimental units), and the experimental design (such as the field blocking design in plant studies) of the experiment. In addition, you need to decide which design factors are fixed effects and which are random. This decision depends on the inference that a researcher will make from the results. Fixed and random factors are discussed in more detail in Chap. 2.

Reserved terms
Do not use the reserved terms given below to name your data fields. They are reserved, meaning that ASReml always interprets them in a particular way as parts of a linear model.

mu the model intercept or the mean
mv missing value estimates
units an extra residual used in some spatial models, described in Chap. 7

If model continues on multiple lines make sure to put comma at the end of non-final lines. Otherwise the programs stops and gives an error. If specifying the residual term, this should occur on a separate line after the model definition, and no comma should be used in the final line of the model specification. For example:

```
height ~ mu ,
        !r family
        residual idv(units)
```

Model terms
General rules:

- Response variable(s) is followed by tilde (~)
- Model terms (factors) are separated by spaces
- Fixed terms are listed directly after ~
- Random terms are listed after !r
- The default variance structure for a random term and for the residual structure is **idv** (homogeneous or identical variances). It can be specified explicitly at the end of the model (new line) but this is not necessary. The following two models are identical.
  ```
  MODEL 1: height ~ mu  !r female  block
  MODEL 2: height ~ mu  !r idv(female)  block
          residual idv(units)
  ```
- The + sign can be used to separate model terms, except at the end of the line. If the model continuous on several lines and if the + sign is used, then the lines after the first line start with a + sign. For example:

```
height ~ mu ,
    + prov ,
    + !r idv(female) block
```

Model formulas use the syntax described below, using the example of 'family', 'site', 'block', and 'prov' as factor variables, 'volume' as a numeric covariate, and 'height' as the response variable:

height ~ mu + family
> **residual idv(units)**
> Fits the overall mean plus a fixed family effect: $Y = \mu + family + \varepsilon$. ε is the residual from the fitted model. Make sure the residual term starts on a new line.

height ~ mu !r family
> **residual idv(units)**
> Fits the overall mean plus a random family effect: $Y = \mu + family + \varepsilon$. Now we assume family effects are drawn from a normal distribution: family $\sim N\left(0, \sigma^2_{family}\right)$.

height ~ mu + volume
> Fits the overall mean plus the effect of a fixed linear regression coefficient (β) that relates changes in height to changes in volume: $Y = \mu + (\beta^* volume) + \varepsilon$.

height ~ mu + site !r family family.site
> Fits a mixed model with site as a fixed effect and family and family-by-site interaction as random: $Y = \mu + site + family + family \times site + \varepsilon$, where family $\sim N(0, \sigma^2_{family})$ and family $\times$ site $\sim N(0, \sigma^2_{family \times site})$.

height ~ mu + site*family
> Fits the main effects of both site and family plus their interaction as fixed effects: $Y = \mu + site + family + family \times site + \varepsilon$. This is equivalent to:
> **height ~ mu + site family family.site.**

height ~ mu + prov/family
> Fits the main effect of prov and the effect of family nested within prov as fixed effects: $Y = \mu + prov + family (prov) + \varepsilon$. This is equivalent to:
> **height ~ mu + prov + prov.family.**

height ~ mu + site !r at(site).block family
> Fits the main effect of site as fixed, a random family effect and **at(site).block** specifies a unique variance component for block at each site: $Y = \mu + site + block(site) + family + \varepsilon$, where family $\sim N(0, \sigma^2_{family})$ and block $\sim N(0, \sigma^2_{block-sitei})$, i.e. there is a different block variance for each site i.

height ~ mu + site !r at(site,1).block family
> Fits the main effect of site as fixed, a random family effect and **at(site,i).block** specifies a random block effect only within the ith level of the site factor, in this example only at the first site.

Variance Header Line and Random Model Terms

Before ASReml *Release 4*, variance specification for models beyond simple *idv* structure (identical or homogenous) required a *variance header line*, followed by specification of variance structures for the residuals and random effects. The syntax could be complicated and hard to learn for beginners. With the new syntax for specifying variance structures in ASReml Release 4, the variance header line and specifications of residual variance and variance for random effects are not needed anymore. However, they can still be used with version 4, and we include the following sections for users who are familiar with the previous releases to help with the transition to Release 4. New users can skip this section.

By default, if a factor is specified as a random term in the model, the distribution of the factor level effects is assumed to follow a simple variance-covariance matrix structure whereby each level effect is a random sample from a common normal distribution described by a variance component $\left(\sigma^2_i\right)$ with no covariances between effects. This is what the notation family $\sim N\left(0, \sigma^2_{family}\right)$ means in the examples in the previous section.

In some cases (many of them very important to breeding applications), we may wish to specify more complex variance-covariance matrices for some random terms in the model (so-called "**G** structures") or for the residual effects (the "**R** structure"). This is done in ASReml by including a variance header line below the model formula.

We provide here two small examples to introduce the variance header line and variance component structure definitions. These examples describe a more verbose way to specify the default variance component structures. There is no need to do this for default structures, but this syntax is required for more complex structures to be encountered later in the book, so we show these examples as an introduction to the syntax. We will delve into the details of such structures and their use in ASReml analyses starting in Chap. 3.

The variance header line follows the model as shown below.

```
height   ~ mu prov  !r  female  block block.female
1 1 0        # variance header line, s = 1, c = 1, g = 0
```

The variance header line consists of three numbers:

---

s c g

s   Refers to the number of sections for the residual variance structure (**R** structure). Each section of the **R** structure is independent of the others and may have a unique variance component associated with it. In multiple environmental trials, each site could correspond to a section.

c   Refers to the number of sub-matrices that define each section of the R structure. This can be 1, in which case each section of the **R** structure is defined by a single matrix, or it can be 2, in which case the **R** structure for a section is the direct product of two sub-matrices. Direct products are introduced in Chap. 3 and their use in defining complex **R** structures is shown in Chap. 7.

g   Refers to the number of random model terms whose **G** structures will be specified in the subsequent code.

If $s > 0$, then the **R** structure definition lines should follow the variance header line first, then the **G** structure definitions are given. If $s = 0$, then c also must equal 0, and the **G** structure definitions would immediately follow the variance header line.

Here is an example where we specify the R structure for Code example 1.1. This will produce exactly the same model fit and results as Code example 1.1 but uses the variance header line to indicate that the residual structure consists of only one section and it is composed of an identity matrix (I) multiplied by a common residual error variance component. In statistical notation, $\varepsilon_i \sim N\left(0, I\sigma_\varepsilon^2\right)$.

Code example 1.3
Specifying the residual and random effects variance-covariance structures with the variance header line
(Code 1-3_Variance header line.as)

```
!OUTFOLDER V:\Book\Book1_Examples\ch01_asreml\outputfiles
Title: Pine_provenance. Variance header line
 treeid !A   female !I   male  *   prov !I   block *  plot *
 height    diameter   volume !*10

!FOLDER V:\Book\Book1_Examples\data
Pine_provenance.csv !SKIP 1   !DOPATH 1

!PATH 1  # 1 R structure, G is not specified
height  ~ mu prov  !r  idv(female) block female.block
1 1 0         #variance header line, s = 1, c = 1, g = 0
treeid 0 IDV !s2=2.7 #R structure definition

!PATH 2  # 1 R structure, G is not specified
height ~ mu prov   !r female*block
   residual idv(units)
```

In PATH 1, we use the variance header line '**1 1 0**' to specify that there is one **R** structure, which is defined by a single matrix, and we are not specifying any special **G** structures. The **R** structure definition line '**treeid 0 IDV**' has three components:

treeid	Specifies the factor whose levels uniquely identify the residual effects. In Code example 1.3, residual effects are associated with levels of '**treeid**'.
0	Specifies the number of error effects in the section of the R structure. We could put 914 here in Code example 1.3, but 0 is handy because it tells ASReml that the number of error effects equals the number of levels in the factor specified in the first term of the R structure definition ('treeid' in this example), so we don't have to know the exact number of levels when writing the code.
IDV	Specifies the form of the variance-covariance matrix for the residuals. In this case, we specify an identity matrix.
! **s2=2.7**	Provides a starting value for the residual error variance component. In this example, we used the variance component estimate already obtained from the previous analysis in Code example 1.1. In practice, one may have to guess at a ballpark initial value.

In PATH 2 the **R** structure is explicit, like in PATH 1 but the we used functional specification (a different syntax) to define **R**.

Here is an example of specifying the variance-covariance structure of the random term 'female' from the model shown in Code example 1.1. Again, the example shown in code example below will produce exactly the same result as Code example 1.1. We used structural and functional specifications to define the G structure for the random model term 'family' effect.

```
...
!PATH 3    #  G is specified, R is not
! Structural specification
height  ~ mu prov   !r  female block  block.female
0 0 1             #variance header line, s = 0, c = 0, g = 1
female 1          #G structure definition part 1
female 0 IDV 0.2 #G structure definition part 2

!PATH 4    #  G is specified, R is not
! Functional specification
height  ~ mu prov   !r  idv(female)  block  block.female
```

Running ASReml

If environment variables are set up inside of ConTEXT text editor correctly as shown in Fig. 1.2, a command script can be submitted to ASReml directly from inside ConTEXT, as follows:

- Open the job file (.as) in ConTEXT that you wish to run ASReml.
- Press the "F9" key or icon 1 to run your command file (.as).
- A plot of residuals will come to the screen.
- Check the job progress in the Console (a window at the bottom of the screen).
- Hit any key to finish the job. Do NOT hit X in the upper-left corner of ConTEXT to close the graphics windows – that would cancel the job.

ASReml can be run as a batch process with minimal user interaction in Windows, Linux, or Macintosh operating systems. Command line instructions to analyze the job command file '*Pine_provenenance.as*' on a Windows machine look something like this:

```
C:\ASReml>asreml  c:\myfolder\Pine_provenance.as
```

The job options can be set *on the command line* or *on the first line* of the job file.

- If the job options are set *on the command line* they can be *concatenated string* in the same format as for the command line. For example, in the following command line a set of job control options are listed before the job file.

```
C:\ASReml>asreml -h11r 1 2 3 4 Pine_provenance.as
```

- They are combined in a single string starting with '–' sign. Command line options are not case sensitive. They can be written as –H11R 1 2 3 4
- In above example, the command file runs four jobs (1, 2, 3, 4) in the job file *Pine_provenance.as*, renames the output files (r) by adding the PART numbers 1 2 3 4 and sets interactive graphics device (h) and suppresses graphics screens. It produces windows meta file images (11).

If the job options are listed on the first line of the job file as follows:

```
!-h22r 1 2 3 4 # or !HARDCOPY !WMF !RENAME !ARGS 1 2 3 4
Title: Pine data
 treeid
 ...
!FOLDER V:\Book\Book1_Examples\data
Pine_provenance.csv !SKIP 1  !DOPATH $A
```

then the command line is simply

```
C:\ASReml>asreml  Pine_provenance.as
```

If qualifiers are listed at the top of the job file *they must be in one line*. Spreading them over two lines is not allowed because ASReml takes the second line as the title and gives an error. ASReml ignores the options on the job line if there are arguments on the command line. See page 194 in ASReml Release4 manual for the list of the job options (Gilmour et al. 2014).

If the ASReml job executes correctly, a number of output files having the same prefix (e.g., "Pine_provenance1") but different suffixes (e.g., ".asr", ".sln") will be created. The contents of these files are detailed in the section "ASReml output files" later in this chapter.

ASReml Output Files

Let's run Code 1-1_Corrected Pine_provenance.as file and examine output file *.asr* (primary output file). The primary output file with extension *.asr* is the most important output file. It includes of information about the data, the qualifiers, model convergence messages, possible errors, variance components estimates, and so on. It is important to read and examine this file carefully. Here is an interpretation of some lines in the primary output file:

```
ASReml 4.1 [28 Dec 2014] Title. Pine_provenance. Corrected job file
    Build 1r [18 Mar 2015]   64 bit  Windows x64
 22 Nov 2015 14:14:32.608     32 Mbyte  01-1_Corrected Pine_provenance
 Licensed to: NCSU Cooperative Tree Improvement Program
 Your ASReml license expires in  8 days
```

```
***************************************************************
* Contact support@asreml.co.uk for licensing and support     *
************************************************************ ARG *
Folder: V:\Book\Book1_Examples\ch01_asreml
!CSKIP  1
female  !I
prov    !I
block   !I
height  diameter  volume !*10
Note: 1 data fields will be skipped.
QUALIFIERS: !SKIP 1
QUALIFIER: !DOPART    1 is active
Reading V:\Book\Book1_Examples\data\Pine_provenance.csv  FREE FORMAT skipping    1 lines

Univariate analysis of height
Summary of 914 records retained of 914 read

 Model term      Size #miss #zero   MinNon0   Mean    MaxNon0  StndDevn
  1 female        36     0     0       1     18.5985    36
 Warning: Fewer levels found in male  than specified
  2 male           2     0   914       0      0.0000     0
  3 prov           4     0     0       1      2.7856     4
  4 block          5     0     0       1      2.9387     5
  5 plot         240     0     0       1    115.8239   240
  6 height    Variate    0     0    4.000    10.48     15.20     1.794
  7 diameter            0     0    5.000    18.31     31.20     4.398
  8 volume              0     0  0.3000E-01  1.181     4.293     0.6611
  9 mu                  1
 10 block.female       180  4 block    :    5 1 female     :    36
 Forming     226 equations:   5 dense.
 Initial updates will be shrunk by factor   0.316
 Notice:      1 singularities detected in design matrix.
```

First always check the summary of data records. The 'Size' column shows the number of levels for factor variables; '#miss' is the number of missing data points in that column, '#zero' is the number of 0's in the column, 'MinNon0' is the smallest value observed in the column aside from any 0's. The 'Mean' column is useful information only for varieties and covariates.

Forming 226 equations: 5 dense.: "Forming 226 equations: 5 dense' refers first to the number of effects being estimated in the model. In this example, the number of equations =1 mu + 36 families +5 blocks +4 provenances +180 plots =226 total. '5 dense' refers to the number of fixed effects in the model: 1 mu + 4 provenances.

Notice: 1 singularities detected in design matrix: Singularities detected in design matrix refer to the fact that we cannot uniquely estimate all 5 of the fixed effects. For example, we cannot uniquely estimate *mu* and *4 prov* effects from the mean values of 4 provenances. However, by constraining the *prov* effect estimates in some way (such as by forcing them to sum to zero, or alternatively by fixing one *prov* effect equal to zero), we can uniquely estimate linear combinations of mu + each prov effect. So, we have one singularity among the fixed provenance effects.

```
     1 LogL=-935.002    S2=  2.4765       910 df
     2 LogL=-934.402    S2=  2.4939       910 df
     3 LogL=-934.158    S2=  2.5138       910 df
     4 LogL=-934.122    S2=  2.5276       910 df
     5 LogL=-934.122    S2=  2.5270       910 df
     6 LogL=-934.122    S2=  2.5270       910 df
```

This section illustrates the log likelihood ('LogL') and residual error variance ('S2') at each iteration as the mixed model equations are solved iteratively.

'910 df' refers to the number of degrees of freedom available to the random part of the model. REML maximizes the likelihood of the random part of the model after absorbing the fixed effects. So, the *df* here do not refer to the df for the residual variance or any other part of the model, but to the 914 observations -3 df for provenances -1 df for mu $= 910$ df remaining after accounting for fixed effects.

If there are a small number of variance components estimated, you may see Gamma values for each random effect in the model at each iteration. The gamma value is usually the ratio of the estimated variance component for a model factor to the error variance estimate. For complex models with many variance components, as in this example, none of the gamma values at each iteration are shown.

```
           - - - Results from analysis of height - - -
     Akaike Information Criterion      1876.24 (assuming 4 parameters).
     Bayesian Information Criterion    1895.50

              Approximate stratum variance decomposition
     Stratum      Degrees-Freedom    Variance  Component Coefficients
     block                  4.01     20.7243    162.8    0.1     3.4      1.0
     female                31.83     8.39947      0.0   25.6     5.2      1.0
     block.female         135.20     3.56543      0.0    0.0     5.3      1.0
     Residual Variance    738.96     2.52703      0.0    0.0     0.0      1.0

     Model_Term                       Gamma        Sigma    Sigma/SE   % C
     block           IDV_V    5   0.425370E-01   0.107492      1.20    0 P
     female          IDV_V   36   0.749901E-01   0.189502      2.26    0 P
     block.female    IDV_V  180   0.782019E-01   0.197618      2.29    0 P
     units                  914 effects
     Residual        SCA_V  914   1.00000        2.52703      19.22    0 P
```

This section shows the variance components estimates. We will not explain the top section ('Approximate stratum variance decomposition'), although one can note that the 'Degrees-Freedom' sum to 914 for the random part of the model. One might expect 35 degrees of freedom for the female term in this analysis, so this section of output might confuse readers and probably should be ignored except by REML experts.

Instead, users will want to focus on the bottom section, which contains the variance components estimates themselves. The gamma values are presented again, and the next column '*Sigma*' shows the actual variance components themselves. So, for this model, the female variance component estimate was 0.189 and the residual variance estimate was 2.527. The column 'Sigma/SE' provides the ratio of the variance component estimate to its standard error. If the factor has sufficient levels, this approximates a t-value, so anything above 2 is probably significant at $\alpha = 0.05$. It is only an approximate rule of thumb, however, and this test should not be used to declare factors non-significant, since it has low power for factors with few levels. See Chap. 2 for details on testing significance of model effects. The column '%' shows the percent change in the variance component between the final iteration and the previous iteration. If the percent change is not small, it suggests that the estimate is fluctuating substantially between iterations and the final value may not be stable. The final column 'C' provides information about constraints on the variance components estimate. In this example, both variance components are constrained by REML to be positive 'P'. Other constraint indicators include 'B' (for variance components that are very close to the boundary of 0 and have been made a very small positive number to avoid having a zero variance component estimate hinder continued iteration) and 'U' (for unconstrained components, such as covariance components that can be positive or negative).

```
                         Wald F statistics
      Source of Variation    NumDF      DenDF     F-inc    P-inc
   9 mu                        ' 1        5.7   3566.99   <.001
   3 prov                        3       32.0     10.01   <.001
Notice: The DenDF values are calculated ignoring fixed/boundary/singular
         variance parameters using algebraic derivatives.

            Solution     Standard Error   T-value   T-prev
   3 prov
        12  -0.623741       0.370135        -1.69
        11  -1.66563        0.374134        -4.45    -4.33
        13  -1.22019        0.376989        -3.24     1.77
   9 mu
         1   11.5121        0.361992               31.80
   4 block                            5 effects fitted
   1 female                          36 effects fitted
  10 block.female                   180 effects fitted
Finished: 22 Nov 2015 14:14:34.004   LogL Converged
```

The last part of the output in the .asr file displays results for fixed effects in the model. Obtaining appropriate denominator degrees of freedom for testing fixed effects in unbalanced mixed models is a very complex problem, and ASReml provides several options for doing this. The default option has the advantage of least computing time, but provides approximate degrees of freedom that may not be very accurate in some cases. "F-inc" and "P-inc" are the F-statistics and associated *p*-values from incremental (or 'sequential') fitting of fixed effects in the model. Users should be aware that the order that terms are fit into the model can affect these F-statistics, and the approximate nature of the degrees of freedom computations also may result in inaccurate p-values. Users with interest in obtaining accurate and appropriate F-tests should consult the User Guide (Gilmour et al. 2014) and read about the various options for F-tests and degrees of freedom computation. In particular, the job qualifier !FCON will provide a particular kind of conditional test and !FOWN permits users to specify the terms used to condition a test.

A note on model convergence message is printed at the end. Look for this message: 'LogL Converge'. If the model has not converged, the analysis may require more iterations to converge, or the model may be poorly specified and require some changes.

Next we examine the solutions output file (.sln).

.sln (Solutions for Fixed /Random effects)
This file does not include a header line, but the columns can be identified as:

Model_Term	Level	Effect	seEffect
prov	10	0.000	0.000
prov	12	-0.6237	0.3701
prov	11	-1.666	0.3741
prov	13	-1.220	0.3770
mu	1	11.51	0.3620
block	1	0.3110	0.1855
block	2	0.1268	0.1858
block	3	0.2056	0.1858
block	4	-0.2919	0.1867
block	5	-0.3516	0.1878
female	191	-0.8257E-01	0.3395
female	192	0.2459	0.3419
female	170	-0.1634	0.3433
...			

BLUE are best linear unbiased estimators of fixed effects; BLUP are best linear unbiased predictors of random effects. Note that the first estimate of provenance effects (level 10) is fixed to zero. These constraints are required to avoid the singularities noted previously. Random effect predictions (BLUPs) of *block* and *female* are centered on zero.

.yht (Predicted and residual values)

Record	Yhat	Residual	Hat
1	11.770	-1.070	0.1839
2	11.770	-0.2702	0.1839
3	11.770	0.3298	0.1839
4	11.770	0.2298	0.1839
5	11.770	0.4298	0.1839

...

Record: Observation number, **Yhat**: Predicted value (Yhat = y − (Xb + Zu), **Residual**: Observed − predicted. **Hat**: Diagonal elements of Hat matrix, the magnitude of these values indicate the influence of the data point on the fitted value (outliers have large Hat values).

.res (Statistics from residuals for model selection)

```
=== === === === Residual statistics for V:\Book\Book1_Examples\ch01_asreml\outputfiles/
Code01-1_Corrected Pine_provenance.asr === === === ===

Convergence sequence of variance parameters

Iteration     1          2          3          4          5          6
LogL   -935.002   -934.402   -934.158   -934.122   -934.122   -934.122
Change %     22         24         17          1          0          0
Adjusted      0          0          0          0          0          0
StepSz     0.316      0.562      1.000      1.000      1.000      1.000
   3 G    0.100000   0.061854   0.047019   0.042197   0.042533   0.042537   0.042537   0.0
   4 G    0.100000   0.090260   0.080426   0.074692   0.074991   0.074990   0.074990   0.0
   5 G    0.100000   0.092618   0.084077   0.078013   0.078203   0.078202   0.078202   0.0

 Covariance \cr Variance \cr Correlation matrix from block.female
[BLUPS]
 0.706E-01-0.198    -0.196E-01 0.161    -0.228
-0.114E-01 0.465E-01-0.446    -0.350    -0.100
-0.123E-02-0.229E-01 0.564E-01 0.100    -0.130
 0.711E-02-0.125E-01 0.393E-02 0.275E-01-0.221
-0.127E-01-0.454E-02-0.649E-02-0.770E-02 0.440E-01

Rescaled Covariance \cr Variance matrix from block.female
 0.78202E-01
-0.15499E-01  0.78202E-01
-0.15311E-02 -0.34911E-01  0.78202E-01
 0.12605E-01 -0.27385E-01  0.77940E-02  0.78202E-01
-0.17869E-01 -0.78428E-02 -0.10186E-01 -0.17286E-01  0.78202E-01
        Trace of W(W'R^W+G^)^W'  174.74017

Plot of Residuals [   -5.2983    4.3232] vs Fitted values [   8.7152   12.2790] _RvE_1
```

```
     -----------------1---------1-------1----1-------------------
     .                     1 1                  1            .
     .               1               11    11    1           .
     .          1    1    1              111    1            .
     .           1 1 3 1     3 3 1   2   2 11  2       1 1    .
     .         11   121 11111 2    1    1122 2 1      1    1.
     .       1  21   1 1 1 411 323 22 1  332223    2     2   1.
     .1     11 1  4 1  2 312 1521913313  14 2 131 1     1     1
     .    1 2  11 1    1 46 14224133247 1557 82343 21 2231     .
     1    2 1 1   1 3 12 421 122341243221 63314 41 32 22331 2    2
     11     2  11 1 31112611112665 322 22214 55353211  1 31 2   1.
     -1-2111-1-211-23-3-61-32145347325-1-86132121113--2222----11
     .  21 3   2 212113 22 1 34*5321 2  491 4421  11   1 1    1.
     1     3 1 111311 21 2 11243444422  24212 11122  1  1    1.
     .   1     2  1    11  23314 4 3      31131 1 1   2 1    .
     .    12 1 113  1 2 211212316112    13111133  11 1 1      .
     1   1 1  11 3 12 1       11  334   13 1 1    1          .
     11        1 1  111  1 213 211121   2   1 1 11          .
     .1   2  1   1 1 1    2 1 1              1           .
     .   1      1 1  1 1 1 1       1 1     1             .
     .     1 12   1      1       1      1 1            .
     .     11          1      1      2    11          .
     .                    1         1                 .
     .           1 1  1 1                             .
     ----1-------------------------1---------------------------
SLOPES FOR LOG(ABS(RES)) on LOG(PV) for Section    1
 -1.16
SLOPES FOR LOG(SDi) on LOG(PVBari) for Section    1
 -1.30

Histogram of RvE_1_A: MaxFreq 55, Range  -5.2983      4.3232
                            *
                            *
                           * *
                          * ***** **
                         * *********
                         ************
                      *  ************* *
                      *  *************
                     ** *************** **
                    *  ** ******************
                   * ******************** *
                   ************************ **
              * *   * * ****************************
     * ***********************************************************
Min Mean Max  -5.2983     -0.39119E-13 4.3232      omitting      0 zeros
```

This file provides a first visualization of the relationship between residual and predicted values. This is a useful diagnostic tool to check that residual errors are independent of the predicted values. (If one sees that errors increase in variance as predicted values increase, a transformation of data may be required).

Next, a histogram of residuals is displayed, which can be useful to check the assumptions of normality. Finally the line starting with 'STND RES' reports data records with large standardized residuals that might be outliers and need to be examined. In this example we do not have any large standardized residuals. If we had one, the output would look like as follows:

```
STND RES  47  5.2000  -3.73
```

Means that observation 47 in the data file (in the order that they exist in the data file) had a value for height of 5.2, and its standardized residual value from the model fit was −3.73 because it is much lower than expected based on the estimated values of its block, provenance, and maternal parent effect estimates.

.aov (ANOVA and conditional F-tests)
The file contains the details of ANOVA and conditional F-tests. Most of what users need is already provided in the .asr file, so this output is mainly to get more details on the sequence of fitting effects.

```
This file reports details concerning the calculation and testing of the
        Wald F-statistics reported in the .asr file.

Table showing the reduction in the numerator degrees of freedom
                for each term as higher terms are absorbed.
    Source              2  1
    1 mu                4  1
    2 prov              3

    Incremental Wald F statistics - calculation of Denominator degrees of freedom
    Source           Size NumDF   F-value  Lambda*F   Lambda   DenDF
    mu                  1    1   3566.9926 3566.9926  1.0000    5.7230
    prov                4    3     10.0066   10.0066  1.0000   32.0240
```

.vvp (Covariance matrix of random effects)
This file contains the approximate variance-covariances of the parameters reported in the .asr file. It is designed to be read by the .pin file for calculation of the standard errors of linear combinations of parameters (e.g., phenotypic variance, heritability and correlations). The matrix is lower triangle row-wise in the order of parameters (variances in this example) given in the .asr file. In this simple example, the order of the three variance components in the .asr file is as follows:

```
    Model_Term                    Gamma      Sigma   Sigma/SE  % C
    block          IDV_V    5  0.425370E-01  0.107492    1.20   0 P
    female         IDV_V   36  0.749901E-01  0.189502    2.26   0 P
    block.female   IDV_V  180  0.782019E-01  0.197618    2.29   0 P
    units                914 effects
    Residual       SCA_V  914  1.00000       2.52703    19.22   0 P
```

The .vvp file is as follows:

```
Variance of Variance components      4
   0.808625E-02
   0.243998E-04   0.705259E-02
  -0.134800E-03  -0.137794E-02   0.743660E-02
  -0.371893E-04  -0.922963E-05  -0.328918E-02   0.172833E-01
```

The first row (0.808625E-02) is the variance of 'block' variance component.

In the second row, 0.243998E-04 is the covariance between the female variance component and block variance component, 0.705259E-02 is the 'female' variance component.

In the third row, $-0.134800E-03$ is the covariance between the block variance component and the plot variance components etc.

.rsv (resetting initial parameter values)
The *.rsv* file holds the initial variance parameter values between runs of ASReml. The file is not normally modified by the user.

```
227 6 3349 270
# This .rsv file holds parameter values between runs of ASReml and
# is not normally modified by the User.  The current values of the
# the variance parameters are listed as a block on the following lines.
# They are then listed again with identifying information
# in a form that the user may edit.
     0.000000        0.000000      0.4253697E-01  0.7499005E-01  0.7820191E-01  1.000000
RSTRUCTURE              1    1    3
 VARIANCE               1    1    0
    6, V, P,    1.0000000        0    0
  STRUCTURE            914    0    0
block                                1    1
    3, G, P,    0.42536968E-01    0    0
female                               1    1
    4, G, P,    0.74990052E-01    0    0
block.female                         1    1
    5, G, P,    0.78201913E-01    0    0
```

Example: The following code runs two models in PATH 1 and PATH 2, renames the output files for each run by adding the run-number to the base file. The .asr output file in the second run produces a message as 'Notice: ReStartValues taken from testPine2.rsv'

```
!RENAME 1 !ARGS 1 2 # Rename the output files for run 1 and 2
Title: Pine_provenance. Variance header line
!DOPATH $1
 treeid !A  female !I  male   *
 prov !I   block !I   plot  !I
 height diameter volume

!PATH 1
Pine_provenance.csv !SKIP 1
height  ~ mu prov  !r  female

!PATH 2
# !CONTINUE !RSV use initial values from previous run
Pine_provenance.csv !SKIP 1 !CONTINUE !RSV
height  ~ mu prov   !r  idv(female) block block.female
```

.tsv (resetting initial parameter values)
The *.tsv* file is created after the initial run of the job. It holds initial variance parameters created by ASReml. If !CONTINUE 2 or !CONTINUE !TSV are used after the data file, then the .tsv file is used instead of the .rsv file.

```
# This .tsv file is a mechanism for resetting initial parameter values
# by changing the values here and rerunning the job with !CONTINUE 2.
# You may not change values in the first 3 fields
#                       or RP fields where RP_GN is negative.

# Fields are:
# GN, Term, Type, PSpace, Initial_value, RP_GN, RP_scale.

   3, "block", G, P,    0.10000000    ,    3,    1
   4, "female", G, P,    0.10000000    ,    4,    1
   5, "block.female", G, P,    0.10000000    ,    5,    1
   6, "Variance 1", V, P,    1.0000000    ,    6,    1

# Valid values for Pspace are F, P, U and maybe Z.

# RP_GN and RP_scale define simple parameter relationships;
# RP_GN links related parameters by the first GN number;
# RP_scale must be 1.0 for the first parameter in the set and
# otherwise specifies the size relative to the first parameter.
# Multivalue RP_scale parameters may not be altered here.

# Notice that this file is overwritten if not being read.
```

.msv (Resetting initial parameter values)

After each iteration the current values of variance parameters are written to *.rsv* and *.msv* files. It is easier to identify each variance parameter in the *.msv* file. This file can be read by rerunning the job with !CONTINUE 3 or !CONTINUE !MSV listed after the data file.

```
# This .msv file is a mechanism for resetting initial parameter values
# by changing the values here and rerunning the job with !CONTINUE 3.
# You may not change values in the first 3 fields
#                       or RP fields where RP_GN is negative.

# Fields are:
# GN, Term, Type, PSpace, Initial_value, RP_GN, RP_scale.

   3, "block", G, P,    0.42536968E-01,    3,    1
   4, "female", G, P,    0.74990052E-01,    4,    1
   5, "block.female", G, P,    0.78201913E-01,    5,    1
   6, "Variance 1", V, P,    1.0000000    ,    6,    1

# Valid values for Pspace are F, P, U and maybe Z.

# RP_GN and RP_scale define simple parameter relationships;
# RP_GN links related parameters by the first GN number;
# RP_scale must be 1.0 for the first parameter in the set and
# otherwise specifies the size relative to the first parameter.
# Multivalue RP_scale parameters may not be altered here.

# Notice that this file is overwritten if not being read.
```

Tabulation

TABULATE statements provide a simple way of summarizing data by groups. In the example below, if the command file were named 'Pine_provenance.as', the descriptive statistics of height for each level of *prov* are written to output file 'Pine_provenance.tab'.

```
Title. Example of TABULATE
 treeid !A    female !I   male   *
 prov  !I   block !I     plot  !I
 height diameter volume

Pine_provenance.csv !SKIP 1

TABULATE   height ~ prov !STATS !DECIMAL 1

height ~ mu prov !r block female block.female
```

The output in the .tab file looks like:

```
Title. Example of TABULATE

       Simple tabulation of height

prov      Mean   StandDevn    Minimum     Maximum Count
  10   11.5104      1.3081     7.3000     14.2000    77
  12   10.8843      1.6383     5.2000     15.2000   306
  11    9.8933      1.8536     4.0000     14.9000   267
  13   10.2989      1.7916     5.1000     14.7000   264
```

- Multiple response variables can be listed after Tabulate:
  ```
  TABULATE height volume ~ prov !STATS
  ```

- Separate summary statistics can be requested for multiple factors:
  ```
  TABULATE height ~ prov block !STATS
  ```

- !COUNT requests counts as well as means:
  ```
  TABULATE height ~ prov !COUNT
  ```

- !SD requests standard deviation for each cell:
  ```
  TABULATE height ~ prov !SD
  ```

- !STATS is shorthand for !COUNT, !SD and !RANGE:
  ```
  TABULATE height ~ prov !STATS
  ```

- !DECIMAL is to control for number of decimals in summary statistics:
  ```
  TABULATE height ~ prov !STATS ! DECIMAL 1
  ```

Prediction

Tabulation works by summarizing the data by groups, as shown above. If data are unbalanced, the mean values of different groups may not be directly comparable because they may be affected differently by other factors in the model. So, tabulation is a useful summary analysis of the data but not best suited for making comparisons among factor levels. Prediction, on the other hand, is ideal for making such comparisons; because predictions are adjusted for all model factors so they are more directly comparable. We can request predictions for both fixed and random model factors, as shown in the example below:

```
Title. Example of prediction
 treeid !A   female !I   male   *
 prov  !I    block !I    plot  !I
 height diameter volume

!FOLDER V:\Book\Book1_examples\data
Pine_provenance.csv !SKIP 1

height ~ mu prov !r female block
predict   prov
predict   prov female   !present prov female
```

You may use multiple prediction statements to obtain predicted values for fixed and random effect factors. For female effect, we need to use `!present prov family` so that we get predictions from only the family x prov combinations that actually exist. If we just use `predict prov family`, we will get the big hypertable that will have predictions for all 36 females for each provenance ($4 \times 36 = 144$ predictions) which does not make sense.

The predictions appear in a file with suffix ".pvs":

```
Title. Pine_provenance. Corrected job file       22 Nov 2015 19:44:24
                                                 cted Pine_provenance
Ecode is E for Estimable, * for Not Estimable

The predictions are obtained by averaging across the hypertable
     calculated from model terms constructed solely from factors
     in the averaging and classify sets.
Use !AVERAGE to move ignored factors into the averaging set.

---- ---- ---- ---- ---- ----    1  ---- ---- ---- ---- ---- ----
Predicted values of height
The ignored set: block female

prov     Predicted_Value Standard_Error     Ecode
 10    11.5121           0.3620              E
 12    10.8883           0.2213              E
 11     9.8464           0.2279              E
 13    10.2919           0.2326              E
SED: Overall Standard Error of Difference   0.3162

---- ---- ---- ---- ---- ----    2  ---- ---- ---- ---- ---- ----
Predicted values of height
The ignored set: block
Warning:    108  non-estimable [empty] cell(s) may be omitted from the table.

prov     female    Predicted_Value Standard_Error Ecode
 10      191       11.4295          0.3392          E
 10      192       11.7580          0.3498          E
 10      170       11.3487          0.3560          E
...
 13      232        9.8923          0.3205          E
 13      247        9.3855          0.3246          E
 13      238       10.6756          0.3225          E
 13      252       10.6467          0.3275          E
SED: Overall Standard Error of Difference   0.4142
```

You can notice that the predicted values for *prov* effect are very similar, but not exactly identical to the mean values from the tabulate command. The predictions are the "best linear unbiased estimates" of the prov effects. The predictions for female effects are called "best linear unbiased predictors" (BLUPs) because they are random effects. The differences in these concepts are described in detail in Chap. 2. "SED Overall Standard Error of Difference" is the mean standard error of a difference between predictors, averaged over all possible comparisons. It can be used as a rough guide to test significance of particular differences.

Processing Multiple Analyses with One Command File

The qualifiers **!CYCLE**, **!DOPART**, **!ARGS**, and **!RENAME** provide users a way to include multiple models or analyses of different traits or parts of the data set in a single job file, and to control how ASReml processes the different parts of the program and produces output files.

Multiple analyses in one job file with outputs combined: The !CYCLE qualifier

As an example, consider an experiment where the traits height, diameter, and volume were measured on each tree in a replicated experiment. We could write three separate .as files to analyze the three traits separately. Or we can combine the three analyses into a single .as file that will produce a single set of output files containing the results for all three analyses using the **!CYCLE** qualifier on the top line of the .as file, before the title line:

Code example 1.4
Example of using !CYCLE qualifier for multiple jobs (*Code 1-4_CYCLE qualifier.as*)

```
!OUTFOLDER V:\Book\Book1_Examples\ch01_asreml\outputfiles
!CYCLE height diameter volume
Title: Example of using CYCLE qualifier
 treeid !A   female !I   male   *
 prov  !I    block !I    plot   !I
 height diameter volume

Pine_provenance.csv !SKIP 1

$I ~ mu !r block*female
```

Notice the model formula: instead of writing a particular trait name, such as 'height' as the independent variable in the model, we write '**$I**'. This indicates where the variables listed after the **!CYCLE** qualifier should be inserted into the program. So, ASReml takes the first variable listed after **!CYCLE**, in this case it is 'height', and inserts it into the model formula, then fits the model to 'height' data. Then, ASReml substitutes the second variable listed, 'diameter', and refits the model to the 'diameter' data, and so on. This is similar to how macros work in some programming languages or in SAS programs.

In this example, the results from each of the different trait analyses are combined together in the relevant output files. For example, a single .asr file is produced, and it contains the results from the three trait analyses. Parts of the .asr file look like:

```
ASReml 3.0 [01 Jan 2009]   Title: Example of using CYCLE qualifier
...
 Cycle 1 value is height
Univariate analysis of height
...
```

```
- - - Results from analysis of height - - -
LogL:    LogL  Residual       NEDF  NIT Cycle Text
LogL: -941.67  2.53118         913    6 height "LogL Converged"
Akaike Information Criterion      1891.34 (assuming 4 parameters).
Bayesian Information Criterion    1910.61

...

Model_Term                   Gamma       Sigma   Sigma/SE   % C
block          IDV_V    5 0.414920E-01  0.105024     1.19  0 P
female         IDV_V   36  0.175159     0.443359     3.16  0 P
block.female   IDV_V  180 0.759541E-01  0.192253     2.24  0 P
Residual       SCA_V  914  1.00000      2.53118     19.21  0 P

...

Finished: 14 Jul 2015 16:40:06.002   LogL Converged
```

...

Cycle 2 value is diameter
Univariate analysis of diameter
...

```
        - - - Results from analysis of diameter - - -
LogL:-1789.00   16.9107         913    6 diameter "LogL Converged"
Akaike Information Criterion      3586.00 (assuming 4 parameters).
Bayesian Information Criterion    3605.27

...

Model_Term                   Gamma       Sigma   Sigma/SE   % C
block          IDV_V    5 0.931124E-02  0.157459     0.82  0 P
female         IDV_V   36  0.102997     1.74176      2.82  0 P
block.female   IDV_V  180 0.392224E-01  0.663277     1.33  0 P
Residual       SCA_V  914  1.00000      16.9107     19.24  0 P

...

Finished: 14 Jul 2015 16:40:06.891   LogL Converged
```

...

Cycle 3 value is volume
Univariate analysis of volume
...

```
        - - - Results from analysis of volume - - -
LogL:  -52.43  0.373735        913    7 volume "LogL Converged"
Local CYCLE LogL Peak at CYCLE:    3 volume LogL:    -52.43 Deviance: 3473.14
Akaike Information Criterion       112.86 (assuming 4 parameters).
Bayesian Information Criterion     132.12

Model_Term                   Gamma           Sigma      Sigma/SE   % C
block          IDV_V    5 0.216182E-01  0.807949E-02     1.08  0 P
female         IDV_V   36  0.116894     0.436872E-01     2.93  0 P
block.female   IDV_V  180 0.423567E-01  0.158302E-01     1.43  0 P
Residual       SCA_V  914  1.00000      0.373735        19.25  0 P

...

Finished: 14 Jul 2015 16:40:07.698   LogL Converged
```

The !CYCLE qualifier does not have to be on the top line of the job file, it can also be placed on a separate line after the data file specification and before the model formula:

```
!OUTFOLDER V:\Book\Book1_Examples\ch01_asreml\outputfiles
Title: Example of using CYCLE qualifier
 treeid !A  female !I  male  *
 prov  !I   block !I   plot  !I
 height diameter volume

!FOLDER V:\Book\Book1_Examples\data
Pine_provenance.csv !SKIP 1

!CYCLE height diameter volume
$I ~ mu !r block*female
```

This produces an identical output.

We can also analyze different models in the same job file by combining the! CYCLE qualifier with the !DOPATH qualifier. We can write different models in the same file, indicating the beginning of each model *i* with '**!PATH i**'. We can select which of these models to fit in the current job run with the qualifier **!DOPATH i** placed after the data file specification. Combining this syntax with the !CYCLE qualifier allows us to cycle through multiple jobs, selecting a different part ('PATH') each time. For example, if we had three different models we wanted to fit in a single job, we could indicate ! CYCLE 1:3 , or equivalently: !CYCLE 1, 2, 3 on the top line of the .as file, then use !DOPATH $I as a qualifier after the data file specification. This would cause the '$I' symbols to be replaced by the variables listed after the !CYCLE qualifier in turn. In this example, it would be like running three jobs sequentially, specifying 'DOPATH 1', then 'DOPATH 2', and finally 'DOPATH 3'.

Code example 1.5
Using CYCLE and DOPATH qualifiers for multiple jobs (*Code 1-5_CYCLE and DOPATH qualifier.as*)

```
!OUTFOLDER V:\Book\Book1_Examples\ch01_asreml\outputfiles
!CYCLE 1:3
Title: Example of using CYCLE and DOPATH qualifiers
 treeid !A  female !I  male  *
 prov !I   block !I   plot  !I
 height diameter volume

!FOLDER V:\Book\Book1_Examples\data
Pine_provenance.csv !SKIP 1  !DOPATH $I

!PATH 1 #no block effect
height ~ mu prov !r  female

!PATH 2 #block effect
height ~ mu prov !r  block female

!PATH 3 # block and plot  effects
height ~ mu prov !r  block*female
```

Again, the output of the three jobs is combined into a single set of output files. For example, the .asr file looks like:

```
...
QUALIFIER: !DOPART     1 is active
Cycle 1 value is 1
Reading Pine_provenance.csv  FREE FORMAT skipping     1 lines

Univariate analysis of height
Summary of 914 records retained of 914 read
...
         - - - Results from analysis of height - - -
LogL:   LogL  Residual         NEDF NIT Cycle Text
LogL: -947.87  2.77622          910   6 1 "LogL Converged"
Akaike Information Criterion      1899.73 (assuming 2 parameters).
Bayesian Information Criterion    1909.36
...
Model_Term                   Gamma        Sigma   Sigma/SE  % C
female         IDV_V   36 0.165272     0.458929       3.37  0 P
Residual       SCA_V  914 1.00000      2.77681       20.95  0 P

                       Wald F statistics
     Source of Variation   NumDF     DenDF    F-inc    P-inc
  10 mu                        1      32.1 11943.35    <.001
   4 prov                      3      32.0     9.44    <.001
...
Finished: 14 Jul 2015 16:53:34.961   LogL Converged

...
QUALIFIER: !DOPART     2 is active
Cycle 2 value is 2
...
Univariate analysis of height
Summary of 914 records retained of 914 read
...
         - - - Results from analysis of height - - -
LogL: -937.89  2.68806          910   6 2 "LogL Converged"
Akaike Information Criterion      1881.78 (assuming 3 parameters).
Bayesian Information Criterion    1896.22

Model_Term                   Gamma        Sigma   Sigma/SE  % C
block          IDV_V    5 0.403228E-01 0.108390       1.24  0 P
female         IDV_V   36 0.831129E-01 0.223412       2.71  0 P
Residual       SCA_V  914 1.00000      2.68806       20.91  0 P
                       Wald F statistics
     Source of Variation   NumDF     DenDF    F-inc    P-inc
  10 mu                        1       6.2  3549.60    <.001
   4 prov                      3      32.0     9.78    <.001
...
Finished: 14 Jul 2015 16:53:35.633   LogL Converged

...
QUALIFIER: !DOPART     3 is active
Cycle 3 value is 3
...
Univariate analysis of height
Summary of 914 records retained of 914 read
...
```

```
      - - - Results from analysis of height - - -
LogL: -934.12   2.52703        910    6 3 "LogL Converged"
Local CYCLE LogL Peak at CYCLE:    3 3 LogL:  -934.12 Deviance:    27.49
Akaike Information Criterion      1876.24 (assuming 4 parameters).
Bayesian Information Criterion   1895.50
...
Model_Term                    Gamma      Sigma   Sigma/SE   % C
  block        IDV_V    5  0.425370E-01  0.107492      1.20   0 P
  female       IDV_V   36  0.749901E-01  0.189502      2.26   0 P
  block.female IDV_V  180  0.782019E-01  0.197618      2.29   0 P
  Residual     SCA_V  914  1.00000       2.52703      19.22   0 P

                       Wald F statistics
      Source of Variation   NumDF    DenDF    F-inc   P-inc
   10 mu                       1      5.7   3566.99   <.001
    4 prov                     3     32.0     10.01   <.001
Finished: 14 Jul 2015 16:53:35.633   LogL Converged
```

In the examples just given, we make one substitution in each iteration of the cycling operation. For example, we changed the trait to be analyzed *or* we changed the path to be analyzed in each step. We can combine up to four substitutions in each cycle step to allow more flexibility in combining multiple analyses in a single command file. This is done by using $I, $J, $K, and $L in the command file as reserved names for the first through fourth strings in the program to be substituted. Then we provide a list of the substitutions to make in each cycle step following the !CYCLE qualifier. For a particular step in the cycle we list the string substitutions for $I, $J, $K, and $L by separating them with semicolons. Spaces then separate the substitutions for other cycle steps.

For example, we can write a single command file that will combine fitting three different models to each trait for three traits, thus combining the two approaches just shown. To do this, we will use $I to indicate the trait to be analyzed in the model formulas and we will use $J to indicate the program path to execute. Then we will provide a list of all nine combinations of traits and paths to the !CYCLE qualifier in the format: height;1 height;2 height;3 diameter;1 . . . volume;3. In the first cycle step, the program will substitute 'height' everywhere that '$I' is found in the program file and will substitute '1' everywhere that $J is found in the program. In the final cycle step, the program will substitute 'volume' for '$I' and '3' for $J. The results will be combined into common output files. Here is the command file to do this example:

Code example 1.6
Example of CYCLE with two substitutions (*Code 1-6_CYCLE with two substitutions.as*)

```
!OUTFOLDER V:\Book\Book1_Examples\ch01_asreml\outputfiles
!CYCLE height;1 height;2 height;3 diameter;1 diameter;2 diameter;3
volume;1 volume;2 volume;3
Title: Example of CYCLE with two substitutions
 treeid !A    female !I   male !I
 prov !I    block !I    plot !I
 height    diameter   volume !*10

!FOLDER V:\Book\Book1_Examples\data
Pine_provenance.csv !SKIP 1   !DOPATH $J

!PATH 1 #no block effect
$I ~ mu prov !r  female

!PATH 2 #block effect
$I ~ mu prov !r  block female

!PATH 3 #block and provenance effects
$I ~ mu prov !r  block*female
```

Now we have nine analyses combined into each of the output files. In the .asr output file, the different cycle steps are indicated and the substitutions made at each step are shown. The .asr file produced by the example program shown above will have nine parts, here we show a bit of the first part of each cycle output:

```
QUALIFIER: !DOPART    1 is active
 Cycle 1 value is height
...
QUALIFIER: !DOPART    2 is active
 Cycle 2 value is height
...
QUALIFIER: !DOPART    3 is active
 Cycle 3 value is height
...
QUALIFIER: !DOPART    1 is active
 Cycle 4 value is diameter
...
QUALIFIER: !DOPART    2 is active
 Cycle 5 value is diameter
...
QUALIFIER: !DOPART    3 is active
 Cycle 6 value is diameter
...
QUALIFIER: !DOPART    1 is active
 Cycle 7 value is volume
...
QUALIFIER: !DOPART    2 is active
 Cycle 8 value is volume
...
QUALIFIER: !DOPART    3 is active
 Cycle 9 value is volume
```

Multiple analyses in one job file with outputs separated: the !ARGS and !RENAME qualifiers

One may wish to create separate output files, each with a name that indicates which trait or model results are included. This can be done by using the !ARGS and !RENAME qualifiers instead of the !CYCLE qualifier. Instead of indicating the variable to change as $I through $L, as we did with !CYCLE, we use either $A or $1 when using !ARGS:

Code example 1.7

Using ARGS and RENAME qualifiers for different traits (*Code 1-7_ARGS and RENAME qualifiers.as*)

```
!ARGS height diameter volume !RENAME !OUTFOLDER V:\Book
Title: Example of using ARGS and RENAME qualifiers
 treeid !A   female !I   male !I
 prov !I   block !I   plot !I
 height   diameter   volume !*10

!FOLDER V:\Book\Book1_Examples\data
Pine_provenance.csv !SKIP 1

$A ~ mu prov !r block*female
```

This creates three sets of output files. For example, it creates three .asr files; the name of each file contains the string that was substituted for $A in that step of the analysis. So, in this case, if we named the .as file as 'Pine.as', the .asr files would be automatically named 'Pineheight.asr', 'Pinediameter.asr', and 'Pinevolume.asr'. Each .asr file contains only the results of the trait that was substituted into the model formula in that step.

We can use the same trick to analyze several different models for a single trait and rename the outputs to correspond to the model analyzed:

Code example 1.7b
Using ARGS and RENAME qualifiers for different models (*Code 1-7b_ARGS and RENAME qualifiers.as*)

```
!ARGS 1 2 3 !RENAME #must specify 1 2 3 rather than 1:3
Title: Example of using ARGS and RENAME qualifiers
 treeid !A    female !I   male !I
 prov !I    block !I    plot !I
 height     diameter    volume !*10

!FOLDER V:\Book\Book1_Examples\data
Pine_provenance.csv !SKIP 1   !DOPATH $1

!PATH 1 #no block effect
height ~ mu prov !r  female

!PATH 2 #block effect
height ~ mu prov !r  block female

!PATH 3 #block and provenance effects
height ~ mu prov !r  block*female
```

If this command file were named "Pine.as", the .asr output files would be named "Pine1.asr", "Pine2.asr", and "Pine3.asr".

!ARGS also can be used to make multiple substitutions in each run, but it works differently than !CYCLE in this regard. Similar to how $I, $J, $K, and $L were used with !CYCLE to indicate where multiple string substitutions would occur in a command file, we can use $A, $B, $C, etc. or $1, $2, $3, etc. in combination with !ARGS to make this happen (but use only the letter designations or the number designations, you cannot mix and match the two styles in the same command file). Importantly, and unlike with !CYCLE, with !ARGS and !RENAME we can only substitute different values for one of the arguments.

We explain with an example. Let's say that we want to simultaneously substitute a path number and a trait name for each of multiple ASReml jobs in a common file using !ARGS and !RENAME. We cannot cycle through each combination of trait and path as we did with the !CYCLE example given above. We can, however, cycle through one of the two variables and while still including both the trait and path number as part of the output file names. This may be convenient to clearly label output files produced from a series of ASReml analyses. In the example code below, we use $1 and $2 to indicate the positions of two string substitutions in the program code. We can only make one substitution for $1, but we can make multiple substitutions for $2. So, we write '**!ARGS 1 height diameter volume**' to indicate that the first string '1' will be substituted for $1 in the subsequent code in every run, but that we will substitute each of 'height', 'diameter', and 'volume' in different runs. We then also include on the same line '!RENAME 2' to indicate that we are going to use two different variable substitutions in the output file names.

Code example 1.7c
Using ARGS and RENAME qualifiers to rename output files (*Code 1-7c_ARGS and RENAME qualifiers.as*).

```
!ARGS 1 height diameter volume !RENAME 2   !OUTFOLDER V:\Book
Title: Example of using ARGS and RENAME qualifiers
   treeid !A   female !I   male !I
   prov !I    block !I    plot !I
   height     diameter    volume !*10

!FOLDER V:\Book\Book1_Examples\data
Pine_provenance.csv !SKIP 1   !DOPATH $1

!PATH 1 #no block effect
$2 ~ mu prov !r  female

!PATH 2 #block effect
$2 ~ mu prov !r  block female

!PATH 3 #block and provenance effects
$2 ~ mu prov !r  block*female
```

This program will produce three output files. If the command file is named 'Pine.as" the output files will be named "Pine1_height.asr", "Pine1_diameter.asr", "Pine1_volume.asr", where the "1" in each file name refers to the first variable substitution (the path number) and the trait analyzed is also indicated in the output file name. In this case, only the path 1 model is analyzed for each trait. We could then change the top line of this command file to "**!ARGS 2 height diameter volume !RENAME 2**" to produce three new outputs corresponding to the path 2 model for each trait.

We can extend this idea to add a third string to the output names, for example:

Code example 1.7d
Using ARGS and RENAME qualifiers to rename output files with three variables (*Code 1-7d_ARGS and RENAME qualifiers.as*).

```
!ARGS 1 female height diameter volume !RENAME 3 !OUTFOLDER V:\Book
Title: Example of using ARGS and RENAME qualifiers
   treeid !A   female !I   male !I
   prov !I    block !I    plot !I
   height    diameter    volume !*10
!FOLDER V:\Book\Book1_Examples\data
Pine_provenance.csv !SKIP 1   !DOPATH $1

!PATH 1 #no block effect
$3 ~ mu prov !r  $2
```

This will produce a set of outputs labelled as "*Pine1_female_height.asr*", "*Pine1_ female_diameter.asr*", "*Pine1_ female_volume.asr*", each containing the path 1 model fitting the random female term as the only effect in the model. Notice that now we had to use '**!RENAME 3**' to indicate that the strings remaining after the first two substitutions would be used in a cycle to iteratively replace '**$3**' in the job run.

!ARGS and !CYCLE working together
Finally, we can combine !ARGS and !CYCLE to combining cycling of some strings and joining of parts of the output along with renaming output files according to the arguments. !ARGS defines an outer loop, in each step of which !CYCLE is run as

an inner loop. Consider how we ran all nine combinations of three models and three traits in a single job using the !CYCLE example previously shown:

```
!CYCLE height;1 height;2 height;3 diameter;1 diameter;2 diameter;3 volume;1 volume;2 volume;3
```

Perhaps we would prefer the output files to be separated and named according to the trait analyzed, with the outputs of three models for a particular trait combined. This can be accomplished by using !ARGS to define the outer loop of traits and ! CYCLE to define the inner loop of models ('paths'). Recall that '$I' through '$J' are used to identify where the string substitutions controlled by !CYCLE are made, whereas $A through $H or $1 through $9 are used to identify where the substitutions made by !ARGS occur:

Code example 1.7e
Using ARGS and CYCLE qualifiers to join outputs from multiple models within output files separated and named according to the trait variable (*Code 1-7e_ARGS and RENAME qualifiers.as*).

```
!ARGS height diameter volume !RENAME !OUTFOLDER V:\Book
!CYCLE 1:3
Title: Example of CYCLE with two substitutions
 treeid    !A    female !I    male    !I
 prov !I    block !I    plot !I
 height diameter volume

!FOLDER V:\Book\Book1_Examples\data
Pine_provenance.csv !SKIP 1    !DOPATH $I

!PATH 1 # no block effect
$A ~ mu prov !r  female

!PATH 2 # block effect
$A ~ mu prov !r  block female

!PATH 3 # block and provenance effects
$A ~ mu prov !r  block*female
```

This produces three sets of output files. If we named this command file "Pine.as" the .asr output files would be named "Pine_height.asr", "Pine_diameter.asr", and "Pine_volume.asr". Each .asr file would have results from the three models for the trait identified in the file name.

Substitutions using !CYCLE and !ARGS qualifiers can also be implemented in the command line rather than inside the .as file. This may be convenient when controlling ASReml using batch jobs or as a subroutine to some other program (e.g., one can write an R program to execute a series of ASReml jobs).

Linear Combinations of Variance Components

One can use the variance components reported in the .asr file to calculate linear combinations of variance components, such as phenotypic variance, ratios of variance components and heritability. The variances and covariances of variance components estimates in the .VVP file can be used to estimate approximate standard errors of such functions of variance components using the Delta method (Lynch and Walsh 1998; Holland et al. 2003).

We can use VPREDICT !DEFINE qualifiers after the model to obtain linear combinations or ratios of variances. We add the qualifiers after the model.

Code example 1.8
Linear combinations of variance components (*Code 1-8_Heritability.as*)

```
!OUTFOLDER V:\Book\Book1_Examples\ch01_asreml\outputfiles
Title: Pine_provenance. PIN file
   treeid !A    female !I   male !I
   prov !I    block !I    plot !I
   height    diameter    volume  !*10

!FOLDER V:\Book\Book1_Examples\data
Pine_provenance.csv   !SKIP 1   !DOPATH 1

!PATH 1
height   ~ mu prov,    # Specify fixed model
           !r  idv(female)*block  # Specify random model
         residual idv(units)

VPREDICT !DEFINE
F GenVar idv(female)*4.0     # Multiply idv(female) component by 4
F PhenVar idv(female)+ idv(female).block + idv(units) # Phen var
H h2   GenVar PhenVar       # Heritability, divide 3rd by 4th term
```

The bold face fonts show how to estimate functions of variance components using the VPREDICT !DEFINE qualifiers. The qualifier will create a new file with the same root name as the .as file but with the extension .pin. When the .as command file is executed the .pin file will also be processed to estimate the functions of variance components and produce the result in a file with extension .pvc.

Alternatively users can also create the .pin file directly if desired. The *.pin* file can have the same name of the job file or can have a different name. It can be run independently.

Each line of the !PIN definition has three components: *Letter, Label, Coefficients*

1. *Letter* Must start from the first field (no space before the letter)
   F is used to calculate linear combinations of variance components
   H is used to calculate ratios, e.g. heritability and standard error of the ratio
   R is used to calculate correlations and their standard errors
   S is used to take the square root
   V is for converting components related to a CORUH or an XFA structure into components related to a US structure
2. *Label* A user-defined name for the function. Labels longer than eight characters can cause errors.
3. *Coefficients* List of arguments/coefficients for the linear function. A function definition where the relevant variance component in the function is referred to by its order in the .asr output

F GenVar 1*4.0	Forms a linear function that we will call 'GenVar' (because it will equal the estimated additive genetic variance component) by multiplying '1', the first component (female variance) in the .asr output file, by 4.0.

Alternatively we can use the name of the random term:

F GenVar idv(female)*4.0.	Multiplies the variance component of the *idv(female)* in the .asr file by 4.0. This syntax is better because the order of the random terms in the .asr file can change.
F PhenVar 1+2	Estimates a term called 'PhenVar' which will be the phenotypic variance by adding variance components '1' and '2' in the .asr output. This is the sum of the first (family variance) and second components (error variance) in the .asr file.

Alternatively we can use the names of the terms to calculate the phenotypic variance as:

F PhenVar idv(female)+idv(units)	Sums the two variance components named *idv(female)* and *idv(units)* in the .asr file to calculate phenotypic variance.
H h2 3 4	Estimates a ratio of variance components named 'h2' by dividing the third term by the fourth term. This will be the heritability estimate because the third term in this case is the 'Additive' estimate and the 4th term is the 'Phenotypic' variance, considering that these terms were defined next in order after the two variance components estimated in the .asr file.

Alternatively we can use the name of the terms to calculate the heritability as

H h2 GenVar PhenVar	Take the ratio of the two variance components previously defined as GenVar and PhenVar in the VPREDICT section to estimate heritability.

The results are reported in the *Pine_provenance.pvc* file:

```
ASReml 4.1 [28 Dec 2014] Title: Heritability and functions
        V:\Book\Book1_Examples\ch01_asreml\outputfiles/Code01-8_Heritability.pvc created
22 Nov 2015 20:31:33.411

        - - - Results from analysis of height - - -




    1 block                        V    5   0.107492    0.895767E-01
    2 idv(female)                   V   36   0.189502    0.838504E-01
idv(units)            914 effects
    3 idv(units);Residual          V  914   2.52703     0.131479
idv(female).block       180 effects
    4 idv(female).block;idv(female) V    1   0.197618    0.862961E-01
    5 GenVar   2          0.75801          0.33592
    6 PhenVar  3          2.9141           0.14973
    Herit        = GenVar     5/PhenVar    6=  0.2601   0.1099
Notice: The parameter estimates are followed by
        their approximate standard errors.
```

The estimated additive variance is 0.758 with an approximate standard error of about 0.3359. The phenotypic variance estimate is 2.9141 with a standard error of about 0.1497. The heritability estimate is 0.26 with a standard error of about 0.11.

A Brief Introduction to ASReml-R

ASReml-R version 3.0 was introduced in 2009 to integrate the flexibility and power of ASReml for mixed models with the advantages of the R environment for managing data, results, and graphical display (Butler et al. 2009). The typical workflow for using ASReml-R has some differences with the use of standalone ASReml (Fig. 1.6). In particular, users can complete all data processing steps (merging and concatenating data on different traits and from different sub-experiments) inside of R, analyze the data with ASReml-R, and manage all of the output from each analysis in R objects, rather than in separate files written directly to the hard drive.

Large data sets and complicated models may require more memory in ASReml-R than in standalone ASReml. If ASReml-R runs out of memory for an analysis, the standalone version of ASReml is recommended, since it manages memory more efficiently (Butler et al. 2009). Here we briefly introduce the use of ASReml-R and focus mainly on how to fit a simple mixed model and access the results. Users should consult the manual for many details not covered here, including the installation for different platforms.

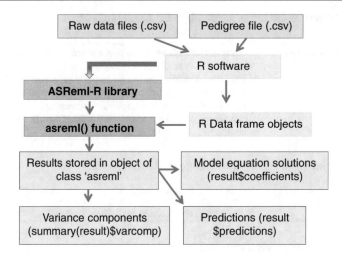

Fig. 1.6 Typical ASReml-R workflow

Data Set Used in the Analysis

We will use the same pine data (Pine_provenance.csv) used previously in this chapter to demonstrate ASReml-R. In this section, all commands are given in R. First, users should set the working directory and read in the data from a CSV file to an R data frame object:

Code example 1.9
Using ASReml R. The code is written using R markdown (rmd), which can be run in RStudio to generate data summary in different formats such as PDF using the KnitR package (see file *Code 1-9_ASReml-R_intro.Rmd* for more details)

```
#set working directory and read data file
setwd("C:/ Google Drive/Book/Book1_Examples/data ")
pine_prov <- read.csv(file = 'Pine_provenance.csv', header = TRUE)
```

The data frame should be checked before doing any analysis. In particular, the read.csv() and read.table() functions assume that numeric data columns should be treated as a numeric variate in any subsequent analyses. That means that, in this example, 'block' will be treated like a covariate (including block in the analysis results in block being fit as a 1 df regression variable). To change this, we need to force the numeric model variables to *factors* in R. First, we check what the 'structure' of the data frame is using str(). This will tell us what kind of variable each column in the data set is considered to be.

```
> str(pine_prov)
## 'data.frame': 914 obs. of 9 variables:
## $ treeid : Factor w/ 914 levels "170_1_3_1","170_1_3_2",..: 24 25 26 27 28 29 53 54 55 56
...
## $ female : int 191 191 191 191 191 191 192 192 192 192 ...
## $ male : int 0 0 0 0 0 0 0 0 0 0 ...
## $ prov : int 10 10 10 10 10 10 10 10 10 10 ...
## $ block : int 1 1 1 1 1 1 1 1 1 1 ...
## $ plot : int 1 1 1 1 1 1 2 2 2 2 ...
## $ height : num 10.7 11.5 12.1 12 12.2 12.5 12.8 12 13.9 13.1 ...
## $ diameter: num 15 22 23.8 22.7 21.5 22.8 24.1 19.8 23.4 22.1 ...
## $ volume : num 0.0722 0.167 0.2056 0.1855 0.1692 ...
```

Notice that *treeid* is correctly considered a factor variable, but the other factors we want to include in the model (*female*, *prov*, and *block*) are considered integer ('int') variables, which are a special type of numeric variable. We coerce these variables to be factors with the as.factor() function:

```
> pine_prov$female <- as.factor(pine_prov$female)
#or, pine_prov[,'female'] <- as.factor(pine_prov[,'female'])
> pine_prov$prov <- as.factor(pine_prov$prov)
> pine_prov$block <- as.factor(pine_prov$block)
```

Now, we re-check the structure of the data frame to be certain these commands did what we want:

```
> str(pine_prov)
## 'data.frame': 914 obs. of 9 variables:
## $ treeid   : Factor w/ 914 levels "170_1_3_1","170_1_3_2",..: 24 25 26 27
28 29 53 54 55 56 ...
## $ female   : Factor w/ 36 levels "170","191","192",..: 2 2 2 2 2 2 3 3 3 3
...
## $ male     : int 0 0 0 0 0 0 0 0 0 0 ...
## $ prov     : Factor w/ 4 levels "10","11","12",..: 1 1 1 1 1 1 1 1 1 1 ...
## $ block    : Factor w/ 5 levels "1","2","3","4",..: 1 1 1 1 1 1 1 1 1 1
...
## $ plot     : int 1 1 1 1 1 1 2 2 2 2 ...
## $ height   : num 10.7 11.5 12.1 12 12.2 12.5 12.8 12 13.9 13.1 ...
## $ diameter : num 15 22 23.8 22.7 21.5 22.8 24.1 19.8 23.4 22.1 ...
## $ volume   : num 0.0722 0.167 0.2056 0.1855 0.1692 ...
```

We can use a few other functions to inspect the data set before proceeding to analysis. head() shows the first rows of the data frame:

```
> head(pine_prov) #inspect the first few rows of the data frame
##   treeid       female male prov block plot height diameter volume
## 1 191_1_1_1    191    0    10   1     1    10.7   15.0     0.0722
## 2 191_1_1_2    191    0    10   1     1    11.5   22.0     0.1670
## 3 191_1_1_3    191    0    10   1     1    12.1   23.8     0.2056
## 4 191_1_1_4    191    0    10   1     1    12.0   22.7     0.1855
## 5 191_1_1_5    191    0    10   1     1    12.2   21.5     0.1692
## 6 191_1_1_6    191    0    10   1     1    12.5   22.8     0.1949
```

names() shows the column names of the data frame:

```
> names(pine_prov) #get the column names
## [1] "treeid" "female" "male" "prov" "block" "plot"
## [7] "height" "diameter" "volume"
```

unique() can be used to shows the names of the levels of a particular factor variable: The following code gets the names of the levels of 'female' in the order they first appear. Note these are returned as a factor vector.

```
> unique(pine_prov$female)

## [1] 191 192 170 210 216 211 212 217 224 219 225 218 213 227 226 207 206
## [18] 201 196 200 202 205 204 197 203 198 233 231 251 253 245 239 232 247
## [35] 238 252
## 36 Levels: 170 191 192 196 197 198 200 201 202 203 204 205 206 207 ... 253
```

levels() can also be used to shows the names of the levels of a particular factor variable. The names of the levels of 'female' in sorted order, are returned as a vector of character strings.

```
> levels(pine_prov$female)
## [1]  "170" "191" "192" "196" "197" "198" "200" "201" "202" "203" "204"
## [12] "205" "206" "207" "210" "211" "212" "213" "216" "217" "218" "219"
## [23] "224" "225" "226" "227" "231" "232" "233" "238" "239" "245" "247"
## [34] "251" "252" "253"
```

summary() returns summary statistics for each column of the data frame, including alphanumeric factors, for which the summaries are not very useful. Basic summary statistics for each column, results don't mean much unless the column is a numeric variable.

```
>    summary(pine_prov)
##   treeid              female          male     prov        block      plot
##   170_1_3_1 :   1     218   :   30    Min.   :0   10:77    1:192     Min.   :1
##   170_1_3_2 :   1     191   :   29    1st Qu.:0   11:267   2:187     1st Qu.:56
##   170_1_3_5 :   1     232   :   29    Median :0   12:306   3:188     Median :112
##   170_1_3_6 :   1     238   :   28    Mean   :0   13:264   4:179     Mean   :116
##   170_2_84_1:   1     202   :   27    3rd Qu.:0   5 :168             3rd Qu.:181
##   170_2_84_3:   1     204   :   27    Max.   :0   Max.   :240
##   (Other)   :908     (Other):744
##  height          diameter         volume
##   Min.   : 4.0    Min.   : 5.0    Min.   :0.0030
##   1st Qu. : 9.4   1st Qu.:15.5    1st Qu. :0.0702
##   Median :10.6    Median :18.5    Median :0.1098
##   Mean   :10.5    Mean   :18.3    Mean   :0.1181
##   3rd Qu. :11.8   3rd Qu.:21.4    3rd Qu. :0.1596
##   Max.   :15.2    Max.   :31.2    Max.   :0.4293
##
```

We can check how many individual progeny trees tested from each female parent using the table() function:

```
> table(pine_prov$female)
##
## 170 191 192 196 197 198 200 201 202 203 204 205 206 207 210 211 212 213
##  23  29  25  22  25  25  23  22  27  24  27  23  24  25  26  25  27  26
## 216 217 218 219 224 225 226 227 231 232 233 238 239 245 247 251 252 253
##  25  25  30  24  27  27  23  21  24  29  26  28  27  27  27  26  26  24
```

It can be easier to read the results of table() by coercing them to a data frame object first:

```
> as.data.frame(table(pine_prov$female))
## Var1 Freq
## 1 170 23
## 2 191 29
## 3 192 25
...
## 36 253 24
```

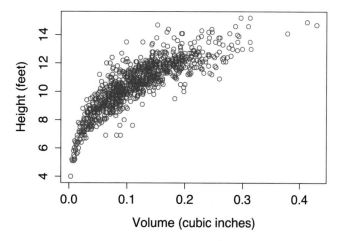

Fig. 1.7 Scatter plot of height and volume

Make a scatterplot of height and volume to check for outliers (Fig. 1.7):

```
plot(x = pine_prov$volume, y = pine_prov$height,
     xlab = "Volume (cubic inches)",
     ylab = "Height (feet)",
     main = "Volume by height", col = "blue")
```

Fitting a Model in ASReml-R

Now that we are assured that the data are formatted correctly, we can fit a linear mixed model using ASReml-R. First we call the ASReml-R library to make its functions available to the current session. Then we fit a model for height considering prov as fixed effect and female parent, block and female by block interaction as random effects.

```
> library(asreml)
## Loading required package: lattice
## Licensed to: North Carolina State University
## Serial Number: 40246062 Expires: 31-dec-2015 (332 days)
```

Mixed models that do not involve complicated covariance structures can be specified using model formula rules common to R linear models functions such as `lm()` in base R and `lmer()` in the lme4 package (Bates et al. 2013) and similar to the rules used for ASReml standalone, in terms of how main effects, interactions, and nested effects are specified.

We will fit the same linear model: $Y = \mu + B_i + P_j + F_{jk} + BF_{ijk} + \varepsilon_{ijkl}$, given in Code example 1.1. Where Y is the measured tree height data, B_i is the random effect of the ith field block, P_j is the fixed effect of the jth provenance, F_{jk} is the random effect of the kth female parent nested in the jth provenance, BF_{ijk} is the random effect of female by block interaction and ε_{ijk} is the random residual error effect of the tree or plot of trees from female parent jk in the ith block.

The model syntax is

```
#specify trait and fixed effects
> HT <- asreml(fixed = height ~ prov,
    random = ~ female*block, #specify random effects
    data = pine_prov,        #specify the data frame
    na.method.X = 'include', #include missing (NAs) in X variables
    na.method.Y = 'include') #include NAs in Y variable
```

This results in the following message output in the R console:

```
ASReml: Sun Nov 22 21:05:29 2015

    LogLik          S2         DF       wall      cpu
   -935.0022      2.4765      910    21:05:29     0.0
   -934.4020      2.4939      910    21:05:29     0.0
   -934.1584      2.5138      910    21:05:29     0.0
   -934.1221      2.5276      910    21:05:29     0.0
   -934.1220      2.5270      910    21:05:29     0.0
   -934.1220      2.5270      910    21:05:29     0.0

Finished on: Sun Nov 22 21:05:29 2015

LogLikelihood Converged
```

The model variance parameter estimates are converged. If the model does not converge we can use the update(HT) statement to rerun the model from its final variance parameters.

This creates an output object called HT (because that is the name we assigned to it in the function call). HT is an object of class 'asreml'. It is a sub-class of type 'List', which is an important object class in R. It contains a number of different components that we can access to get different parts of the results. We can use the str() function again to look at the overall structure of the asreml object first. We present only a small part of the output from the str() function below, because it is very long – there are 43 components inside of this object!

```
> str(HT)
List of 43
$ monitor :'data.frame':    7 obs. of 8 variables:
..$ 1 : num [1:7] -935 2.48 910 0.1 0.1 ...
..$ 2 : num [1:7] -934.402 2.4939 910 0.0903 0.0618 ...
..$ 3 : num [1:7] -934.1584 2.5138 910 0.0804 0.047 ...
...
$ factor.names : chr [1:5] "prov" "(Intercept)" "female" "block" ...
$ fixed.formula :Class 'formula' length 3 height ~ prov
.. ..- attr(*, ".Environment")=<environment: R_GlobalEnv>
$ random.formula :Class 'formula' length 2 ~female * block
.. ..- attr(*, ".Environment")=<environment: R_GlobalEnv>
$ sparse.formula :Class 'formula' length 2 ~NULL
.. ..- attr(*, ".Environment")=<environment: 0x10b28a230>
- attr(*, "class")= chr "asreml"
```

Since HT is a list, we can access the components by name using the '$' operator. For example, the tests of the fixed effects are in the component named 'aovTbl':

```
> HT$aovTbl

id df denDF Finc Fcon M Fprob
(Intercept) 2 1 NA 3566.99258 0 0 0
prov 3 3 NA 10.00657 0 0 0
```

This result matches the result of the standalone ASReml analysis, it is just formatted a little differently. Here is the log likelihood:

```
> HT$loglik
## [1] -934.122
```

The solutions to fixed and random effects are in the list component "coefficients", but this component is itself a list:

```
> str(HT$coefficients)
## List of 3
$ fixed : Named num [1:5] 0 -1.666 -0.624 -1.22 11.512
..- attr(*, "names")= chr [1:5] "prov_10" "prov_11" "prov_12" "prov_13" ...
$ random: Named num [1:221] -0.1634 -0.0826 0.2459 -0.4606 0.3556 ...
..- attr(*, "names")= chr [1:221] "female_170" "female_191" "female_192" "female_196" ...
$ sparse: NULL
- attr(*, "Terms")= chr [1:226] "prov" "prov" "prov" "prov" ...## $ fixed :
Named num [1:10] 0 -1.653 -0.627 -1.226 0 ...
```

We access the fixed and random effect solutions from sub-lists of 'coefficients' list, this is done by specifying both the top layer and lower layer list components using two '$' characters in the command:

```
> HT$coefficients$fixed

  prov_10     prov_11     prov_12     prov_13   (Intercept)
0.0000000  -1.6656326  -0.6237406  -1.2201920   11.5120637

> HT$coefficients$random

female_170     female_191     female_192     female_196     female_197
-0.163374395  -0.082566563    0.245940958   -0.460593732    0.355640427
female_198     female_200     female_201     female_202     female_203
-0.450571771    0.267092595   -0.190573032   -0.028973086    0.384757027
...
```

Again, the formatting is different than from the ASReml standalone output in the .sln file, but results are the same. For example, for both analyses we estimated 0 and -0.6272 for prov levels 10 and 12, respectively. And we predicted a value of -0.1002 for female 191 in both analyses.

You may notice that the variance components for random effects are not given directly in the asreml object HT. Instead, the HT object has the residual error variance estimate in the component 'sigma2', and gives estimates of the model random variance components as ratios relative to the residual error variance in the 'gammas' component:

```
> HT$sigma2

[1] 2.527026
> HT$gammas
female!female.var     block!block.var     female:block!female.var     R!variance
0.07499005            0.04253697          0.07820191                  1.00000000
```

We can get the variance components estimates by multiplying the gamma estimates by the residual error variance:

```
> HT$gammas * HT$sigma2

female!female.var      block!block.var    female:block!female.var      R!variance
0.1895018              0.1074920          0.1976183                    2.5270259
```

Now you can see that we get the same variance components estimates as we obtained earlier from standalone ASReml. This is somewhat non-obvious, but we can more easily obtain a more user-friendly by applying the summary() function to the asreml object:

```
  > summary(HT)
$call
asreml(fixed = height ~ prov, random = ~female * block, data = pine_prov,
na.method.Y = "include", na.method.X = "include")

$loglik
[1] -934.122

$nedf
[1] 910

$sigma
[1] 1.589662

$varcomp

                               gamma       component    std.error    z.ratio    constraint
female!female.var          0.07499005    0.1895018    0.08397968    2.256520    Positive
block!block.var            0.04253697    0.1074920    0.08992360    1.195371    Positive
female:block!female.var    0.07820191    0.1976183    0.08623574    2.291605    Positive
R!variance                 1.00000000    2.5270259    0.13146615   19.221875    Positive

attr(,"class")
[1] "summary.asreml"## $call
```

Now we can get the variance components directly from the summary of the asreml object.

Finally, if you pass the asreml object to the generic plot() function of R you get a nice set of residual diagnostic plots with no effort (Fig. 1.8):

We can compute heritability as a function of the variance components by extracting the relevant pieces from the output and doing the computation in R:

```
# Variance components
> HTvcs <- summary(HT)$varcomp$component

# Genetic variance= female*4
> Va <- HTvcs[1]*4

> Vp = sum(HTvcs[-2]) # Phenotypic variance
> h2 <- Va/Vp       # Heritability
> print(h2)

## [1] 0.26
```

Fig. 1.8 Residual diagnostic
plots of the model

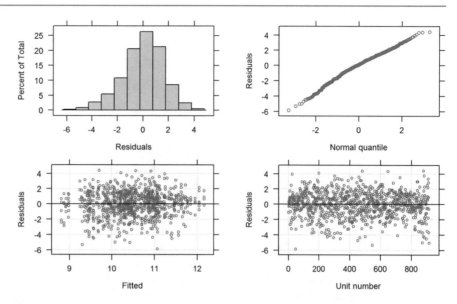

In calculation of phenotypic variance (Vp), we exclude the component explained by the block effect using HTvcs[−2], where −2 is the second element (block variance) being excluded. We get the same result as we did from using the ASReml standalone VPREDICT !DEFINE method. Unfortunately, ASReml-R does not provide the standard error of the estimate automatically. The user will have to perform the computations for the standard errors separately.

```r
# predict prov levels
HTprov_pred <- predict(HT,classify="prov")

# Not adjusted for prov effect
HTprov_female <- predict(HT, classify = "female")

# adjusted for prov effect
HTprov_fem.prov <- predict(HT,
        classify = "prov:female",
        present = c("prov", "female") )
```

Finally, we can get predictions using the `predict()` function in ASReml-R. There are three predictions in the code. The first one is for prov effect, the second one is for female effect, which does not adjust for prov effect and the third one is for female effect after adjusting for prov effect.

```r
# prediction for prov effect
predict.prov <- HTprov_pred$predictions

# not adjusted for prov effect
predict.female <- HTprov_female$predictions

#includes non-estimable combinations
HTprov_fem.prov$predictions

#only the estimable predictions
predict.female.prov <-
HTprov_fem.prov$predictions$pvals[HTprov_fem.prov$predictions$pvals$es
t.status == "Estimable",]

#Print out
round(predict.female.prov[c('prov','female',
        'predicted.value','standard.error')], 3)
```

Each of these `predict()` function calls results in the model being re-run and each also produces an asreml object output, very similar to the original HT output object. The difference is that the component `$predictions` in HT is set to NULL, whereas we get predictions for prov or female in the output of the `predict()` function calls:

```
> predict.prov <- HTprov_pred$predictions

$pvals

Notes:
- The predictions are obtained by averaging across the hypertable
calculated from model terms constructed solely from factors in
the averaging and classify sets.
- Use "average" to move ignored factors into the averaging set.

- The ignored set: female block

  prov predicted.value standard.error est.status
1   10       11.512064      0.3619918    Estimable
2   11        9.846431      0.2278751    Estimable
3   12       10.888323      0.2212824    Estimable
4   13       10.291872      0.2325509    Estimable

$avsed
overall
0.3162359
```

The `classify="prov:female"`, options creates a hyper table of prov and female effect. The `present=` option in ASReml-R sets the value of the 'extra' combinations to NA and labels them as 'Aliased'. We can use a filter to drop them.

```
#Print
> round(predict.female.prov[c('prov','female',
          'predicted.value','standard.error')], 3)

Notes:
- The predictions are obtained by averaging across the hypertable
  calculated from model terms constructed solely from factors in
  the averaging and classify sets.
- Use "average" to move ignored factors into the averaging set.

- The ignored set:  block
- Empty cells:  108.000000

    prov female predicted.value standard.error
1     10    170          11.349          0.356
2     10    191          11.429          0.339
3     10    192          11.758          0.350
40    11    196           9.386          0.336
41    11    197          10.202          0.329
42    11    198           9.396          0.329
43    11    200          10.114          0.335
...
```

Notice that the predictions match exactly those from the ASReml standalone analysis.

Electronic supplementary material: The online version of this chapter (doi:10.1007/978-3-319-55177-7_2) contains supplementary material, which is available to authorized users.

Abstract

In this chapter we will introduce the basic concepts and matrix algebra methods used to perform linear mixed models analysis. For readers who are more familiar with traditional analysis of variance (ANOVA) based on ordinary least squares methods, we first will review the ANOVA and compare ANOVA to mixed models analysis to help introduce this topic. We will show that under certain conditions, results from ANOVA and mixed models analysis are largely equivalent, but that when data are unbalanced or when we want to relax certain assumptions of the ANOVA, the mixed model analysis has properties that make it preferable. In this section, we focus on hypothesis testing and estimation using an empirical data set to show how these analyses are conducted for different methods and for different software packages. Only a few details of the mathematical machinery involved in the mixed models analysis will be covered here. A more detailed description of mixed model theory will be covered in later sections of the book. For readers interested in a more formal treatment of the argument details, they can be found in Sorensen and Gianola (Likelihood, Bayesian, and MCMC Methods in Quantitative Genetics. Springer Science & Business Media, 2007). Details of ANOVA for balanced and unbalanced data can be found in Rawlings et al. (Applied regression analysis: A research tool. New York: Springer, 2001) and Milliken and Johnson (Analysis of messy data, Designed Experiments (Vol. 1). Boca Raton: Chapman & Hall/CRC, 2004).

Mixed Models Compared to Traditional ANOVA

Maize Recombinant Inbred Lines Data

Example data (MaizeRILs.csv) were obtained by testing 62 recombinant inbred line (RIL) progeny from the cross between inbred maize lines B73 and MO17. RILs were grown in experimental units (plots) of 20 plants each using a randomized complete block design with two replications at each of four locations. The mean height for five plants within each plot is the dependent variable used for this experiment. Some data were missing from the actual data set, these were filled in with simulated data to create a balanced data set for demonstration.

location	rep	block	plot	RIL	pollen	silking	ASI	height
ARC	1	1	1	RIL-53	74	77	3	184.8
ARC	1	1	2	RIL-40	75	75	0	225.2
ARC	1	1	4	RIL-41	74	74	0	174.4
ARC	1	1	5	RIL-28	69	71	2	147.6
ARC	1	1	6	RIL-11	69	71	2	181.6

location	Location of the progeny test
rep	Replication number
block	Block number. There were 2 blocks at each location
plot	Plot number
RIL	Recombinant inbred line ID
pollen	days to pollen shed
silking	days to silking
ASI	anthesis-silk interval (*silking – pollen*)
height	Mean height of five plants in each plot

The linear model for this experiment is:

$$y_{ijk} = \mu + L_i + B(L)_{ij} + G_k + GL_{ik} + \varepsilon_{ijk} \tag{2.1}$$

Where μ = overall mean, L_i= effect of location i, $B(L)_{ij}$= effect of block j nested within location i (replication effect), G_k= effect of genotype k (RIL effect), GL_{ik}= effect of interaction between genotype k and location i, ε_{ijk}= residual (experimental error) effect of the plot containing genotype k in block j of location i. We will assume that all effects except the overall mean are random.

Balanced Data: ANOVA with SAS Proc GLM

The ANOVA layout for this experiment can be generated before data analysis based on the linear model using the names of the factors in the data set ("rep" = block, "RIL" = genotype), following the rules described by (Steel et al. 1997), where n_l is the number of locations, n_b is the number of blocks (reps) per location, and n_g is the number of genotypes (RILs):

Source	df	Expected mean squares
Location (L)	n_l-1	$\sigma_\varepsilon^2 + n_b\sigma_{GL}^2 + n_g\sigma_{B(L)}^2 + n_b n_g\sigma_L^2$
Rep(Location) (B)	$(n_b-1)\,n_l$	$\sigma_\varepsilon^2 + n_g\sigma_{B(L)}^2$
RIL (G)	n_g-1	$\sigma_\varepsilon^2 + n_b\sigma_{GL}^2 + n_b n_l\sigma_G^2$
RIL*Location (GL)	$(n_g-1)(n_l-1)$	$\sigma_\varepsilon^2 + n_b\sigma_{GL}^2$
Error	$(n_g-1)(n_b-1)n_l$	σ_ε^2
Total	$n_l n_b n_g-1$	

We can obtain a traditional ANOVA with SAS Proc GLM using this code:

Code example 2.1
Ordinary least squares ANOVA with SAS GLM procedure using the maize RIL data (see Code 2-1_Mixed models with SAS.sas **for details)**

```
/* if a sas file, then set a library (working directory) */
libname ril "V:\Book\Book1_Examples\data" ;

/* Print the first 5 lines */
proc print data=ril.maizeril (obs=5) ; run;

/* Code example 2.1: Traditional ANOVA with GLM - Balanced data */
title "GLM ANALYSIS OF MAIZE RILS BALANCED DATA SET";
proc glm data = ril.maizeril;
class location rep RIL;
model height = location rep(location) RIL RIL*location;
random location rep(location) RIL RIL*location/test;
lsmeans RIL;
run;
```

The results found in the SAS output are:

```
The GLM Procedure
Dependent Variable: height
                            Sum of
Source            DF       Squares    Mean Square   F Value    Pr > F
Model            251   264462.4916     1053.6354     16.24    <.0001
Error            244    15832.2400       64.8862
Corrected Total  495   280294.7316
```

```
R-Square       Coeff Var         Root MSE      height Mean
0.943516       4.423030          8.055199         182.1195

Source              DF      Type I SS    Mean Square    F Value    Pr > F
location             3    84931.3312     28310.4437      436.31    <.0001
rep(location)        4     3594.2244       898.5561       13.85    <.0001
RIL                 61   154937.5322      2539.9595       39.14    <.0001
location*RIL       183    20999.4038       114.7508        1.77    <.0001

Source              DF    Type III SS    Mean Square    F Value    Pr > F
location             3    84931.3312     28310.4437      436.31    <.0001
rep(location)        4     3594.2244       898.5561       13.85    <.0001
RIL                 61   154937.5322      2539.9595       39.14    <.0001
location*RIL       183    20999.4038       114.7508        1.77    <.0001
```

Notice that the default F-tests for each factor shown here are correct only for the model in which all effects except residuals are fixed. The default F-tests use the residual error variance as the denominator in all cases. Since we have assumed that all effects are random, the correct form of the F-test depends on the expected mean squares. So, for example, to test the null hypothesis that the variation among RILs is zero, the correct F-test is:

$$F = MS(RIL)/MS(location*RIL)$$

$$= 2539.959/114.7508 = 22.13 \text{ with } 61 \text{ and } 183 \text{ df}.$$

We obtain the correct F-tests by using the "random" specification in the SAS code shown above, and the result is a table of modified F-tests:

```
Tests of Hypotheses for Random Model Analysis of Variance
Dependent Variable: height

Source              DF     Type III SS    Mean Square    F Value    Pr > F
location             3          84931          28310       29.85    0.0022
Error           4.4543    4224.553277     948.420722
Error: MS(rep(location)) + MS(location*RIL) - MS(Error)

Source              DF     Type III SS    Mean Square    F Value    Pr > F
rep(location)        4    3594.224445     898.556111       13.85    <.0001
location*RIL       183          20999     114.750840        1.77    <.0001
Error: MS(Error)   244          15832      64.886230

Source              DF     Type III SS    Mean Square    F Value    Pr > F
RIL                 61         154938    2539.959544       22.13    <.0001
Error              183          20999     114.750840
Error: MS(location*RIL)
```

We can also estimate variance components for random effects from the ANOVA using the method of moments, by which we equate the observed mean squares to their expectations and solve for the variance components. For example, the expected mean square for the error variance is equal to the error variance component, but the expected mean square for the location*RIL variance component is the sum of the error variance plus twice the location*RIL variance component. So, we can solve for the estimated variance components as follows:

$$\widehat{\sigma}_{\varepsilon}^2 = MS(error) = 64.89$$

$$\widehat{\sigma}_{GL}^2 = \frac{MS(location*RIL) - MS(error)}{n_b} = \frac{114.75 - 64.89}{2} = 24.93$$

$$\widehat{\sigma}_{G}^2 = \frac{MS(RIL) - MS(location*RIL)}{n_l n_b} = \frac{2539.96 - 114.75}{8} = 303.15$$

(2.2)

Finally, we can estimate the predicted marginal mean value of each RIL using the "lsmeans" statement in the SAS code above, resulting in least squares means estimates of each genotype. The first eight RILs have the following mean values in the Proc GLM output:

```
Least Squares Means
          height
RIL            LSMEAN
RIL-1       182.100000
RIL-11      182.875000
RIL-12      185.200000
RIL-14      194.250000
...
```

Balanced Data: ANOVA with R

We can also obtain the same results using R and `lm()` function:

Code example 2.2
Ordinary least squares ANOVA with R using balanced data (see **Code 2-2_Mixed models with R.R** for more details)

```
#ANOVA
rils <- read.csv("MaizeRILs.csv")
rils$rep <- as.factor(rep) # make sure rep number is coded as a factor
m0 <- lm(height~location*RIL + rep:location, data = rils)
anova(m0)
```

R results from this output:

```
Analysis of Variance Table
Response: height
               Df    Mean Sq    F value    Pr(>F)
location        3    28310.4    436.3090   < 2.2e-16   ***
RIL            61     2540.0     39.1448   < 2.2e-16   ***
location:RIL  183      114.8      1.7685   1.643e-05   ***
location:rep    4      898.6     13.8482   3.408e-10   ***
Residuals     244       64.9
---
Signif. codes:  0 '***' 0.001 '**' 0.01 '*' 0.05 '.' 0.1 ' ' 1
Full sample code is available in ''example2a.r'' file.
```

Again, the default F-tests are correct only if we assume all effects are fixed. If we considered these effects to be random, we would have to construct F-tests some other way (or more appropriately, just do a proper mixed models analysis)!

The summary() function will present the estimates of coefficients for each term in the model, but in the case of factors in analysis of variance, these coefficients may be hard to interpret. The lm() function arbitrarily sets the coefficient for the first level of each factor to zero to avoid singularity because we are fitting more parameters than we have degrees of freedom to estimate independently. In this case, the output of the summary() call on the m0 object is:

```
> (m0_summary <- summary(m0))
Coefficients:
                        Estimate Std. Error t value Pr(>|t|)
(Intercept)              179.797      5.742  31.315  < 2e-16  ***
locationCLY               29.122      8.120   3.587  0.000405 ***
locationPPAC             -15.965      8.120  -1.966  0.050413 .
locationTPAC              -3.289      8.120  -0.405  0.685770
RILRIL-11                  4.200      8.055   0.521  0.602559
RILRIL-12                 13.400      8.055   1.664  0.097492 .
RILRIL-14                 26.600      8.055   3.302  0.001103 **
...
locationCLY:RILRIL-11     -7.700     11.392  -0.676  0.499727
locationPPAC:RILRIL-11   -10.200     11.392  -0.895  0.371465
locationTPAC:RILRIL-11     4.200     11.392   0.369  0.712681
locationCLY:RILRIL-12    -25.400     11.392  -2.230  0.026679 *
locationPPAC:RILRIL-12   -10.800     11.392  -0.948  0.344041
locationTPAC:RILRIL-12    -5.000     11.392  -0.439  0.661113
locationCLY:RILRIL-14    -28.600     11.392  -2.511  0.012701 *
locationPPAC:RILRIL-14   -16.800     11.392  -1.475  0.141569
locationTPAC:RILRIL-14   -12.400     11.392  -1.089  0.277446
...
locationARC:rep2          -8.395      1.447  -5.802  2.02e-08 ***
locationCLY:rep2          -1.839      1.447  -1.271  0.204968
locationPPAC:rep2          5.535      1.447   3.826  0.000165 ***
locationTPAC:rep2          3.384      1.447   2.339  0.020147 *
```

Notice that although there are four locations, there are coefficients estimated only for three of the four. The first level of location is "ARC" which has a coefficient set to zero. Similarly, there is no coefficient estimate for RIL-1, it is set to zero.

To obtain the least square means of each RIL, we need to understand how the least square means are estimated as functions of the model coefficient estimates. The estimate of the mean value of RIL-1 is:

$$LSMean(RIL1) = \bar{y}_{..1} = \mu + \overline{L.} + \overline{B(L)}_{..} + G_1 + \overline{GL}_{.1} \tag{2.3}$$

This is simply the linear model we started with, but replacing the generic k subscript for genotypes with 1, averaging over all other terms in the model, and dropping the term for residual effects since they do not contribute to a marginal mean prediction. Realizing that the lm() function sets the coefficients for G_1 and G_{i1}'s to zero, we are left with this equation for the first RIL:

$$LSMean(RIL1) = \bar{y}_{..1} = \mu + \overline{L.} + \overline{B(L)}_{..} \tag{2.4}$$

We can obtain this from the results of the ANOVA by assigning the 'coefficients' element of the m0_summary object to a new object called 'coefs' and pulling out the pieces we need by indexing the row by name and the column by number 1 (since the coefficients are the first column of the object):

```
>coefs <- m0_summary$coefficients

>#Estimate mean value for RIL1 as mu + mean(Loc) + mean(Rep(Loc))
>(RIL1_mn <- coefs["(Intercept)",1] + (sum(coefs[c("locationCLY",
"locationPPAC", "locationTPAC")],1])/4) +
  (sum(coefs[c("locationARC:rep2", "locationCLY:rep2",
"locationPPAC:rep2", "locationTPAC:rep2")],1])/8))

[1] 182.1
```

This agrees with the 'lsmean' estimate from SAS Proc GLM. Notice that we averaged over four location effects even though we have estimates for only three (because the ARC location effect is zero); similarly, we averaged over 8 rep effects, even though we have coefficients for four of them – the other four are set to zero, but nonetheless are part of the average value for rep effects.

To obtain the least square means of a different RIL, we need to also include the RIL coefficient and the average of its interactions with locations (and again, notice that this is an average over four interaction terms including one coefficient that is set to zero):

```
>#Estimate mean value for RIL11 as mu + mean(Loc) + mean(Rep(Loc)) +
RIL11 + mean(RIL11*Loc)
>(RIL11_mn <- RIL1_mn + coefs["RILRIL-11",1] +
(sum(coefs[c("locationCLY:RILRIL-11", "locationPPAC:RILRIL-11",
"locationTPAC:RILRIL-11")],1])/4))

[1] 182.875
```

Of course, this would be a pain to repeat for all of the RILs, so in practice we can use the lsmeans() function of the lsmeans package, or the LSMeans() function of the doBy package. Here we illustrate applying the lsmeans() function on the 'm0' linear model object:

```
>library(lsmeans)
>lsmeans(m0, 'RIL')
NOTE: Results may be misleading due to involvement in interactions
 RIL      lsmean       SE  df lower.CL upper.CL
 RIL-1  182.1000 2.847943 244 176.4903 187.7097
 RIL-11 182.8750 2.847943 244 177.2653 188.4847
 RIL-12 185.2000 2.847943 244 179.5903 190.8097
 RIL-14 194.2500 2.847943 244 188.6403 199.8597
 ...
```

Balanced Data: ANOVA with ASReml

The same basic ANOVA results can be obtained using the following ASReml code:

Code example 2.3
Ordinary least squares ANOVA with ASReml using balanced data (see **Code 2-3_Mixed models with ASReml.
as** for more details)

```
!OUTFOLDER  V:\Book1_Examples\ch02_mm\outputfiles
!CYCLE 1 2
Title: Maize RILs balanced.
 location  !A
 rep   *
 block   *
 plot   !I
 RIL    !A
 pollen  !I
 silking  !I   ASI   *   height

!FOLDER  V:\Book1_Examples\data     # data folder
MaizeRILs.csv  !SKIP 1    !DOPART $I

!PART 1 # fixed effects model
height  ~ mu location rep.location RIL location.RIL
predict RIL

!PART 2 # random effects model
height  ~ mu ,
     !r location rep.location RIL location.RIL
predict RIL
```

This produces the following ANOVA output in ASReml:

```
        - - - Results from analysis of height - - -

LogL:      LogL    Residual      NEDF    NIT Cycle Text
LogL:    -723.88    64.8862       244     2 1 "LogL Converged"
Akaike Information Criterion        1449.76 (assuming 1 parameters).
Bayesian Information Criterion      1453.26

Model_Term               Gamma       Sigma   Sigma/SE   % C
Residual      SCA_V 496  1.00000     64.8862     11.05   0 P

                    Wald F statistics
     Source of Variation   NumDF    DenDF     F-inc    P-inc
 10  mu                       1     244.0    0.25E+06   <.001

  1  location                 3     244.0     436.31    <.001
 11  rep.location             4     244.0      13.85    <.001
  5  RIL                     61     244.0      39.14    <.001
 12  location.RIL           183     244.0       1.77    <.001
```

ASReml treats all the terms as fixed, as SAS GLM procedure type 1 sum of squares. Notice that the F statistic for RIL is 39.14, the same as the F test obtained from GLM. The incremental F-tests are not correct. The predictions for RILs are lsmeans and are produced in the .pvs file.

Balanced Data: Mixed Models Analysis with SAS Proc MIXED

We can analyze these data using a mixed model with this SAS Proc MIXED code:

Code example 2.1
(continued)

```
proc mixed data=ril.maizeril covtest;
title "MIXED MODELS ANALYSIS OF MAIZE RILS BALANCED DATA" ;
class location rep RIL;
model height = /solution;
random location rep(location) RIL RIL*location/solution;
ods output solutionR = random solutionF = fixed;
```

Note that the class statement is identical to Proc GLM, but now we include only fixed effects in the model statement (in this case the overall mean or intercept is the only fixed term, and since it is implicit, we can write model "trait = ;"), and random effects in the random statement.

The relevant output from this analysis in SAS is:

```
                              The Mixed Procedure

                    Covariance Parameter Estimates
                                    Standard        Z
Cov Parm              Estimate        Error      Value      Pr Z
location                220.66       186.48       1.18     0.1184
rep(location)          13.4463      10.2484       1.31     0.0948
RIL                     303.15      57.5088       5.27     <.0001
location*RIL           24.9323       6.6787       3.73     <.0001
Residual               64.8862       5.8745      11.05     <.0001

          Fit Statistics
-2 Res Log Likelihood         3833.2
AIC (smaller is better)       3843.2
AICC (smaller is better)      3843.3
BIC (smaller is better)       3840.1
```

Some differences between the Proc MIXED and Proc GLM outputs are immediately obvious. First, Proc MIXED output does not provide degrees of freedom, sum of squares, mean squares, or F-tests for random terms in the model. Instead, the variance components estimates are provided directly. Note that they are equivalent to the variance components estimates in this case. Proc MIXED by default estimates variance components and random effects with restricted maximum likelihood (REML). One of the nice properties of REML is that it gives variance components estimates equal to method of moments estimates when data are balanced and all components estimates are greater than zero, as in this case. REML restricts variance components to be greater than or equal to zero, but this is not true for the method of moments. If the method of moments estimate for one variance component is negative, then all of the variance components estimates may differ between REML and method of moments methods.

Balanced Data: Mixed Models Analysis with R

We can fit the same mixed model with the *lme4* package in R with this code.

**Code example 2.2
(continued)**

```
require(lme4)
mm <-lmer(height ~1 + (1|RIL) + (1|location/rep) + (1|location:RIL))
summary(mm)
```

which produces this output:

```
Linear mixed model fit by REML
Formula: height ~ 1+(1|RIL)+(1|location/rep)+ (1|location:RIL)
   Data: rils
  AIC  BIC logLik deviance REMLdev
 3845 3870  -1917     3839     3833

Random effects:
 Groups         Name          Variance   Std.Dev.
 location:RIL   (Intercept)    24.932     4.9932
 RIL            (Intercept)   303.151    17.4112
 rep:location   (Intercept)    13.446     3.6669
 location       (Intercept)   220.661    14.8547
 Residual                      64.886     8.0552
Number of obs: 496, groups: location:RIL, 248; RIL, 62; rep:location, 8; location, 4

Fixed effects:
               Estimate Std.  Error t value
 (Intercept)    182.119        7.869   23.14
```

The variance components estimates from *lme4* are identical to the estimates from Proc MIXED, but the AIC and BIC values are different between *lme4* and the SAS Proc MIXED output as well. These differences arise due to details in some of the assumptions made in models fit with the *R lmer* function versus SAS Proc MIXED or ASReml and will not be explained in detail here. For the reader interested in a broader treatment of the subject, see (Bates et al. 2013). In addition, the "Std.Dev." values in this output are not the standard errors of the variance components estimates, instead they are simply the square roots of the variance components estimates (e.g., $\overline{\sigma}_G$, not $SE(\overline{\sigma}_G^2)$).

Measures of uncertainty in the variance components estimated by lme4() can be obtained as confidence intervals, however:

```
>confint(mm)
Computing profile confidence intervals ...
                  2.5 %       97.5 %
.sig01         3.570868     6.281389
.sig02        14.574225    21.154625
.sig03         1.908606     9.279741
.sig04         6.928737    32.399264
.sigma         7.390111     8.826378
(Intercept)  164.924810   199.314146
```

The labeling of terms in the output is not totally clear, but the confidence intervals are labeled numerically to match their order in the summary of the *lme4* object. Thus, '.sig02' refers to the second variance component in the previous summary output, that is, the RIL term. Also, the confidence intervals are reported on the standard deviation scale rather than the variance scale. So, by squaring the interval end points, we can obtain the confidence interval for the variance component as $(14.574^2, 21.155^2) = (212.4, 447.5)$. This can be compared to the approximate 95% confidence interval constructed from the Proc MIXED estimate of the variance component +/− two times its standard error = $303.15 +/- 2(57.5088) = (188.1, 418.2)$. The latter estimate assumes that the estimators are normally distributed, which also implies that they are symmetrically distributed around the point estimate. The confidence interval from `lme4()` is based on evaluating the model likelihood at different values of the variance component and does not assume symmetry, which is why it is more conservative at the high end and less conservative at the low end compared to the approximate confidence interval based on the standard error. This likelihood profile based confidence interval is more accurate than the approximation based on the standard error of the variance component.

Balanced Data: Mixed Models Analysis with ASReml

Finally, we can also fit this mixed model in ASReml with part 2 of the following code.

Code example 2.3
(continued)

```
!OUTFOLDER  V:\Book\Book1_Examples\ch02_mm\outputfiles
!CYCLE 1 2 !JOIN
Title: Example 2A maize B73 Mo17 RILs.
#location,rep,block,plot,RIL,pollen,silking,ASI,height
 location   !A
 rep   2
 block   8
 plot   !I
 RIL    !A
 pollen   silking   ASI *   height

!FOLDER  V:\Book\Book1_Examples\data
MaizeRILs.csv  !SKIP 1 !DOPART $I

!part 1 # fixed effects model
height ~ mu location rep.location RIL location.RIL
predict RIL

!part 2    # random effects model
height  ~ mu ,    # Specify fixed effect
     !r location rep.location RIL location.RIL  # random effects
predict RIL
```

The ASReml output follows:

Model_Term			Gamma	Sigma	Sigma/SE	%	C
location	IDV_V	4	3.40074	220.661	1.18	0	P
rep.location	IDV_V	8	0.207229	13.4463	1.31	0	P
RIL	IDV_V	62	4.67204	303.151	5.27	0	P
location.RIL	IDV_V	248	0.384246	24.9323	3.73	0	P
Residual	SCA_V	496	1.00000	64.8862	11.05	0	P

The standard error of the RIL variance component can be obtained from the following relationship: SE = Sigma/(Sigma/SE) = 303.151/5.27 = 57.52, very close to the estimated SE from SAS Proc MIXED.

Hypothesis Testing with Mixed Models

Hypothesis testing for the variance components can be based on the "Z value" obtained by using the "covtest" option in the Proc MIXED statement. The Z value is the ratio of the variance component to its standard error, and this is also given in the ASReml output as 'Sigma/SE'. This test has low power, particularly for variance components estimated with few degrees of freedom. Thus, in the SAS and ASReml basic output tables it is clear that the variance components for RIL and location*RIL are significant, but location variation appears not to be significant based on this test, even though it has a rather large variance component.

Hypothesis testing with higher power can be implemented with the likelihood ratio test. This test requires one to fit an additional mixed model for each factor to be tested, in which one removes the factor of interest from the model. The likelihood of this "reduced" model can be compared to the likelihood of the "full" model to form a test of the null hypothesis that the variance component for the dropped term is zero. If removing the term causes a large decrease in the likelihood of the reduced model, then there is more evidence that the variance component for the term is greater than zero.

In this example, we can test the null hypothesis of no variation among locations by removing location from the model and comparing its likelihood to the full model with this code in SAS:

(See **Code 2.4_Mixed models with SAS.sas** for more details).

```
proc mixed data=ril.MaizeRIL;
class location rep RIL;
model height =;
random rep(location) RIL RIL*location;
```

All we need from the output of this model to conduct this test is the likelihood:

```
          Fit Statistics
-2 Res Log Likelihood           3841.3
```

The term in the output is actually -2 times the natural log of the likelihood. The reason this number is provided is to make the likelihood ratio test (LRT) easy to compute. The LRT is:

$$\begin{aligned}
\text{LRT} &= (-2) \ln \text{ (likelihood of reduced model/likelihood of full model)} \\
&= -2^* \ln \text{ (likelihood of reduced model)} - (-2)^* \ln \text{ (likelihood of full model)} \\
&= 3841.3 - 3833.2 = 8.1
\end{aligned}$$

This value is distributed approximately as a chi-square value with one degree of freedom because the models differ for one parameter. The p-value for a chi-squared value of 8.1 with 1 df is 0.004. We need to adjust the p-value to half of the tabular value because we are testing the null hypothesis of a variance component equal to zero; see (Self and Liang 1987) for details. So, the adjusted p-value is 0.002. As you can see, the LRT provides strong evidence that the location effect is significant even when the Z-test did not. Further, the adjusted p-value from the LRT is equivalent to the p-value from the F-test for location using ANOVA in this example with balanced data (shown previously).

For completeness, we show the results of LRTs for the RIL and location*RIL components:

Model	−2 RLL	LRT	Raw p-value	Adjusted p-value
Full	3833.2			
No RIL	4092.7	259.5	<0.0001	<0.0001
No location*RIL	3850.4	17.20	<0.0001	<0.0001

Prediction: BLUE and BLUP

As noted before, the predicted marginal mean value of a fixed effect is obtained as a least square mean. This prediction is obtained as a linear combination of fixed effects estimates from the model. In the ordinary least squares setting, the only true random effect is the residual error, and these do not contribute to the least square means estimates. The least square mean for a particular level of a factor is a 'marginal mean' because it is averaged over levels of other fixed factors in the model and also over interaction effects involving the factor level. In the example used in this section, the least square mean of an RIL involves the linear combination of the overall mean, the particular RIL effect estimate, the average of all environment effects, and the average of all RIL-by-environment interaction effects that involve the particular RIL.

As we move to the mixed model, we can use the same concept to estimate the predicted marginal means of fixed effects. These are also called 'least square means' in SAS and elsewhere. In the mixed model framework, however, effects for a given random factor are assumed to have mean values of zero, so that when we average over levels of a random factor, the average is zero. Thus, random factors do not contribute to the least square mean of a fixed effect. Furthermore, the interaction effects involving a random effect with one level of a fixed factor also sum to zero; therefore interactions with any random factors in the model can be ignored in the least square mean of a fixed factor level.

These predicted marginal means or least square means for fixed effects are called best linear unbiased estimators (BLUEs). They are 'best' because they maximize correlations between true values and predicted values, they are 'linear' because they are linear combinations of model effect estimates, and they are 'unbiased' because their expected value is equal to the true value.

To predict the marginal value of random effects in the models, we use an analogous value called a best linear unbiased predictor (BLUP). To obtain the marginal prediction BLUP, we would again average over levels of fixed effects in the model, average over interactions between the random effect level of interest and all levels of other fixed effects, ignore other random effects, and include the intercept (mu) plus the effect prediction for the particular level of the random factor. A marginal predicted BLUP value is not the only type of BLUP used, however. By convention, animal breeders typically use the random genotype effect predictions directly as BLUPs for making selections and comparisons, whereas crop breeders often use marginal effect predictions. In some cases, predictions are made for random effects at a particular level of some other effect (perhaps a specific level of a fixed treatment or at a particular environment). Any linear combination of effect estimates from a mixed model analysis that includes random effects is a BLUP (Robinson 1991).

A sometimes more useful BLUP in plant breeding is a 'conditional' BLUP, which will have a smaller standard error than the marginal BLUP because it is predicted for a specific set of random factor levels. The specific set can be chosen as the average of the random effects, in which case the conditional and marginal BLUPs are usually equal, but the variability in the other random effects does not contribute to the standard error of the conditional BLUP. This is convenient in situations such as multi-environment crop variety trials because it relates to the precision of comparisons of variety BLUPs within the set of tested environments, and the standard error of the conditional predictions relate directly to the typical concept of heritability relating to selection among variety values evaluated across multiple environemnts. We will demonstrate this difference in Chap. 7.

To compare the prediction of random RIL values from a mixed model to estimation of RIL means from a fixed model, consider how this is done in SAS Proc MIXED vs. SAS Proc GLM. The first difference one will notice is that an error message will result if one includes the statement "lsmeans RIL" as part of the Proc MIXED analysis if RIL is included in the random effects part of the model. This happens because random effects are 'predicted' by BLUPs instead of estimated by BLUEs, so there is no such thing as a 'least square mean' of a random effect.

In practice, we obtain the RIL effect predictions by requesting the solutions for the random effects in the model using the "/solution" option on the random statement in Proc MIXED. We can then construct marginal BLUPs in this case by simply adding the estimated overall mean effect (μ, obtained with the "/solution" option on the model statement) to each RIL effect prediction. (Recall from above that if we considered all the other model effects random, they do not contribute to the BLUPs for RILs).

The estimate of μ is 182.12, obtained from the SAS Proc MIXED output:

```
                    Solution for Fixed Effects
                         Standard
 Effect         Estimate    Error      DF     t Value    Pr > |t|
 Intercept      182.12      7.8719     3      23.14      0.0002
```

The SAS code provided saves the random effect estimates into a new data set, where they can be processed by extracting just the RIL effects and adding them to the mean estimate to get the RIL BLUPs. The SAS code is included in *Code 2-1_ANOVA with SAS GLM procedure.sas* file; here we are interested in the result of that processing for the first eight RILs for comparison to the *LSmeans* obtained from Proc GLM:

RIL	Random effect predictor	mu	BLUP	LSMEAN	Fixed effect estimate	Fixed effect*h^2
RIL-1	-0.02	182.12	182.10	182.10	-0.02	-0.02
RIL-11	0.72	182.12	182.84	182.88	0.76	0.72
RIL-12	2.94	182.12	185.06	185.20	3.08	2.94
RIL-14	11.58	182.12	193.70	194.25	12.13	11.58
RIL-15	13.04	182.12	195.16	195.78	13.66	13.04
RIL-16	-8.87	182.12	173.25	172.83	-9.29	-8.87
RIL-20	26.38	182.12	208.50	209.75	27.63	26.38
RIL-21	-15.53	182.12	166.59	165.85	-16.27	-15.53

Notice that the BLUPs are shrunk back toward the overall mean (182.12) compared to the LSmeans. For example, RIL-20 has an LSmean of 209.75 but a BLUP of 208.50, whereas RIL-21 has an LSmean of 165.85 and BLUP of 166.50. The amount of shrinkage depends on the heritability. In this case the heritability of entry means is:

$$\widehat{h}^2 = \frac{\widehat{\sigma}^2_{RIL}}{\widehat{\sigma}^2_{RIL} + \frac{\widehat{\sigma}^2_{RIL*location}}{4} + \frac{\widehat{\sigma}^2_{\varepsilon}}{8}} = 0.954 \tag{2.5}$$

Notice that in this case of balanced data, the random effect predictor for RILs is equal to the deviation of the line's least square mean from the overall mean times the heritability.

To summarize, ANOVA and mixed models analysis provide equivalent variance components estimates in the case of balanced data and no negative variance components estimates, although we have to use some different methods to test hypotheses about the variance components. Random effects are predicted by BLUPs from the mixed model, which are 'shrunk' toward the overall mean relative to least squares means (BLUEs). When data are unbalanced either due to a balanced experimental design with missing data or due to unbalanced designs, ANOVA and mixed models results diverge in more important ways, as described in the next section.

Unbalanced Data

ANOVA with SAS Proc GLM

To demonstrate the effects of unbalanced data on analysis of variance, we now consider the true original data set of maize RILs, which includes about 4.5% missing data (MaizeRIL_miss.csv). We will show that the pattern of missing data has important effects on estimation; note that RIL-5 has data from only two of four locations and RIL-51 has data only from three locations. When we fit the linear model to these data with SAS Proc GLM, we obtain the following results:

Code example 2.4
Using SAS GLM procedure for unbalanced data (See **Code 2-4_Mixed models with SAS unbalanced data.sas** for more details):

```
* ANOVA with GLM-Unbalanced data
title "GLM ANALYSIS OF MAIZE RILS UNBALANCED DATA SET";
proc glm data = maize_miss;
 class location rep RIL;
 model mean_height = location rep(location) RIL RIL*location;
 random location rep(location) RIL RIL*location/test;
 lsmeans RIL;
run;
```

Output from the code

```
The GLM Procedure
Dependent Variable: height
```

Source	DF	Sum of Squares	Mean Square	F Value	Pr > F
Model	248	251616.8381	1014.5840	15.06	<.0001
Error	225	15157.9644	67.3687		
Corrected Total	473	266774.8025			

R-Square	Coeff Var	Root MSE	height Mean
0.943181	4.509687	8.207846	182.0048

Source	DF	Type I SS	Mean Square	F Value	Pr > F
location	3	79345.6274	26448.5425	392.59	<.0001
rep(location)	4	3693.2364	923.3091	13.71	<.0001
RIL	61	150287.3376	2463.7268	36.57	<.0001
location*RIL	180	18290.6367	101.6146	1.51	0.0018

Source	DF	Type III SS	Mean Square	F Value	Pr > F
location	3	77171.8286	25723.9429	381.84	<.0001
rep(location)	4	3677.7950	919.4488	13.65	<.0001
RIL	61	149644.4648	2453.1879	36.41	<.0001
location*RIL	180	18290.6367	101.6146	1.51	0.0018

Comparing this output to the Proc GLM output from the balanced data set, a few key differences should be noted. First, the degrees of freedom for RIL are still 61 but the degrees of freedom for location*RIL are now 180 instead of 183. The reason for this is that we have no data on two of the location*RIL interactions involving RIL-5 (because we have no data on this RIL from two locations) and one of the location*RIL interactions involving RIL-51 (as it is missing in one location). Second, note that now the Type I and Type III sums of squares (SS) and mean square (MS) results are different from each other in this case. This occurs because the Type I statistics are computed by fitting the effects in the order given in the model and computing the sums of squares accounting for each term sequentially, whereas the Type III statistics are computed by calculating the sums of squares attributable to each term after accounting for all other terms in the model. In the case of balanced data, all of the model terms are orthogonal to each other such that the order of fitting factors does not affect how much variation they are associated with. In contrast, with unbalanced data, the different model factors become correlated and the variation associated with any one term may also be partly associated with a different term, such that the order of fitting terms affects the sums of squares for the term. Because of this, Type III statistics are preferred since they indicate the amount of variation attributable to each factor after accounting for the other factors in the model As a result, the sum of Type III statistics will be less than the total sums of squares for the model: in this example the sum of the Type III SS = 248784.7251, whereas the total SS for the model is 266774.8025.

Variance components can be estimated by the method of moments from ANOVA Type III MS, but two complications arise in the case of unbalanced data: First, such estimates are reasonable estimates if the data are not too badly balanced, but the statistical properties of such estimators are unknown, so it can be difficult to know how reliable they are for a given data set. Second, the expected mean squares shown for the balanced data set above are not correct for the unbalanced data case, as the coefficients on the variance components are affected by the data structure. The computation of the coefficients can be horribly complex (Milliken and Johnson 2004; Rawlings et al. 2001), but we can get the coefficients using SAS Proc GLM with the random statement, resulting in this output:

```
Source              Type III Expected Mean Square
location            Var(Error) + 1.8769 Var(location*RIL) + 57.243
                    Var(rep(location)) + 114.49 Var(location)

rep(location)       Var(Error) + 57.25 Var(rep(location))
RIL                 Var(Error) + 1.9149 Var(location*RIL) + 7.6105Var(RIL)
location*RIL        Var(Error) + 1.9348 Var(location*RIL)
```

The complex coefficients in the expected mean squares result in the standard F-tests being incorrect. So, for example, with balanced data we could test the null hypothesis of no RIL variation by the F-test of **MS(RIL)/MS(location*RIL)** which had the expectation:

$$F_{RIL} = \frac{MS(RIL)}{MS(location * RIL)} = \frac{\sigma_\varepsilon^2 + 2\sigma_{location*RIL}^2 + 8\sigma_{RIL}^2}{\sigma_\varepsilon^2 + 2\sigma_{location*RIL}^2} = 1 + \frac{8\sigma_{RIL}^2}{\sigma_\varepsilon^2 + 2\sigma_{location*RIL}^2} \tag{2.6}$$

The extent to which the F-value is greater than one provides evidence for variation due to RILs. However, consider the expectation of this same F-test in the case of unbalanced data:

$$F_{RIL} = \frac{MS(RIL)}{MS(location * RIL)} = \frac{\sigma_\varepsilon^2 + 1.9149\sigma_{location*RIL}^2 + 7.6105\sigma_{RIL}^2}{\sigma_\varepsilon^2 + 1.9348\sigma_{location*RIL}^2} \tag{2.7}$$

As this expectation does not equal 1 when the null hypothesis is true (RIL variance component $=0$), the F-test is not correct. Instead, more complicated forms of the F-test are required, and Proc GLM computes these forms when the random statement is given:

```
The GLM Procedure
Tests of Hypotheses for Random Model Analysis of Variance
Dependent Variable: height

Source              DF      Type III SS     Mean Square    F Value    Pr > F
location             3            77172           25724      27.00    0.0031

Error        4.2929     4089.266951      952.566239
Error: 0.9999*MS(rep(location)) + 0.9701*MS(location*RIL)
 - 0.9699*MS(Error)

Source              DF      Type III SS     Mean Square    F Value    Pr > F
rep(location)        4     3677.795036      919.448759      13.65    <.0001
location*RIL       180            18291      101.614649       1.51    0.0018

Error: MS(Error)   225            15158       67.368731

Source              DF      Type III SS     Mean Square    F Value    Pr > F
RIL                 61           149644     2453.187948      24.23    <.0001

Error         182.48            18479      101.262345
Error: 0.9897*MS(location*RIL) + 0.0103*MS(Error)
```

Similarly, the variance components can be estimated by method of moments, but the estimation is more complex:

$$\widehat{\sigma}_\varepsilon^2 = MS(error) = 67.37$$

$$\widehat{\sigma}_{GL}^2 = \frac{MS(location^*RIL) - MS(error)}{1.9348} = \frac{101.61 - 67.37}{1.9348} = 17.7$$

$$\widehat{\sigma}_G^2 = \frac{MS(RIL) - 1.9149\widehat{\sigma}_{GL}^2 - \sigma_\varepsilon^2}{8} = \frac{2539.96 - 1.9149(17.7) - 67.37}{8} = 304.84$$

(2.8)

Furthermore, as mentioned above, the statistical properties of these estimates are not known, so, for example, exact estimates of their standard errors are unknown.

Estimation of least square means can also be problematic. Recalling that RILs 5 and 51 were completely missing at least one location, note the results when we request the least square means for RILs with Proc GLM:

```
                height
RIL                LSMEAN

RIL-1           182.100000
RIL-11          182.875000
RIL-12          185.200000
...
RIL-49          176.975000
RIL-5              Non-est
RIL-50          200.275000
RIL-51            Non-est
RIL-53          174.425000
```

"Non-est" indicates that the LSmeans for RILs 5 and 51 are non-estimable. To understand why this is the case, consider the formula for computing the LSmean of genotype k:

LSmean $(Y_{.k}) = (Y_{..k}) = \mu + \bar{L}_. + \overline{B(L)}_{..} + G_k + \overline{GL}_{.k}$, where $\bar{L}_.$ and $\overline{B(L)}_{..}$ are the averages over all location and replication effects, and $\overline{GL}_{.k}$ is the average over all locations of the interaction between RIL k and each location. In the fixed model, the interaction of a genotype-by-location interaction effect is non-estimable if there are no data on that combination of genotype and location. Then if some interaction effects included in the LSmean equation are non-estimable, the whole LSmean is non-estimable.

Unbalanced Data: Mixed Models Analysis with SAS

For comparison, we perform the mixed models analysis of the same data set and obtain the following results from SAS Proc MIXED. The code is the same as in the previous section "Balanced Data: Mixed Models Analysis with SAS Proc MIXED", but now we apply it to the data set with missing observations:

Code example 2.4
(continued)

```
proc mixed data = ril.maizeril_miss covtest;
title "MIXED MODELS ANALYSIS OF MAIZE RILS UNBALANCED DATA SET";
class location rep RIL;
model height =/solution;
random location rep(location) RIL RIL*location/solution;
ods listing exclude solutionR;
ods output solutionR = random solutionF = fixed;
run;
```

which produces the following output:

```
                        Covariance Parameter Estimates
                                   Standard            Z
Cov Parm               Estimate      Error          Value          Pr Z
location                217.69       184.69          1.18         0.1193
rep(location)           15.0591      11.4593         1.31         0.0944
RIL                     309.19       58.5756         5.28         <.0001
location*RIL            17.1466      6.4757          2.65         0.0041
Residual                67.9184      6.4110         10.59         <.0001
```

Notice that the variance components are similar to, but not the same as, those obtained from the ANOVA. More importantly these estimators have known asymptotic statistical properties (permitting estimation of their standard errors) and have the properties of maximum likelihood – among all possible estimates they have the highest probabilities given the observed data and the assumption that the effects of the different levels of random factor are normally distributed.

The BLUPs for RILs are computed as before, and we compare them to the RIL LSmeans shown above from the ANOVA:

RIL	Random effect predictor	mu	BLUP	LSMEAN
RIL-1	0.39	181.70	182.08	182.10
RIL-11	1.13	181.70	182.83	182.88
RIL-12	3.37	181.70	185.06	185.20
...				
RIL-49	−4.53	181.70	177.16	176.98
RIL-5	−28.74	181.70	152.96	Non-est
RIL-50	17.84	181.70	199.54	200.28
RIL-51	0.53	181.70	182.22	Non-est
RIL-53	−6.98	181.70	174.71	174.43

The mixed model assumes that the various levels of each effect were sampled from a random normal distribution with a mean of zero. Thus, when a genotype-location combination is missing from the data the expectation of its value is zero. We can still predict both the specific genotype-location combination value and the overall genotype value. Comparing BLUPs to LSmeans is further complicated because the shrinkage effect for each BLUP can be different as it depends on the amount and pattern of data for each genotype. Thus, the shrinkage effect on the RILs with only a few data points is greater than that for RILs with more data points, reflecting the higher confidence one has in deviations from the mean based on more data. Similar results can be obtained with ASReml and lme4 in R.

Box 2.1. A Brief Introduction to Matrices

In the next section of this chapter, we introduce some computational details of mixed models using matrix algebra, as this provides the most succinct way to represent the mixed model equations and their solutions. A good understanding of matrix forms and matrix algebra will help the reader understand the material that follows throughout the rest of the book. We provide this sidebar introduction for readers not well versed in matrix representations and matrix algebra. More advanced readers can skip this section.

A **Matrix** is a two-dimensional rectangular array of numbers with the following properties:

- A_{ij} denotes the **element** in the i-th row and in the j-th column of matrix A
- A matrix is indexed by its rows and columns; the combination of row number and column number uniquely specifies each element.
- The dimensions of the matrix are defined by the number of rows and number of columns, in that order ($r \times c$)

Box 2.1 (continued)

- A matrix with one of its dimensions equal to one is a **vector**. A vector with dimension $(r \times 1)$ is called a 'column vector' because it represents a single column. A vector with dimension $(1 \times c)$ is called a 'row vector' because it represents a single row.

We use bold lower case Roman letters (e.g., **a**), for scalars or vectors and use bold capital letters (e.g., **A**) for rectangular or square matrices.

Examples of vectors are:

$\mathbf{a} = (4)$ scalar (matrix with dimensions $r = 1$, $c = 1$)

$\mathbf{c} = (4 \quad 3)$ row vector (matrix with dimensions $r = 1$, $c = 2$)

$\mathbf{d} = \begin{pmatrix} 1 \\ 4 \\ 2 \end{pmatrix}$ column vector (matrix with dimensions $r = 3$, $c = 1$)

Matrices can take various forms:

$$
\begin{array}{ccc}
\text{Square} & \text{Symmetric} & \text{Diagonal} \\
\mathbf{A} = \begin{pmatrix} 1 & 2 & 3 \\ 4 & 9 & 5 \\ 6 & 7 & 7 \end{pmatrix}, & \mathbf{S} = \begin{pmatrix} a & b & c \\ b & k & e \\ c & e & f \end{pmatrix} & \mathbf{D} = \begin{pmatrix} a & 0 & 0 \\ 0 & b & 0 \\ 0 & 0 & 0 \end{pmatrix} \\
r = c & S_{ij} = S_{ji} & D_{ij} = 0 (i \neq j)
\end{array}
$$

$$
\begin{array}{ccc}
\text{Identity}(\mathbf{I}) & \text{Unity} & \text{Null}(\mathbf{0}) \\
\mathbf{I}_{3x3} = \begin{pmatrix} 1 & 0 & 0 \\ 0 & 1 & 0 \\ 0 & 0 & 1 \end{pmatrix} & \mathbf{J}_{3,2} = \begin{pmatrix} 1 & 1 \\ 1 & 1 \\ 1 & 1 \end{pmatrix} & \mathbf{0}_{3,2}\begin{pmatrix} 0 & 0 & 0 \\ 0 & 0 & 0 \end{pmatrix} \\
I_{ij} = 0(i \neq j) & & \\
I_{ii} = 1 & J_{ij} = J_{ij} = 1 & 0_{ij} = 0_{ii} = 0
\end{array}
$$

Matrix algebra

Matrix algebra is a set of rules for adding, subtracting, multiplying, and dividing matrices. The rules of matrix algebra differ from those for scalar numbers.

Summation

$$\mathbf{A} + \mathbf{B} = \begin{pmatrix} a & c & e \\ b & d & f \end{pmatrix} + \begin{pmatrix} g & j & n \\ h & k & m \end{pmatrix} = \begin{pmatrix} a+g & c+j & e+n \\ b+h & d+k & f+m \end{pmatrix}$$

Subtraction

$$\mathbf{A} - \mathbf{B} = \begin{pmatrix} a & c & e \\ b & d & f \end{pmatrix} - \begin{pmatrix} g & j & n \\ h & k & m \end{pmatrix} = \begin{pmatrix} a-g & c-j & e-n \\ b-h & d-k & f-m \end{pmatrix}$$

Multiplication

Multiplying a matrix by a scalar, z, results in a matrix where each element is the original matrix element times the scalar:

$$z\mathbf{S} = z\begin{pmatrix} a & b & c \\ b & k & e \\ c & e & f \end{pmatrix} = \begin{pmatrix} za & zb & zc \\ zb & zk & ze \\ zc & ze & zf \end{pmatrix}$$

Box 2.1 (continued)

- In order to multiply $\mathbf{A}$ ($r_1 \times c_1$) by $\mathbf{B}$ ($r_2 \times c_2$), their inner indices must be equal. That is, the number of columns of $\mathbf{A}$ must be equal to the number of rows of $\mathbf{B}$ ($c_1 = r_2$). The resulting matrix has dimensions equal to the outer indices (r_1, c_2) of $\mathbf{A}$ and $\mathbf{B}$.

$$\mathbf{A}_{r1 \times c1} \times \mathbf{B}_{r2 \times c2} = \mathbf{C}_{r1 \times c2}$$

- The order of matrices is important for multiplication, as the commutative property of scalar multiplication does not hold for matrices. In general, $\mathbf{AB} \neq \mathbf{BA}$
- The element of the resulting product matrix ($\mathbf{C}_{ij}$) is obtained by multiplying **row i of A** by **column j of B**, which is done by multiplying corresponding elements (where the column index of the element in row i matches the row index of the element in column j) and then summing the results:

$$\mathbf{A}_{2x3}\mathbf{B}_{3x2} = \begin{pmatrix} a & c & e \\ b & d & f \end{pmatrix} \begin{pmatrix} g & j \\ h & k \\ i & n \end{pmatrix} = \begin{pmatrix} ag+ch+ei & aj+ck+en \\ bg+dh+fi & bg+dh+fi \end{pmatrix}$$

This operation requires the inner indices to be equal ($r_1 = c_1 = 3$) and produces a resulting matrix with dimensions $r_1 \times c_2$ (2×2 in this example)

Identity matrix (I)

- The elements on the diagonal (i = j) are 1.
- The elements off the diagonal (i ≠ j) are 0.
- The identity matrix plays a role similar to 1 in regular algebra. It is like multiplying a scalar by 1:

$$\mathbf{a} \times 1 = 1 \times \mathbf{a} = \mathbf{a}$$

$$\mathbf{A} = \mathbf{AI} = \mathbf{A}$$

The **trace** of a square matrix is the sum of its diagonal elements:

$$Tr(\mathbf{S}) = Tr \begin{pmatrix} a & b & c \\ b & k & e \\ c & e & f \end{pmatrix} = a+k+f$$

Transposition

- The transpose of a matrix is another matrix with rows equal to the columns of the first matrix and columns equal to the rows of the first matrix:

$$\mathbf{B} = \begin{pmatrix} a & c & e \\ b & d & f \end{pmatrix}, \quad \mathbf{B}^T = \begin{pmatrix} a & b \\ c & d \\ e & f \end{pmatrix}$$

- The transpose of a matrix $\mathbf{B}$ can be denoted with as $\mathbf{B}^{\mathbf{T}}$ or as $\mathbf{B}'$ and is called 'B prime'.
- If two matrices are square, the transpose of their product is equal to the product of the transpose of each.

$$(\mathbf{AB})^T = \mathbf{A}^T \mathbf{B}^T$$

Box 2.1 (continued)

- Transposition can be used to convert a row vector $(1 \times c)$ into a column vector $(r \times 1)$ and vice versa:

$$a = (a_1 \quad \ldots \quad a_n)$$

$$a^T = \begin{pmatrix} a_1 \\ \vdots \\ a_n \end{pmatrix}$$

$$\left(a^T \right)^T = a$$

- We can combine transposition and matrix product operations to define the 'inner' and 'outer' products of two vectors. Let **a** and **b** be column vectors with elements $a_i, \ldots a_n$ and $b_i, \ldots b_n$, or equivalently, matrices each with dimension $n \times 1$.
- The 'inner product' of the two vectors is the sum of the cross-products of their corresponding elements, which is a scalar. We can obtain this result by transposing **a** and multiplying **a** by **b**:

$$(a_i \ldots a_n) \begin{pmatrix} b_i \\ . \\ b_n \end{pmatrix} = \mathbf{a}^T \mathbf{b} = \sum_{i=1}^{n} a_i b_i$$

- The 'outer product' of the two vectors is a square matrix of dimension $n \times n$, of which each element ij is the product of $a_i \times b_j$. This is obtained by multiplying **a** by the transpose of **b**:

$$\begin{pmatrix} a_1 \\ \vdots \\ a_n \end{pmatrix} (b_1 \quad \cdots \quad b_n) = \mathbf{a}\mathbf{b}^T = \begin{pmatrix} a_1 b_1 & \cdots & a_1 b_n \\ \vdots & \cdots & \vdots \\ a_n b_1 & \cdots & a_n b_n \end{pmatrix}$$

Inverse of a matrix

Multiplication of a matrix by its inverse (in either order) produces an identity matrix:

$$\mathbf{A}^{-1}\mathbf{A} = \mathbf{A}\mathbf{A}^{-1} = \mathbf{I}$$

$$\mathbf{A}\mathbf{A}^{-1} = \mathbf{I} = \begin{pmatrix} 1 & 1 & 1 \\ 1 & 2 & 3 \\ 1 & 3 & 4 \end{pmatrix} \begin{pmatrix} 1 & 1 & -1 \\ 1 & -3 & 2 \\ -1 & 2 & -1 \end{pmatrix} = \begin{pmatrix} 1 & 0 & 0 \\ 0 & 1 & 0 \\ 0 & 0 & 1 \end{pmatrix}$$

Thus, a matrix inverse is analogous to the reciprocal of a scalar value, in that any number divided by its reciprocal equals 1. Also, in scalar algebra, multiplying some number x by the reciprocal of a number y is equal to dividing x by y:

$$xy^{-1} = \frac{x}{y}$$

And by analogy, multiplying by the inverse of a matrix is something like dividing by that matrix. There is no 'division' operation defined for matrices, so matrix inverses play an important role to accomplish something analogous to division. For example, if we have the following equation:

$$Ax = b$$

we can solve the value of the vector x by multiplying both sides of the equation by the inverse of **A**:

$$(Ax)A^{-1} = bA^{-1}$$
$$x = bA^{-1}$$

An important difference between matrix inverses and scalar reciprocals is that whereas all non-zero scalars have a reciprocal value, not all matrices have an inverse. In particular, only square matrices have an inverse. But not all square matrices can be inverted. Matrices that cannot be inverted are called '**singular**' matrices. We will avoid details of how

Box 2.1 (continued)

one determines if a square matrix is invertable or singular, but will simply note that a matrix **A** is singular when there is no unique solution to the system of equations represented by:

$$Ax = b$$

For example, a singularity occurs in fitting statistical models to data when we include confounded parameters in the model. This could happen if one collected data from an experiment conducted over several years at a different site in each year, and then tried to fit a model to simultaneously estimate the effects of years and sites. In linear regression or mixed models analyses, this would result in a singularity in a matrix that must be inverted in order to solve for the model effects, and no unique solution can be given.

The direct sum and direct product of matrices

These two operators will be used frequently in later chapters to form the structures of the variance-covariance matrices for residual effects and model effects (these will be called **R** and **G** structures, respectively).

Direct Sum - The $\oplus$ symbol denotes a direct sum of two matrices. A direct sum adds square matrices as independent blocks along the diagonal:

$$D = \begin{bmatrix} 4 & 3 \\ 6 & 8 \end{bmatrix} \quad F = \begin{bmatrix} 1 & 2 \\ 3 & 5 \end{bmatrix}$$

$$D \oplus F = \begin{bmatrix} D & 0 \\ 0 & F \end{bmatrix} = \begin{bmatrix} 4 & 3 & 0 & 0 \\ 6 & 8 & 0 & 0 \\ 0 & 0 & 1 & 2 \\ 0 & 0 & 3 & 5 \end{bmatrix}$$

In statistical models, a direct sum can only be used for independent effects. In this example the **D** and **F** are assumed independent.

Direct (Kronecker) Product - The $\otimes$ symbol denotes a Kronecker product.

A direct product multiplies each element of the 1st matrix element-wise by the 2nd matrix.

$$D = \begin{bmatrix} 4 & 3 \\ 6 & 8 \end{bmatrix} \quad F = \begin{bmatrix} 1 & 2 \\ 3 & 5 \end{bmatrix}$$

$$D \otimes F = \begin{bmatrix} 4xF & 3xF \\ 6xF & 8xF \end{bmatrix} = \begin{bmatrix} 4x1 & 4x2 & 3x1 & 3x2 \\ 4x3 & 4x5 & 3x3 & 3x5 \\ 6x1 & 6x2 & 8x1 & 8x2 \\ 6x3 & 6x5 & 8x3 & 8x5 \end{bmatrix} = \begin{bmatrix} 4 & 8 & 3 & 6 \\ 12 & 20 & 9 & 15 \\ 6 & 12 & 8 & 16 \\ 18 & 30 & 24 & 40 \end{bmatrix}$$

This structure will be useful later to model correlations between effects of different random factors.

Matrix algebra with R

R has nice facilities for encoding and manipulating matrices. The following functions are key for matrix algebra:

`c()`: if scalar values are given as arguments, they are concatenated into a numeric vector.

`matrix()`: forms a matrix from the vector(s) of scalar values given as arguments

`diag(n)`: when n is a scalar value, diag(n) forms an identity matrix of dimension n × n

`diag(X)`: when X is a matrix, diag(X) returns a vector of the diagonal elements of X.

`t(X)`: returns the transpose of X

`solve(X)`: returns the inverse of X

`A %*% B`: returns the product of matrices A and B

`rbind(A, B)`: returns a new matrix formed by stacking the rows of matrix **A** on top of the rows of matrix **B** (**A** and **B** must have the same number of columns).

`cbind(A, B)`: returns a new matrix formed by stacking the columns of matrix **A** to the left of the columns of matrix **B** (**A** and **B** must have the same number of rows).

Code example 2.5

Some examples of matrix algebra using R (see **Code 2-5_Matrix algebra with R.R** for more details)

```
> #c() forms a vector from scalar inputs:
> a <- c(1,0,4)   # vector a
> b <- c(2,1,5)   # vector b


> # Inner product of a and b. t() returns the transpose of a.
> # %*% is used for matrix multiplication
> atb <-t(a)%*%b
> atb
     [,1]
[1,]   22


> # Outer product of a and b.
> abt <-a%*%t(b)
> abt
     [,1] [,2] [,3]
[1,]    2    1    5
[2,]    0    0    0
[3,]    8    4   20


> # Create a square matrix of 4x4 from a sample of 16 random normal
variates
> set.seed(1)    #this will make the example reproducible
> A <- matrix(rnorm(16),nrow=4)

> # Multiply each element by 5 and round off to have integers
> A <- round(A*5)
> A
     [,1] [,2] [,3] [,4]
[1,]   -3    2    3   -3
[2,]    1   -4   -2  -11
[3,]   -4    2    8    6
[4,]    8    4    2    0

> # Inverse of A is obtained using 'solve' function
> Ainv <- solve(A)
> round(Ainv, 2)
       [,1]   [,2]   [,3]   [,4]
[1,] -0.12   0.05   0.03   0.09
[2,]  0.26  -0.15  -0.15   0.04
[3,] -0.05   0.10   0.16   0.05
[4,] -0.10  -0.05   0.03  -0.02

> # Product of A with its inverse gives an identity matrix
> AinvA <- Ainv%*%A
> round(AinvA)
     [,1] [,2] [,3] [,4]
[1,]    1    0    0    0
[2,]    0    1    0    0
[3,]    0    0    1    0
[4,]    0    0    0    1
```

```
> # Take the diagonal and name it d
> d <- diag(A)
> d
[1] -3 -4  8  0
> sum(d) #this is the trace of A
[1] 1

> # Use diag() function to create a 4 x 4 identity matrix
> I <- diag(4)
> I
     [,1] [,2] [,3] [,4]
[1,]    1    0    0    0
[2,]    0    1    0    0
[3,]    0    0    1    0
[4,]    0    0    0    1

> # Matrix A times I (or I times A) returns A
> I%*%A
     [,1] [,2] [,3] [,4]
[1,]   -3    2    3   -3
[2,]    1   -4   -2  -11
[3,]   -4    2    8    6
[4,]    8    4    2    0

> A%*%I
     [,1] [,2] [,3] [,4]
[1,]   -3    2    3   -3
[2,]    1   -4   -2  -11
[3,]   -4    2    8    6
[4,]    8    4    2    0

> # Make a new matrix that combines A and I row-wise:
> B = cbind(A, I)
> B
     [,1] [,2] [,3] [,4] [,5] [,6] [,7] [,8]
[1,]   -3    2    3   -3    1    0    0    0
[2,]    1   -4   -2  -11    0    1    0    0
[3,]   -4    2    8    6    0    0    1    0
[4,]    8    4    2    0    0    0    0    1

> # Make a new matrix that combines A and I row-wise:
> C = rbind(A, I)
> C
     [,1] [,2] [,3] [,4]
[1,]   -3    2    3   -3
[2,]    1   -4   -2  -11
[3,]   -4    2    8    6
[4,]    8    4    2    0
[5,]    1    0    0    0
[6,]    0    1    0    0
[7,]    0    0    1    0
[8,]    0    0    0    1
```

```
##############################################
### Kronecker product of matrices.
# The order of the matrices produces different outcomes

# upload library for matrices
library(Matrix)
library(reshape2)

# Create 2 x 2 identity matrix
I = diag(2)
I
     [,1] [,2]
[1,]    1    0
[2,]    0    1


# Create 3 x 3 G matrix
G = matrix(c(5.55, 0.44, 0.43,
             0.44, 8.88, 0.42,
             0.43, 0.42, 3.33),   nrow=3, ncol=3, byrow=T)
G
     [,1] [,2] [,3]
[1,] 5.55 0.44 0.43
[2,] 0.44 8.88 0.42
[3,] 0.43 0.42 3.33




# The order of matrices while taking the kronecker product matters
GI <-kronecker(G, I)
GI


     [,1] [,2] [,3] [,4] [,5] [,6]
[1,] 5.55 0.00 0.44 0.00 0.43 0.00
[2,] 0.00 5.55 0.00 0.44 0.00 0.43
[3,] 0.44 0.00 8.88 0.00 0.42 0.00
[4,] 0.00 0.44 0.00 8.88 0.00 0.42
[5,] 0.43 0.00 0.42 0.00 3.33 0.00
[6,] 0.00 0.43 0.00 0.42 0.00 3.33

IG <-kronecker(I,G)
IG


     [,1] [,2] [,3] [,4] [,5] [,6]
[1,] 5.55 0.44 0.43 0.00 0.00 0.00
[2,] 0.44 8.88 0.42 0.00 0.00 0.00
[3,] 0.43 0.42 3.33 0.00 0.00 0.00
[4,] 0.00 0.00 0.00 5.55 0.44 0.43
[5,] 0.00 0.00 0.00 0.44 8.88 0.42
[6,] 0.00 0.00 0.00 0.43 0.42 3.33
```

Mixed Models in a Nutshell: Theory and Concepts

Mixed linear models are a particular class of models containing both fixed and random effects. Loosely speaking, a mixed model is a model where some terms remain constant over repeated sampling and some other terms vary at random according to some distribution. For simplicity from now on we will drop the notation linear and refer to these models simply as mixed models. For any mixed model we can identify three main components: the equation of the model, the expectations and variance-covariance for the random effects, and all the remaining assumptions regarding the model.

The Model

A model is a mathematical representation of our understanding of the biological process that explains our observations. We can think of each observation as a single equation (and in this case we are confining ourselves to linear equations), containing the trait of interest on one side and factors on the other side that explain the observations. For example:

$$a_1 x_1 + a_2 x_2 + \ldots + a_n x_n = b \tag{2.9}$$

A system of equations is then a set of these linear equations and a solution for the system must satisfy all equations. With n unknown parameters (factors) the system takes form:

$$\begin{aligned} a_{11} x_1 + a_{12} x_2 + \ldots + a_{1n} x_n &= b_1 \\ a_{21} x_1 + a_{22} x_2 + \ldots + a_{2n} x_n &= b_2 \\ a_{31} x_1 + a_{32} x_2 + \ldots + a_{3n} x_n &= b_3 \\ &\vdots \\ a_{m1} x_1 + a_{m2} x_2 + \ldots + a_{mn} x_n &= b_m \end{aligned} \tag{2.10}$$

with the first subscript referring to the equation number and the second to the variable number. The same set of equations can be rewritten in a more convenient matrix notation:

$$\mathbf{Ax=b}$$

where

$$\mathbf{A} = \begin{bmatrix} a_{11} + & a_{12} + & \ldots + & a_{1n} \\ a_{21} + & a_{22} + & \ldots + & a_{2n} \\ a_{31} + & a_{32} + & \ldots + & a_{3n} \\ \cdot & & & \\ \cdot & & & \\ \cdot & & & \\ a_{m1} + & a_{m2} + & \ldots + & a_{mn} \end{bmatrix}, \mathbf{x} = \begin{bmatrix} x_1 \\ x_2 \\ x_3 \\ \cdot \\ \cdot \\ \cdot \\ x_m \end{bmatrix}, \mathbf{b} = \begin{bmatrix} b_1 \\ b_2 \\ b_3 \\ \cdot \\ \cdot \\ \cdot \\ b_m \end{bmatrix} \tag{2.11}$$

From now on we will assume that our model contains both fixed and random effects. Traditionally, mixed models have been represented in matrix form as follows (Henderson 1990):

$$\mathbf{y} = \mathbf{Xb} + \mathbf{Zu} + \mathbf{e} \tag{2.12}$$

where $\mathbf{y}$ is the vector of the observations, $\mathbf{b}$ is a vector of fixed effects, $\mathbf{u}$ is a vector of random effects (unknown), $\mathbf{e}$ is a vector of random residuals (whatever we cannot explain with our model), and $\mathbf{X}$ and $\mathbf{Z}$ are incidence matrices that assign each element of $\mathbf{b}$ and $\mathbf{u}$ to their corresponding element in $\mathbf{y}$.

Fixed and Random Effects

In most breeding applications (and in all cases in this book) elements of $\mathbf{y}$ can be assumed as drawn from a normal distribution. Keep in mind that although this is a rather convenient assumption that facilitates the analysis of the data, it is not always tenable. We have mentioned in the previous section how we will consider both fixed and random effects in our models. We will now provide a short explanation on what we might consider fixed vs. random. The distinction between fixed and random applies to the unknown model components. A fixed effect is a known constant that will remain the same over conceptual repeated sampling, while a random effect is a random variable that arises from the subsampling and random selection of "treatment" levels.

Imagine a very simple fixed effect model similar to the following:

$y_i = b_0 + e_i$ where we are assuming $e \sim NIID(0, \sigma^2)$.

Where NIID stands for normally, independently and identically distributed. If we were to simulate data points for such a model we will proceed as follow:

(a) Set the value for b_0 (let's say 10)
(b) Set the value for σ^2 (let's say 5)
(c) Draw a sample of size n of random deviations from a standard normal distribution $N(0,1)$ (z values)
(d) Form a vector of residuals:

$$e_1 = 5z_1$$
$$e_2 = 5z_2$$
$$.$$
$$e_n = 5z_n$$

(e) Then your vector of observations $\mathbf{y}$ will be:

$$y_1 = 10 + 5z_1$$
$$y_2 = 10 + 5z_2$$
$$.$$
$$y_n = 10 + 5z_n$$

You should notice that all the variability in your sample comes from $\mathbf{e}$ and this would remain the same were you to simulate new data (a new experiment).

Now consider the same model but assume $b_0 \sim NIID(a, \sigma_b^2)$, $e \sim NIID(0, \sigma^2)$ and $Cov(b_0, e_i) = 0$

In this case you would simulate your data like this:

(a) Set the value for the mean of the b_0 distribution ('a'; let's say 2)
(b) Set the value of σ_b^2 (let's say 3)
(c) Draw 1 (z) normal deviate from $N(0,1)$
(d) Form $\beta_0 = \alpha + \sigma_\beta z$
(e) Set the value for σ^2 (let's say 5)
(f) Draw n (z) random deviations from a standard normal distribution $N(0,1)$
(g) Form a vector of residuals:

$$e_1 = 5z_1$$
$$e_2 = 5z_2$$
$$.$$
$$e_n = 5z_n$$

Your vector of observations $\mathbf{y}$ will in this case be:

$$y_1 = \beta_0 + 5z_1$$
$$y_2 = \beta_0 + 5z_2$$
$$y_3 = \beta_0 + 5z_3$$
$$.$$
$$y_n = \beta_0 + 5z_n$$

Note that in this case your observed values will be dependent on the realized value of b_0 for that particular realization (experiment) so that the variance for **y** will be $\sigma_b^2 + \sigma^2$.

If you were to repeat the experiment m times you would find that in the limit (as m approaches infinity) your estimates of both a and σ_b^2 will get closer and closer to their true value.

In reality the distinction between fixed and random effects often depends on the practical use and interpretation of parameter estimates. When the investigator is interested in comparing specific levels of a certain factors (let's say amount of fertilizer for a plant or concentrate for a cow) then it is sensible to consider them as fixed effect. When a parameter is not of relevance for the analysis but rather a nuisance that we want to account for, more often than not we end up treating that effect as random.

If the researcher believes that the levels of a particular design factor represent random samples from some larger reference population of effects, and that the distribution of those effects follows normality, then the factor can be considered random. For random factors, the researcher wants to make inference to the distribution of the population by estimating variance components. In breeding and genetics studies, researchers often wish to make inference to larger reference populations of animals or plants, such that the individuals, families, or genotypes represent random effects. Estimations of heritability or genetic correlation between traits are typical examples where the genetic samples are considered random. In contrast, factors are considered fixed when inference is made only to the particular levels of the factor studied in the experiment. An example of a fixed factor might be a specific nutrition treatment provided to some animals (or similarly, specific levels of fertilizer application in a plant study).

Sometimes genetic factors can also be fixed. This most commonly occurs in plant variety trials where a small sample of elite cultivars is compared for the purpose of recommending the best cultivar for a particular growing region. Of course, there are situations where the distinction between random and fixed effects is less clear cut, for example when plant families are evaluated at a relatively small number of test sites, in which case it may be uncertain whether the test environments should be considered random or fixed effects. The researcher may hope that the environments represent a random sample of target production environments, but the researcher may be limited to a few managed research farms that are not really random samples. On the other hand, if the experiment is replicated over years, the yearly component of environmental variation may be more random. In such cases, the researcher should decide primarily based on what inferences are to be made, and if the factor is to be considered random, have enough levels been sampled to adequately estimate a variance component? Finally, if inference is primarily aimed at the genetic factors in the study, and environments are used mainly to replicate the genetic evaluations over a reasonable number of environments, the researcher can declare environments as fixed simply to make the analysis more efficient.

The distinction between fixed and random effects is a prerogative of assuming a frequentist point of view. People employing Bayesian statistics to analyze mixed models will argue that in reality all the effects are random.

Expectations and Variance-Covariance for the Random Effects

Since we have assumed that random effects come from some large population we need to define location and dispersion for these parameters. Note that we have assumed that our observations are normally distributed and we will always assume that our residuals are normally distributed. Keeping the general matrix notation that we have seen before, the expectations of **u**, **e** and **y** are.

$$
\begin{aligned}
E(\mathbf{u}) &= \mathbf{0} \\
E(\mathbf{e}) &= \mathbf{0} \\
E(\mathbf{y}) &= E(\mathbf{Xb} + \mathbf{Zu} + \mathbf{e}) \\
&= E(\mathbf{Xb}) + E(\mathbf{Zu}) + E(\mathbf{e}) \\
&= \mathbf{X}E(\mathbf{b}) + \mathbf{Z}E(\mathbf{u}) + E(\mathbf{e}) \\
&= \mathbf{Xb} + \mathbf{0} + \mathbf{0} \\
&= \mathbf{Xb}
\end{aligned}
\tag{2.13}
$$

Also, the variances of $\mathbf{u}$ and $\mathbf{e}$ are

$$\mathbf{V} = \begin{pmatrix} \mathbf{u} \\ \mathbf{e} \end{pmatrix} = \begin{pmatrix} \mathbf{G} & \mathbf{0} \\ \mathbf{0} & \mathbf{R} \end{pmatrix} \tag{2.14}$$

The actual structure of $\mathbf{G}$ and $\mathbf{R}$ is flexible (and much of this book is concerned with fitting complex $\mathbf{G}$ and $\mathbf{R}$ structures), but in the simplest cases $\mathbf{G} = \mathbf{I}\sigma_u^2$ and $\mathbf{R} = \mathbf{I}\sigma_e^2$.

We usually assume that $\text{cov}(\mathbf{u},\mathbf{e}) = 0$, so that

$$
\begin{aligned}
V(\mathbf{y}) &= V(\mathbf{Xb} + \mathbf{Zu} + \mathbf{e}) \\
&= V(\mathbf{Zu} + \mathbf{e}) \\
&= \mathbf{Z}V(\mathbf{u})\mathbf{Z}' + V(\mathbf{e}) + \mathbf{Z}Cov(\mathbf{u},\mathbf{e}) + Cov(\mathbf{e},\mathbf{u})\mathbf{Z}' \\
&= \mathbf{ZGZ}' + \mathbf{R}
\end{aligned}
\tag{2.15}
$$

$$Cov(\mathbf{y},\mathbf{u}) = \mathbf{ZG}$$
$$Cov(\mathbf{y},\mathbf{e}) = \mathbf{R}$$

If we call $\mathbf{V} = \mathbf{ZGZ'} + \mathbf{R}$, we can summarize the distribution of the data and model factors as:

$$
\begin{aligned}
\mathbf{y} &\sim N(\mathbf{Xb}, \mathbf{V}); \\
\mathbf{u} &\sim N(0, \mathbf{G}); \\
\mathbf{e} &\sim N(0, \mathbf{R})
\end{aligned}
\tag{2.16}
$$

A Trivial Example: Daughters Lactation Yield

We will follow the motivational example provided by Robinson (1991). Let's assume we have collected data on lactation yields of dairy cows in three different herds. Let's assume that the cows' sire genetic merit is treated as a random effect while herds are treated as fixed.

Herd	Sire	Yield
1	ZA	110
1	AD	100
2	BB	110
2	AD	100
2	AD	100
3	CC	110
3	CC	110
3	AD	100
3	AD	100

Then we can write the system of linear equations as.

$$110 = herd_1 + sire_{ZA} + e$$
$$100 = herd_1 + sire_{AD} + e$$
$$110 = herd_2 + sire_{BB} + e$$
$$100 = herd_2 + sire_{AD} + e$$
$$100 = herd_2 + sire_{AD} + e$$
$$110 = herd_3 + sire_{CC} + e$$
$$110 = herd_3 + sire_{CC} + e$$
$$100 = herd_3 + sire_{AD} + e$$
$$100 = herd_3 + sire_{AD} + e$$

And in matrix form:

$\mathbf{y} = \mathbf{Xb} + \mathbf{Zu} + \mathbf{e}$, where:

$$\mathbf{y} = \begin{bmatrix} 110 \\ 100 \\ 110 \\ 100 \\ 100 \\ 110 \\ 110 \\ 100 \\ 100 \end{bmatrix}$$

is our (n x 1) vector of phenotypic observations,

$$\mathbf{b} = \begin{bmatrix} h_1 \\ h_2 \\ h_3 \end{bmatrix}$$

is a (p x 1) vector of fixed herd effects,

$$\mathbf{u} = \begin{bmatrix} S_{ZA} \\ S_{BB} \\ S_{CC} \\ S_{AD} \end{bmatrix}$$

is a (q x 1) vector of random sire effects,

$$\mathbf{e} = \begin{bmatrix} e_1 \\ e_2 \\ e_3 \\ e_4 \\ e_5 \\ e_6 \\ e_7 \\ e_8 \\ e_9 \end{bmatrix}$$

is a (n x 1) vector of residuals and the design matrices are:

$$\mathbf{X} = \begin{bmatrix} 1 & 0 & 0 \\ 1 & 0 & 0 \\ 0 & 1 & 0 \\ 0 & 1 & 0 \\ 0 & 1 & 0 \\ 0 & 0 & 1 \\ 0 & 0 & 1 \\ 0 & 0 & 1 \\ 0 & 0 & 1 \end{bmatrix}, \mathbf{Z} = \begin{bmatrix} 1 & 0 & 0 & 0 \\ 0 & 0 & 0 & 1 \\ 0 & 1 & 0 & 0 \\ 0 & 0 & 0 & 1 \\ 0 & 0 & 0 & 1 \\ 0 & 0 & 1 & 0 \\ 0 & 0 & 1 & 0 \\ 0 & 0 & 0 & 1 \\ 0 & 0 & 0 & 1 \end{bmatrix}$$

$\mathbf{X}$ (n x p) and $\mathbf{Z}$ (n x q) are incidence matrices that relate phenotypic observations to herd and sire effects.

Let's now write the assumptions of the model.

$$\begin{aligned} &\mathbf{V_u} = \mathbf{G} = \mathbf{I}\sigma_u^2 \\ &\mathbf{V_e} = \mathbf{R} = \mathbf{I}\sigma_e^2 \\ &\mathbf{V_y} = \mathbf{Z'GZ} + \mathbf{R}, \text{so that :} \\ &\mathbf{y} \sim \text{NID}(\mathbf{Xb}, \mathbf{V}); \mathbf{u} \sim \text{NID}(0, \mathbf{G}); \quad \mathbf{e} \sim \text{NID}(0, \mathbf{R}) \end{aligned}$$ (2.17)

For this example, we assume that we know that $\sigma_e^2 = 1$ and $\sigma_u^2 = 0.1$ and that sires are not related. We can obtain a matrix representation of the model using the following R code:

Code example 2.6
Mixed models using matrix algebra in R (see **Code 2-6_Mixed models using matrices.R** for more details)

```
y=c(110,100,110,100,100,110,110,100,100)   # the y vector
X=matrix(c(1,1,0,0,0,0,0,0,0,
           0,0,1,1,1,0,0,0,0,
           0,0,0,0,0,1,1,1,1),  9,byrow=F)   # X matrix
Z=matrix(c(1,0,0,0,0,0,0,0,0,
           0,0,1,0,0,0,0,0,0,
           0,0,0,0,0,1,1,0,0,
           0,1,0,1,1,0,0,1,1),  9,byrow=F)   # Z matrix

Iu=diag(4)
Ie=diag(9)                                   # identity matrix
se=1                                         # error variance
su=0.1                                       # sire variance
G=Iu*su                                      #G
R=Ie*se                                      #R
V=Z%*%(G)%*%t(Z)+(R)                         #V
```

Solving the Model

Here we demonstrate the matrix algebra involved in solving the mixed model equations. For this small example, we will not actually estimate the variance components from the data but instead will use fixed values for the variance components as if they were estimated (or known) *a priori* ($\sigma_e^2 = 1$ and $\sigma_u^2 = 0.1$ in this example). This is seldom the case in practice and later we will see how mixed models provide a powerful tool to simultaneously obtain both solutions for fixed and random effects and estimates of variance components.

From the model outlined above, the solutions for fixed effects (BLUEs) are:

$$\widehat{\mathbf{b}} = \left(\mathbf{X}'\mathbf{V}^{-1}\mathbf{X}\right)^{-1}\mathbf{X}'\mathbf{V}^{-1}\mathbf{y} \tag{2.18}$$

which are the generalized least squares estimates for **b**.

For the random effects, the solutions for random effects (BLUPs) are:

$$\widehat{\mathbf{u}} = \mathbf{G}\mathbf{Z}'\mathbf{V}^{-1}\left(\mathbf{y} - \mathbf{X}\widehat{\mathbf{b}}\right) \tag{2.19}$$

Solving for our example gives us

$$\widehat{\mathbf{b}} = (105.64, 104.28, 105.46)'$$
$$\widehat{\mathbf{u}} = (0.40, 0.52, 0.76, -1.67)'$$

We can do this with matrix algebra in R:

```
Xt=t(X)                                          # X matrix transpose
Zt=t(Z)                                          # Z matrix transpose
Vinv=solve(V)                                    # the inverse of the variance
b=solve(Xt%*%Vinv%*%X)%*%(Xt%*%Vinv%*%y)         #BLUE
u=G%*%Zt%*%Vinv%*%(y-X%*%b)                       #BLUP
```

You should notice that using the formulas above involves finding the inverse of the variance matrix. While for this little example this is easily done, that is seldom the case with large data and complex models. We need a more convenient form.

The Mixed Model Equations

For the general mixed linear model described above, a particular set of equations can be used to find the solutions of each effect. These are the Henderson's mixed model equations and were developed for animal breeding by Henderson (1949).

$$\begin{pmatrix} \mathbf{X'R^{-1}X} & \mathbf{X'R^{-1}Z} \\ \mathbf{Z'R^{-1}X} & \mathbf{Z'R^{-1}Z+G} \end{pmatrix} \begin{pmatrix} \widehat{\mathbf{b}} \\ \widehat{\mathbf{u}} \end{pmatrix} = \begin{pmatrix} \mathbf{X'R^{-1}y} \\ \mathbf{Z'R^{-1}y} \end{pmatrix}$$

(2.20)

We usually call this the Left
Hand Side (LHS)

We usually call this the
Right Hand Side (RHS)

If we assume that residual variance is IID (identical and independent for all observations), the **R** matrix can be factored out. In most of the applications we will see from now on, the following form of the equations will be more convenient:

$$\begin{pmatrix} \mathbf{X'X} & \mathbf{X'Z} \\ \mathbf{Z'X} & \mathbf{Z'Z+I}\alpha \end{pmatrix} \begin{pmatrix} \widehat{\mathbf{b}} \\ \widehat{\mathbf{u}} \end{pmatrix} = \begin{pmatrix} \mathbf{X'y} \\ \mathbf{Z'y} \end{pmatrix}$$

(2.21)

where:

$$\alpha = \frac{\sigma_e^2}{\sigma_u^2}$$

(2.22)

The solutions to these equations are the best linear unbiased estimators of **b** and the best linear unbiased predictors of **u**:

BLUE (**b**)

$$[\mathbf{X'X} \quad \mathbf{X'Z}]^{-1} \begin{bmatrix} \mathbf{X'y} \\ \mathbf{Z'y} \end{bmatrix}$$

(2.23)

BLUP (**u**)

$$[\mathbf{Z'X} \quad \mathbf{Z'Z+I}\alpha]^{-1} \begin{bmatrix} \mathbf{X'y} \\ \mathbf{Z'y} \end{bmatrix}$$

(2.24)

which give us the same solutions as before:

$$\widehat{b} = (105.64, 104.28, 105.46)'$$
$$\widehat{u} = (0.40, 0.52, 0.76, -1.67)'$$

We can obtain the solutions using the following small R script:

```
alpha=se/su                                          # alpha
XpX=crossprod(X)                                      #X'X
XpZ=crossprod(X,Z)                                    #X'Z
ZpX=crossprod(Z,X)                                    #Z'X
ZpZ=crossprod(Z)                                      #Z'Z
Xpy=crossprod(X,y)                                     #X'y
Zpy=crossprod(Z,y)                                     #Z'y
LHS=rbind(cbind(XpX,XpZ),cbind(ZpX,ZpZ+diag(4)*alpha)) #LHS
RHS=rbind(Xpy,Zpy)                                     #RHS
sol=solve(LHS)%*%RHS                                  #Solutions
```

Estimability in Models with Multiple Fixed Effects

While we have said we will not spend too much time dealing with fixed effects, a note on estimability is necessary. When there are multiple effects in the model it is often impossible to obtain unique BLUE for each level of the fixed effects.

Let's follow the previous example.

Herd	Sire	Yield
1	ZA	110
1	AD	100
2	BB	110
2	AD	100
2	AD	100
3	CC	110
3	CC	110
3	AD	100
3	AD	100

But let's consider for this example both herd and sire as fixed so that we can rewrite the **X** matrix as.

$$\mathbf{X} = \begin{bmatrix}
1 & 0 & 0 & 1 & 0 & 0 & 0 \\
1 & 0 & 0 & 0 & 0 & 0 & 1 \\
0 & 1 & 0 & 0 & 1 & 0 & 0 \\
0 & 1 & 0 & 0 & 0 & 0 & 1 \\
0 & 1 & 0 & 0 & 0 & 0 & 1 \\
0 & 0 & 1 & 0 & 0 & 1 & 0 \\
0 & 0 & 1 & 0 & 0 & 1 & 0 \\
0 & 0 & 1 & 0 & 0 & 0 & 1 \\
0 & 0 & 1 & 0 & 0 & 0 & 1
\end{bmatrix}$$

You should notice that there are dependencies among rows and columns. For example, in this simple case the fourth column (in red) is equal to the difference of the other columns.

As a consequence $\mathbf{X'X}$ is not full rank since its dimension is 7 x 7 yet there are only 6 independent rows and columns. In this case a unique inverse of the coefficient matrix ($\mathbf{X'X}$) does not exist. Therefore we cannot obtain the BLUE estimates for herd and sire. This is the same problem we encountered in the previous example in this chapter where some genotypes (RILs) were missing from some environments. Nonetheless some (useful) linear functions of the solutions are still estimable.

Let's look at the following small R code that uses the same data.

```
#sires
sire<-c("ZA","AD","BB","AD","AD","CC","CC","AD","AD")

#herds
herd<-c("one","one","two","two","two","three","three","three","three")

#yields
yield<-c(110,100,110,100,100,110,110,100,100)

# putting everything in a dataframe
new_data<-as.data.frame(cbind(yield,herd,sire))

# making sure that yield is treated as a numeric value
new_data$yield<-as.numeric(as.character(new_data$yield))
# fitting a linear model omitting the intercept (-1)
fm<-lm( yield~ herd + sire -1,data=new_data)
summary(fm)
```

The code produces the following output.

```
Call:
lm(formula = yield ~ herd + sire - 1, data = new_data)

Residuals:
1 2 3 4 5 6 7 8 9
0 0 0 0 0 0 0 0 0

Coefficients:
          Estimate Std. Error t value Pr(>|t|)
herdone        100          0     Inf   <2e-16 ***
herdthree      100          0     Inf   <2e-16 ***
herdtwo        100          0     Inf   <2e-16 ***
sireBB          10          0     Inf   <2e-16 ***
sireCC          10          0     Inf   <2e-16 ***
sireZA          10          0     Inf   <2e-16  ***
```

A few things are apparent:

If we look at the output of the estimates we notice that what R did was to set to 0 the first level of the sire fixed effect (sire AD in this case). In this way the model was reparametrized to be full rank and the solutions presented are an estimable function of the (unknown) BLUEs. Specifically the functions estimated are.

$$Sire_{AD} - Sire_{BB}$$

$$Sire_{AD} - Sire_{CC}$$

$$Sire_{AD} - Sire_{ZA}$$

Each of the sires other than AD has daughters that yield 10 units more (on average) than daughters of sire AD. The same principle can be applied to construct other meaningful estimable functions. Furthermore, treating sire effects as fixed is equivalent to omitting the $\mathbf{G}^{-1}$ matrix from the mixed model equations of the previous example.

The LHS of the MME equations obtained by R looks like this:

```
LHS
      [,1] [,2] [,3] [,4] [,5] [,6] [,7]
[1,]    2    0    0    1    0    0    1
[2,]    0    3    0    0    1    0    2
[3,]    0    0    4    0    0    2    2
[4,]    1    0    0   11    0    0    0
[5,]    0    1    0    0   11    0    0
[6,]    0    0    2    0    0   12    0
[7,]    1    2    2    0    0    0   15
```

Omitting $\mathbf{G}^{-1}$ effectively means reducing the last 4 diagonal elements of the LHS matrix by 10.

```
LHSm=rbind(cbind(XpX,XpZ),cbind(ZpX,ZpZ))
```

```
LHSm
      [,1] [,2] [,3] [,4] [,5] [,6] [,7]
[1,]    2    0    0    1    0    0    1
[2,]    0    3    0    0    1    0    2
[3,]    0    0    4    0    0    2    2
[4,]    1    0    0    1    0    0    0
[5,]    0    1    0    0    1    0    0
[6,]    0    0    2    0    0    2    0
[7,]    1    2    2    0    0    0    5
```

The solution given in this case is the least-squares solution for both sires and herds. If we compare what was produced by this analysis with what obtained through BLUP we get a glimpse of how BLUP works. BLUP solutions take into account the fact that sire information has less variability than the variance of lactation yield of a single sire's daughter. Effectively the sires estimates are shrunk toward the mean (assumed 0). The amount of shrinkage is dependent on the amount of information available for the sire. For example, if we take the predictions for the 4 sires obtained by BLUP (0.40, 0.52, 0.76, −1.67) we notice that prediction for sire CC is better than the one for sires ZA and BB even if lactation yield for his daughters are the same (110) than those for sires ZA and BB. This is because we have more information available for that sire.

Standard Errors and Accuracy of the Estimates

Accuracy refers to the correlation between true and predicted random genetic effects $\left(r_{\hat{g},g}\right)$. In some cases, 'reliability' of the predictions is reported; this is the squared correlation between true and predicted random effects $\left(r_{\hat{g},g}^2\right)$, such that accuracy is simply the square root of reliability (Mrode 2014). Reliabilities are related to prediction error variances (PEV) as follows:

$$PEV = \text{var}(\hat{u} - u) = \left(1 - r_{\hat{u},u}^2\right)\sigma_u^2 \tag{2.25}$$

For breeding value predictions, the u's in this equation refer to the genetic effects or breeding values of the individuals, such that:

$$PEV = \text{var}(\hat{g} - g) = \left(1 - r_{\hat{g},g}^2\right)\sigma_A^2 \tag{2.26}$$

The closer the predictions are to the true values, the closer the reliability is to one, and the smaller the prediction error variance is (Mrode 2014). The prediction error variances may differ for different individuals, as individuals with more information will have smaller prediction errors. We can obtain the prediction error variances using the inverse elements of the mixed model equations.

Let:

$$\begin{pmatrix} \mathbf{X'X} & \mathbf{X'Z} \\ \mathbf{Z'X} & \mathbf{Z'Z} + \mathbf{I}\alpha \end{pmatrix} = \begin{pmatrix} \mathbf{C_{11}} & \mathbf{C_{12}} \\ \mathbf{C_{21}} & \mathbf{C_{22}} \end{pmatrix} \tag{2.27}$$

Then let the inverse of the left hand matrix be

$$\begin{pmatrix} \mathbf{C_{11}} & \mathbf{C_{12}} \\ \mathbf{C_{21}} & \mathbf{C_{22}} \end{pmatrix}^{-1} = \begin{pmatrix} \mathbf{C^{11}} & \mathbf{C^{12}} \\ \mathbf{C^{21}} & \mathbf{C^{22}} \end{pmatrix} \tag{2.28}$$

Now the prediction error variance is $PEV = V(\hat{u} - u) = \mathbf{C^{22}}\sigma_e^2$. So for each level of a random effect i, $PEV_i = \left(d_i\sigma_e^2\right)$ where d_i is the diagonal element of $\mathbf{C^{22}}$. Returning to our first example let's find the PEV for each sire. If we take the inverse of the LHS:

```
round(solve(LHS), digit=3)
         [,1]     [,2]     [,3]     [,4]     [,5]     [,6]     [,7]
[1,]    0.547    0.030    0.024   -0.050   -0.003   -0.004   -0.044
[2,]    0.030    0.383    0.031   -0.003   -0.035   -0.005   -0.057
[3,]    0.024    0.031    0.297   -0.002   -0.003   -0.050   -0.045
[4,]   -0.050   -0.003   -0.002    0.095    0.000    0.000    0.004
[5,]   -0.003   -0.035   -0.003    0.000    0.094    0.000    0.005
[6,]   -0.004   -0.005   -0.050    0.000    0.000    0.092    0.008
[7,]   -0.044   -0.057   -0.045    0.004    0.005    0.008    0.083
```

The bottom right corner (bold fonts in diagonal) is $\mathbf{C^{22}}$.

In this case we have assumed a value of $\sigma_e^2 = 1$ so that the prediction error variances are simply the diagonal elements of $\mathbf{C^{22}}$:

$$PEV = \begin{bmatrix} 0.095 \\ 0.094 \\ 0.092 \\ 0.083 \end{bmatrix}$$

From these it is possible to obtain the standard errors of predictions (SEP) as the square roots of the prediction variances. Furthermore, we can compute the reliabilities for each prediction given the PEVs and the estimated additive variance component by re-arranging Eq. 2.1 to obtain:

$$r^2_{\hat{u},u} = 1 - \frac{PEV}{\sigma_u^2} \tag{2.29}$$

Interestingly, the ratio of the variances of the true values to the error variance $\left(\frac{\sigma_u^2}{\sigma_e^2}\right)$ can be obtained from the sum of a row or from the sum of a column of $\mathbf{C}^{22}$ (the inverse of the LHS). In this example they are all equal to 0.1, but there is arounding error:

```
0.095 + 0.000 + 0.000 + 0.004 ~ 0.099
0.000 + 0.094 + 0.000 + 0.005 ~ 0.099
0.000 + 0.000 + 0.092 + 0.008 ~ 0.099
0.004 + 0.005 + 0.008 + 0.083 ~ 0.099
0.099   0.099   0.099   0.099
```

Putting everything together we obtain the following summary, including the prediction (BLUP), prediction variance (PEV), standard error of the prediction (PEV), and reliability (REL) for each of the sires:

Sire	BLUP	PEV	SEP	REL
ZA	0.40	0.095	0.308	0.05
BB	0.52	0.094	0.306	0.06
CC	0.76	0.092	0.303	0.08
ZD	−1.67	0.083	0.288	0.17

```
REL(Sire ZA) = (0.1 - 0.095)/ 0.1 = 0.05
```

You should notice in this case that sires with no information (and no relationship to observed individuals) would have a PEV of 0.1 and therefore a reliability of 0.

A Brief Note on REML

Prior to this point in this section we have only been concerned with describing the model used and we assumed that the variance components were known without error. In practice we must estimate (co)variance components from the data. The most common method for variance components estimation (and the one used by default by SAS, R and ASReml) is the restricted maximum likelihood (REML). REML estimates are often ideal for the analysis of complex breeding data sets. We will describe the principles of REML estimation without formality REML here; See Lynch and Walsh (Lynch and Walsh 1998) for a more thorough explanation.

REML is based on the maximization of the likelihood function (the probability of observing the data given a set of parameters). In other words, the estimates of variances and covariances chosen by REML are those that would have been most likely to give us the observed data. While REML and ML (maximum likelihood) share the same principle, REML attempts to account for the degrees of freedom used in estimating the fixed effects. In other words, in ML we estimate variance components conditionally on the solutions for fixed effects, essentially treating the fixed effects as known without error. This is clearly not the case for most of our analyses where we substitute the true effects with their estimates. REML partially accounts for that uncertainty.

Since we are breeders, in the rest of the book we will deal with the estimation of genetic variances for traits under selection which could be potentially biased by the very selection process. For most applications we would be interested in estimating the genetic variances in the "unknown" base population rather than the observed variance in our sample of related individuals. REML estimates obtained using a relationship matrix in the mixed model to account for genetic relationships among the individuals (more on this in Chap. 11) are not (or at least less) influenced by selection in the base population. More importantly, REML (and ML) estimates of variance components restrict the possible parameter space so that variances are always positive (or 0). This avoids the embarrassment of having to report negative estimates of a squared quantity as can happen with methods of moments estimators!

REML estimates of (co)variance components must be obtained through iteration and many algorithms can be used. Among the most popular are derivative free approaches that do not require any likelihood derivative (for example, those implemented in the software MTDFREML), methods that require the first derivative of the likelihood function (e.g., the expectation maximization (EM) algorithm), and finally, methods that require the second derivative of the likelihood function (the Newton-Raphson algorithm, Fisher scoring, and average information algorithm). SAS proc. MIXED implements the Newton-Raphson algorithm to solve the mixed model equations, whereas ASReml uses the average information algorithm (Gilmour et al. 2009; SAS Institute, Inc. 2011a). The more recently developed SAS Proc HPMIXED also implements the average information algorithm and some other optimization techniques (including sparse matrix representation) to reduce computational time and memory demands compared to SAS Proc MIXED, but at this time only has limited modeling flexibility (e.g., only a small set of G and R structures are available) (SAS Institute, Inc. 2011b). Finally, the *lmer* function in the R package *lme4* uses yet a different method than the other software, including a penalized iteratively reweighted least squares method for updating estimates at each iteration (Bates et al. 2013).

One last word of caution: While REML has several nice properties many REML algorithms do not guarantee that the iterative procedure will converge to a global maximum. Thus, especially with small datasets and multiple trait models (discussed in Chap. 7), it is always a good idea to repeat analyses with several different starting values for variance-covariance parameters.

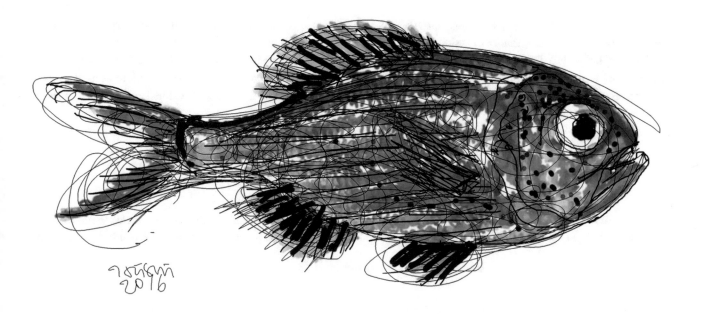

Electronic supplementary material: The online version of this chapter (doi:10.1007/978-3-319-55177-7_3) contains supplementary material, which is available to authorized users.

F. Isik et al., *Genetic Data Analysis for Plant and Animal Breeding*, DOI 10.1007/978-3-319-55177-7_3

Abstract

Understanding the matrix representations of variance-covariance models is important to be able to fit mixed models with complex variance structures. In particular, ASReml makes use of a notation for direct products of matrices to form some complex variance structures. The direct product notation can be applied both to the residual errors from the model (in the '**R** structure') and to random model factors (in the '**G** structure'). In this chapter we introduce the major variance models to form more complex **R** and **G** structures with some examples, but more detailed applications of variance modeling will be covered in later chapters.

Variance Model Specifications

Gamma and Sigma Parameterization in ASReml

As we saw in Chap. 2, the variance of response variable $\mathbf{y}$ is $\mathbf{Var(y)} = \mathbf{V} = \mathbf{ZGZ'} + \mathbf{R}$. By default the variance of random effects is $\mathbf{G} = \mathbf{I}\sigma_u^2$ and the variance of residuals is $\mathbf{R} = \mathbf{I}\sigma_e^2$. For a simple mixed model with one random term, the variance of $\mathbf{y}$ is $\mathbf{V} = \sigma_u^2 \mathbf{ZZ'} + \mathbf{I}\sigma_e^2$. The model has two variance parameters or *sigmas*, one for the random model effect $\left(\sigma_u^2\right)$ and one for the residual term $\left(\sigma_e^2\right)$. This is called *sigma parameterization*. The other parameterization used in ASReml is the *gamma parameterization*. In this parameterization the variance of $\mathbf{y}$ is formulated as the ratios of residual variance σ_e^2 as shown below.

$$\mathbf{V} = \sigma_e^2 \sum_{i=1}^{b} \left(\gamma_g \mathbf{ZZ'} + \mathbf{I}_n\right) \quad \text{or}$$
$$\mathbf{V} = \sigma_e^2 \left[\mathbf{ZG}\left(\gamma_g\right)\mathbf{Z'} + \mathbf{R}_c(\gamma_r)\right] \tag{3.1}$$

$\mathbf{Z}$ and $\mathbf{Z'}$ are the design matrix and its transpose, respectively, for the random term. The gamma value (scaled variance) of the random term (γ_g) is the ratio of the variance component for the random term and the variance of the residual error term:

$$\gamma_g = \sigma_u^2/\sigma_e^2 \tag{3.2}$$

The gamma value of the residual error term (γ_r) is the ratio of the residual variance component to itself: $\gamma_r = \sigma_e^2/\sigma_e^2 = 1$. This is the default parameterization for a simple univariate mixed model in ASReml because it is easier to guess starting values for the unitless *gammas* than for the variances (*sigmas*), which depends on the scale of the trait measurement units. Gamma parameterization can also speed up model convergence (Butler et al. 2009).

For more complex variance structures, such as correlated residuals, the default parameterization is the *sigma parameterization*. The user can switch from gamma to sigma parameterization in ASReml Release 4, using the !SIGMAP qualifier. The qualifier must be placed after the response variable as follows:

```
height   !SIGMAP ~ mu location ,
                  !r block
residual idv(units)
```

There are many variance-covariance functions used for different models. A summary of the most common variance functions is given below (Table 3.1).

Correlation variance models can be appended with v e.g. idv() to add a common (homogeneous) variance or with h (e.g., idh()) to add a separate (heterogeneous) variance for each level of the factor.

Table 3.1 Variance-covariance structures in ASReml

Function name	Type	Description
id()	Correlation	IID with variance 1
idv()	Variance	IID with common variance
idh()	Variance	Independent with heterogeneous variance
diag()	Variance	Same as idh()
ar1()	Correlation	Auto regressive correlation structure of order 1
cor()	Correlation	Unstructured correlation matrix
giv()	Known correlation	User defined correlation, or inverse of correlation matrix
giv(,ped=T)	Known correlation	User defined inverse correlation matrix derived with a factor argument
us()	Variance	General unstructured
fa(,k)	Variance	Factor analytic model of order k

See the ASReml User Guide (Gilmour et al. 2014, Chapter 7) for details

Homogenous Variance Models

We will use the maize recombinant inbred lines (RIL) data with missing observations (MaizeRILs_miss.csv) introduced in Chap. 2 to show the most common variance modeling in ASReml Release 4 (standalone). Let's start with a simple linear model.

$$y_{ijk} = \mu + L_i + R_j + e_{ijk} \tag{3.3}$$

where y_{ijk} is the k-th observation of the j-th RIL at i-th location, μ is the intercept (fixed), L_i is the i-th location effect (fixed), R_j is the j-th RIL effect (random), e_{ijk} is the random error term associated with the k-th observation.

We can write the same model in matrix form as $\mathbf{y} = \mathbf{Xb} + \mathbf{Zu} + \mathbf{e}$ where $\mathbf{y}$ is the vector of observations, $\mathbf{b}$ and $\mathbf{u}$ are vector of fixed and random effects, respectively; $\mathbf{X}$ and $\mathbf{Z}$ are incidence matrices of fixed and random effects and $\mathbf{e}$ is the vector of residuals. The usual model assumptions are that the residual effects (e_{ijk}) are normally and independently distributed $\mathbf{e} \sim \text{NID}(0, \mathbf{R})$ and random effects are normally and independently distributed $\mathbf{u} \sim \text{NID}(0, \mathbf{G})$.

There are 474 observations in this data set (recall that the data are not completely balanced because of missing data on some plots). The variance-covariance matrix of residual effects ($\mathbf{R}$) is a square matrix with 474 rows by 474 columns. The diagonal elements are variances of residual effects, all equal to σ_e^2. The off-diagonal elements are covariances of residual effects for different observations, which are all 0 because we assume residuals are independent of each other.

$$\mathbf{R} = \sigma_{\mathbf{e}}^2 \mathbf{I}_{474} = \sigma_{\mathbf{e}}^2 \begin{bmatrix} 1 & \cdots & 0 \\ \vdots & \ddots & \vdots \\ 0 & \cdots & 1 \end{bmatrix} = \begin{bmatrix} \sigma_e^2 & \cdots & 0 \\ \vdots & \ddots & \vdots \\ 0 & \cdots & \sigma_e^2 \end{bmatrix} \tag{3.4}$$

The variance for the residuals is set up as a correlation matrix (in this case, the identity matrix $\mathbf{I}$) scaled by a variance $\left(\sigma_e^2\right)$.

Similarly, we can write the variance-covariance matrix of the 62 random RIL effects (the $\mathbf{G}$ structure) as follows:

$$\mathbf{G} = \sigma_{\mathbf{u}}^2 \, \mathbf{I}_{62} = \sigma_{\mathbf{u}}^2 \begin{bmatrix} 1 & \cdots & 0 \\ \vdots & \ddots & \vdots \\ 0 & \cdots & 1 \end{bmatrix} = \begin{bmatrix} \sigma_u^2 & \cdots & 0 \\ \vdots & \ddots & \vdots \\ 0 & \cdots & \sigma_u^2 \end{bmatrix} \tag{3.5}$$

The variance for the random RIL effect is set up as a correlation matrix (here the identity matrix $\mathbf{I}$) scaled by the variance component for RIL effects $\left(\sigma_u^2\right)$.

The model terms in ASReml would be as follows:

Code example 3.1
Default variance modeling in ASReml. In the first part (PART 1) the R and G structures are not defined (implicit) but they are explicit in model 2 (PART 2). (Code 3-1_Default variance modeling.as)

```
!OUTFOLDER  V:\Book1_Examples\ch03_cov\outfiles # write output files
Title: Maize RILs unBalanced data.
 location  !A   rep *    block *    plot !I   RIL !A
 pollen
 silking
 ASI *
 height


!FOLDER V:\Book1_Examples\data   # data folder
MaizeRILs_miss.csv  !SKIP 1

!CYCLE   1 2

!PART 1 # Variance modeling is implicit
height !SIGMAP ~ mu location ,
      !r  RIL   rep.location


!PART 2 # The R and G structures are fully defined (explicit)
height !SIGMAP ~ mu location ,
        !r  idv(RIL)  id(rep).location
           residual idv(units)
```

- The model defined in PART 2 is *equivalent* to the model given in PART 1. They will produce the same results.
- The term `residual idv(units)` tells ASReml that the residual variance structure is an identity (ID) matrix multiplied by a uniform variance (V) for all the data points, $\mathbf{R} \sim N(0, \sigma_e^2 \mathbf{I}_{474})$. This is the default error variance structure, so if the error variance structure is not otherwise specified (as in the model in Part 1 in the example), it will be used.
- Similarly, the **G** variance structure `idv(RIL)` indicates that the variance associated with the RIL effect is an identity (ID) matrix multiplied by a uniform variance (V) for all the data points, $\mathbf{G} \sim N(0, \sigma_u^2 \mathbf{I}_{62})$. Again, this is the default, so it does not need to be specified (as in the model in Part 1).
- The variance structure for the nested `rep` effect is an identity matrix $N(0, \mathbf{I}_8 \sigma_r^2)$
- The `!SIGMAP` qualifier forces ASReml to use a sigma parameterization rather than the default gamma parameterization used for simple models.
- The `units` is a reserved term that refers to independent residuals of individual observations.

A subset of the output is given below:

```
...
Cycle 1 value is 1
          - - - Results from analysis of height - - -
LogL:    LogL  Residual       NEDF  NIT Cycle Text
LogL:-1389.77   82.6113         470    9 1 "LogL Converged"
Akaike Information Criterion    2785.54 (assuming 3 parameters).
Bayesian Information Criterion  2798.00

Model_Term                        Gamma       Sigma   Sigma/SE   % C
rep.location          IDV_V   8  0.180426     14.9052     1.29   0 P
RIL                   IDV_V  62  3.77412     311.785      5.32   0 P
Residual              SCA_V 474  1.00000      82.6113    14.23   0 P
```

```
...
Cycle 2 value is 2
...
            - - - Results from analysis of height - - -
  LogL:-1389.77   82.6113         470    9 2 "LogL Converged"
  Akaike Information Criterion      2785.54 (assuming 3 parameters).
  Bayesian Information Criterion    2798.00

  Model_Term                        Gamma        Sigma  Sigma/SE   % C
  rep.location        IDV_V    8  0.180426      14.9052      1.29   0 P
  RIL                 IDV_V   62  3.77412      311.785       5.32   0 P
  Residual            SCA_V  474  1.00000       82.6113     14.23   0 P
```

Heterogeneous R Variance Structures

Sections May Have Different Residual Variances

In the context of residual variance structures, a "section" refers to a group of observations that all have a common error variance structure and are independent of observations in other groups. In traditional analyses of variance and in the example in the previous section, we assumed that there is only one section in the data: all the observations have a common error variance structure. Breeding and genetics experiments often have natural groupings of observations (such as environments or ages of individuals) that may differ in terms of their error variation. If there are important differences among residual variances among sites, but these are ignored in the analysis by using an IDV $\mathbf{R}$ structure, one may observe some trends in the residual diagnostic plots, such as a positive relationship between predicted and residual values or distinct clusters of error effects representing sections with different residual variances. These distributions suggest that the model is not a good fit because the assumption of homogeneous error variance is wrong. If this is the case, we need to consider a block diagonal $\mathbf{R}$ structure (one section for each location) instead of a common IDV $\mathbf{R}$ structure for all observations. In the maize RIL example, it is reasonable to model the residuals as being independent but having a unique variance within each location. The $\mathbf{R}$ matrix will have 4 sections and we can define the $\mathbf{R}$ structure using the direct sum as follows:

$$\mathbf{R} = \sigma_{e1}^2 \mathbf{I}_{121} \oplus \sigma_{e2}^2 \mathbf{I}_{118} \oplus \sigma_{e3}^2 \mathbf{I}_{116} \oplus \sigma_{e4}^2 \mathbf{I}_{119} \tag{3.6}$$

The dimensions of sub-matrices of $\mathbf{R}$ (corresponding to sections) are given by the number of observations at each location. For example, there were 121 observations in location 1, so the dimension of the section for location 1 is 121×121 and this section is the product of a residual error variance specific to location 1 and the identity matrix $\left(\sigma_{e1}^2 \mathbf{I}_{121}\right)$. The overall dimension of $\mathbf{R}$ is 474×474, or the sum of all four sections. The block-diagonal $\mathbf{R}$ matrix is:

$$\mathbf{R} = \oplus_{j=1}^{s} R_j = \begin{bmatrix} R_1 & 0 & 0 & 0 \\ 0 & R_2 & 0 & 0 \\ 0 & 0 & R_3 & 0 \\ 0 & 0 & 0 & R_4 \end{bmatrix} = \begin{bmatrix} \sigma_{e1}^2 \mathbf{I}_{121} & 0 & 0 & 0 \\ 0 & \sigma_{e2}^2 \mathbf{I}_{118} & 0 & 0 \\ 0 & 0 & \sigma_{e3}^2 \mathbf{I}_{116} & 0 \\ 0 & 0 & 0 & \sigma_{e4}^2 \mathbf{I}_{119} \end{bmatrix}$$

We can zoom in on the $\mathbf{R}$ structure for section 1 to show its elements:

$$R_1 = \sigma_{e1}^2 \mathbf{I}_{121} = \begin{bmatrix} \sigma_{e1}^2 & 0 & \cdots & 0 \\ 0 & \sigma_{e1}^2 & \cdots & 0 \\ 0 & 0 & \cdots & 0 \\ 0 & 0 & \cdots & \sigma_{e1}^2 \end{bmatrix}$$

The ASReml syntax for a block-diagonal R structure follows:

Code example 3.2
Block diagonal R structure or heterogeneous residual variance modeling (see Code 3-2_Variance modeling Block Diagonal.as **for more details**)

```
!OUTFOLDER V:\Book1_Examples\ch03_cov\outfiles
Title: Maize RILs unBalanced data.
 location  !A   rep 2   block 8   plot !I  RIL !A
 pollen
 silking
 ASI  *
 height

!FOLDER V:\Book1_Examples\data
MaizeRILs_miss.csv   !SKIP 1 !DOPART 1



!PART 1 # Block-diagonal R
height  ~ mu location  !r rep.location  RIL
        residual sat(location).idv(units)
```

Here we used the `sat()` function to divide the **R** matrix into sections defined by levels of the factor location. The model term 'residual sat(location).idv(units)' means that the residual **R** structure is divided into sections defined by the levels of the factor location, and that within each section the residual effects are distributed identically and independently ('IID' or 'IDV'). ASReml forms this structure as a direct sum (see 'A brief introduction to matrices' in Chap. 2) of the individual section submatrices (**R**$_i$). The term `idv(units)` tells ASReml that the residuals with each section are uniform and independent, with 'units' being a reserved term that refers to individual measurement observations.

```
        - - - Results from analysis of height - - -
LogL:    LogL  Residual       NEDF  NIT Cycle Text
LogL:-1378.47   1.00000        470    8 1 "LogL Converged"
Akaike Information Criterion    2768.94 (assuming 6 parameters).
Bayesian Information Criterion  2793.86

Model_Term                          Sigma       Sigma  Sigma/SE  % C
rep.location            IDV_V    8  15.4426     15.4426     1.31  0 P
RIL                     IDV_V   62  310.573     310.573     5.35  0 P
sat(location,01).idv(units)    121 effects
units                   ID_V    1  54.1740      54.1740     6.35  0 P
sat(location,02).idv(units)    118 effects
units                   ID_V    1  139.725     139.725     7.06  0 P
sat(location,03).idv(units)    116 effects
units                   ID_V    1  78.3899      78.3899     6.57  0 P
sat(location,04).idv(units)    119 effects
units                   ID_V    1  61.3601      61.3601     6.59  0 P
```

In the output above (a subset of the .asr file) we highlighted the separate residual error variance estimated for each location, ranging from 54 to 140, suggesting that there are indeed important differences in the error variance among environments. We can formally test the null hypothesis that the error variances are equal among environments using a likelihood ratio test (the log(Likelihood) is also highlighted in the output above):

$$\text{LRT} = -2(\text{Log } L(\text{reduced model}) - \text{Log } L(\text{full model})) = -2(-1389.77 + 1378.47) = -2*(-11.3) = 22.6$$

This statistic is distributed approximately as a chi-square variable with 3 degrees of freedom. The degrees of freedom for this statistic are obtained by comparing the number of variance component parameters fit in the full model (6 components) to that for the reduced model (3 components). We can get the p-value for this statistic using a little R code:

```
> raw_p = 1 - pchisq(22.6,df = 3, lower.tail = T)
> (adj_p = raw_p/2)
[1] 2.446329e-05
```

Recall that we need to divide the raw p-value by two for this likelihood ratio test. In this case, modeling heterogeneous error variances significantly improves the fit of the model to the data.

Error Effects May Not Be Independent

Modeling heterogeneous error variances as shown in the previous section allows us to have models that do not require the 'identical' part of the assumption about identical and independent (IID) residual effects. We can also generalize the model to allow for non-independent residual effects. This will result in an **R** matrix structure that does not have 0 values for all of the off-diagonal elements. Exactly what those off-diagonal covariance values are will depend on some other assumptions we can make about what patterns of correlations may exist among the residual along with estimation of the correlations from the actual data.

Before discussing the mathematical details on non-independent residual effects, it may be helpful to consider what kinds of experimental conditions might lead to correlations among residual effects. A common scenario that occurs in many crop and tree field experiments is that the experimental field is never completely homogeneous. Even before the plants are established in the field plots, there is some level of variability in soil properties. After the experiment is established, additional variation among the experimental units may occur due to management effect, for example fertilizer or irrigation is not applied equally among plots. Under these conditions, experimental units (the field plots or plant positions) that are closer together tend to be more similar than plots that are separated by greater distances. Correlations among residual effects may also arise in other scenarios, for example when measurements on experimental units are taken at different times and measurement pairs taken at similar times have more similar residual effects than pairs measured at more widely spaced intervals. In some cases, the variability in experimental plots can be fit into the model using some known factor that is related to the variability. Often in field experiments, the plots are arranged in a rectangular grid pattern of rows and columns, such that a linear or non-linear function of row or column position can be fit as a covariate in the model to account for some of the field variability. Such covariates may not model the experimental error variability well if the variation is 'patchy' (highly non-linearly and not regularly related to the field position coordinates). Fitting a model that allows some correlation between plots as a function of their physical distance and estimating the correlation from the data often better models the data with such patchy variability.

How do we model the relationship between distance among experimental units (in space or time) and the correlation between their residual effects? In fact, many models are possible and they vary in how they relate the correlation between residuals to the distance between observations: the relationship may be linear or non-linear, one- or two-dimensional, involve a moving mean or not, and so forth. ASReml provides a wide array of possible models that can be fit to the data (Table 7.6 of Gilmour et al. 2014). In practice, we most often just fit autoregressive first-order (AR1) correlation structures to field data, as this model has proven robust across many different data sets. This is not to say that errors are really correlated in an AR1 fashion, but it means that the AR1 model is a useful approximation to a wide variety of error effect distributions.

The AR1 structure for residuals means that there are two parameters that we need to estimate to describe the **R** structure. We have a variance component for residual effects as always, but in addition there is a correlation coefficient that needs to be estimated. For any pair of residuals, their covariance is the product of the error variance component and the correlation coefficient raised to a power equal to the distance between the two observations. Since a correlation coefficient is never larger than one, the resulting covariance between observations decreases as the distance increases and the correlation

$$R = \begin{array}{c} \begin{array}{ccccccccccc} & \overbrace{}^{\text{row 1}} & & & \overbrace{}^{\text{row 2}} & & & \overbrace{}^{\text{row 10}} & & \text{row indices} \end{array} \\ \begin{array}{ccccccccccc} \text{c1} & \text{c2} & \dots & \text{c10} & \text{c1} & \dots & \text{c10} & \dots & \text{c1} & \dots & \text{c10} & \text{column indices} \end{array} \end{array}$$

Fig. 3.1 R structure for residuals with autoregressive first order correlations in row direction and independent in column direction

coefficient is raised to higher powers. Negative correlation coefficients are also possible, and in plant field data may indicate inter-plot competition, perhaps mediated by differences in plant heights. The units of distance obviously impact the powers of the correlation coefficient that are fit in the elements of the $\mathbf{R}$ structure, but since the correlation coefficient is estimated from the data, its value will change accordingly. So if we only care about the final $\mathbf{R}$ structure and not the specific value of the correlation coefficient, we do not need to worry about the units of distance.

Next, we could use linear distance between plots as the relevant distance for the AR1 model, but more often in two-dimensional field grid situations, we consider that distances in the row direction may have a different effect on the residual covariance as distances in the column direction. Thus we often model an $\mathbf{R}$ structure that has 'separable' correlations in the row and column direction. At one extreme, we may assume that distances in only the column direction and not in the row direction affect the covariance between residuals. To show what this $\mathbf{R}$ structure looks like, assume we have 100 observations measured on a grid of 10 rows $\times$ 10 columns. If we sort the observations by row position and column within row and fit an AR1 correlation (ρ) only in the row direction, we have an $\mathbf{R}$ structure as described in Fig. 3.1:

Observations in different field columns are independent of each other and have covariance of zero. Observations in a common row have a covariance that depends on their distance in row units. For example, observations in row 1 and row 2 of column 1 have a distance of 1 unit and a correlation of $\rho^1 = \rho$, resulting in a covariance of $\rho\sigma_e^2$ (Fig. 3.1). Observations in row 1 and in row 10 of a common column have a distance of 9 units, so their covariance is $\rho^9\sigma_e^2$, which will be near zero unless ρ is very close to one.

We can write this matrix in a much more concise form using a direct product between an identity matrix with the dimensions equal to the number of columns by the matrix of spatial correlations among row residuals. The row residual correlation matrix for a grid with r rows is:

$$\Sigma\rho_r = \begin{bmatrix} 1 & \rho & \rho^2 & \rho^3 & \cdots & \rho^{r-1} \\ \rho & 1 & \rho & \rho^2 & \cdots & \rho^{r-2} \\ \rho^2 & \rho & 1 & \rho & \cdots & \rho^{r-3} \\ \vdots & \vdots & \vdots & \vdots & \ddots & \vdots \\ \rho^r & \rho^{r-1} & \rho^{r-2} & \rho^{r-3} & \cdots & 1 \end{bmatrix}$$

Then the complete $\mathbf{R}$ structure is efficiently written as a Kronecker product:

$$\mathbf{Var}(e) = \mathbf{R} = \left(\sigma_e^2\mathbf{I}_c \otimes \Sigma\rho_r\right) \tag{3.7}$$

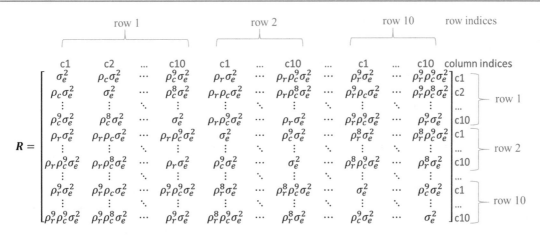

$$R = $$

Fig. 3.2 R structure for residuals with separate autoregressive first order correlations in both row and column directions

The dimensions of this R matrix are $rc \times rc$. Writing the R structure using this direct product formulation is helpful for understanding how to fit the models using ASReml because ASReml allows the user to specify the components of the direct product, which is the easiest way to represent this structure in terms of data set factors. For example, the R structure shown here would be represented as a direct product of two structures in ASReml: `idv(column)` and `ar1(row)`.

If residual errors are correlated both in row and column directions, we have two additional parameters, ρ_r and ρ_c, representing the correlation coefficient in the row and column, directions, respectively. If observations are sorted in row order and column within row order, the **R** structure is shown in Fig. 3.2:

Again, this "AR1 × AR1" **R** structure can be written as a Kronecker product involving two spatial correlation matrices

$$\mathbf{R} = \sigma_e^2 \begin{bmatrix} 1 & \rho_c & \rho_c^2 & \cdots & \rho_c^{c-1} \\ \rho_c & 1 & \rho_c & \cdots & \rho_c^{c-2} \\ \rho_c^2 & \rho_c & 1 & \cdots & \rho_c^{c-3} \\ \vdots & \vdots & \ddots & \cdots & \vdots \\ \rho_c^{c-1} & \rho_c^{c-2} & \rho_c^{c-3} & \cdots & 1 \end{bmatrix} \otimes \begin{bmatrix} 1 & \rho_r & \rho_r^2 & \cdots & \rho_r^{r-1} \\ \rho_r & 1 & \rho_r & \cdots & \rho_r^{r-2} \\ \rho_r^2 & \rho_r & 1 & \cdots & \rho_r^{r-3} \\ \vdots & \vdots & \ddots & \cdots & \vdots \\ \rho_r^{r-1} & \rho_r^{r-2} & \rho_r^{r-3} & \cdots & 1 \end{bmatrix}$$

$$= \left(\sigma_e^2 \, \Sigma \rho_c \otimes \Sigma \rho_r \right)$$

We will cover more details about the spatially correlated residual structures in Chap. 7 (Spatial analysis). To demonstrate briefly how the Kronecker products are coded in ASReml, we show how to fit the models with residual correlations only in row direction or in both row and column directions in Code example 3.3. The data set for this example is a single-site replicated experiment of barley varieties where the field was arranged as a rectangular grid of rows and columns. The data file (barley.asd) is provided with the ASReml program code, and on a Windows computer will be installed in a folder similar to "C:\Program Files\ASReml4\Examples\Functional". This example is also discussed in the ASReml User Guide (Gilmour et al. 2014) and in Gilmour et al. (1997).

Code example 3.3
Correlated residual variance modeling (see *Code 3-3_Correlated Residuals.as* **for more details**)

```
!outfolder V:\Book1_Examples\ch03_cov\outfiles
#Rep,RowBLK,ColBlk,row,column,variety,yield
Title: Correlated residuals
  Rep 6
  RowBlk 30     ColBlk 30     row 10     column 15
  variety 25
  yield
!folder V:\Book1_Examples\data
barley.asd !skip 1 !dopart  1

!PART 1   # residuals correlated in row direction
yield ~ mu variety !r Rep !f mv
       residual idv(column).ar1(row)
```

$$\boxed{\sigma_e^2 I} \nearrow \qquad \nwarrow \boxed{\Sigma\rho_r}$$

```
!PART 2   # residuals correlated in row and column directions
yield ~ mu variety !r    !f mv
       residual ar1v(column).ar1(row)
```

$$\boxed{\sigma_e^2\Sigma\rho_c} \nearrow \qquad \nwarrow \boxed{\Sigma\rho_r}$$

In both cases, the **R** structure is indicated as a direct product of one covariance matrix and one correlations matrix, using a period to separate the two components. Recall that we want the R structure to have variances on the diagonal and this is produced from a direct product between a covariance matrix (which has variances on the diagonal) and a correlation matrix (which usually has ones on the diagonal). A direct product between two correlation matrices would not work because it would be scaled to ones on the diagonal instead of the residual variance. Similarly, a direct product between two covariance matrices would not work because it would result in variances squared on the diagonals. So, for model 1, in principle, we want to use `residual idv(column).ar1(row)` instead of `residual id(column).ar1(row)`; the difference is simply the character 'v' appended to 'id' to make it a covariance rather than correlation structure. In practice, however, ASReml will, if possible, adjust the components of the residual structure to produce an appropriate R structure. So, one can 'get away' with writing `residual id(column).ar1(row)` in this case, but we recommend writing the correct structures as a habit, since there may be instances where one has to get the structure exactly correct, for example, when providing starting values for covariances vs correlations.

Output from model 1

```
      - - - Results from analysis of yield - - -
Akaike Information Criterion      1468.03 (assuming 2 parameters).
Bayesian Information Criterion    1473.68

Model_Term                         Gamma      Sigma    Sigma/SE   % C
idv(column).ar1(row)           150 effects
Residual               SCA_V   150  1.00000    43374.9       6.16   0 P
row                    AR_R      1  0.579377   0.579377      8.12   0 P
...
```

In this example, the residuals in column direction are identical and independent (not correlated), whereas residual errors are correlated in the row direction. The correlation in the row direction is is $\rho = 0.579$. This means that the covariance between residuals on plots that are in a common column but separated by two rows is $\rho_r^2 \sigma_e^2 = (0.579^2)43374.9 = 14541$.

Output from model 2

```
           - - - Results from analysis of yield - - -
     Akaike Information Criterion     1406.64 (assuming 3 parameters).
     Bayesian Information Criterion   1415.13

     Model_Term                          Gamma      Sigma    Sigma/SE    % C
     ar1(column).ar1(row)          150 effects
     Residual              SCA_V   150   1.00000    38754.3       5.00   0 P
     column                AR_R      1   0.683769   0.683769     10.80   0 P
     row                   AR_R      1   0.458594   0.458594      5.55   0 P
```

The residuals from model 2 are correlated both in column ($\rho_c = 0.683$) and row directions ($\rho_r = 0.458$) with the common residual variance of 38754.3. So, residuals on plots separated by two rows and three columns would have a covariance of $\rho_r^2 \rho_c^3 \sigma_e^2 = (0.459^2)(0.684^2)38754.3 = 3819.9$.

We can also create models that have spatially correlated residuals with unique variances and correlations across different sets of observations using the sat() function to model a separate and correlated residual structure for sections (e.g., experimental locations). The ASReml manual provides example code for fitting heterogeneous spatially correlated residuals for a field experiment replicated across three sites (Gilmour et al. 2014). To fit three separate AR1 × AR1 structures at each site, we can use:

```
residual sat(site).ar1v(column).ar1(row)
```

If we find after fitting the model that the column correlation was not significant at site 2, then we can modify the structure to specify an AR1 structure for rows only specifically at site 2, while fitting row and column correlations at the other two locations:

```
residual sat(site,1).ar1v(column).ar1(row),
         sat(site,2).idv(column).ar1(row) ,
         sat(site,3).ar1v(column).ar1(row)
```

Or, more concisely:

```
residual sat(site,1,3).ar1v(column).ar1(row),
         sat(site,2).idv(column).ar1(row)
```

Residual effects are also often correlated when we have repeated measures on the same subjects (e.g., measuring the same trees or animals at different ages) or when we measure multiple traits on each experimental unit and want to fit multivariate models. A trait measured at different ages can be treated in some cases as just a specific form of a multivariate model with different traits. For example, tree height could be measured at ages 2, 4 and 6 and recorded as Height2, Height4, Height6. The algebraic expression of residual variance matrix for repeated measure and for multivariate models is

$$I_n \otimes \Sigma \tag{3.8}$$

where I is the identity matrix with n units (random residuals), Σ is usually an *unstructured covariance matrix* with t x t dimensions and t is the number of traits (Gilmour et al. 2014). Traits are ordered *within* units. This structure means that residuals are assumed to be independent between experimental units, but residuals for different traits measured on the same experimental unit can be correlated. The unstructured covariance matrix means that we allow each trait to have a unique error variance and each pair of trait residuals to have its own covariance.

We can specify the residual R structure for a multivariate analysis of height and yield in the maize RIL data set using:

```
height yield ~ Trait Trait.location ,
               !r us(Trait).id(RIL)
               residual id(units).us(Trait)
```

The `id(units).us(Trait)` structure for repeated and multivariate models is a default. ASReml 'knows' we are fitting a multivariate model if we have more than one dependent variable listed on the left hand side of the model statement. Even if we do not define the residual term in the model ASReml will automatically add it. Again, it is a useful habit to write the residual structures explicitly because the user may need to add starting parameter values in some cases.

Heterogeneous G Variance Structures

Block Diagonal G Structure

For a univariate linear mixed model with one random term the variance of the response variable $\mathbf{y}$ is $\mathbf{Var}(\mathbf{y}) = \sigma_u^2 \mathbf{ZZ}' + \sigma_e^2 \mathbf{I_n}$. If there is more than one random term, the random effect vectors $\boldsymbol{u}$ can be partitioned into sub-vectors as $\mathbf{u} = \left[\mathbf{u}_r^T, \ldots, \mathbf{u}_b^T\right]^T$ and the design matrix for random effects can be partitioned into sub-matrices as $\mathbf{Z} = [\mathbf{Z}_r, \ldots, \mathbf{Z}_b]$.

The $\mathbf{G}$ structure for multiple random terms becomes a block diagonal, with one section and one variance component for each term. Let's say there are two random terms, such as the genotype (RIL) and rep effects from the maize RIL example; we use the direct sum $\oplus$ to define the $\mathbf{G}$ matrix as

$$\mathbf{G} = \oplus_{j=1}^{L} \sigma_{uj}^2 = \sigma_g^2 \mathbf{I_g} \oplus \sigma_r^2 \mathbf{I_r} \tag{3.9}$$

$$\mathbf{G} = \oplus_{j=1}^{L} \sigma_{uj}^2 = \begin{bmatrix} \sigma_g^2 \mathbf{I_g} & 0 \\ 0 & \sigma_r^2 \mathbf{I_r} \end{bmatrix} \tag{3.10}$$

where $\mathbf{I_g}$ is identity matrix with dimensions $g \times g$ (g being the number of RIL genotypes), $\mathbf{I_r}$ is the corresponding identity matrix for the block effect (r being the number of blocks). To show this structure directly, assume we took a small subset of the data that included only three RIL genotypes and two reps. The $\mathbf{G}$ structure would have dimensions 7×7:

$$\mathbf{G} = \oplus_{j=1}^{L} \sigma_{uj}^2 = \begin{bmatrix} \sigma_g^2 & 0 & 0 & & & \\ 0 & \sigma_g^2 & 0 & & 0 & \\ 0 & 0 & \sigma_g^2 & & & \\ & & & \sigma_r^2 & 0 & \\ & 0 & & 0 & \sigma_r^2 & \end{bmatrix} \tag{3.11}$$

The ASReml code for multiple random terms is:

```
!PART 1 # Block-Diagonal G
height  ~ mu location  !r RIL  rep
         residual sat(location).idv(units)
```

We can be more explicit about the **G** structure while fitting the same model as follows:

```
!PART 2 # Block-Diagonal G explicit
height  ~ mu location  !r idv(RIL)  idv(rep)
          residual sat(location).idv(units)
```

The model terms `idv(RIL)` and `idv(rep)` are *simple terms*. The *RIL* and *rep* effects are assumed to have IID variance structures, meaning that variance associated with the RIL is the same (identical) across different environments and the rep effects have the same IID variance structure at different locations. These may not be realistic assumptions.

Nested and Interaction Terms in the G Structure

Interaction effects are formed by joining two terms with a dot or colon (A.B or A:B). Model terms with more than one factor are called *compound terms*. In the following ASReml code, a compound model term is given for RIL nested within location.

```
!PART 3 # Block-Diagonal G. RIL nested in locations
height  ~ mu location  ,
          + !r location.RIL
          residual sat(location).idv(units)
```

The program understands the compound term `location.RIL` is a nested term because the location main effect is included, but the RIL term is never listed as a main effect and appears only as an interaction with location. ASReml creates a compound term with the number of levels equal to the number of locations × number of RIL levels. If there were 3 locations and 20 RIL varieties in each location, the nested (interaction) term would have 60 levels.

The variance of the compound model term can still be IID. However, we can change this assumption and instead fit a model where RILs have different variances at each location. In ASReml, we create a *consolidated* model term by combining the compound term ($location.RIL$) and associating it with different variances according to location. For example, we may want a G structure that assume RILs have a different variance at each location and that RILs are independent within locations and across locations:

$$\mathbf{G_{L.R}} = \Sigma \otimes \mathbf{I_{nr}} = \begin{bmatrix} \sigma_{g1}^2 \mathbf{I_g} & 0 & 0 & 0 \\ 0 & \sigma_{g2}^2 \mathbf{I_g} & 0 & 0 \\ 0 & 0 & \sigma_{g3}^2 \mathbf{I_g} & 0 \\ 0 & 0 & 0 & \sigma_{g4}^2 \mathbf{I_g} \end{bmatrix} \tag{3.12}$$

We can write this structure as a direct sum of four IDV matrices that each have a separate genotypic variance:

$$\mathbf{G_{L.R}} = \sigma_{g1}^2 \mathbf{I_g} \oplus \sigma_{g2}^2 \mathbf{I_g} \oplus \sigma_{g3}^2 \mathbf{I_g} \oplus \sigma_{g4}^2 \mathbf{I_g} \tag{3.13}$$

We can also write the same structure as a direct product of a matrix (Σ) representing the variance-covariance relationships of any one genotype across the four locations and an identity matrix with dimensions equal to the number of RILs:

$$\mathbf{G_{L.R}} = \Sigma \otimes \mathbf{I_g} = \begin{bmatrix} \sigma_{g1}^2 & 0 & 0 & 0 \\ 0 & \sigma_{g2}^2 & 0 & 0 \\ 0 & 0 & \sigma_{g3}^2 & 0 \\ 0 & 0 & 0 & \sigma_{g4}^2 \end{bmatrix} \otimes \mathbf{I_g} \tag{3.14}$$

The ASReml code for this G structure along with a heterogeneous error variance structure across locations is:

```
!PART 4   # Block-diagonal R, IID G, IID G*L
height   ~ mu location  !r RIL    location:RIL
          residual sat(location).idv(units)


!PART 5   # Block-diagonal R and G
height   ~ mu location  !r  idh(location).id(RIL)   # idh = diag
residual sat(location).idv(units)
```

In this example, we use the term idh(location) to create an 'identity with heterogeneous variances' matrix (in other words, a diagonal matrix), which represents the $\sum$ matrix. Part of the output from this model follows:

```
Model_Term                            Sigma      Sigma    Sigma/SE   % C
  sat(location,01).idv(units)    121 effects
  units                  ID_V    1   84.7896    84.7896       5.43   0 P
  sat(location,02).idv(units)    118 effects
  units                  ID_V    1   107.056    107.056       5.34   0 P
  sat(location,03).idv(units)    116 effects
  units                  ID_V    1   69.9691    69.9691       5.31   0 P
  sat(location,04).idv(units)    119 effects
  units                  ID_V    1   66.1969    66.1969       5.34   0 P
  diag(location).id(RIL)         248 effects
  location              DIAG_V    1   320.142    320.142       4.82   0 P
  location              DIAG_V    2   381.756    381.756       4.73   0 P
  location              DIAG_V    3   241.861    241.861       4.69   0 P
  location              DIAG_V    4   323.672    323.672       4.95   0 P
```

In the output, the first four sigmas are residual variances at each of the four locations and the last four sigmas are RIL variances, one for each location. DIAG_V indicates that the G variance structure is block diagonal.

Effects in G Can Be Correlated

Is the assumption of *independent RIL effect* across sites realistic? Probably not, because the same RIL varieties are tested across sites and the performance of a variety in one environment is likely to be related to its performance in another environment. This implies a correlation between the effects of a common genotype in different environments. We can no longer use a direct sum $\bigoplus$ to model such a **G** matrix; instead we need to use a direct product $\bigotimes$ and include the pairwise environment correlations in one of the component matrices.

As a simplified example, consider the variance structure for three genotypes nested within two locations, allowing each genotype within location effect to have a correlation with the effect of the same genotype at a different location. The **G** matrix is the product of a variance-covariance matrix of genotype effects across levels of location (this time being unstructured instead of diagonal) and an identity matrix with $g \times g$ (3×3) dimensions:

$$G_{L.R} = \sum \otimes I_g \tag{3.15}$$

$$= \begin{bmatrix} \sigma_{g1}^2 & \sigma_{g12} \\ \sigma_{g12} & \sigma_{g2}^2 \end{bmatrix} \otimes \begin{bmatrix} 1 & 0 & 0 \\ 0 & 1 & 0 \\ 0 & 0 & 1 \end{bmatrix} = \begin{bmatrix} \sigma_{g1}^2 & 0 & 0 & \sigma_{g12} & 0 & 0 \\ 0 & \sigma_{g1}^2 & 0 & 0 & \sigma_{g12} & 0 \\ 0 & 0 & \sigma_{g1}^2 & 0 & 0 & \sigma_{g12} \\ \sigma_{g12} & 0 & 0 & \sigma_{g2}^2 & 0 & 0 \\ 0 & \sigma_{g12} & 0 & 0 & \sigma_{g2}^2 & 0 \\ 0 & 0 & \sigma_{g12} & 0 & 0 & \sigma_{g2}^2 \end{bmatrix} \tag{3.16}$$

Since we have only two sections in this example, there are two variance components $\left(\sigma_{g1}^2 \text{ and } \sigma_{g2}^2\right)$ and one covariance (σ_{g12}) between RIL effects in different locations. We can also parameterize this in terms of correlations instead of covariances:

$$
G_{L.R} = \sqrt{\sigma_{g1}^2 \sigma_{g2}^2}
\begin{bmatrix}
\dfrac{\sigma_{g1}^2}{\sigma_{g2}^2} & r_{g12} \\[2ex]
r_{g12} & \dfrac{\sigma_{g2}^2}{\sigma_{g1}^2}
\end{bmatrix}
\otimes
\begin{bmatrix}
1 & 0 & 0 \\
0 & 1 & 0 \\
0 & 0 & 1
\end{bmatrix}
=
\begin{bmatrix}
\sigma_{g1}^2 & 0 & 0 & \sigma_{g12} & 0 & 0 \\
0 & \sigma_{g1}^2 & 0 & 0 & \sigma_{g12} & 0 \\
0 & 0 & \sigma_{g1}^2 & 0 & 0 & \sigma_{g12} \\
\sigma_{g12} & 0 & 0 & \sigma_{g2}^2 & 0 & 0 \\
0 & \sigma_{g12} & 0 & 0 & \sigma_{g2}^2 & 0 \\
0 & 0 & \sigma_{g12} & 0 & 0 & \sigma_{g2}^2
\end{bmatrix}
$$

Using a correlation matrix parameterization sometimes makes it easier to provide initial values and to converge on a correct result. We can fit this kind of structure to the maize RIL example data using the following ASReml code:

```
!PART 6    # Block-diagonal R and G, G=CORUH
height   ~ mu location   !r   coruh(location).id(RIL)
residual sat(location).idv(units)
```

The variance function `coruh(location).id(RIL)` means that RIL effects nested within location have a uniform correlation between pair of sites (**coru**h) but their variances is heterogeneous across locations (coru**h**). The second part of the consolidated term, `id(RIL)` means that RIL effects are independent of each other. The order of the effects does not matter as long as we assign the appropriate variance function to the compound terms. For example, `coruh(location).id(RIL)` and `id(RIL).coruh(location)` produces the same results.

Output from this model is:

```
Model_Term                          Sigma      Sigma     Sigma/SE   % C
sat(location,01).idv(units)    121 effects
units                   ID_V   1    79.1657    79.1657    6.27     0 P
sat(location,02).idv(units)    118 effects
units                   ID_V   1   128.281    128.281     6.13     0 P
sat(location,03).idv(units)    116 effects
units                   ID_V   1    79.3361    79.3361    6.14     0 P
sat(location,04).idv(units)    119 effects
units                   ID_V   1    61.8943    61.8943    5.98     0 P
coruh(location).id(RIL)        248 effects
location                COR_R  1     0.991999   0.991999  49.49    0 P
location                COR_V  1   325.759    325.759     4.87     0 P
location                COR_V  2   367.645    367.645     4.70     0 P
location                COR_V  3   237.983    237.983     4.71     0 P
location                COR_V  4   324.171    324.171     4.98     0 P
```

It is obvious from this result that RIL effects are strongly correlated across locations, almost perfectly so. The previous model where we fit a diagonal variance-covariance matrix for the effects within and across locations is analogous for forcing the correlation coefficient to be zero, which in this case is obviously a poor model.

Correlated Effects Due to Genetics

In the examples up to now, we assumed that the levels of random genotypic effects are independent. In reality, individuals in a population can have varying levels of genetic similarity. If we have a way to estimate the relatedness for each pair of individuals in the data set (using pedigree information or marker data), we can account for these relationships in the **G** structure of the mixed model. We will cover details in later chapters but give a simple demonstration here.

Let's say there are 8 individuals in the pedigree. Individual A is the mother of progenies 4 and 7, each from a different father. Individual B has progenies 5 and 8, each from different fathers, and individual C has progeny 6. The matrix of additive genetic relationships derived from pedigree looks like:

	A	B	C	4	5	6	7	8
A	1	0	0	0.5	0	0	0.5	0
B	0	1	0	0	0.5	0	0	0.5
C	0	0	1	0	0	0.5	0	0
4	0.5	0	0	1	0	0	0.25	0
5	0	0.5	0	0	1	0	0	0.25
6	0	0	0.5	0	0	1	0	0
7	0.5	0	0	0.25	0	0	1	0
8	0	0.5	0	0	0.25	0	0	1

Individuals A, B, and C in the parental generation are assumed to be unrelated and independent, so we assign them as having 0 genetic relationships. The additive genetic covariance between individual 1 and its progeny 4 and 7 is 0.5; the additive genetic covariance between half-sibs 4 and 7 is 0.25. The algebraic form of the $\mathbf{G}$ matrix for the random term is $\mathbf{G} = \sigma_A^2 \mathbf{A}$, where σ_A^2 is the variance component associated with the additive genetic effects and $\mathbf{A}$ is the additive genetic relationship matrix derived from the pedigree. We can fit this relationship matrix to the data on a series of individuals by supplying a special pedigree information file to ASReml, declaring the appropriate factor (e.g., individual) as associated with the pedigree using the !P field definition qualifier, and using the model function nrm(). For example:

```
height   ~ mu location   ,
           + !r nrm(individual)
```

Initial Values

As described in Chap. 2, ASReml uses the Average Information algorithm to obtain maximum likelihood estimates of the variance parameters from the data. This requires iteration beginning from some starting values, at each step obtaining maximum likelihood estimates for one parameter at a time dependent on current values of the other parameters. These steps are repeated until the parameter estimates change only very little from step to step. This is referred to as 'convergence' on the maximum likelihood solution. Specifically, ASReml stops iterating if the successive REML log-likelihood (LogL) values of two iterations do not change more than 0.002 and if the parameter estimates are stable.

ASReml by default uses the phenotypic variance of the response to obtain initial values for the variance structure parameters. For most simple models, the user does not need to provide initial values in the code. For complex models, such as multivariate models, the ASReml algorithm can have difficulty converging when the starting values are not reasonably close to the REML solution (Gilmour et al. 2014). When you see 'Convergence failed' or 'LogL did not converge' messages after running a model, it is likely that the initial values based on the phenotypic variance are not good enough to finish the job. In such cases, especially for complex models, user need to supply good starting values to achieve convergence.

There are multiple ways to provide initial values. We will use MaizeRILs_miss.csv data

1. **Insert explicit initial values in the job file (.as) using the** `!INIT` **qualifier**.

```
!PART 2   # Initial values for random effects
height   ~ mu location   !r idv(RIL !INIT 300)   idv(block !INIT 1.2)
              residual sat(location).idv(units)
```

In the code, 300 is the initial variance for RIL effect and 1.2 is the initial variance for the block effect.

When !INIT is used to supply initial values for more complex variance structures, the number of parameters and their order must be correct. In example below a bivariate model (height and silking) is fit to estimate genetic/residual correlations and covariances between two traits.

```
!PART 8 # initial values for correlated residuals
height silking ~ Trait Trait.location   !r  Trait.RIL
                    residual idv(units).us(Trait !INIT 92 0.1 3.2)
```

We used `!INIT` qualifier and listed the residual variance for height (92), the covariance between height and silk (0.1) and the variance for silking (3.2). Notice the order of the initial values. It follows the order of traits in the model.

We can also use `!ASSIGN` and `!INIT` qualifiers to supply initial values for the correlated residual effects models. In the example below, the initial values for the unstructured `us()` residual variance structure are supplied in the order lower-triangle row-wise:

```
######## R and G structures for MULTIVARIATE MODELS
!PART 7     # initial values for correlated residuals
!ASSIGN USe   !< !INIT
92                # Error variance of height
0.1 3.2           # Error covariance, error variance of silk
!>
height silking ~ Trait Trait.location   !r  Trait.RIL
              #   residual idv(units).us(Trait)
                    residual idv(units).us(Trait $USe)
```

Since the initial parameters continue on multiple rows, we use !< *string* !> to enclose the term. With the !ASSIGN qualifier we created a file called *USe* and used it in the code to supply initial values.

Initial values for the RIL variance can be supplied in a similar way:

```
!PART 8 # initial values
!ASSIGN USg !INIT 300 0.2 5   # Genetic var-cov-var
!ASSIGN USe !INIT 92 0.1 3    # Error var-cov-var
height silking ~ Trait Trait.location ,
              !r us(Trait $USg).id(RIL)
                    residual idv(units).us(Trait $USe)
```

USg is the initial values for RIL effect and USe is the initial values for the residuals. Since the values for residuals or for RIL effect are in their respective rows there is no need to use !< !> to enclose the string.

2. **Using .tsv and .msv files to provide initial values**.

An example of .tsv file is given below for *Code 3-1_Default variance modeling.as.*

```
!OUTFOLDER    V:\Book\Book1_Examples\ch03_cov\outfiles
Title: Maize RILs unBalanced data.
 location !A   rep 2    block 8    plot !I    RIL  !A
 pollen
 silking
 ASI   *
 height

!FOLDER   V:\Book\Book1_Examples\data
MaizeRILs_miss.csv   !SKIP 1

height ~ mu location   !r RIL
```

In each run ASReml writes the initial values of the variance parameters to a file with *Code 3-1_Default variance modeling. tsv* extension (template-start-value).

```
# This .tsv file is a mechanism for resetting initial parameter values
# by changing the values here and rerunning the job with !CONTINUE 2.
# You may not change values in the first 3 fields
#                  or RP fields where RP_GN is negative.

# Fields are:
# GN, Term, Type, PSpace, Initial_value, RP_GN, RP_scale.
    4, "rep.location", G, P,    0.10000000    ,    4,    1
    5, "RIL",          G, P,    0.10000000    ,    5,    1
    6, "Variance 1",   V, P,    1.0000000     ,    6,    1

# Valid values for Pspace are F, P, U and maybe Z.

# RP_GN and RP_scale define simple parameter relationships;
# RP_GN links related parameters by the first GN number;
# RP_scale must be 1.0 for the first parameter in the set and
# otherwise specifies the size relative to the first parameter.
# Multivalue RP_scale parameters may not be altered here.
```

After each iteration, the current values of the variance parameters are written to two files with extensions *.rsv* (re-start values) and *.msv*.

The *Code 3-1_Default variance modeling.rsv* file

```
78 6 2173 121
# This .rsv file holds parameter values between runs of ASReml and
# is not normally modified by the User.  The current values of the
# the variance parameters are listed as a block on the following lines.
# They are then listed again with identifying information
# in a form that the user may edit.
```

```
     0.000000          0.000000          0.000000         0.1804261        3.774121        1.000000
 RSTRUCTURE                     1    1    3
  VARIANCE                      1    1    0
    6, V, P,    1.0000000        0    0
  STRUCTURE                   474    0    0
 rep.location                            1    1
    4, G, P,    0.18042608       0    0
 RIL                                     1    1
    5, G, P,    3.7741206        0    0
```

The *Code 3-1_Default variance modeling.msv* file is easier to understand, as it clearly identifies the fields.

```
# This .msv file is a mechanism for resetting initial parameter values
# by changing the values here and rerunning the job with !CONTINUE 3.
# You may not change values in the first 3 fields
#                      or RP fields where RP_GN is negative.

# Fields are:
# GN, Term,          Type, PSpace, Initial_value, RP_GN, RP_scale.
   4, "rep.location", G,   P,        0.18042608   ,    4,    1
   5, "RIL",          G,   P,        3.7741206    ,    5,    1
   6, "Variance 1",   V,   P,        1.0000000    ,    6,    1
# Valid values for Pspace are F, P, U and maybe Z.
# RP_GN and RP_scale define simple parameter relationships;
# RP_GN links related parameters by the first GN number;
# RP_scale must be 1.0 for the first parameter in the set and
# otherwise specifies the size relative to the first parameter.
# Multivalue RP_scale parameters may not be altered here.

# Notice that this file is overwritten if not being read.
```

Notice the difference between initial values in the .tsv and .msv files. The initial values in the .msv file are generated after the last iteration are closer to actual values. You may edit the *PSpace* and *Initial_value* fields.

We can use the !CONTINUE f qualifier to use one of these starting value files.

If !CONTINUE 2 or !TSV is used then the **.tsv** file is used instead of the .rsv file.

if !CONTINUE 3 or !MSV is used then the **.msv** file is used instead of the .rsv file.

```
 !OUTFOLDER    V:\Book\Book1_Examples\ch03_cov\outfiles
 Title: Maize RILs unBalanced data.
  location !A    rep 2    block 8    plot !I    RIL   !A
  pollen
  silking
  ASI   *
  height

 !FOLDER   V:\Book\Book1_Examples\data
 MaizeRILs_miss.csv   !SKIP 1    !CONTINUE !MSV

 height ~ mu location   !r RIL
```

We provide the specific file names after !CONTINUE or after !MSV as "Code 3-1_Default variance modeling"

```
...
!FOLDER   V:\Book\Book1_Examples\data
MaizeRILs_miss.csv   !SKIP 1 !CONTINUE !MSV "Code03-1_Default variance
modeling.msv"

height ~ mu location   !r RIL
```

ASReml looks for "Code 3-1_Default variance modeling.msv" file, scans it for parameter values related to current model, replaces the values obtained from .as file before iteration resumes. If a file name is not provided, ASReml looks for an .rsv file with the same base name used for the output files, i.e., the .as file suffix name, possibly appended by arguments.

Electronic supplementary material: The online version of this chapter (doi:10.1007/978-3-319-55177-7_4) contains supplementary material, which is available to authorized users.

Abstract

In this chapter we cover the basics of estimating breeding values from field progeny test data for half-sib family selection. The individual level or 'animal' model is introduced to demonstrate prediction of individual breeding values across generations. Modifications to the basic model are considered, such as maternal effects and genetic group effects.

Family Selection

Many breeders are interested in family selection for species that can easily produce large half-sib or full-sib families, particularly for traits that cannot be measured on living individuals. Family selection is favored for a number of reasons. Traits (such as meat quality in livestock species) that cannot be measured on living individuals can be measured in progeny or siblings of selection candidates, and selections made based on the mean phenotypes of the relatives. Many traits of interest are controlled by many genes with small effects, and show low to moderate heritability. The phenotype of an individual is often a poor predictor of its genetic merit as a breeding parent because of large environmental effects, so phenotypic selection on single individual (mass selection or strict within family selection) is not effective. Testing of inbred lines is one solution to this problem, but for those species in which inbred lines are not feasible, selection based on family mean phenotype is an alternative solution. Mass production of progeny of a specific cross may also be a goal, to exploit favorable specific combining ability (non-additive genetic effects). For example, in forest trees, seed orchard managers prefer parental selection to establish seed orchards. This is sometimes called backward selection (selection of individuals of a previous generation) or **among-family selection**. Breeders use phenotypic data from many progeny to calculate the breeding values of families or parents with high accuracy.

Depending on the mating design and the experimental design, there are different types of family selection. Half-sib family selection is often used to select superior individuals for their general combining ability. In contrast, full-sib family selection is preferred if breeders want to market seed with a known father and mother to increase genetic gain and capture specific combining ability and hybrid vigor between two parents.

In general, selection units and schemes are more complex when information is available on multiple individuals in some families (Falconer et al. 1996; White et al. 2007). Selection can be **strictly within family** (Lynch and Walsh 1998). Individuals are ranked within each family and then the best individual(s) within each family are selected. In another scenario, individual estimated values are adjusted by subtracting their family means and the individuals with highest within-family deviations are selected regardless of their family. This is called **selection on within-family deviation**. Another type of selection is **family index selection**. This is sometimes called **combined family and within family selection**. Different weights are assigned to family and individuals within family to rank them. In general, the objective is to increase genetic gain while putting constraints on coancestry of selected individuals to avoid inbreeding as much as possible in future generations.

In practice, breeders either use a **family model** or general combining ability (GCA) model to predict parental breeding values. **General Combining Ability** (GCA_i) is the deviation of the mean value of progeny of a particular individual ($y_{i.}$) from the population mean (μ) : $GCA_i = y_{i.} - \mu$. **Breeding value** (BV) is the value of an individual measured by the mean phenotype of its progeny obtained by random mating with the population (Falconer and Mackay 1996). Breeding value measures the average effect of an individual's alleles as they affect progeny performance, rather than the performance of the individual itself (which is its genotypic value). In diploid species and in the absence of epistasis, an individual's GCA value is half of its BV because a parent transmits exactly half of its alleles to any individual progeny. Compared to animal models, GCA models do not require as much computer memory to solve mixed model equations, because mixed model equations are solved only for the parents.

Moving to the progeny generation, we can include parental BV's in a linear model for the progeny j as the average of its parental BVs along with a fixed overall mean and random error effect (e_i) : $y_j = \mu + 0.5(u_m + u_f) + e_j$, where u_m and u_f are male and female parental BVs, respectively. Alternatively, breeders can fit **animal models** to simultaneously predict parental and progeny breeding values on the same scale and carry out various selection strategies.

In this chapter we cover half-sib family selection, a commonly used approach in many plant and animal breeding programs. We first briefly cover some definitions, interpretations of observed variance components and genetic interpretations, then introduce several examples of the mixed model equations for GCA models and the syntax in ASReml. We also give an example on how to derive half-sib family means and the variance of family means as the selection unit and introduce two animal models. The first example uses the same pine data with shallow pedigree (parent and progeny) used for the GCA model. The second example uses data from a pig breeding population with deep pedigree.

Causal Variance Components and Resemblance

Variance components estimated from data are *observed variance components*. We partition the total phenotypic variance into groups, such as between families and within families. Using the observed variance components and genetic covariances among relatives, we can calculate the *causal variance components*. Additive genetic variance is the *causal variance component* arising from additive effects of genes that cause resemblance between relatives. Falconer and MacKay (1996) denote observed variance component by the symbol 'σ^2' and the causal components by the symbol 'V'. For most models we use observed variance components to estimate causal variances or to estimate heritability and other population parameters. The relationships between the observed and causal variance components is dictated by the degree of covariances among relatives. Relatives can be classified as *ancestral* (e.g. parent-offspring) or *collateral* (e.g. sibs) (Lynch and Walsh 1998). Relatives resemble each other more than they do other individuals in the population, for traits that have non-zero heritability. The more closely related the individuals are, the more they resemble each other. The resemblance between individuals is measured by the genetic covariance, a concept introduced by Fisher (1918) and Wright (1922). We use genetic covariances to estimate causal variance components from observed estimates.

Covariance of half-sibs
In half-sib families (each consisting of a group of progeny that all share one parent in common, while the other parent is randomly chosen from the population), the expected value of the estimated variance component explained by half-sib family effects is 1/4 of the additive genetic variance (Falconer and Mackay 1996). Where does this relationship come from?

We can use a simple model of genotypic values for individuals based on their additive and dominance effects at one locus. We assume the population is in linkage equilibrium and for now will assume no epistasis, such that our single locus model can be simply generalized to a multi-locus model for the whole genome by summing up effects and variances over loci. Obviously, this assumption is probably never completely true in real populations, but nevertheless this kind of model often is an adequate approximation to the genetic architecture of real populations. The genotypic value of individual A is the sum of the overall population mean (μ), the average effects of the two alleles that it carries, and the dominance interaction between those two alleles:

$$G_A = \mu + \alpha_{A1} + \alpha_{A2} + \delta_{A12} \tag{4.1}$$

The phenotypic value of this individual is the sum of its genotypic value and a combination of environmental effects and random residual term that we will indicate as ε:

$$P_A = G_A + \varepsilon_A = \mu + \alpha_{A1} + \alpha_{A2} + \delta_{A12} + \varepsilon \tag{4.2}$$

The phenotypic variance in the population is the variance of the phenotypic values, which is the sum of the genotypic variance and environmental/residual effect variance, since we will assume that there is no covariance between genotypic values and environmental values:

$$V_P = V_G + V_\varepsilon = V(\alpha_{A1} + \alpha_{A2} + \delta_{A12}) + V(\varepsilon) \tag{4.3}$$

We can further sub-divide the genotypic variance into components due to the additive effects and the dominance effect. We assume the population is in Hardy-Weinberg equilibrium, so that the two alleles at a locus within any individual are not correlated. This lack of correlation results in the variance of the sum of the genetic effects being equal to the sum of the variances of those effects:

$$V_G = V(\alpha_{A1} + \alpha_{A2} + \delta_{A12}) = V(\alpha_{A1}) + V(\alpha_{A2}) + V(\delta_{A12}) = 2V(\alpha) + V(\delta) = V_A + V_D \tag{4.4}$$

Notice that, by definition, the additive variance (V_A) is twice the variance of individual allele additive effects: $V_A = 2V(\alpha)$. This definition also follows from the idea that the breeding value of an individual is the sum of its allelic additive effects, so for one locus $BV_A = \alpha_{A1} + \alpha_{A2}$ and the variance of breeding values is:

$$V(BV_A) = V(\alpha_{A1} + \alpha_{A2}) = 2V(\alpha) = V_A \tag{4.5}$$

Now, we use this model to find the covariance between half-sibs. Let's say individual A mates with individuals B and C, and produces progenies X (from A × B) and Y (from A × C). We can write the covariance between the genotypic values of these two offspring as:

$$Cov(G_X, G_Y) = C((\alpha_{X1} + \alpha_{X2} + \delta_{X12}), (\alpha_{Y1} + \alpha_{Y2} + \delta_{Y12})) \tag{4.6}$$

We use the subscripts X1 and X2 to denote the two alleles in individual X. We know that one of the two alleles was inherited from its parent A and the other from its parent B. So, we can switch the notation to refer to allele X1 as allele Ai, where Ai has an equal probability of being either allele carried by parent A. We can make similar substitutions for the other alleles and we will refer to allele Y1 as the allele that individual Y inherited from A as Ai', to indicate that it may or may not be the same as Ai inherited in individual X:

$$\begin{aligned}
Cov(G_X, G_Y) &= C((\alpha_{Ai} + \alpha_{Bi} + \delta_{AiBi}), (\alpha_{Ai'} + \alpha_{Ci} + \delta_{Ai'Ci})) \\
&= C(\alpha_{Ai}, \alpha_{Ai'}) + C(\alpha_{Ai}, \alpha_{Ci}) + C(\alpha_{Ai}, \delta_{Ai'Ci}) \\
&\quad + C(\alpha_{Bi}, \alpha_{Ai'}) + C(\alpha_{Bi}, \alpha_{Ci}) + C(\alpha_{Bi}, \delta_{Ai'Ci}) \\
&\quad + C(\delta_{AiBi}, \alpha_{Ai'}) + C(\delta_{AiBi}, \alpha_{Ci}) + C(\delta_{AiBi}, \delta_{Ai'Ci})
\end{aligned} \tag{4.7}$$

The assumption of random mating among unrelated individuals means that alleles inherited from different parents cannot be identical by descent (IBD), so they are independent. Independence among alleles implies that all of the covariances involving parents B and C are zero. Random-mating also means that there are no covariances between additive effects and dominance effects and that the dominance effects have no covariance even if they have one allele in common (both alleles must be identical by descent for there to be a covariance). Thus, Eq. 4.7 simplifies greatly because all but one covariance component is zero by definition under random mating:

$$\boldsymbol{Cov(G_X, G_Y) = C(\alpha_{Ai}, \alpha_{Ai'})} \tag{4.8}$$

The probability that the allele inherited by X from parent A is the same as the allele inherited by Y from parent A is ½. So the covariance equals half of the covariance between an additive effect and itself, or in other words, half of the variance of one additive effect, which equals a quarter of the overall additive genetic variance:

$$Cov(G_X, G_Y) = \frac{1}{2}C(\alpha_{Ai}, \alpha_{Ai}) = \frac{1}{2}V(\alpha_{Ai}) = \frac{1}{4}V_A \tag{4.9}$$

If we have randomized the progenies with respect to environmental and error effects, then there is no covariance between those effects on different progenies, and the phenotypic variance is equal to the genotypic variance:

$$Cov(P_X, P_Y) = Cov(G_X + \varepsilon_X, G_Y + \varepsilon_Y) = Cov(G_X, G_Y) = \frac{1}{4}V_A \tag{4.10}$$

The degree of resemblance between X and Y is measured using the coefficient of coancestry θ_{xy}, which is simply the probability that an allele in individual X is identical by descent to an allele at the same locus in individual Y. The coancestry coefficient determines the covariance between breeding values of individuals, denoted here as A_X and A_Y:

$$Cov(A_X, A_Y) = 2\theta_{XY}V_A \tag{4.11}$$

More generally, the covariance between genotypic values of individuals X and Y is a function of the additive variance scaled by twice the coancestry coefficient and the dominance variance scaled by the 'double coancestry coefficient', Δ_{XY}. In the absence of inbreeding, the double coancestry coefficient is the probability that the *pair* of alleles at a locus in one individual is IBD with the pair of alleles in the other individual:

$$Cov(G_X, G_Y) = 2\theta_{XY}V_A + \Delta_{XY}V_D \tag{4.12}$$

To summarize (Lynch and Walsh 1998):

```
Genetic covariance between half-sibs

Case                          Probability    Contribution
X and Y have 0 allele IBD 1/2                0
X and Y have 1 allele IBD 1/2               V_A/2

Giving the genetic covariance between half-sibs as
```
$$Cov(G_{X_1}, G_{Y_1}) = \tfrac{1}{2}V_A/2 = V_A/4$$

Covariance of full-sibs (both parents are common to all progeny)

Two parents (B and C) are mated to produce full-sib offspring (X and Y). At one locus, the genotype values (G) of parents are;

$$G_B : \alpha_{B1} + \alpha_{B2} + \delta_{B12}$$

$$G_C : \alpha_{C1} + \alpha_{C2} + \delta_{C12}$$

Each full sib receives one paternal and one maternal allele. The probability that each sib receives the same paternal allele or maternal allele is ½. The probability that a randomly sampled allele from X is IBD to a randomly sampled allele from its full-sib Y is the probability that we sample from X and Y sibs inherited from a common parent (1/2) times the probability that two alleles sampled from that parent are IBD (1/2 for non-inbred parents). Thus, the coancestry coefficient for full sibs is $\theta_{XY} = 1/4$. To determine the double coancestry coefficient, consider that three cases are possible for the number of alleles per locus that are IBD between full sibs. They can share 0, 1, or 2 alleles IBD (Lynch and Walsh 1998):

$$Pr(0 \; alleles \; IBD) = Pr(paternal \; allele \; not \; IBD)^{*}Pr(maternal \; allele \; not \; IBD)$$

$$= \frac{1}{2} * \frac{1}{2} = \frac{1}{4}$$

$$Pr(2 \; alleles \; IBD) = Pr(paternal \; allele \; IBD)^{*}Pr(maternal \; allele \; IBD)$$

$$= \frac{1}{2} * \frac{1}{2} = \frac{1}{4}$$

$$Pr(1 \; allele \; IBD) = 1 - (Pr(2 \; alleles \; IBD) + Pr(0 \; allele \; IBD))$$

$$= 1 - \left(\frac{1}{4} + \frac{1}{4}\right) = \frac{1}{2}$$

The probability that the allele pair is IBD between the full-sibs is the probability that both alleles are IBD between them, which is shown above as 1/4. In the case where the allele pair is IBD, their genetic covariance is:

$$Cov(\alpha_{A1} + \alpha_{B1} + \delta_{A1B1}, \alpha_{A1} + \alpha_{B1} + \delta_{A1B1})$$
$$V(\alpha_{A1} + \alpha_{B1} + \delta_{A1B1}) = V_A + V_D \tag{4.13}$$

Putting the pieces together (Lynch and Walsh 1998);

Genetic covariance between full-sibs

Case	Probability	Contribution
X and Y have 0 allele IBD	1/4	0
X and Y have 1 allele IBD	1/2	$V_A/2$
X and Y have 2 alleles IBD	1/4	$V_A + V_D$

$$Cov(G_X, G_Y) = \frac{1}{2}\frac{V_A}{2} + \frac{1}{4}[V_A + V_D]$$

$$= \frac{1}{2}V_A + \frac{1}{4}V_D$$

Or, using the coancestry and double coancestry coefficients directly, we get the same result:

$$Cov(G_X, G_Y) = 2\theta_{XY}V_A + \Delta_{XY}V_D = 2\left(\frac{1}{4}\right)V_A + \frac{1}{4}V_D = \frac{1}{2}V_A + \frac{1}{4}V_D \tag{4.14}$$

See Lynch and Walsh (1998, Chapter 7) and Falconer and MacKay (1996, Chapter 9) for more details about the genetic covariances among relatives.

The GCA (Family) Model

There are multiple ways to estimate variance components and breeding values from progeny test data. We might be interested in parental breeding values. In this case, the GCA models (parental models) may be preferred because they are easier to fit. All we need to do is to solve linear mixed model equations for the parents, not for the progeny. For example, a linear mixed model to predict GCA values of parents based on their progeny values measured at a particular age is $y_i = \mu + age + GCA_i + e_i$, where *age* refers to the age of individual progeny (a fixed effect). We can set up the mixed model equations to obtain solutions (GCA estimates) for parents, as follows:

$$\begin{bmatrix} \mathbf{X'X} & \mathbf{X'Z} \\ \mathbf{Z'X} & \mathbf{Z'Z} + \mathbf{I}\alpha \end{bmatrix} \begin{bmatrix} \hat{\mathbf{b}} \\ \hat{\mathbf{u}} \end{bmatrix} = \begin{bmatrix} \mathbf{X'y} \\ \mathbf{Z'y} \end{bmatrix} \tag{4.15}$$

where **b** and **u** are vectors of fixed and random effects, respectively. In this example, **b** includes only the effect of the fixed covariate for age, and **u** includes the parental GCA values. **Z** and **X** are design matrices for the fixed and random effects, respectively. **I** is the identity matrix with dimensions equal to the number of parents, α is a ratio of residual variance and genetic variance explained by the random family effect $\alpha = \sigma_e^2/\sigma_u^2$. By rearranging the mixed model equations we can obtain solutions for the fixed (**b**) and random effects (**u**) as follows:

$$\begin{bmatrix} \hat{\mathbf{b}} \\ \hat{\mathbf{u}} \end{bmatrix} = \begin{bmatrix} \mathbf{X'X} & \mathbf{X'Z} \\ \mathbf{Z'X} & \mathbf{Z'Z} + \mathbf{I}\alpha \end{bmatrix}^{-1} \begin{bmatrix} \mathbf{X'y} \\ \mathbf{Z'y} \end{bmatrix} \tag{4.16}$$

$$BLUE(\mathbf{b}) = [\mathbf{X'X} \quad \mathbf{X'Z}]^{-1} \begin{bmatrix} \mathbf{X'y} \\ \mathbf{Z'y} \end{bmatrix} \tag{4.17}$$

$$BLUP(\mathbf{u}) = [\mathbf{Z'X} \quad \mathbf{Z'Z} + \mathbf{I}\alpha]^{-1} \begin{bmatrix} \mathbf{X'y} \\ \mathbf{Z'y} \end{bmatrix} = \begin{bmatrix} u_1 \\ u_2 \\ \vdots \\ u_n \end{bmatrix} \tag{4.18}$$

$\alpha = \sigma_e^2 / \sigma_u^2$ is sometimes called the shrinkage factor and has a range of 0 to 1. Notice what happens in the mixed model equations at the extreme values of heritability for $h^2 = 0$ or $h^2 = 1$. As $h^2 \to 0$, then $\alpha = \sigma_e^2 / \sigma_u^2 \to \infty$ and the diagonal components of the $\mathbf{Z'Z + I\alpha}$ part of the BLUP formula become infinitely large. Since that part of the formula gets inverted, it is like dividing the observed differences in family means by infinity, or like multiplying them all by zero. So, the result is that all breeding value predictions shrink to zero (there are no genetic differences among parents). As $h^2 \to 1$, then $\alpha = \sigma_e^2 / \sigma_u^2 \to 0$, with the result that the observed progeny values would be perfect indicators of the parental breeding values, so breeding values would be equal to the mean phenotypes of their progeny (adjusted only by differences in the fixed covariate).

The identity matrix, $\mathbf{I}$, used in the mixed model equation above is appropriate when we assume no relationships among the parents. If we know or can estimate the relationships among the parents, based either on recorded pedigrees or genetic marker data, we can estimate their coancestries and replace the $\mathbf{I}$ matrix by a matrix, $\mathbf{A}$, of additive genetic correlations among the parents. Each element of the $\mathbf{A}$ matrix is two times the coancestry between a pair of individuals ($2\theta_{XY}$) as we saw before:

$$\mathrm{BLUP}(\mathbf{u}) = \begin{bmatrix} \mathbf{Z'X} & \mathbf{Z'Z + A\alpha} \end{bmatrix}^{-1} \begin{bmatrix} \mathbf{X'y} \\ \mathbf{Z'y} \end{bmatrix} = \begin{bmatrix} u_1 \\ u_2 \\ \vdots \\ u_n \end{bmatrix} \tag{4.19}$$

The genetic covariance matrix is traditionally derived from the pedigree, although we will show in later chapters how DNA markers can be used to estimate genetic relationships. One effect of replacing the $\mathbf{I}$ matrix by an $\mathbf{A}$ matrix is that the BLUP for a parent is influenced not only by the phenotype values of its own offspring, but also by the phenotypes of the offspring of any parents that have a pedigree relationship with the individual ($\theta > 0$ between the parents). In this case, the effect of the shrinkage factor, α, on the BLUPs is to weigh the influence of the offspring of the parent being predicted versus the information from offspring of related parents. As heritability increases and α decreases, not only is the overall shrinkage reduced, but the relative influence of information from direct progeny (which comes from the $\mathbf{Z'Z}$ part of the BLUP equation) versus the progenies of related parents (which comes from the $\mathbf{A\alpha}$ part of the equation) increases. With lower heritabilities, the information from relatives becomes relatively more important. Again, however, if heritability is zero, the BLUPs will all be shrunk back to zero.

Among the advantages of the BLUP method are that it:

- Accounts for fixed effects (e.g. site, age) while calculating breeding values.
- Is an efficient method to use information from all relatives while calculating breeding values, allowing breeding values to be predicted even for individuals that do not have a measured phenotype.
- Accounts for trends in the data and founder effects
- Allows models to account for genotype by environment interactions (correlations)
- Makes calculation of genetic gain straightforward

Analysis of Half-Sib Progeny Data Using GCA Model

We will use the *Pine_provenance.csv* data introduced in Chap. 1 to demonstrate an example of prediction of half-sib family breeding values. To recap, the data are on trees from 36 half-sib families, each sampled from one of four provenances. The field experimental design was a randomized complete block design with five replications and from two to six trees measured within each plot. When analyzing data, it is important to write the linear model first:

$$y_{ijkl} = \mu + B_i + P_j + F_{k(j)} + BF_{ik(j)} + \varepsilon_{ijkl} \tag{4.20}$$

where

y_{ijkl} is the lth observation of the ith block, jth provenance and kth family;
μ is the overall mean;

B_i is the random jth block effect $\sim N\left(0, \sigma_b^2\right)$;

P_j is the fixed provenance effect ($j = 1,..,4$)

$F_{k(j)}$ is the random kth female parent effect within its provenance group $\sim N\left(0, \sigma_f^2\right)$

$BF_{ik(j)}$ is the random block by female interactions (plot effect), $\sim N\left(0, \sigma_{bf}^2\right)$

ε_{ijkl} is the tree within plot residual term $\sim N\left(0, \sigma_\varepsilon^2\right)$

Depending on the experimental design or mating design, the linear model can change. For example, if we had one tree per female per block (single-tree plots), then the plot term (BF) would be dropped from above model and would be included in the residual term.

For this specific example, here is why factors in the model are considered fixed or random:

- We would like to explain the sources of variation for height among the entire population from which trees in the experiment were sampled. How much of the phenotypic variation is due to genetics and how much is due to environment? To answer this question, we treat families and their female parents as **a random sample** of the breeding population. Since female parents represent a random sample of the population, any interaction effects involving female parents (block by female interaction, BF) are also random. The residual effect is always considered random.
- The blocks in this experiment represent some larger population of potential block (or microenvironment) effects, and we want to make inference beyond the specific sample of blocks used in this study, so we treat the block effect as **random.**
- Female parents were sampled from four provenances. 'Provenance' in forest trees means the region where female parents were originally selected. Provenance effect was treated as a fixed effect in this example. One reason is that there are too few (only four) to estimate a variance for the provenance effect. Another reason might be that we are specifically interested in differences among these four provenances.

Our first objective is to estimate variance components to make some inferences about genetic variation in the population. We can fit the linear model in ASReml as follows:

Code example 4.1
Half-sib family progeny test data analysis (*Code 4-1_HSfamily.as*)

```
!OUTFOLDER    T:\Book\Book1_Examples\ch04_gca\outputfiles
Title: Pine_provenance.
#treeid,female,male,prov,block,plot,height,diameter,volume
 treeid  !A
 female  !I
 male     *
 prov      !I
 block *    plot  *
 height     diameter     volume  !*10

!FOLDER    T:\Book\Book1_Examples\data
Pine_provenance.csv  !SKIP 1   !DOPART 3

!PART 1
height   ~ mu prov ,     # Specify fixed model
         !r idv(female) block block.female  # Specify random model
         residual units
```

Variance components ('Sigma') and the ratio of Sigma's to their standard errors ('Sigma/SE') are reported in the *Code 4-1_HSfamily.asr* file:

```
Model_Term                    Gamma        Sigma    Sigma/SE   % C
  block         IDV_V    5  0.425370E-01  0.107492      1.20    0 P
  idv(female)   IDV_V   36  0.749901E-01  0.189502      2.26    0 P
  block.female  IDV_V  180  0.782019E-01  0.197618      2.29    0 P
  units                914  effects
  Residual      SCA_V  914  1.00000       2.52703      19.22    0 P
```

Variance Components and Their Linear Combinations

One of the main objectives of progeny testing is to partition observed phenotypic variance into genetic and environmental components. Additive genetic variance, phenotypic variance, heritability and genetic gain predictions are calculated using linear combinations of variance components. Here is an example for estimation of additive and phenotypic variances and their standard errors from the output for half-sib progeny test data.

Variance components in the *.asr* output file are observed variance components. For simplicity we will use rounded numbers. The family variance component ($\sigma^2_f = 0.189$) is an estimate so it has an error (variance) associated with it. The variances and covariances of variance components estimates are reported in the ASReml output file *Code 4-1_HSfamily.vvp*.

The standard error of the *family* variance component is:

$$SE\left(\sigma^2_f\right) = \sqrt{Var\left(\sigma^2_f\right)} = \sqrt{0.705259E\text{-}02} = 0.084$$

Notice that the ratio of the estimated variance component itself to its standard error is $0.189/0.084 = 2.26$, matching the value in the column labeled '**Sigma/SE**' in the .asr output above.

The phenotypic variance is the sum of the female variance, plot variance, and within-plot residual variance components:

$$\sigma^2_P = \sigma^2_f + \sigma^2_{bf} + \sigma^2_e$$

$$\sigma^2_P = 0.1895 + 0.1976 + 2.527 = 2.91$$

The variance of a sum of estimators is the sum of the variances of each term, plus two times their covariances (using a Taylor series approximation). Using the information from the *Code 4-1_HSfamily.vvp* file, we get:

$$Var\left(\sigma^2_P\right) = Var\left(\sigma^2_f\right) + Var\left(\sigma^2_{bf}\right) + Var\left(\sigma^2_\varepsilon\right) +$$

$$2\left[Cov\left(\sigma^2_f, \sigma^2_{bf}\right) + Cov\left(\sigma^2_f, \sigma^2_\varepsilon\right) + Cov\left(\sigma^2_{bf}, \sigma^2_\varepsilon\right)\right]$$

$$Var\left(\sigma^2_P\right) = 0.00705 + 0.00744 + 0.0173 + 2^*[-0.00138 - 0.00009 - 0.00329]$$

$$Var\left(\sigma^2_P\right) = 0.02227$$

The standard error of σ^2_P is the square root of the variance:

$$SE\left(\sigma^2_P\right) = \sqrt{Var(\sigma^2_P)} = Sqrt(0.02227) = 0.149$$

Additive genetic variance

The variance component associated with the female parent (or half-sib family) effect is 1/4 of the additive genetic variance. Therefore, the additive genetic variance estimate is four times the family variance component:

$$\sigma_A^2 = 4\sigma_f^2 = 4{*}0.1895 = 0.758 \quad \text{Additive genetic variance}$$

$$\text{Var}\left(\sigma_A^2\right) = \text{Var}\left(4\sigma_f^2\right) = 16\text{Var}\left(\sigma_f^2\right) \quad \text{Variance of additive genetic variance}$$

$$\text{SE}\left(\sigma_A^2\right) = \sqrt{16\text{Var}\left(\sigma_f^2\right)} = 4\sqrt{\text{Var}\left(\sigma_f^2\right)} = 4{*}0.00705259 = 0.3359 \quad \text{Standard error of additive genetic variance}$$

Narrow-sense heritability

Narrow-sense heritability is a ratio of additive ($\sigma^2{}_A$) and phenotypic variances ($\sigma^2{}_P$). It is a population parameter with a range of 0–1. For this half-sib family example, heritability is

$$h_i^2 = \frac{\sigma_A^2}{\sigma_P^2} = \frac{4\sigma_f^2}{\sigma_f^2 + \sigma_{bf}^2 + \sigma_\varepsilon^2} = 0.758/2.91 = 0.26$$

The approximate variance of a heritability estimator $\text{Var}\left(h_i^2\right)$ can be obtained in several ways (Piepho and Möhring 2007), including the following two methods:

1. Assuming phenotypic variance $\left(\sigma_P^2\right)$ is a constant, we can use the Dickerson approximation (Dickerson 1969) to calculate variance of heritability:

$$Var\left(h_i^2\right) = \frac{16Var\left(\sigma_f^2\right)}{\left(\sigma_P^2\right)^2} = \frac{16{*}0.00705}{\left(2.91\right)^2} = 0.01335$$

$$SE\left(h_i^2\right) = \sqrt{0.01335} = 0.1155$$

2. The Delta method is a good approximation to obtain the variance of a ratio of estimators because it uses all the information of moments. See Lynch and Walsh (1998, Appendix 1) and Holland et al. (2003) for a detailed explanation of the Delta method. Conveniently, ASReml provides Delta method estimates of the standard error of heritability estimators by default when the VPREDICT !DEFINE statement is used to define a heritability (or correlation) estimate. We show below how to use this statement to define various functions of the variance components estimates, including linear combinations or ratios:

Code example 4.1
Half-sib family progeny test data analysis, continued (*Code 4-1_HSfamily.as*)

```
...
height   ~ mu prov ,     # Specify fixed model
         !r  idv(female) block block.female   # Specify random model
         residual units
VPREDICT !DEFINE
# Make sure that function F (sum, extraction, multiplication)
# comes before function H (ratio)
F Pheno        idv(female) + block.female + units   # sum of variances
F Additive     idv(female)*4         # Multiply component by 4
H fam_ratio    idv(female) Pheno     # Ratio of female var in pheno var
H plot_ratio   block.female Pheno    # Ratio of plot var in pheno var
H error_ratio  units Pheno           # Ratio of residual var in pheno var
H Herit        Additive  Pheno       # Heritability, divide A by P
```

Alternatively, the block of code following VPREDICT !DEFINE can be placed in a separate file with the same root name as the .as command file but using the extension *.pin*. Either way, the output from these estimates of functions of variance components are found in a file with extension *.pvc*, in this case *Code 4-1_HSfamily.pvc*:

```
ASReml 4.1 [28 Dec 2014] Title: Pine_provenance.
          Code4-3_HSfamily.pvc created 25 Aug 2015 16:33:26.811


          - - - Results from analysis of height - - -

    1 idv(female)        V    36  0.194272  0.830222E-01
    2 block.female       V   180  0.158139  0.859451E-01
  units                      914 effects
    3 units;Residual     V   914      2.53125   0.131836
    5 Pheno   2              2.9141     0.14973
    6 Additive  2            0.75801    0.33592
      fam_ratio    = idv(fema   2/Pheno  2   5= 0.0650  0.0275
      plot_ratio   = block.fe   3/Pheno  2   5= 0.0678  0.0290
      error_ratio  = units;Re   4/Pheno  2   5= 0.8672  0.0341
      Herit        = Additive   6/Pheno  2   5= 0.2601  0.1099
  Notice: The parameter estimates are followed by
            their approximate standard errors.
```

The highlighted block contains the estimate of heritability (0.2601) and its standard error (0.1099) obtained with the Delta method.

Variation Among Family Means

Narrow-sense heritability is useful to predict genetic gains from selection of *individuals* in a population. This is sometimes called mass selection. However, in many plant and animal breeding programs the selection units are families rather than individuals. Similarly, in some animal breeding programs, sires are selected on the basis of their progeny values. In this case, breeders would be mostly interested in heritability of family means and genetic gain from family (or sire) selection. For example, in tree breeding, seed orchards are established to produce large quantities of seeds from a small subset of the best families based on progeny tests.

In the example of the pine provenance data set, we could use half-sib family means as the selection units. After adjusting out the provenance effects, each family mean is averaged over plots and individual trees (residual effects) that contain that family effect:

$$\bar{Y}_{..k.} = \mu + F_k + \frac{\sum_{i=1}^{b} BF_{ik}}{b} + \frac{\sum_{i=1}^{b}\sum_{l=1}^{n} \varepsilon_{ikl}}{bn} \tag{4.21}$$

where, b = number of blocks, n = number of trees per family per block. The expected variance of half-sib family means is:

$$V(\bar{Y}_{..k.}) = E\left(\mu + F_k + \frac{\sum_{i=1}^{b} BF_{ik}}{b} + \frac{\sum_{i=1}^{b}\sum_{l=1}^{n} \varepsilon_{ikl}}{bn} - \mu \right)^2 \tag{4.22}$$

Since the different model effects are uncorrelated because of randomization in the experimental design, this simplifies to:

$$V(\bar{Y}_{..k.}) = E(F_k)^2 + E\left(\frac{\sum_{i=1}^{b} BF_{ik}}{b}\right)^2 + E\left(\frac{\sum_{i=1}^{b}\sum_{l=1}^{n}\varepsilon_{ikl}}{bn}\right)^2 = E(F_k)^2 + bE\left(\frac{BF_{ik}}{b}\right)^2 + bnE\left(\frac{\varepsilon_{ikl}}{bn}\right)^2 =$$

(4.23)

$$= \sigma_F^2 + b\frac{\sigma_{BF}^2}{b^2} + bn\frac{\sigma_\varepsilon^2}{(bn)^2} = \sigma_F^2 + \frac{\sigma_{BF}^2}{b} + \frac{\sigma_\varepsilon^2}{bn}$$

We can estimate this variance of half-sib family means using the observed variance components estimates and the number of blocks and plants per block:

$$\mathrm{Var}(\bar{Y}..k.) \sim 0.189 + (0.198/5) + (2.527/(5*5.2)) = 0.326$$

where $b = 5$ is the number of blocks, and $n = 5.2$ is the average number of individuals per female family per block. Note that when data are not balanced, the divisors of the variance components should be the harmonic mean of the number of relevant effects that are averaged over for each family, see Holland et al. (2003).

Heritability of half-sib family means is a ratio of the family variance component (σ_f^2) to the phenotypic variance of family means:

$$h_{HS}^2 = \sigma_f^2/\mathrm{Var}(\bar{Y}..k.) \sim 0.189/0.326 = 0.58$$

Assuming a subset of the families is selected, predicted genetic gain from selection would be the product of family mean heritability and the selection differential: $\Delta = h_{HS}^2 S_f$. The selection differential, S_f, is the difference between mean of the selected families and the population mean.

Within-Family Variation

Breeders might be interested in estimating the predicted genetic gains from family index and within-family selection schemes. For within-family selection we need within-family phenotypic variance and within-family heritability. We can derive these two terms from the variance components estimates. The difference between total phenotypic variance and family mean phenotypic variance is the within-family phenotypic variance (σ_w^2):

$$\sigma_w^2 = \left(\sigma_f^2 + \sigma_{fb}^2 + \sigma_\varepsilon^2\right) - \left(\sigma_f^2 + \sigma_{bf}^2/b + \sigma_\varepsilon^2/bn\right)$$

(4.24)

$$= 0 + \frac{(b-1)}{b}\sigma_{fb}^2 + \frac{(bn-1)}{bn}\sigma_\varepsilon^2$$

Since the variance among half-sib families, σ_f^2, is 1/4 of V_A, the remainder of the additive genetic variance (3/4 V_A) occurs within families and is contained in the within-family phenotypic variance. Within-family heritability is simply the ratio of $3/4V_A = 3\sigma_f^2$ to the within-family phenotypic variance:

$$h_w^2 = \frac{3\sigma_f^2}{\sigma_w^2} = \frac{3*(0.1895)}{\frac{5-1}{5}(0.198) + \left(\frac{(5*5.2)-1}{5*5.2}\right)2.527} = \frac{0.5685}{0.158 + 2.43} = 0.22$$

Genetic gain from selection among individuals within families can be predicted as the product of within-family heritability and the mean selection differential within families (the difference between the mean of selected progeny and the family mean).

We can estimate the heritabilities for these additional within-family selection schemes by adding some additional code for functions of the variance components to the VPREDICT !DEFINE part of the .as file:

Code example 4.1
Half-sib family progeny test data analysis, continued (*Code 4-1_HSfamily.as*)

```
# Make sure that function F (sum, extraction, multiplication)
# comes before function H (ratio)
VPREDICT !DEFINE
F Pheno        idv(female) + block.female + Residual # sum of variances
F Additive     idv(female)*4         # Multiply component by 4
H fam_ratio    idv(female) Pheno     # Ratio of female var in pheno var
H plot_ratio   block.female Pheno    # Ratio of plot var in pheno var
H error_ratio  Residual Pheno        # Ratio of residual var in pheno
F Phen_hs      idv(female) + block.female*0.2 + Residual*0.039
F Phen_w       block.female*0.80 + Residual*0.96 # within female phen
F fam3         idv(female)*3.0       # 3/4 of Additive var
H h2i          Additive Pheno        # Heritability
H h2fm         idv(female) Phen_hs   # Family mean heritability
H h2w          fam3  Phen_w          # Within family heritability
```

Output file is *Code 4-1_HSfamily.pvc*

```
- - - Results from analysis of height - - -
1 block             V     5  0.107492     0.895767E-01
2 idv(female)       V    36  0.189502     0.838504E-01
3 block.female      V   180  0.197618     0.862961E-01
 units                 914 effects
4 units;Residual    V   914  2.52703      0.131479
5 Pheno  2                  2.9141     0.14973
6 Additive  2               0.75801    0.33592
  fam_ratio    = idv(fema   2/Pheno  2   5=  0.0650  0.0275
  plot_ratio   = block.fe   3/Pheno  2   5=  0.0678  0.0290
  error_ratio  = units;Re   4/Pheno  2   5=  0.8672  0.0341
7 Phen_hs  2                0.32758    0.82299E-01
8 Phen_w  3                 2.5840     0.12504
9 fam3  2                   0.56851    0.25194
  h2i          = Additive   6/Pheno  2   5=  0.2601  0.1099
  h2fm         = idv(fema   2/Phen_hs   7=  0.5785  0.1174
  h2w          = fam3  2    9/Phen_w    8=  0.2200  0.0992
```

Family effect predictions can be obtained from the solution file *Code 4-1_HSfamily.sln*. The solutions for the family effect also represent the BLUP of GCA values of the female parent of each family, and they are distributed around zero. A subset of the solution file is given below:

Model_Term	Level	Effect	seEffect
prov	10	0.000	0.000
prov	12	-0.6237	0.3701
prov	11	-1.666	0.3741
prov	13	-1.220	0.3770
mu	1	11.51	0.3620

```
block              1      0.3110      0.1855
block              2      0.1268      0.1858
block              3      0.2056      0.1858
block              4     -0.2919      0.1867
block              5     -0.3516      0.1878
idv(female)      191     -0.8257E-01  0.3395
idv(female)      192      0.2459      0.3419
idv(female)      170     -0.1634      0.3433
...
```

The GCA values of parents 191 and 192 are −0.0826 and 0.2459, respectively, after adjusting for provenance effects. These GCA values are appropriate for comparing parental breeding values within the same provenance, but to compare across provenances, one would want to add the provenance effect to the breeding value. To obtain GCA values centered on the mean, one may also add the intercept "mu" value.

The Accuracy of Breeding Values

The standard error of a BLUP can be used to assess the reliability of the prediction. However, the standard error is not a useful estimate to compare the prediction reliability of different traits or for the same trait with different scales because it depends on the measurement unit. As an alternative, animal breeders developed 'accuracy' estimates derived from standard errors to assess the confidence in predictions. The correlation between true and predicted breeding values is called the accuracy of the BV (Mrode 2014), and this value is comparable among traits and measurement units because it is scaled to between 0 and 1. As discussed in Chap. 2, reliabilities are the squares of accuracies, and the accuracies of BLUPs can be obtained as a function of the standard error of predictions (available in the .sln file). BLUP accuracies for families are: (Gilmour et al. 2014):

$$r = \sqrt{1 - \frac{S^2}{(1+F)\sigma_F^2}} \tag{4.25}$$

where S is the standard error of the family effect prediction reported in the .*sln* file, and σ_F^2 is the family variance component and F is the inbreeding coefficient of individuals. We show below how to compute accuracies of BLUPs within provenances; this would be an appropriate breeding value to use to select among families within provenances.

Code example 4.2
R script to estimate breeding values within provenances and their accuracies (Code 4-2_HSfamily.R)

```
#~~~~~~~~~~~~~~~~~~~~~~~~~~~~~~~~~~~~~~~~~~~~~~~~~~
# Calculate breeding values and their accuracy
#~~~~~~~~~~~~~~~~~~~~~~~~~~~~~~~~~~~~~~~~~~~~~~~~~~

# Remove everything from memory
rm(list=ls())

# Set the path of the working directory
path="/Users/fisik/Google
 Drive/Book/Book1_Examples/ch04_gca/outputfiles"
setwd(path)
```

```
# Read in the data file and call it 'dat'.
dat=read.table(file='Code04-1_HSfamily.sln',header=T)

# Get the intercept term value from ASReml output file
# and call it 'mu'
mu.idx=which(dat$Model_Term=='mu')
mu=dat$Effect[mu.idx]

# Get the half-sib GCA values from ASReml output file
# and store them in array 'half.fam.bv'
fam.idx=dat$Model_Term=='idv(family)'
family=dat$Level[fam.idx]      # select the levels
gca=dat$Effect[fam.idx]        # select the GCA estimates
se.gca=dat$seEffect[fam.idx] # select the SE of GCA

# Calculate breeding values on the mean scale
bv = mu + 2*gca
# the denominator (family variance=0.194) is taken from the .asr file
accuracy <- sqrt(1- (se.gca^2/0.194272))

# merge all in a data frame
solution <- data.frame(cbind(family,gca,se.gca,mu,bv, accuracy))
round(head(solution),3)
```

Breeding values of the six five families and their accuracies are printed here:

female	gca	se.gca	mu	bv	accuracy
170	-0.163	0.343	11.51	11.183	0.627
191	-0.083	0.340	11.51	11.345	0.638
192	0.246	0.342	11.51	12.002	0.631
196	-0.461	0.307	11.51	10.589	0.717
197	0.356	0.302	11.51	12.221	0.729
198	-0.451	0.302	11.51	10.609	0.729

Individual ("Animal") Model

To predict BVs of grandparents, progeny and parents simultaneously, we can use the so-called animal model, which may also be referred to as the individual model when applied to plant species. In the individual model, we write the phenotypic value of an individual as a linear combination of overall mean (μ), individual effect (u) and the error (e):

$$y_i = \mu + u_i + e_i \tag{4.26}$$

In the mixed model equations for the individual model, the vector $\mathbf{u}$ includes predictions for progeny and parents (and, further, any individuals with known relationships to the observational units can be included):

$$\begin{bmatrix} \mathbf{X'X} & \mathbf{X'Z} \\ \mathbf{Z'X} & \mathbf{Z'Z}+\dfrac{\sigma_e^2}{\sigma_u^2}\mathbf{A}^{-1} \end{bmatrix} \begin{bmatrix} \mathbf{b} \\ \mathbf{u} \end{bmatrix} = \begin{bmatrix} \mathbf{X'y} \\ \mathbf{Z'y} \end{bmatrix} \longrightarrow \mathbf{u} = \begin{bmatrix} u_1 \\ u_2 \\ u_3 \\ u_4 \\ u_5 \\ u_6 \\ u_7 \\ u_8 \end{bmatrix} \begin{matrix} \text{Parents} \\ \\ \\ \\ \text{Progeny} \\ \\ \end{matrix}$$

In the animal model, individuals are no longer simply observational units, they (rather than the family units) are **factors** in the model representing effects that need to be estimated. As with the GCA models, the animal model requires a genetic relationship matrix, derived either from a pedigree file or markers.

Why use the individual (animal) model and pedigree information?

- It provides predictions (BV) for grandparents, parents and progeny in one step.
- Breeding values are comparable across all the generations because they are on the same scale
- Facilitates more complex genetic models, such as additive, dominance, paternal and clonal effects
- Uses all the information from relatives. Each relative contributes to the estimate of each individual's BV depending on how genetically close they are.
- It provides a prediction even if an individual does not have a measured phenotype, as long as one or more relatives are measured.

Animal Model for Half-Sib Family Data

To demonstrate a simple example of an individual model, we return to the *pine_provenance.csv* dataset previously used for half-sib GCA analysis. We will compare the individual model with the GCA model fit previously in this Chapter. Our objective is to simultaneously predict parental and progeny breeding values.

We assume that female parents are random independent samples from the reference population, so they have no relationships. The pedigree for the parents is quite simple:

```
female     parent1 parent2
170           0        0
191           0        0
192           0        0

. . .

191.1        191       0
191.2        191       0

. . .
```

Individuals 191.1 and 191.2 are half-sibs as they share the same female parent and different (unknown) male parents. The additive genetic relationship matrix (**A**) is derived from the pedigree with dimension 951 × 951 (915 progeny +36 female parents). We present as small subset of the **A** matrix for the three parents and five offspring with pairwise additive genetic relationship coefficients (2θ) for individuals (Fig. 4.1):

Fig. 4.1 Additive genetic relationship matrix for three selected female parents and selected open-pollinated progenies from the pine_provenance.csv data set. The complete relationship matrix includes 36 parents and 915 progenies and has dimension 951 × 951

	Parents			Progeny				
	191	192	170	191.1	191.2	192.1	192.2	170.1
191	1	0	0	0.5	0.5	0	0	0
192	0	1	0	0	0	0.5	0.5	0
170	0	0	1	0	0	0	0	0.5
191.1	0.5	0	0	1	0.25	0	0	0
191.2	0.5	0	0	0.25	1	0	0	0
192.1	0	0.5	0	0	0	1	0.25	0
192.2	0	0.5	0	0	0	0.25	1	0
170.1	0	0	0.5	0	0	0	0	1

	Parents			Progeny				
	191	**192**	**170**	191.1	191.2	192.1	192.2	170.1
	0	0	0	1		0	0	0
$\mathbf{Z} =$	0	0	0	0	1	0	0	0
	0	0	0	0	0	1	0	0
	0	0	0	0	0	0	1	0
	0	0	0	0	0	0	0	1

Fig. 4.2 Incidence (Z) matrix relating breeding value effects for three selected female parents and five selected open-pollinated progenies to phenotypic observations from the pine_provenance.csv data set

The coefficient of relationship is 0.5 between parent and offspring and 0.25 for half-sibs. The diagonal elements are 1+F, where F is the inbreeding coefficient, assumed to be 0 for all individuals in this case.

The linear model (individual tree model) for the data is:

$$y_{ijk} = \mu + B_i + P_j + T_k + \varepsilon_{ijk} \tag{4.28}$$

The main difference from the GCA model is that we have a Tree effect in the model instead of Female (or Family) effect, where T_k is the random kth tree effect $\sim N(0, \sigma_u^2)$. Another subtle difference also occurs since we model the effects of individual trees instead of families: we cannot easily model the plot error effect (which was block by family interaction previously). In this model, we allow the residual to absorb the plot error effects. In matrix form the model is $y = Xb + Zu + e$. The Z matrix relates observations to trees in the pedigree (whether or not they have an observed height record). In this example, trees 191, 192, and 170 are female parents and they do not have height measurements, but trees in the offspring generation do have phenotype records (Fig. 4.2).

The ASReml syntax for the model is:

Code example 4.3
Individual model for half-sib progeny test data (*Code 4-3_TreeModel.as*)

```
!OUTFOLDER    V:\Book\Book1_Examples\ch04_gca\outputfiles
Title: Pine_provenance.
#treeid,female,male,prov,block,plot,height,diameter,volume
 treeid  !P
 female  !I
 male    *
 prov    !I
 block *
 plot  *
 height    diameter    volume   !*10


!FOLDER   V:\Book\Book1_Examples\data
Pine_provenance.csv  !SKIP 1   !ALPHA
Pine_provenance.csv  !SKIP 1

height  ~ mu prov ,     # Specify fixed model
         !r block nrm(treeid)   # Specify random model
         residual units
```

- Notice that the data file is used as a pedigree file. ASReml keeps the first three fields in the data file as pedigree and disregards the rest. This works only for a shallow pedigree, as in pine data where female parents are genetically independent.
- The *treeid* field is a random factor and associated with the pedigree (!P).

- The variance structure for *treeid* is $\mathbf{G} = \sigma_u^2 \mathbf{A}$, a product of the variance component explained by the tree effect and additive genetic relationship matrix derived from the pedigree.

Variance components are reported in the primary output file (*Code 4-3_TreeModel*.asr)

```
Model_Term                      Gamma        Sigma     Sigma/SE   % C
  block          IDV_V   5  0.537164E-01  0.108390        1.24   0 P
  nrm(treeid)    NRM_V 950  0.442879      0.893649        2.71   0 P
  units                914 effects
  Residual       SCA_V 914  1.00000       2.01782         7.06   0 P


                         Wald F statistics
       Source of Variation   NumDF    DenDF   F-inc    P-inc
   10 mu                        1      6.2  3550.83    <.001
    4 prov                      3     32.0     9.78    <.001
```

The variance explained by the *treeid* effect (0.8936) is a bit more than the expected four times larger than the variance component (0.189) explained by the *female* effect in GCA model (Code Example 4.1). The variance explained by the *treeid* is the total additive genetic variance. Also notice that the residual variance in the animal model (2.018) is smaller than the residual variance in the GCA model (2.527). The differences originate from having different terms in the models. Heritability estimation from the animal model is straightforward because all we need to do is divide the variance component explained by the tree effect and the sum of variance components for tree effect and the residual.

$$h_i^2 = \frac{\sigma_A^2}{\sigma_P^2} = \frac{\sigma_t^2}{\sigma_t^2 + \sigma_\varepsilon^2}$$
$$= 0.893649/(0.893649 + 2.0178) = 0.3069$$

We can estimate the heritability and its standard error using the VPREDICT function in ASReml:

Code example 4.3
Individual model for half-sib progeny test data, continued (*Code 4-3_TreeModel.as*)

```
# Estimate linear combinations of variances
VPREDICT !DEFINE
F Additive nrm(treeid)
F Pheno   nrm(treeid) + Residual
H h2i Additive Pheno
```

The results of these functions of variance components are reported in file Code 4-3_TreeModel.pvc:

```
          Code4-4_TreeModel.pvc created 04 Sep 2015 08:46:41.045


          - - - Results from analysis of height - - -


1 block               V    5      0.108390      0.874113E-01
2 nrm(treeid)         V  950      0.893649      0.329760
  units             914 effects
3 units;Residual      V  914      2.01782       0.285810
```

```
4 Additive  2        0.89365     0.32977
5 Pheno  2           2.9115      0.14845
   h2i            = Additive   4/Pheno  2    5= 0.3069   0.1063
Notice: The parameter estimates are followed by
         their approximate standard errors.
```

Phenotypic variance is 2.91, very close to phenotypic variance estimated from the GCA model (2.88). Heritability is slightly higher because of having different terms in two models.

The solution file (*Code 4-3_TreeModel.sln*) now includes effect predictions for progeny as well for parents:

```
Model_Term          Level      Effect     seEffect
prov                   10        0.000       0.000
prov                   11       -1.649       0.3742
prov                   12       -0.6260      0.3702
prov                   13       -1.224       0.3771
mu                      1       11.51        0.3623
block                   1        0.3187      0.1782
block                   2        0.1299      0.1787
block                   3        0.2106      0.1786
block                   4       -0.3040      0.1795
block                   5       -0.3552      0.1808
nrm(treeid)     170             -0.3788      0.7042
nrm(treeid)     170.1            0.2431E-01  0.7793
nrm(treeid)     170.2           -0.8484      0.7793
nrm(treeid)     170.3           -0.4245      0.7793
..
nrm(treeid)     191             -0.1974      0.6936
nrm(treeid)     191.1           -0.3564      0.7794
nrm(treeid)     191.2           -0.1569      0.7794
nrm(treeid)     191.3           -0.7346E-02  0.7794
...
nrm(treeid)     192              0.5762      0.7002
nrm(treeid)     192.1            0.4575      0.7793
nrm(treeid)     192.2            0.2581      0.7793
nrm(treeid)     192.3            0.7318      0.7793
...
```

Notice that the solutions for parents (in bold font) are about two times that of solutions from the GCA model. The standard errors are also about two times larger. This is because the animal model solutions are breeding values not GCA values. Using this file we can easily rank all the individuals (parents and progeny) on the same scale and make selections (again, if we are interested in selecting with equal weight among provenances rather than selecting for overall best predictions that include provenance effects). Also notice that the standard error of breeding values for parents (191, 192, and 170) are smaller than standard errors of breeding values for progeny. This is expected because parental breeding values use data from their progeny whereas progeny have one data point each. This fact also causes another effect that can be observed in the residual diagnostic plot of residual values plotted against the predicted values (Fig. 4.3).

On the left hand side of Fig. 4.3, the residual values are independently distributed relative to the predicted values from the GCA model. On the right hand side of Fig. 4.3, the residual values are not independent of the predicted values from the individual model, instead they are positively correlated, as is obvious from the trend in the scatterplot. This is an effect of predicting the values of individuals, which have only one observation each, such that the *genetic value of the individual is*

Fig. 4.3 Scatterplots of residual versus predicted values for pine provenance data based on two different models: (**A**) GCA model, and (**B**) individual tree model

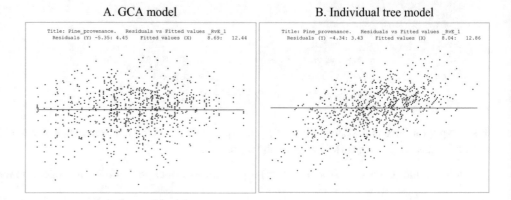

A. GCA model

B. Individual tree model

entangled with the residual effect of that observation. The mixed model nevertheless can make a prediction for each individual by combining the information from relatives (in this case the half-sib family mean) and an estimate of the proportion of an individual's deviation from the family mean is due to genetics (from the ratio of genetic to total variance within families). In contrast, the predicted value of the individuals in the GCA model is simply the family mean.

The accuracy of BLUPs for individuals from the animal model is computed following Gilmour et al. (2014) as:

$$r = \sqrt{1 - \frac{S^2}{(1+F)\sigma_A^2}} \tag{4.29}$$

where S is the standard error of the prediction reported in the *.sln* file, F is the inbreeding coefficient of the individual being predicted, and σ_A^2 is the additive variance for individual predictions (see next section on the individual or 'animal' model). Estimates of F for each individual can be obtained by using the !DIAG qualifier after the pedigree file and are reported in an output file with extension *.aif*. If the pedigree does not imply inbreeding, all of the F coefficients will be zero.

We can compute the accuracy of breeding values by reading this .sln file into R, and performing the computations with the following R code:

Code example 4.4
Breeding values from .sln output file processed by R code (*Code 4-4_TreeModel.R*)

```
#~~~~~~~~~~~~~~~~~~~~~~~~~~~~~~~~~~~~~~~~~~~~~~~~
# Calculate breeding values and their accuracy
#~~~~~~~~~~~~~~~~~~~~~~~~~~~~~~~~~~~~~~~~~~~~~~~~
# Remove everything from memory
rm(list=ls())

# Set the path of the working directory
path="/Users/fisik/Google
 Drive/Book/Book1_Examples/ch04_gca/outputfiles"
setwd(path)

# Read in the data file and call it 'dat'.
dat=read.table(file='Code04-3_TreeModel.sln',header=T)

# Get the intercept term value from ASReml output file
# and call it 'mu'
mu.idx=which(dat$Model_Term=='mu')
mu=dat$Effect[mu.idx]
```

```
# Get the BV values from ASReml output file
# and store them in array 'fam.idx'
fam.idx=dat$Model_Term=='nrm(treeid)'
ID=dat$Level[fam.idx]  # select the levels
BV=dat$Effect[fam.idx] + mu     # select the BV estimates
BV=round(BV,2)
se.BV=dat$seEffect[fam.idx] # select the SE of BV

# the denominator (tree variance=0.88) is taken from the .asr file
accuracy <- round(sqrt(1- (se.BV^2/0.88349)),2)

# merge all in a data frame
solution <- data.frame(ID=ID, BV=BV, se.BV=se.BV, accuracy=accuracy)
head(solution,10)
```

This R code outputs breeding values (BV) centered on the overall mean, their standard errors (se.BV), and their accuracies:

	ID	BV	se.BV	accuracy
1	**191**	**11.31**	**0.6936**	**0.67**
2	191.1	11.15	0.7794	0.56
3	191.2	11.35	0.7794	0.56
4	191.3	11.50	0.7794	0.56
...				
7	191.6	11.60	0.7794	0.56
8	**192**	**12.09**	**0.7002**	**0.67**
9	192.1	11.97	0.7793	0.56
10	192.2	11.77	0.7793	0.56
11	192.3	12.24	0.7793	0.56
...				

Breeding values of parents 191 and 192 from the previous GCA model were 11.34 and 12.00, respectively. Those values are approximately the same as the estimates from the individual model (in bold font above; there are small differences introduced by handling the residual error differently). As we expect, some progeny are better than their parents, e.g. progeny 191.3 has a breeding value of 11.5, whereas its parent (191) has a BV of 11.31.

The Animal Model with Deep Pedigrees and Maternal Effects

An animal model for livestock is in concept the same as the individual tree model just shown. However, the previous example based on the pine_provenance.csv data set was somewhat limited in that the pedigree was quite shallow, covering only two generations (only one of which was phenotyped). Furthermore, since the progeny were derived from open-pollination, we had no information on the fathers of the tree. As an example of a more extensive pedigree that will allow us to combine information across more generations and incorporate information from fathers, we now turn to the data in the "pig_data.txt" file:

pig	sire	dam	year	sex	pen	weanage	weanwt	adg	weight	loinarea
133	2	1	2004	1	52	21	13.25	2.0	264	5.34
654	2	1	2004	1	58	21	12.45	2.0	266	6.62

```
655    2   1   2004   1   54      21   13.55   1.6      215        5.50
153    2   1   2004   1   56      21   15.60   1.9      267        7.04
656    4   3   2004   2   59      21   10.40   1.5      210        3.94
657    4   3   2004   2   57      21   10.10   0.9      153        3.74
```

The first column is an identifier for the individual pig, and the next two columns represent the identifiers of the sire and dam of each pig. Subsequent columns indicate the year the data were collected on the individual, the sex of the individual, its pen number, the age in weeks at which it was weaned, its weight at weaning, and three other traits that we will not use in the subsequent examples.

We have pedigree information on the pigs in this data set extending across seven generations, corresponding to years in the data set. The first generation for which we have phenotype data is from year 2004, and we have pedigree information on the direct parents of those pigs. That previous parental generation is a base generation for making inferences about genetic parameters. We assume that each parent in the first generation of the pedigree was sampled at random from the reference population, so we assume that all of those parents have no genetic relationships among themselves (all coefficient of ancestry values, θ, = 0 for that generation). The pedigree data are available in the file "pig_pedigree.txt":

```
pig       sire    dam
1         0       0
2         0       0
3         0       0
4         0       0
5         0       0
6         0       0
...
2702      547     652
2703      641     653
2704      641     653
2705      641     653
2706      641     653
2707      641     653
```

In the first progeny generation (G_0 in 2004), there are unrelated pairs of individuals ($2\theta = 0$), half-sibs ($2\theta = 0.25$), and full sibs or parent-offspring ($2\theta = 0.5$, Fig. 4.4). By the later generations, there are more complex relationships because of distant common ancestors. For example, individuals 211 and 322 have a small relationship ($2\theta = 0.0625$) because of common ancestry in preceding generations (Fig. 4.4).

To analyze these data and estimate the breeding values of all the individuals in the pedigree, we can include effects of the year to account for some temporal variability, the fixed effect of the sex of the individual, and the age at which the pig has been weaned as a covariate:

$$y_{ijkl} = \mu + Y_i + S_j + W_k + P_l + \varepsilon_{ijkl} \tag{4.30}$$

where Y_i is the fixed effect of year, S_j is the fixed effect of the pig's sex, W_k is the fixed covariate for weaning age, P_l is the random effect of the pig's breeding value, and ε_{ijkl} is the random residual effect.

We can fit this simple animal model in ASReml as shown in Code example 4.5:

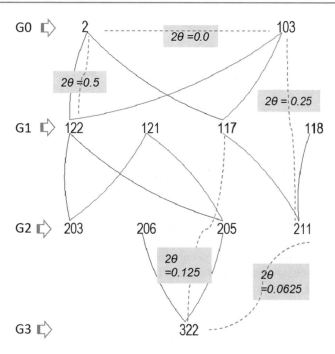

Fig. 4.4 Pedigree of selected individuals in pig_data.txt data set. The pedigree extends over six generations. Only small sample of individuals extending over four generations are presented here. Additive genetic relationship values (2θ) are given for some pairs of individuals connected by *dashed lines*

Code example 4.5
Animal model for pig data including only direct genetic effects (*Code 4-5_AnimalModel.as*)

```
Title: animal1_wewt
 pig    !P
 sire   !P
 dam    !P
 year   !A
 sex    !A
 pen    !A
 weanage   weanwt   adg   weight   loinarea
# The Pedigree file
pig_pedigree.txt   !SKIP 1 !SORT !ALPHA
# The data file
pig_data.txt   !SKIP 1 !MAXIT 100 !EXTRA 5   !NOGRAPH !DOPART 1

!PART 1
weanwt ~ year sex weanage !r nrmv(pig)
VPREDICT !DEFINE
F Additive    nrmv(pig)
P Pheno       Residual + nrmv(pig)
H h2i         Additive Pheno
```

Once again we can obtain functions of the estimated variance components (h^2 in this case) through the VPREDICT ! DEFINE qualifiers to create a .pin file.

The resulting variance components are reported in file *Code 4-5_AnimalModel.pvc* file:

```
            - - - Results from analysis of weanwt - - -

   1 nrmv(pig)        V 2707      2.08202         0.221964
   2 Residual         V 2556      2.83718         0.147693
   3 Additive  1                  2.0820          0.22197
   4 Pheno   2                    4.9192          0.16444
     h2i    = Additive    3/Pheno  2    4= 0.4232  0.0358
```

While the previous model seems reasonable as a first approximation, it is common in livestock species to include maternal effects in the model. These effects are especially important for growth traits and recognize the importance of the maternal environment for the performance of an individual. Maternal effects from the progeny's standpoint are environmental but have a genetic component in the mother; further, the genetic component of the maternal effect may have a covariance with the progeny genetic effect. This double structure can efficiently be accommodated in ASReml. Using the same data just seen, we can expand the previous model as follows:

Code example 4.5
Animal model for pig data, adding maternal genetic effects (*Code 4-5_AnimalModel.as*)

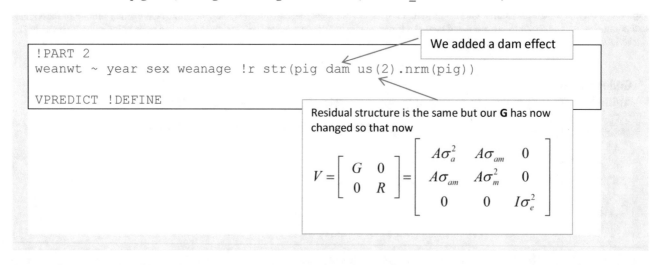

```
!PART 2
weanwt ~ year sex weanage !r str(pig dam us(2).nrm(pig))

VPREDICT !DEFINE
```

We added a dam effect

Residual structure is the same but our **G** has now changed so that now

$$V = \begin{bmatrix} G & 0 \\ 0 & R \end{bmatrix} = \begin{bmatrix} A\sigma_a^2 & A\sigma_{am} & 0 \\ A\sigma_{am} & A\sigma_m^2 & 0 \\ 0 & 0 & I\sigma_e^2 \end{bmatrix}$$

Here we use the str() function in ASReml to indicate that the terms inside the parentheses use the same **Z** matrix to relate observations to parameter values. In this case, we specify two model terms, pig and dam, that share a common structure. Since we have specified both of these terms as pedigree-associated factors using the !P qualifier in the field definitions, ASReml associates both of these terms with the pedigree information we supplied. That information, in turn, is used to model the variance-covariance structure of the pig and dam effects based on the additive genetic relationship matrix. In the previous model, where we just specified pig alone as a random pedigree-associated effect, we also estimated both progeny and dam breeding values as random effects from a common distribution. What is new here is that we are separating the effects of each pig into its own direct genetic effect on weaning weight and also into a maternal genetic effect (if it was female and was a mother to some phenotyped offspring), where the variance-covariance structure of the effects is proportional to the **A** matrix. In other words, we are modeling both the breeding value of the pig itself and the effect of its mother on its weight, where both effects have covariances among individuals determined by their coancestries.

Finally, we allow the direct genetic effect of the pig itself and its mother's maternal genetic effect to have a covariance via the third term in the str() function, which provides the structure information for the effects: str(pig dam us(2). nrm(pig)). The term 'us(2).nrm(pig)' indicates that the two previously identified model factors (pig and dam) have a variance-covariance structure that is the direct product of two terms: an unstructured variance-covariance matrix with dimension 2 and the additive relationship matrix based on the pedigree information from pigs:

$$\mathbf{G} = us(2).nrm(pig) = \begin{bmatrix} V_A & C_{A,Am} \\ C_{A,Am} & V_{Am} \end{bmatrix} \otimes \begin{bmatrix} 2\theta_{11} & 2\theta_{12} & \cdots & 2\theta_{1n} \\ 2\theta_{21} & 2\theta_{22} & \cdots & 2\theta_{2n} \\ \vdots & \vdots & \ddots & \vdots \\ 2\theta_{n1} & 2\theta_{n2} & \cdots & 2\theta_{nn} \end{bmatrix}$$

(4.31)

$$\mathbf{G} = us(2).nrm(pig) = \begin{bmatrix} V_A & C_{A,Am} \\ C_{A,Am} & V_{Am} \end{bmatrix} \otimes \mathbf{A}$$

Here, V_A (or σ_a^2) is the additive variance for direct genetic effects, V_{Am} (or σ_m^2) is the additive variance for maternal genetic effects, and $C_{A,Am}$ (or σ_{Am}) is the covariance between the direct and maternal genetic effects. The dimension of the A matrix in this examples is $2{,}707 \times 2707$ since there are 2,707 animals included in the pedigree. We are estimating 2,707 breeding values for direct genetic effects, including all of the pigs that were phenotyped and also including generation 0 pigs that do not have phenotypes. Similarly, we are estimating 2,707 breeding values for maternal effects. This may seem strange, since male pigs cannot express a maternal effect. However, since we are dealing with breeding values, a male pig transmits alleles that influence the maternal effects that its daughters can express on *their* offspring. So, even male pigs have a breeding value for maternal genetic effects, and they can be predicted using the phenotype information from related females in the same way that we predict breeding values of unphenotyped parents based on their phenotyped relatives.

The resulting variance (correlation) components from this model are in file Code 4-5_AnimalModel.asr:

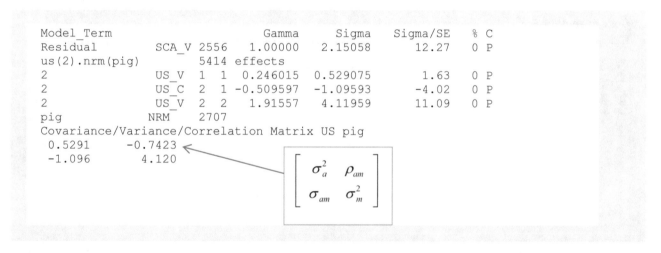

```
Model_Term                      Gamma      Sigma    Sigma/SE   % C
Residual        SCA_V 2556    1.00000    2.15058       12.27   0 P
us(2).nrm(pig)        5414 effects
2               US_V 1  1    0.246015   0.529075        1.63   0 P
2               US_C 2  1   -0.509597  -1.09593        -4.02   0 P
2               US_V 2  2    1.91557    4.11959        11.09   0 P
pig             NRM     2707
Covariance/Variance/Correlation Matrix US pig
   0.5291       -0.7423
  -1.096         4.120
```

$$\begin{bmatrix} \sigma_a^2 & \rho_{am} \\ \sigma_{am} & \sigma_m^2 \end{bmatrix}$$

The estimated variance component for pig effects (direct genetic effects) is 0.529. The dam (maternal genetic) variance component estimate is 4.12. They have a covariance of -1.09 and a correlation of -0.74. It is striking how strong the maternal effect is relative to the pig's own genotype effect on its weaning weight.

Direct and maternal heritabilities can be defined, keeping in mind that $\sigma_p^2 = \sigma_a^2 + \sigma_m^2 + \sigma_{am} + \sigma_e^2$ in this case:

Phenotypic variance is $0.529 + 4.11 + (-1.09) + 2.15 = 5.7$
h^2 for direct effects is $= 0.529/5.7 = 0.09$
h^2 for maternal effects is $= 4.11/5.7 = 0.72$

We can obtain these heritability estimates using the VPREDICT !DEFINE command directly in the ASReml program. For complex models like the one in this example, however, it is often not obvious how the variance components are named or ordered in the output. When in doubt, simply write the VPREDICT !DEFINE command with no subsequent definition lines, and a .pvc file will be created that includes the covariance components with the names that should be used in the definitions of functions of the variance components. For example, if no variance component functions are defined after the VPREDICT !DEFINE directive in !PART 2 of *Code 4-5_AnimalModel.as*, the resulting .pvc output file is created:

Code 4-5_AnimalModel.pvc

```
            - - - Results from analysis of weanwt - - -

1 Residual                         V 2556        2.15066     0.175278
  us(2).nrm(pig)          5414 effects
2 us(2).nrm(pig);us(2)             V  1  1       0.529062    0.324578
3 us(2).nrm(pig);us(2)             C  2  1      -1.09595    -4.02000
4 us(2).nrm(pig);us(2)             V  2  2       4.11973     0.371482
  pig                      NRM    2707
 Notice: The parameter estimates are followed by
         their approximate standard errors.
```

We can then define the phenotypic variance and the heritability by referring to the direct additive genetic variance either as component '2' or as 'us(2).nrm(pig);us(2)[1]'. The suffix '[1]' distinguishes the first variance component of this complex term in the output. Similarly, the covariance between direct additive and maternal effects is referred to as either component '3' or 'us(2).nrm(pig);us(2)[2]'.

The heritability definitions can then be added after the VPREDICT !DEFINE command and the *.as* file can be re-run to get the estimates:

Code example 4.5
Heritability estimation from animal model for pig data with maternal genetic effects (*Code 4-5_AnimalModel.as*)

```
VPREDICT !DEFINE
F Add_pig    us(2).nrm(pig);us(2)[1]
#equivalently:
#F Add_pig    2
F Add_dam    us(2).nrm(pig);us(2)[3]
#equivalently:
#F Add_dam    4
P Pheno Residual + us(2).nrm(pig);us(2)[1] + us(2).nrm(pig);us(2)[2] +
us(2).nrm(pig);us(2)[3]
#equivalently:
#P Pheno = Residual + 2 + 3 + 4
H h2pig    Add_pig Pheno
H h2dam    Add_dam Pheno
```

The results are reported in the *Code 4-5_AnimalModel.pvc* file:

```
V:\Book\Book1_Examples\ch04_gca\outputfiles/Code04-5_AnimalModel.pvc   created   26   Nov   2015
15:31:02.576

            - - - Results from analysis of weanwt - - -

  1 Residual                        V 2556        2.15058     0.175271
us(2).nrm(pig)             5414 effects
  2 us(2).nrm(pig);us(2)            V  1  1       0.529075    0.324586
  3 us(2).nrm(pig);us(2)            C  2  1      -1.09593    -4.02000
  4 us(2).nrm(pig);us(2)            V  2  2       4.11959     0.371469
```

```
pig                      NRM     2707
    5 Add_pig   2                0.52907         0.32555
    6 Add_dam   4                4.1196          0.37156
    7 Pheno   1                  5.7033          0.27376
      h2pig      = Add_pig    5/Pheno  1   7=  0.0928      0.0572
      h2dam      = Add_dam    6/Pheno  1   7=  0.7223      0.0423
  Notice: The parameter estimates are followed by
          their approximate standard errors.
```

A common extension of the maternal effect model is to include non-genetic ('environmental') maternal effects in addition to the genetic maternal effect in the animal model. Environmental maternal effects refer to any maternal differences that are not due to genetics. For example, differences in diet among mothers can affect their ability to nourish their offspring, but these differences may be entirely due to management rather than genetics.

To accomplish this, we simply add a term `ide(dam)` to the random part of the model. The `ide()` function specifies that a factor that we have previously defined as a pedigree factor (and so normally has a variance-covariance structure proportional to the **A** matrix computed from the pedigree information) should be fit *without* the pedigree information. So, we fit the dam effect twice in this model, once where the random dam effects have breeding values with covariances determined by coancestries, and a second time where the random dam effects have no covariances and an IDV structure. This second term corresponds to the maternal environment effects, since it includes any maternal effects that cannot be associated with the pedigree relationships:

Code example 4.5
Animal model for pig data, with maternal genetic and maternal environmental effects
(*Code 4-5_AnimalModel.as*)

```
!PART 3
weanwt ~ year sex weanage ,
!r  str(pig dam us(2).nrm(pig)) ,
    us(2 !GUUU).nrm(pig) ,|
    ide(dam)
```

The variance structure is

$$
V = \begin{bmatrix}
A\sigma_a^2 & A\sigma_{am} & 0 & 0 \\
A\sigma_{am} & A\sigma_m^2 & 0 & 0 \\
0 & 0 & I\sigma_{me}^2 & 0 \\
0 & 0 & 0 & I\sigma_e^2
\end{bmatrix}
$$

Notice that we also included in the model the term 'us(2 !GUUU).nrm(pig)' to permit the variance covariance structure of the estimates for this term to be unconstrained (i.e., correlations based on the variance and covariance components are allowed to go out of bounds, greater than 1.0 or less than −1.0). This only needs to be done if we want to add qualifiers to the model term specified in the str() function. We cannot include the qualifiers directly in the str() function, i.e., 'str (pig dam us(2 !GUUU).nrm(pig))' will not be read as a valid random term, so we specify the same variance structure already given in the str() function a second time and include the qualifiers in this second specification.

OUTPUT PART 3: This model produces the following result:

```
Model_Term                     Gamma         Sigma    Sigma/SE   % C
ide(dam)        IDV_V 2707  0.683986       1.59023       5.65   0 P
Residual        SCA_V 2556  1.00000        2.32494      14.63   0 P
us(2).nrm(pig)        5414 effects
2               US_V  1  1  0.449416E-01   0.104487      0.36   0 U
2               US_C  2  1 -0.250977      -0.583506     -2.28   0 U
2               US_V  2  2  0.621264       1.44440       3.36   0 U
pig             NRM   2707
Covariance/Variance/Correlation Matrix US pig
  0.1045        -1.502
 -0.5835         1.444
```

The maternal environment effect variance component estimate is 1.59, the maternal genetic effect variance component is 1.444, and the direct genetic effect variance is 0.104. You will notice something strange about this result: the correlation between direct and maternal genetic effects is −1.502, which is out of theoretical bounds for a correlation coefficient. This can occur if we do not constrain the unstructured variance-covariance matrix for these two effects (we used the !GUUU qualifier on the model term). If we try to constrain the model in this case, it does not converge. This result suggests that we do not have a reliable estimate of one or more of the variance-covariance terms in this model. Notice that the direct genetic effect variance is close to zero (and is smaller than its standard error). If a variance component really is zero, it cannot have a covariance with any other term. So, the trouble is probably occurring because this direct genetic variance component is approaching zero. We could simplify the model by dropping the direct genetic effects and compare the AIC values of the reduced model to this one to see if the reduced model could be accepted.

Accounting for Genetic Groups Effect in Predictions

Genetic groups are a common phenomenon in plant and animal breeding programs. Rather than descending from a common gene pool, individuals in a program may descend from different groups of founders or different breeds. For example, in forest trees, geographic differences in seed sources (provenances) within a single species can be associated with very large phenotypic differences in progenies (Isik et al. 1999). Subpopulations within a species often have adapted to very different elevations or climates, and this can result in substantial genetic differences among individuals descended from parents sampled across different geographic areas. If the population is composed of distinct genetic groups, such structure should be accounted for in predictions of breeding values.

By default, breeding values are centered on zero (with negative and positive values). This is based on the assumptions that all base (ancestral) individuals are equal and they mated at random to produce the population under study. This is not a safe assumption when the population comprises a mixture of different genetic groups. If the group effect is ignored, additive genetic variance estimates can be biased upward, and selection may favor a particular provenance at the risk of introducing some inbreeding in future generations. The provenance or founder effect should be accounted for when estimating breeding values to avoid such problems.

There are twso ways to handle the genetic group effects in the predictions of breeding values: (1) treating genetic groups as fixed effects in a GCA model or (2) using pedigree information and the *genetic groups* model.

Treating Genetic Groups as a Fixed Effect in GCA model

In some cases, the group effect can be considered random to estimate between and within provenance variances. Such information is useful to study effect of geographical variation among groups for adaptation-related traits. In cases where we have few groups in the study, it may be better to treat the group effect as fixed since otherwise the variance component for groups might be poorly estimated. Groups might also be considered fixed effects if provenances were sampled to represent certain climatic/elevation transects rather than random samples.

We will use the *Pine_provenance.csv* data again to demonstrate modeling genetic group effects. The GCA model that we used previously already treated provenance as a fixed effect, so the model is the same as previously shown:

$$y_{ijkl} = \mu + B_i + P_j + F_{k(j)} + BF_{ik(j)} + \varepsilon_{ijkl} \tag{4.32}$$

Variance components estimates and tests of the fixed effect of provenance are exactly as before. At this point, however, we shift the focus of genetic value prediction from breeding values within provenances (how good a parent/family is compared to the overall provenance mean) to 'net breeding values' which are appropriate to compare breeding values of individuals across provenances if we want to select on both the provenance and the individual within provenance values. Recall that the effect solutions are found in file (*Code04-1_HSfamily.sln*) includes the BLUEs of fixed effects (e.g., prov) and BLUPs of family effects (or parental GCA effects). As a reminder, a subset of the prediction file is given below:

Model_Term	Level	Effect	seEffect
prov	10	0.000	0.000
prov	12	-0.6237	0.3701
prov	11	-1.666	0.3741
prov	13	-1.220	0.3770
mu	1	11.51	0.3620
block	1	0.3110	0.1855
block	2	0.1268	0.1858
block	3	0.2056	0.1858
block	4	-0.2919	0.1867
block	5	-0.3516	0.1878
idv(female)	191	-0.8257E-01	0.3395
idv(female)	192	0.2459	0.3419
idv(female)	170	-0.1634	0.3433
idv(female)	210	-0.3584	0.2979

...

Calculation of net breeding values of families

The net breeding values (NBV) of families from the above model can be calculated as follows:

$$A_{jk} = \mu + P_j + 2GCA_{jk} \tag{4.33}$$

where A_{ij} is the NBV of family k from provenance j, μ is the intercept, P_j is the Best Linear Unbiased Estimate of jth provenance and GCA_{jk} is the GCA prediction of parent k in provenance j.

Parent 191 comes from PROV 10. The adjusted NBV of Parent 191 is

$$= \mu + P_{10} + 2^*GCA_{191}$$
$$= 11.51 + 0.0 + 2^*(-0.0826) = \mathbf{11.34}$$

Parent 210 comes from PROV 12. The adjusted NBV of Parent 210 is

$$= 11.51 + -0.6237 + 2^*(-0.3584) = \mathbf{10.17}$$

Notice that these values are different than the predictions that can be obtained from the same model in ASReml with:

```
PREDICT   prov female !PRESENT prov female
```

The !PRESENT qualifier prevents reporting predicted values for provenance-family combinations that do not appear in the data set, in this case predictions for families in some other provenance than their own actual provenance.

This `PREDICT` statement produces predicted values found in the output file *Code 4-1_HSFamily.pvs*:

prov	female	Predicted_Value	Standard_Error	Ecode
10	191	11.4295	0.3392	E
10	192	11.7580	0.3498	E
10	170	11.3487	0.3560	E
12	210	10.5299	0.3253	E
...				

The predictions in this .pvs file refer to the family mean genetic values (not the parental breeding values), and are obtained as $\widehat{Y}_{ij} = \mu + P_j + F_{jk}$. The family mean genetic value predictions can be obtained from the effect estimates as:

Family 191 from PROV 10:

$$= \mu + P_{10} + F_{191}$$
$$= 11.51 + 0.0 - 0.0826 = \textbf{11.43}$$

Parent 210 from PROV 12:

$$= 11.51 + -0.6237 - 0.3584 = \textbf{10.53}$$

Fitting Genetic Groups as Pedigree Information in Individual Model

An alternative to the GCA model with fixed group effects is an individual model with genetic groups accounted for in the pedigree rather than explicitly as a factor in the model. This alternative will produce directly net breeding values that includes the group effect. Furthermore, this alternative method is useful for later generations of progenies that may trace ancestry to more than one founder group. We need to create a pedigree file in order to include the genetic group information as part of the pedigree. This is done by including the genetic group identifiers at the beginning of the file, then indicating the group identifier as the male and female parents of the individuals in the first sampled generation from the genetic groups. The pedigree file ("pine_provenance_pedigree.txt") appears as:

treeid	mother	father
10	0	0
11	0	0
12	0	0
13	0	0
170	**10**	**10**
191	**10**	**10**
192	**10**	**10**
...	...	...
191.1	191	**10**
191.2	191	**10**
196.9	196	**11**

The first four lines (10, 11, 12, 13) in the pedigree file are genetic group IDs. Zeroes are used to indicate the mother and father of the genetic groups. To distinguish the genetic group identifiers from individuals with no known ancestors, we will also use the !GROUPS *n* qualifier on the pedigree file name line (in the job file .as) to indicate that the first *n* records correspond to groups rather than individuals.

In above pedigree file after the genetic groups, the parental plants are listed in the *treeid* column, followed by the progeny generation plants. Notice that every parent generation individual has a genetic group indicated, whereas the progenies derived from those parents by open-pollination have the known mother plant identified but the genetic group label for the father. If progeny came from full-sib families, however, only the parents would be assigned to genetic groups (provenances).

The individual tree model is:

$$y_{ik} = \mu + B_i + T_k + \varepsilon_{ik} \tag{4.34}$$

In this model, individual tree values are modeled directly, according to the pedigree structure. We now modify this individual model to reflect that the pedigree of each tree will trace back to one or more of the founder groups, and the breeding value of the individual is modeled as the weighted mean of its founder group ancestry effects plus the deviation of the individual value from the expectation of that mean (Mrode 2014):

$$Y_{ijk} = \mu + B_i + \sum_{j=1}^{n} t_{jk} g_k + a_k + \varepsilon_{ijk} \tag{4.35}$$

In this model, each individual may have more than one genetic group effect in its ancestry, and these effects are modeled as the summation of the product of the group effect (g_k) times the additive genetic relationship between individual k and the ancestral group j over $j = 1$ to n groups. Then, for the mixed model, we introduce a matrix, $\mathbf{Q}$, which corresponds to an additive relationship matrix between individuals and the ancestral groups, and the mixed model equation (MME) takes the following form (Mrode 2014; Quaas and Pollak 1981).

$$\begin{bmatrix} \mathbf{X'X} & \mathbf{X'Z} & 0 \\ \mathbf{Z'X} & \mathbf{Z'Z} + \mathbf{A}_{nn}^{-1}\alpha & \mathbf{A}_{np}^{-1}\alpha \\ 0 & \mathbf{A}_{pn}^{-1}\alpha & \mathbf{A}_{pp}^{-1}\alpha \end{bmatrix} \begin{bmatrix} \widehat{\mathbf{b}} \\ \widehat{\mathbf{u}} + \widehat{\mathbf{Qg}} \\ \widehat{\mathbf{g}} \end{bmatrix} = \begin{bmatrix} \mathbf{X'y} \\ \mathbf{Z'y} \\ 0 \end{bmatrix} \tag{4.36}$$

- The inverse of the numerator relationship matrix $\mathbf{A}^{-1}$ is now partitioned into four sub-matrices corresponding to the inverses of the sub-matrix for genetic groups ($\mathbf{A}_{pp}$), the sub-matrix for treeid ($\mathbf{A}_{nn}$), and the sub-matrix for covariances between genetic groups and treeid ($\mathbf{A}_{np}$ or $\mathbf{A}_{pn}$).
- Solving the MME will give the solutions for vectors $\mathbf{u}$ (individuals in the pedigree) and $\mathbf{g}$ (genetic groups).
- The ranking of trees will be based on both the prediction value of the ancestral genetic contribution ($\mathbf{Qg}$) and the prediction for the individual's additive effect deviation from that prediction: $u_i = \mu + Q_i g + a_i$ for tree i, where $\mathbf{Q}_i$ is the row of the $\mathbf{Q}$ matrix corresponding to the ith individual (and containing the additive relationship coefficients between individual i and each of the ancestral groups).

The ASReml syntax for the model is quite simple.

Code example 4.6
Genetic group effect for half-sib progeny test data (*Code 4-6_TreeGeneticGroups.as*)

```
!OUTFOLDER    V:\Book\Book1_Examples\ch04_gca\outputfiles
Title: Pine_provenance.
#treeid,female,male,prov,block,plot,height,diameter,volume
 treeid   !P
 female   !I
 male     *
 prov     !I
 block *
 plot  *
 height    diameter    volume   !*10


!FOLDER   V:\Book\Book1_Examples\data
Pine_provenance_pedigree.txt   !SKIP 1   !ALPHA !GROUPS 4 #groups
Pine_provenance.csv   !SKIP 1

height   ~ mu ,
         !r block   nrm(treeid)        # Specify random model
         residual units
```

The *treeid* field is a random factor and associated with the pedigree (!P). The pedigree file has the qualifier **!GROUPS** 4 to indicate that the first 4 lines of the pedigree file identify genetic groups (with zero in both the mother and father fields).

Variance components are reported in the primary output file (*Code 4-6_TreeGeneticGroups*.asr)

```
Model_Term                    Gamma        Sigma   Sigma/SE   % C
  block         IDV_V   5  0.537163E-01  0.108397     1.24    0 P
  nrm(treeid)   NRM_V 954  0.442879      0.893708     2.71    0 P
  units               914  effects
  Residual      SCA_V 914  1.00000       2.01795      7.07    0 P
```

Again, the variance component for the *treeid* effect is (0.89) is about four times of family variance component in the GCA model, because it represents total additive genetic variance.

The genetic groups model produces predictions for all the levels of genotypes (prov, female and treeid) in a single run. The population mean (intercept) is automatically added to the BV to get net BV.

Predictions (*Code 4-6_TreeGeneticGroups*.sln)

```
Model_Term      Level    Effect    seEffect
mu                1       0.000      0.000
block             1       0.3187     0.1782
block             2       0.1299     0.1786
block             3       0.2106     0.1786
block             4      -0.3040     0.1795
block             5      -0.3552     0.1808
nrm(treeid)  10          11.51       0.3623
nrm(treeid)  11           9.864      0.2282
nrm(treeid)  12          10.89       0.2216
nrm(treeid)  13          10.29       0.2329
nrm(treeid)  170         11.13       0.5826
nrm(treeid)  191         11.32       0.5500
nrm(treeid)  192         12.09       0.5705
...
```

Notice that the net breeding values of parents are close to what we obtained earlier from the GCA model where we treat the genetic groups (*prov*) as fixed. For example, the breeding value prediction for parent 191 from the genetic groups model above is 11.32, compared to 11.31 estimated by a linear function of effect estimates. The prediction file from the genetic groups model produces BV predictions for all members of the pedigree on the same scale. No additional computation (such as multiplying GCA values by two) is needed.

Effect of Self-Fertilization on Variance Components

When one of the parents of progeny is unknown, and the species can self-fertilize, we need to account for the possibility of selfing in the genetic relationship matrix. For example, in eucalyptus species, it is quite common to assume up to 0.3 probability of selfing (Grattapaglia et al. 2004; Patterson et al. 2004). If the selfing is not accounted for, the additive genetic variance will be biased upward because the progeny are not all half-sibs – they are more closely related, and therefore more similar, than the model assumes.

Mendelian sampling is the effect due to random segregation and recombination of genes coming from mother and father in the progeny. Selfing will affect the diagonal elements of the **A** matrix, which represent twice the coancestry of individuals with themselves. Individuals derived from one generation of selfing have an inbreeding coefficient of 0.5. Additional generations of selfing will increase the inbreeding coefficient to:

$$F = 1 - (0.5)^t \tag{4.37}$$

for t generations of recurrent selfing. This quickly approaches 1 as t becomes large.

The coancestry of individuals with themselves is:

$$\theta_{XX} = \frac{1}{2}(1 + F_X) \tag{4.38}$$

The elements of the **A** matrix are twice the coancestries, so the diagonal elements of A are:

$$\mathbf{A}_{ii} = 2\theta_{ii} = (1 + F_i) \tag{4.39}$$

The diagonal elements can be greater than one under inbreeding. For one generation where a father is unknown and the probability that the unknown father is also the mother is denoted s, the diagonal element of the **A** matrix is (Mrode 2014):

$$\mathbf{A}_{ii} = 2\theta_{ii} = (1 - s)(1) + s(1 + 0.5) = 1 + 0.5s \tag{4.40}$$

In ASReml, the probability of selfing is declared after the pedigree file using the !SELF s, qualifier where s is the probability from 0 to 1. In the following example, the qualifier !SELF s is used to tell ASReml that probability of selfing is 0.1 in the pine provenance data. The same model was run for selfing probabilities 0.2, 0.3, 0.4, 0.5 and 1 to demonstrate how the selfing probability affects the variance components estimates.

Code example 4.5
Selfing effect on genetic variance for pine half-sib progeny test data *(Code 4-7_Selfing.as)*

```
!OUTFOLDER    V:\Book\Book1_Examples\ch04_gca\outputfiles
Title: Pine_provenance.
 treeid   !P
 female   !I
 male     *
 prov     !I   block *  plot  *
 height   diameter    volume  !*10

!FOLDER    V:\Book\Book1_Examples\data
Pine_provenance.csv   !SKIP     !SELF 0.1 # selfing. Change to .2, .3 ...
Pine_provenance.csv   !SKIP 1

height   ~ mu prov !r block nrm(treeid)
```

Variance components for different levels of selfing are reported below. With inbreeding, the observed variance in the partly inbred population represents more than the additive variance in the non-inbred reference population. Thus, as the probability of selfing increases, the estimate of the genetic variance in the non-inbred reference population decreases, whereas residual variance increases as shown below.

Selfing	Genetic	Residual
0.0	0.89	2.02
0.1	0.74	2.14
0.2	0.62	2.23
0.3	0.53	2.30
0.4	0.46	2.36
0.5	0.40	2.42
...		
1.0	0.22	2.58

Electronic supplementary material: The online version of this chapter (doi:10.1007/978-3-319-55177-7_5) contains supplementary material, which is available to authorized users.

F. Isik et al., *Genetic Data Analysis for Plant and Animal Breeding*, DOI 10.1007/978-3-319-55177-7_5

Abstract

Some mating designs, such as factorials and diallels produce full-sib families. These mating designs allow decomposing the observed genetic variance into additive and non-additive genetic effects. Selection among full-sib families is common in some plant and animal breeding programs. If full-sib families can be propagated, non-additive genetic effects can contribute to selection gain. In addition, selection among progeny within crosses with known pedigree can be used to establish the next breeding population. In this chapter we demonstrate one diallel and one factorial example to partition the observed variance into additive and dominance genetic components. We then show an example of cloned progeny test data analysis for within-fullsib family selection.

Specific Combining Ability (SCA) and Genetic Values

An individual's phenotype can be defined as a linear combination of additive (A), dominance (D) and environmental (e) effects:

$$y_i = \mu + A_i + D_i + e_i \tag{5.1}$$

Models based on parental general combining ability (GCA) effects, which are a function of additive genetic effects, were introduced in Chap. 4. In this chapter we introduce **specific combining ability** (SCA) effects that result from dominance effects (allelic interactions within loci). The mean genotypic value of offspring from a particular cross may deviate from value expected considering the population mean and the sum of the parental GCA effects (Falconer and MacKay 1996). This deviation is the specific combining ability for that cross. We can define the mean genotypic value (G_{AB}) for the full-sib family produced by crossing parents A and B as the sum of the overall mean (μ), the GCAs of the two parents and the SCA value:

$$G_{AB} = \mu + GCA_A + GCA_B + SCA_{AB} \tag{5.2}$$

The SCA effect for the cross can be derived by subtraction:

$$SCA_{AB} = G_{AB} - (\mu + GCA_A + GCA_B) \tag{5.3}$$

A classical method to estimate dominance genetic variance (D) is to estimate the variance associated with SCA effects of many crosses. The expected value of the observed SCA variance component is 1/4 of the dominance genetic variance in the reference population.

There are several commonly used mating designs used to partition the observed variance into additive and dominance genetic effects, such as factorial mating designs, nested designs and diallels (Hallauer et al. 2010). Diallels are commonly used in plant breeding applications where individual plants can be used as both male and female parents; in addition to using diallels to estimate additive and dominance variances, certain diallel designs may allow estimation of reciprocal cross effects. Diallels cannot be used in dioecious species (female and male flowers occur in different plants), including mammals. However, factorial designs can be used in dioecious species to estimate dominance genetic variance. We will demonstrate one diallel and one factorial example then show an example of cloned progeny test data analysis for within-family selection.

Diallel Mating Designs

In monoecious plant species when the same individuals are used as females and males in breeding, the mating design is a diallel. Here are some commonly used diallel mating designs in plant breeding:

Full diallels involve all the possible combinations of crosses among parents, including reciprocals and self-fertilization of the parents. For a sample of *n* parents, the full-diallel requires $n \times n$ progenies, a number that quickly becomes unmanageable as more parents are sampled.

Half diallels: Each parent is mated with every other parent, excluding selfs and reciprocals. This requires making $n(n-1)/2$ crosses for n parents (Fig. 5.1a).

Partial diallel: Not all the crosses are made. There are no reciprocals or selfs. The goal is to reduce the breeding workload for a given sample of parents by making less than $n(n-1)/2$ crosses for n parents (Fig. 5.1b).

Connected diallels: Two groups (1–6 and 7–12) of individuals are used to form two diallels but they are connected by crossing 4×9, 7×1, 9×3 and 10×2. In the example below, the second diallel also includes some selfs (S) and reciprocals (R) (Fig. 5.2):

If there are no connections between groups of parents, the design is a diallel in sets. Diallel mating designs provide good evaluation of parents and full-sib families. They also provide estimates of both additive and dominance genetic effects, and genetic gains due to additive and dominance genetic effects *if we assume the sample of parents used is sufficient to represent the reference population* (Baker 1978; Holland et al. 2003). One disadvantage of diallels is that the breeding and progeny evaluations can be costly due to large number of crosses required. For a full diallel with 6 parents, 36 crosses are required;

Fig. 5.1 A half-diallel mating without selfs (*left*) and partial mating designs (*right*). There are many variations of diallels used in plant breeding

a)

F/M	1	2	3	4	5	6
1	.	X	X	X	X	X
2		.	X	X	X	X
3			.	X	X	X
4				.	X	X
5					.	X
6						.

b)

F/M	1	2	3	4	5	6
1	.	X		X		X
2		.	X			
3			.		X	
4				.	X	
5					.	X
6						.

Fig. 5.2 A connected diallel mating design of 12 parents. In the diagram crossings between different individuals are designed by X, reciprocals by R and selfing by S

F/M	1	2	3	4	5	6	7	8	9	10	11	12
1	.	X		X		X						
2		.	X									
3			.	X								
4				.					X			
5					.							
6						.						
7	X						S	X		X		X
8							R	S	X		X	
9			X					R	S	X		
10		X					R		R	.		
11								R				S
12							R					S

with 10 parents the number of crosses required is 100. While these numbers may be manageable, it may be questioned if a sample of 10 or so parents can really provide a useful estimate of the reference population genetic variances (Baker 1978). There are many other combinations of diallel mating designs. See White et al. (2007) and Hallauer and Miranda (1988) for details.

Diallel Example

Six pine trees were mated using a half-diallel mating design. There were no selfs or reciprocals in the mating. Pines are monoecious species, meaning that male (pollen catkins) and female (strobili) reproductive organs are found on the same tree. A tree can be used as male, female or both in the mating. One of the parental combinations was not made, so the half diallel was not complete. In total, 14 crosses (full-sib families) were obtained. The field experimental design was a randomized complete block design in which the experimental unit was a plot containing on average 4.3 trees per full-sib family. The experiment was replicated in four locations and there were six blocks per location. Wood density of trees was measured at age 10. The first five observations of the data set (in file '*diallel.csv*') are given below:

Tree	Female	Male	Cross	Location	Block	density
1A11	590	626	590626	1	1	424.9
1A12	590	626	590626	1	1	414.5
1A13	590	626	590626	1	1	394.3
1A14	590	626	590626	1	1	352.4
1A15	590	626	590626	1	1	438.8

Linear mixed model
It is imperative to write the statistical model correctly first before writing the model in ASReml or in another software. Given the mating design and the experimental design, we can use the following linear model to decompose phenotypic variance of a trait into genetic and environmental components:

$$y_{ijklm} = \mu + L_i + B_{j(i)} + G_k + G_l + S_{kl} + LG_{ik} + LG_{il} + LS_{ikl} + \varepsilon_{ijkl} + w_{ijklm} \tag{5.4}$$

where

y_{ijklm} is the observation on the mth tree in the jth block of the ith location for the klth cross;
μ is the overall mean;
L_i is the ith fixed location (environment) effect, $i = 1$ to t;
$B_{j(i)}$ is the random effect of the jth block nested within the ith location, $j = 1$ to b;
G_k, G_l are the random general combining ability (GCA) effects of the kth female and the lth male, respectively. These are normally and independently distributed G_k and $G_l \sim NID(0, \sigma_G^2)$;
S_{kl} is the random specific combining ability (SCA) effect of the kth and the lth parents $(k \neq l) \sim NID(0, \sigma_S^2)$;
LG_{ik}, LG_{il} are the random female and male parent GCA by location interactions $\sim NID(0, \sigma_{LG}^2)$;
LS_{ikl} is the random SCA by location interaction effect $\sim NID(0, \sigma_{LS}^2)$;
ε_{ijkl} is the experimental error term for residual variation among plot mean values $\sim NID(0, \sigma_\varepsilon^2)$;
W_{ijklm} is the experimental error term for residual variation among individual trees within a plot $\sim NID(0, \sigma_w^2)$.

We show in the next sections how to fit the model for the diallel experiment established at multiple locations (Eq. 5.8) in ASReml.

The GCA model
We start with a simple reduced model to estimate genetic variance due to GCA effects. Unlike half-sib family data we saw in Chap. 4, writing the GCA effect for diallels is trickier, because the same parents are used as both female and male.

We are going to model the parental GCA effects *as identical* whether they are transmitted through the male or female side of the parent. So, a progeny from parent A is modeled as having the GCA effect of parent A whether A was its male or female parent. We accomplish this restriction on the model by 'overlaying' the design matrix $\mathbf{Z}$ for male and female parents in the linear mixed model. This will estimate one common GCA variance and one GCA estimate for each parent regardless of how the parents were used.

The reduced linear model ignoring SCA effects is:

$$y_{ijkm} = \mu + L_i + B(L)_{j(i)} + G_k + G_l + \varepsilon_{ijkl} + w_{ijklm} \tag{5.5}$$

Code example 5.1
GCA model for half diallel. (*Code 5-1_diallels.as*)

```
Title: Diallel.
   tree       !A   1480
   female     !I
   male !AS female  # Integrated coding
   cross      !I
   location   *
   block      *
   density

diallel.csv  !SKIP 1  !DOPATH 1

!PATH 1      # GCA effect model
density  ~ mu location ,    # fixed model
      !r  location.block ,  # Random nested Block effect
          female and(male)  # Random GCA effect
          location.block.female.male # Plot level experimental error
          residual units # Variation among individuals within a plot
```

!AS	is required to indicate that the level codes for *female* and *male* parent effects are identical.
and(male)	overlays the design matrix for model term *male* on the design matrix for the immediately preceding model term (in this case, *female*).
!AS and **and()**	used together ensure that we get a single consistent GCA effect estimate for each parent whether a parent is used as female, male or both.

In the model, the block effect is nested within location effect because the block effect appears only with the location as *location.block*. There is no block main effect across locations.

A partial output of *Code 5-1_diallels.asr* file is given below:

```
Notice: ASReml assumes female and and(male)
        have the same levels in the same order.
...

        - - - Results from analysis of density - - -

Model_Term                        Gamma      Sigma    Sigma/SE   % C
   location.block       IDV_V   24  0.296193  125.872     2.83   0 P
   female               IDV_V    6  0.221045   93.9369    1.55   0 P
   location.block.female.m IDV_V 864 0.134900  57.3279    4.38   0 P
   units                      1440  effects
   Residual             SCA_V 1440  1.00000   424.967    23.87   0 P
```

- Notice the message in the output file about female and male having the same levels in the same order. ASReml assumes that a female parent coded '1' is the same individual as a male parent also coded '1'. We used the !AS qualifier and the and() function to make sure this is the case.

It is important to note that although only one parental GCA component of variance ('female') is reported in this output, our model includes GCA variances for both male and female parents. We restricted the model to force them to be equal, which is why only one of the two parental GCA variances is reported. The female variance component is expected to equal 1/4 of the additive genetic variance. The (unreported) male variance component is identical and also is expected to equal 1/4 of the additive genetic variance. Together, these two GCA variances represent half of the additive genetic variance, which is the proportion of additive genetic variance expected among full-sib families (Holland et al. 2003). The residual variance in this model is due to variation among trees within a plot, which includes both within-family genetic variation and extraneous non-genetic error variation. The within-family genetic variation includes the other half of the additive genetic variance and all of the non-additive genetic (SCA) variance.

The solution (GCA estimates) for the parents used in the design are reported in the *Code 5-1_diallels.sln* file:

Model_Term	Level	Effect	seEffect
location	1	0.000	0.000
location	2	-11.86	7.100
location	3	14.37	6.751
location	4	-13.59	6.759
mu	1	410.1	9.242
location.block	1.001	0.1467E-01	5.395
...			
female	590	-0.6103	4.139
female	626	-5.062	4.164
female	634	-12.13	4.140
female	652	-1.542	4.143
female	612	2.686	4.173
female	649	16.66	4.138

- A partial output of the solution file is presented here. Parent 649 has a large positive GCA effect estimate (16.66 with a standard error of 4.138), whereas parent 634 has a large negative value of -12.13 with a standard error of 4.14.

Using pedigree in the GCA model

In the model above, we assumed that the parents are unrelated, such that GCA effects are independent with no covariances: G_k and $G_l \sim NID(0, \sigma_G^2)$. If there are known pedigree relationships among the parents, however, we should include that information in the analysis to account for correlations among effects of related parents.

Pedigree file Among the six parents used in the mating design, trees 590 and 626 have one common parent (333). The remaining parental pairs are unrelated. The first nine rows of the pedigree file are given below:

TREE	FEMALE	MALE
333	0	0
590	333	0
626	0	333
634	0	0
652	0	0
612	0	0
649	0	0
1A11	590	626
1A12	590	626
...		

Although we do not have any observations for grandparent 333, we expand the vector **a** containing additive effects of individuals to include all ancestors back to the base population. Thus we will see a solution (GCA value) for 333 in the solution output file (*.sln*) for models that use this pedigree information.

The ASReml command file used to incorporate the pedigree relationships among the parents takes the following form:

Code example 5.2
GCA model for half diallel using pedigree information on parents (*Code 5-2_diallels_with_pedigree.as*)

```
Title: Diallel with pedigree.
 tree     !A 1480
 female   !P  # Determine levels from pedigree
 male     !P  # Determine levels from pedigree
 cross    !I
 location  *
 block    *
 density

diallel_pedigree.txt  !SKIP 1  # Pedigree
diallel.csv  !SKIP 1  !DOPART 1

!PART 1    # GCA effect
density  ~ mu location ,        # Specify fixed model
    !r  location.block ,        # random Block effect
        female and(male,1),     # random GCA effect
        location.block.female.male # plot error variance
```

A partial output from the primary output file is given below (*Code 5-2_diallels_with_pedigree.asr*). Before the variance components estimates, the .asr file contains information about the reading of the pedigree file. Notice that ASReml correctly identified that individual 333 was used as both father and mother in the pedigree and prints a warning (since this should not happen in a mammalian pedigree). Also, the *.asr* file reports seven unique identities (the six parents plus one known common grandparent) in the pedigrees of the phenotyped progenies.

```
diallel_pedigree.txt  !SKIP 1
 Reading pedigree file diallel_pedigree.txt: skipping       1 lines
 Pedigree Header Line: tree mother father
 Pedigree check: fath 333  previously occurred as a moth Now at line 3: 626  0      333
       7 identities in the pedigree over 1 generations.
     For first parent labelled moth, second labelled fath
   moth moth_of_moth fath_of_moth     fath moth_of_fath fath_of_fath
      1         0            0         1         0            0
 Using an adapted version of  Meuwissen & Luo GSE 1992 305-313:
 PEDIGREE [diallel_pedigree.txt ] has    7 identities, 9 Non zero elements
 GIV0  NRM          7       7      -0.58
```

The variance components estimates are:

Model_Term			Gamma	Sigma	Sigma/SE	%	C
location.block	IDV_V	24	0.296185	125.868	2.83	0	P
female	**NRM_V**	7	0.221136	93.9749	1.55	0	P
location.block.female.m	IDV_V	1176	0.134908	57.3310	4.38	0	P
Residual	SCA_V	1440	1.00000	424.964	23.87	0	P

- The variance explained by the GCA effect is 93.97. The female factor (overlaid with male factor) is associated with the pedigree, which models the variance-covariance structure of the parental effects according to the 'numerator relationship matrix' (identified as 'NRM_V' in the *.asr* output). The female variance component estimate from this pedigree-based model is slightly larger than in the previous model where we assumed no relationships among the parental effects. Some increase is expected since the GCA effects of parents with some pedigree relationship are not independent, so when we treated them as independent in the previous model, we were underestimating the true variance among the GCAs in the reference population.

The solutions for parent GCA effects are given below (*Code 5-2_diallels_with_pedigree.sln*):

Model_Term	Level	Effect	seEffect
...			
female	333	-2.437	8.219
female	590	-0.8351	4.281
female	626	-5.257	4.307
female	634	-12.33	4.270
female	652	-1.742	4.271
female	612	2.489	4.302
female	649	16.46	4.268

Now the model predicts a GCA value for ancestor 333; otherwise the GCA predictions for the parents are similar to the model without pedigree.

Adding GCA by location interaction term

Since the crosses were evaluated at multiple locations, we can include GCA by location interaction effects LG_{ik} to the model:

$$y_{ijklm} = \mu + L_i + B(L)_{j(i)} + G_k + G_l + LG_{ik} + LG_{il} + \varepsilon_{ijkl} + w_{ijklm} \tag{5.6}$$

Modeling the GCA by location interaction for the diallel in ASReml requires some additional tricks. We want to overlay the design matrix for female by location interaction on to that for male by location interaction using the and() model specification. However, we cannot do this directly, because the interaction term 'location.male' must be first defined before it is inserted inside the and() specification. This is best explained by example; the model in ASReml is as follows:

Adding GCA-by-environment interaction to model for half diallel using pedigree information on parents (Continued *Code 5-2_diallels_with_pedigree.as*)

```
!PART 2            # GCA and GxE effect model
density  ~ mu location ,      # fixed model
   !r location.block ,        # Random nested Block effect
      female and(male) ,      # Random GCA effect
      loc.female -loc.male and(loc.male), # GCA by Location effect
      location.block.female.male # plot error variance
```

- loc.female and(loc.male) would be the obvious way to write the last line of model code, but this would not work because the term inside the and() function is parsed before the interaction term is defined. So, any terms inside the and() function must be defined directly by a field definition or be explicitly declared previously.
- The previous line of model code, female and (male) works because 'male' is defined via the field definitions. However, loc.female and(loc.male) fails because the loc.male term that has not yet been defined.
- To work around the strict parsing of the and() function, we write: loc.female -loc.male and(loc.male): -loc.male defines a term for the interaction of location and male without including it in the model so that it is recognized when the and() term is parsed.

- Notice that *location* is shortened as `loc`. ASReml understands it is `location` by matching it to the first three letters of defined field names. Model terms should only be shortened when there is no chance of matching the wrong field.

Variance components from the model are below:

```
Model_Term                              Gamma       Sigma   Sigma/SE   % C
location.block          IDV_V   24   0.295285     125.461      2.84   0 P
female                  NRM_V    7   0.220009     93.4776      1.52   0 P
loc.female              IDV_V   28   0.210848E-01 8.95855      1.51   0 P
location.block.female.m IDV_V 1176   0.112914     47.9751      3.80   0 P
Residual                SCA_V 1440   1.00000      424.881     23.87   0 P
female                  NRM      7
```

- Notice that the plot-level experimental error variance for this model (`location.block.female.male` = 47.97) is smaller than the plot error variance (57.33) for the model without the G × E interaction term; the GCA-by-location variance has been modeled as a random effect instead of being combined with the plot variance. In contrast, the residual within-plot variance has hardly changed.

Specific Combining Ability (SCA) Effect

With the diallel mating design we can decompose total genetic variance into additive genetic variance (explained by the GCA effect) and dominance genetic variance (explained by the SCA effect). In the model below, the term `cross` defines SCA. In the data set, the levels of cross should code for a particular combination of parents without respect to which parent is male or female. In other words, the level for cross should be identical for reciprocals. SCA can also be defined as an interaction between female and male effects, but only if we are not treating male and female as pedigree-associated factors (since the pedigree information forces a specific variance-covariance matrix on the main effects that cannot be simply generalized to their interaction). Furthermore, to be sure we treat the effect of female k × male l as identical to female k × male l, we need to use the `and()` function again. Since there is no field defined as `male.female`, we would have to first define it in the model without fitting it, writing the part of the model coding for SCA as:

```
female.male -male.female and(male.female)
```

This would produce a single variance component estimate for `female.male`, and its value would be exactly half of the variance component estimate for cross, since we are implicitly partitioning the cross variance equally into `female.male` and `male.female` pieces. Considering these complications, it is generally easier to code the model and understand results if we just create a field for 'cross' in the data set.

Finally, we also add a term for SCA by location interaction (*loc.cross*) in the model:

Adding SCA effects to model for half diallel using pedigree information on parents (*Code 5-2_diallels_with_pedigree.as* continued).

```
!PART 3      # GCA, GxE and SCA effect model
density  ~ mu location,     # fixed model
   !r location.block ,      # Random nested Block effect
      female and(male),     # Random GCA effect
      loc.female -loc.male and(loc.male),   # GCA by Loc effect
      cross,                # Random SCA
      loc.cross,            # Random SCA x Loc effects
      location.block.cross
```

Variance components from the primary output file (.asr) are given below:

Model_Term			Gamma	Sigma	Sigma/SE	%	C
cross	IDV_V	14	0.151500E-02	0.643664	0.18	0	P
location.block	IDV_V	24	0.295446	125.524	2.84	0	P
female	NRM_V	7	0.219946	93.4463	1.51	0	P
loc.female	IDV_V	28	0.211815E-01	8.99918	1.51	0	P
loc.cross	IDV_V	56	0.101193E-06	0.429929E-04	0.00	0	B
location.block.cross	IDV_V	336	0.111973	47.5728	3.73	0	P
Residual	SCA_V	1440	1.00000	424.861	23.87	0	P
female	NRM	7					

- SCA and SCA by location interaction variances are both very small. The ratios of the estimates to their standard errors are less than 1, suggesting that the dominance variance is likely not significant. There is almost no difference in log likelihoods between this model (-5179.14) and the previous GCA model (-5179.16). Therefore we cannot reject the null hypothesis that the SCA and SCA by location variances are both zero.

The SCA effects for each level of 'cross' are included in the solution file (.sln), and the effects are all very small, as expected if the SCA variance is close to zero:

Model_Term	Level	Effect	seEffect
cross	**590626**	**0.1036**	0.7806
cross	590634	0.5447E-02	0.7807
cross	590652	0.1325	0.7790
...			
cross	649652	0.2667	0.7802
cross	652626	0.1141	0.7810
cross	652634	-0.3850	0.7823
...			
female	333	-2.414	8.234
female	**590**	**-0.7736**	4.475
female	**626**	**-5.262**	4.500
female	634	-12.04	4.464
female	652	-1.775	4.464
female	612	2.365	4.494
female	649	16.28	4.462

What are the breeding value (BV) and genotypic value (GV) of cross 590×626? Parent 590 has a GCA value of -0.7736. Parent 626 has the GCA value of -5.262. The expected BV of 590×626 is simply the sum of its parental GCA values: BV $(590 \times 626) = (-0.7736 - 5.262) = -6.04$. The GV is simply the BV plus the SCA estimate for the cross. GV $(590 \times 626) = -6.04 + 0.1036 = -5.94$.

Reciprocal Effects

As discussed in the preceding section, the SCA effect for the cross between parents A and B is common between the two reciprocal ways of making the cross SCA(A $\times$ B) = SCA(B $\times$ A). If we have reciprocal crosses in the data, we can model the reciprocal SCA effect, which is due to differences between A $\times$ B and B $\times$ A. Furthermore, some part of the reciprocal effect may be due to differences in the GCA effect of a parent when used as male and female. We can augment the diallel model to include reciprocal effects of both GCA and SCA. To do this, we need to add a column to the data set for an indicator variable that distinguishes reciprocal crosses. This indicator variable can be a covariate set to 1 for crosses in one direction and -1 for crosses in the other direction, and it is arbitrary as to which cross direction is coded as 1.

For example, the file *maize_diallel.csv* contains data from a partial diallel in sets in which 55 parent maize plants were crossed in many ways in 11 sets of five parents each. All possible crosses were attempted, although not all crosses succeeded. A sample of 3,500 progenies were measured in randomized single plant plots across 2 years. Some progeny data rows in the file are shown below:

Year	Progeny	female	male	cross	r	DTA	PltHt
2013	1	171_11	170_1	170_1 × 171_11	−1	72	270
2013	27	171_11	170_1	170_1 × 171_11	−1	73	225
2013	44	170_1	171_11	170_1 × 171_11	1	73	240
2013	76	170_1	171_11	170_1 × 171_11	1	81	220
2013	131	170_1	167_6	167_6 × 170_1	−1	73	265
2013	147	170_1	167_6	167_6 × 170_1	−1	72	255

In the data set, you can see that when parents 170_1 and 171_11 were crossed, the progeny's cross field level is set to 170_1 × 171_11, irrespective of which parent was used as female. The variable 'r' is the covariate that distinguishes the two reciprocals of this one combination. It is set at −1 when 171_11 was used as female and set at 1 when 170_1 was used as female. It is arbitrary which reciprocal is labeled as $r = -1$ or 1 within a particular combination. For example, when 170_1 is used as a female in a cross with parent 167_6, the progeny value for r is set at −1.

Having set up the data this way, we can then write the ASReml code to define the field for the 'r' variable as a numeric covariate, and then fit interactions between GCA and SCA effects with this covariate to capture the reciprocal effects (Holland et al. 2003; Möhring et al. 2011; Piepho and Möhring 2007). The full model should also include reciprocal by environment interactions, which are really three-way interaction between genetic effects, the reciprocal covariate, and environment:

Code example 5.3
Model for diallel with reciprocal crosses. (see details in *Code 5-3_reciprocal_effects.as*)

```
 Title: Maize diallel
  Year   !I
  Progeny  !I
  female   !A
  male   !A !AS female
  cross  !A
  r      # make sure this is coded as a numeric covariate!
  DTA    # days to anthesis
  PltHt  # plant height

 maize_diallel.csv  !SKIP 1 !MAXITER 30 !DOPART 1

 !PART 1
 PltHt  ~ mu Year,      # Specify fixed model
    !r female and(male),  # GCA
    Year.female -Year.male and(Year.male),  #GCA x Year
    female.r -male.r and(male.r),  # reciprocal GCA
    Year.female.r -Year.male.r and(Year.male.r),  #reciprocal GCA x Year
    cross Year.cross,     # SCA and SCA x Year
    cross.r Year.cross.r  # reciprocal SCA and reciprocal SCA x Year
```

The variance components estimates from this analysis are:

Model_Term			Gamma	Sigma	Sigma/SE	% C
female	IDV_V	53	0.107210	64.1576	2.93	0 P
Year.female	IDV_V	106	0.497229E-02	2.97556	0.78	0 P
female.r	IDV_V	53	0.297211E-03	0.177859	0.14	0 P
Year.female.r	IDV_V	106	0.101193E-06	0.605567E-04	0.00	0 B
cross	IDV_V	89	0.799283E-01	47.8314	2.07	0 P
cross.r	IDV_V	89	0.101193E-06	0.605567E-04	0.00	0 B
Year.cross	IDV_V	178	0.122755E-01	7.34601	0.86	0 P
Year.cross.r	IDV_V	178	0.101193E-06	0.605567E-04	0.00	0 B
Residual	SCA_V	3196	1.00000	598.429	38.86	0 P

- These results show little evidence for reciprocal effects, the variance components corresponding to reciprocal effects are either very small (*female.r*, corresponding to reciprocal GCA) or essentially zero.

Interpretation of Observed Variances from Diallels

Assuming that epistatic genetic effects are negligible, the variance component associated with the *female* effect (GCA) is 1/4 of the additive genetic variance, and the variance component associated with the *cross* effect (SCA) is 1/4 of the dominance variance. If the experiment were established in one environment, these variances may also include contributions due to confounded genetic-by-environment variances. Variance associated with reciprocal GCA effects (*female.r*, σ^2_{GCAr}) can be due to *maternal effects and cytoplasmic genetic effects that are shared among all crosses with a common female parent, but not shared by crosses where that parent is used as male.* The reciprocal SCA variance (*cross.r*, σ^2_{SCAr}) is due to any effects that are specific to a particular direction of a specific cross; this could occur due to interactions between cytoplasm of the maternal parent and nuclear alleles of the paternal parent.

The interpretation of observed GCA and SCA variance components as equal to their expected values in terms of additive and dominance variance in the reference population has to be done with caution. As already mentioned, a major limitation to the use of diallels for estimating genetic variance components is their generally limited sampling of parents, resulting in potentially large variation among estimates from different parental samples (Isik et al. 2005). Small samples can also cause the assumption of independence among loci to be violated even when the reference population is in linkage equilibrium, and this will affect the estimated variance components (Baker 1978). In general, it is safer to treat the parents as a fixed factor and avoid estimating variance components (Hallauer and Miranda 1988), but the fixed effects approach may result in difficulties in estimability of GCA effects if some crosses are missing.

We showed previously that relationships among the parents can be modeled by including pedigree information, which results in the GCA variance being properly adjusted to reflect the variance among unrelated samples of the reference population. ASReml internally makes this adjustment by using the numerator relationship matrix to model the additive effects of the parents. However, ASReml does not automatically adjust the SCA variance to account for the possibility that some progenies that appear to be half-sibs based on the immediate generation pedigree may in fact have a closer relationship and share dominance effects. Returning to the pine diallel example, progenies from the cross of parent 590 × 634 are closer than half-sibs to progenies from 626 × 634 ($2\theta = 0.3125$) because parents 590 and 626 are both progenies of ancestor 333. Furthermore, progenies within the cross of 590 × 626 are more closely related than full-sibs, their additive relationship is $2\theta = 0.625$. ASReml accounts for the increased additive relationships in these cases when it uses the pedigree information on the parents. It does not, however, account for the small covariance between the dominance effects of families 590 × 624 and 626 × 624 ($\Delta = 0.0625$), or the slightly increased dominance covariance between progenies within the 590 × 626 family ($\Delta = 0.28125$). Ignoring these relationships leads to biased estimates of the dominance variance based on the diallel SCA variance estimate. In the example shown, the bias is expected to be small since the changes in affected coefficients are small.

Linear Combinations of Variances from Diallels

Keeping in mind the caveats stated in the previous section, we still may want to use the estimated variance components from the diallel model to estimate heritability. Given the family structure inherent in the diallel design, we can estimate several different heritability values, which can be interpreted as being related to response to selection among individuals, half-sib families, or full-sib families, or combined selection among and within families (Holland et al. 2003).

If we have data on individuals (rather than just mean or total values from plots containing multiple individuals from a common family), we can estimate phenotypic variance among individuals $\left(\sigma_{phen}^2\right)$ and the narrow-sense heritability $\left(h_i^2\right)$ from the diallel analysis results. The phenotypic variance among individuals within a block from a diallel without reciprocals replicated across several locations and with multiple individuals per experimental unit is:

$$\sigma_{phen}^2 = 2\sigma_G^2 + \sigma_S^2 + 2\sigma_{LG}^2 + \sigma_{LS}^2 + \sigma_\varepsilon^2 + \sigma_w^2 \tag{5.7}$$

Narrow-sense heritability is the ratio of additive genetic variance and phenotypic variance among individuals within a block:

$$h_i^2 = \sigma_A^2/\sigma_{phen}^2 = 4\sigma_G^2/\sigma_{phen}^2 \tag{5.8}$$

Broad-sense individual heritability ($h_{bs}{}^2$) is similar but includes both additive and dominance variance in the numerator:

$$h_{bs}^2 = 4\left(\sigma_G^2 + \sigma_S^2\right)/\sigma_{phen}^2 \tag{5.9}$$

The selection unit in diallels could be **half-sib family means,** in which case the relevant heritability to predict selection response would be a ratio of the amount of additive variance among families to the variance of half-sib family means. The variance of half-sib family means can be derived from the linear model and the observed variance components, as follows: the mean of half-sib family k is:

$$\bar{Y}_{..k.} = \mu + \bar{L}_. + \bar{B}_{..} + G_k + \frac{\sum\limits_{l \neq k}^{p-1} G_l}{p-1} + \frac{\sum\limits_{l \neq k}^{p-1} S_{kl}}{p-1} + \frac{\sum\limits_{i}^{e} LG_{ik}}{e} + \frac{\sum\limits_{l \neq k}^{p-1}\sum\limits_{i}^{e} LG_{il}}{e(p-1)} + \frac{\sum\limits_{l \neq k}^{p-1}\sum\limits_{i}^{e} LS_{ikl}}{e(p-1)} + \frac{\sum\limits_{l \neq k}^{p-1}\sum\limits_{i}^{e}\sum\limits_{j}^{r} \varepsilon_{ijkl}}{er(p-1)} + \frac{\sum\limits_{l \neq k}^{p-1}\sum\limits_{i}^{e}\sum\limits_{j}^{r}\sum\limits_{m}^{n} w_{ijklm}}{er(p-1)n}$$

$$= \frac{p\sigma_{GCA}^2}{p-1} + \frac{\sigma_{SCA}^2}{p-1} + \frac{p\sigma_{LG}^2}{e(p-1)} + \frac{\sigma_{LS}^2}{e(p-1)} + \frac{\sigma_\varepsilon^2}{er(p-1)} + \frac{\sigma_w^2}{ern(p-1)}$$

$$\tag{5.10}$$

In this equation, p is the number of parents of the diallel, e is the number of environments (locations in this example), r is the number of replications or blocks per environment, and n is the number of individuals per plot. The half-sib family mean includes G_k, the GCA of the common female parent k, and averages over $p - 1$ GCA effects of male parents, $p - 1$ SCA effects of combinations between parent k and the other parents, and over the relevant genetic-by-environment interaction effects, plot effects, and plant within plot effects. If p is large (say, $p > 20$) then $p/(p-1)$ is approximately 1 and the variance of family means ($Var(\bar{Y}_{..k.})$) is approximately:

$$\sigma_{HS}^2 = \sigma_{GCA}^2 + \frac{\sigma_{SCA}^2}{p-1} + \frac{\sigma_{LG}^2}{e} + \frac{\sigma_{LS}^2}{e(p-1)} + \frac{\sigma_\varepsilon^2}{er(p-1)} + \frac{\sigma_w^2}{ern(p-1)} \tag{5.11}$$

If the number of individuals per experimental unit varies, then we replace n in the equation above (which refers to a constant number of individuals per unit), with the harmonic mean of the number of individuals per experimental unit (n_h) (Holland et al. 2003; Nyquist 1991). This equation will also hold for half-sib families derived from male parents.

Heritability of half-sib family means is simply a ratio of the variance component due to GCA and the phenotypic variance of half-sib family means:

$$h^2_{HS} = \frac{\sigma^2_G}{\sigma^2_{HS}} = \frac{\frac{1}{4}\sigma^2_A}{\sigma^2_{HS}} \qquad (5.12)$$

Alternatively, breeders may select among **full-sib family** means. The mean of half-sib family kl is:

$$\bar{Y}_{..k.} = \mu + \bar{L}_. + \bar{B}_{..} + G_k + G_l + S_{kl} + \frac{\sum\limits_{i}^{e} LG_k}{e} + \frac{\sum\limits_{i}^{e} LG_l}{e} + \frac{\sum\limits_{i}^{e} LS_{ikl}}{e} + \frac{\sum\limits_{i}^{e}\sum\limits_{j}^{r} \varepsilon_{ijkl}}{er} + \frac{\sum\limits_{i}^{e}\sum\limits_{j}^{r}\sum\limits_{m}^{n} w_{ijklm}}{ern} \qquad (5.13)$$

and the variance of full-sib family means $(Var(\bar{Y}_{..k.})$ is:

$$\sigma^2_{FS} = \sigma^2_{GCA} + \sigma^2_{GCA} + \sigma^2_{SCA} + \frac{\sigma^2_{LG}}{e} + \frac{\sigma^2_{LG}}{e} + \frac{\sigma^2_{LS}}{e} + \frac{\sigma^2_{\varepsilon}}{er} + \frac{\sigma^2_w}{ern} \qquad (5.14)$$

$$= 2\sigma^2_{GCA} + \sigma^2_{SCA} + \frac{2\sigma^2_{LG}}{e} + \frac{\sigma^2_{LS}}{e} + \frac{\sigma^2_{\varepsilon}}{er} + \frac{\sigma^2_w}{ern} \qquad (5.15)$$

If the breeder recombines individuals from selected full-sib families to create a new population, only the additive portion of the variance among full-sib families contributes to selection gain in the next generation. The dominance variance does not contribute to gain because the dominance interactions are disrupted by random mating among the selected families:

$$h^2_{FS} = 2\sigma^2_G/\sigma^2_{FS} \qquad (5.16)$$

In some species, it may be possible to reproduce seeds or progenies of a superior cross in sufficient quantity to serve as a unit of production, then the selected families could be used directly in production. The gain predicted from growing these selected families in independent environments compared to growing the original population involves the complete variation among full-sib family genotype values in the numerator, as the response units share dominance deviations with the selection units (Holland et al. 2003):

$$H^2_{FS} = \frac{2\sigma^2_G + \sigma^2_S}{\sigma^2_{FS}} = \frac{0.5\sigma^2_A + 0.25\sigma^2_D}{\sigma^2_{FS}} \qquad (5.17)$$

Breeders might also be interested in selection of the best crosses and the best progeny within crosses. This is simply an index of family and within family selection. The phenotypic variance within families is the difference between the total variance among individuals and the phenotypic variance among full-sib family means:

$$\begin{aligned}
\sigma^2_{fsw} &= \sigma^2_{phen} - \sigma^2_{FS} \\
&= \left(2\sigma^2_G + \sigma^2_S + 2\sigma^2_{LG} + \sigma^2_{LS} + \sigma^2_{\varepsilon} + \sigma^2_w \right) \\
&\quad - \left(2\sigma^2_G + \sigma^2_S + \frac{2\sigma^2_{LG}}{e} + \frac{\sigma^2_{LS}}{e} + \frac{\sigma^2_{\varepsilon}}{er} + \frac{\sigma^2_w}{ern} \right) \\
&= \frac{(e-1)2\sigma^2_{LG}}{e} + \frac{(e-1)\sigma^2_{LS}}{e} + \frac{(er-1)\sigma^2_{\varepsilon}}{er} + \frac{(ern-1)\sigma^2_w}{ern} \\
&= \frac{(e-1)}{e}\left(2\sigma^2_{LG} + \sigma^2_{LS} \right) + \frac{(er-1)\sigma^2_{\varepsilon}}{er} + \frac{(ern-1)\sigma^2_w}{ern}
\end{aligned} \qquad (5.18)$$

The broad-sense and narrow sense within-family heritability estimates are:

$$h^2_w = 2\sigma^2_G/\sigma^2_{fsw} \qquad (5.19)$$

$$H_w^2 = \left(2\sigma_G^2 + 3\sigma_S^2\right)/\sigma_{fsw}^2 \qquad\qquad (5.20)$$

We can obtain estimates of these kinds of heritabilities as functions of the variance components from ASReml. For example, we can obtain the heritability estimators for the tree diallel example, by substituting the coefficients $e = 4$, $r = 6$, and $p = 6$ into the equations above, and using the following VPREDICT !DEFINE function after the model including GCA and SCA and their interactions with environments. There is a slight complication that we also need to compute the harmonic mean of individuals per plot because the number of individuals per plot was not constant. We can compute the harmonic mean easily using a little bit of R code:

Code example 5.4
R script to compute the harmonic mean of the number of trees per plot in the diallel.csv data set
(*Code 5-4_Harmonic_mean.R*):

```
# harmonic mean of number of trees per plot in diallel.csv data
diallel <- read.csv("diallel.csv")
#create a data frame with the counts of trees within each plot
trees_per_plot <- aggregate(density ~ MALE + FEMALE + CROSS + SITE +
BLOCK, data = diallel, length)
#harmonic mean of number of trees per plot
(nh = 1/mean(1/trees_per_plot$density))
```

This returns 4.0017. So, we can use 4 as the value of n_h in the family mean heritability computations.

Code example calculating linear combinations of variance components from diallels (*Code 5.2_diallels_with_pedigree. as*).

```
VPREDICT !DEFINE
F SCA        cross
F GCA        female
F GCAxL      loc.female
F SCAxL      loc.cross
F Error      location.block.cross   #plot error variance
F Within     Residual # within-plot variance
F A_var        female*4.0      # additive variance
F D_var        cross*4.0       # dominance variance
F GCA2         female*2.0      # 2x of GCA variance or 1/2A
F SCA.GCA2   SCA + GCA*2.0      # Numerator of BS heritability for FS
F SCA3.GCA2 SCA*3.0 + GCA*2.0 # Numerator of BS heritability within FS
F phen_i   GCA*2.0 + SCA + GCAxL*2.0 + SCAxL + Error + Within #
pheno_var
F phen_hs   GCA*1.2 + SCA*0.2 + GCAxL*0.3 + SCAxL*0.05 + Error*0.0083 +
Within*0.002 #Variance of HS means
F phen_fs   GCA*2.0 + SCA + GCAxL*0.5 + SCAxL*0.25 + Error*0.041667 +
Within*0.0104 #Variance of FS means
F phen_wfs GCAxL*1.5 + SCAxL*0.75 + Error*0.958333 + Within*0.99
#Variance within FS families
H h2_i      A_var   phen_i        # NS individual heritability
H h2_hs     GCA phen_hs           # NS half-sib family heritability
H h2_fsns   GCA2    phen_fs       # NS Full-sib family heritability
H H2_fsbs   SCA.GCA2   phen_fs  # BS Full-sib family heritability
H h2_wfs    GCA2    phen_wfs     # NS Within Full-sib family heritability
H H2_wfs    SCA3.GCA2 phen_wfs # BS Within Full-sib family heritability
```

Output of the functions of variance components defined by **VPREDICT** is stored in the *Code 5-2_diallels_with_pedigree.pvc* file:

```
            - - - Results from analysis of density - - -
1 cross                  V    14    0.643664        3.57591
2 location.block         V    24    125.524         44.1986
3 female                 V     7    93.4463         61.8850
4 loc.female             V    28    8.99918         5.95972
5 loc.cross              V    56    0.429929E-04    0.00000
6 location.block.cros    V   336    47.5728         12.7541
7 Residual               V  1440    424.861         17.7990
female              NRM       7
8 SCA   1            0.64366           3.6528
9 GCA   3            93.446            61.768
10 GCAxL   4         8.9992            5.9421
11 SCAxL   5         0.42993E-04       0.18008E-05
12 Error   6         47.573            12.766
13 Within   7        424.86            17.796
14 A_var   3         373.79            247.07
15 D_var   1         2.5747            14.611
16 GCA2   3          186.89            123.54
17 SCA.GCA2   8      187.54            123.55
18 SCA3.GCA2   8     188.82            123.91
19 phen_i   9        677.97            125.11
20 phen_hs   9       116.21            74.104
21 phen_fs   9       198.44            123.53
22 phen_wfs 10       479.70            19.794
   h2_i       = A_var   3   14/phen_i    19=   0.5513   0.2648
   h2_hs      = GCA   3      9/phen_hs   20=   0.8041   0.0232
   h2_fsns    = GCA2   3    16/phen_fs   21=   0.9418   0.0429
   H2_fsbs    = SCA.GCA2   17/phen_fs    21=   0.9451   0.0373
   h2_wfs     = GCA2   3    16/phen_wfs  22=   0.3896   0.2582
   H2_wfs     = SCA3.GCA  18/phen_wfs    22=   0.3936   0.2591
Notice: The parameter estimates are followed by
        their approximate standard errors.
```

- The estimate of dominance variance is much smaller than its standard error, so it is likely not different from zero. This is expected in this case, as we already noted that the SCA variance was not significant in these data.
- As a consequence, the 'broad-sense' heritability estimators that include dominance variance in their numerators are about equal to the corresponding 'narrow-sense' heritability estimators.
- The estimate of heritability among individual plants (h2_i) has a large standard error, but the heritabilities among family means are very high and have relatively small standard errors.
- We caution the reader again that these estimates of heritability should not be considered reliable because of the very small sample of six parental plants used to establish the diallel.

Factorial Mating Designs

In factorial mating designs, a set of individuals are used as females and another set of individuals are used as males for mating. Like diallels, there are multiple versions of factorial mating designs (Hallauer and Miranda 1988; White et al. 2007). Single and double pair mating designs depicted in Fig. 5.3 are often used because the number of crosses required are greatly reduced compared to a complete factorial.

Fig. 5.3 Single pair (*left*) and double pair factorial mating designs (*right*). A full factorial would produce N = f × m, where f and m are the numbers of females and males, respectively

Male parents \ Female parents	5	6	7	8
1	X			
2		X		
3			X	
4				X

Male parents \ Female parents	5	6	7	8
1	X	X		
2		X	X	
3			X	X
4				X

A linear model example for a factorial mating design

A factorial mating design was used to cross three unrelated female and three unrelated male individuals. Nine full-sib families (crosses) were produced. A randomized complete block design with six blocks were used to establish a field test. The experiment was replicated in another location. Tree height is the response variable measured after six growing seasons. The statistical model is

$$y_{ijpqk} = \mu + L_i + B_{j(i)} + F_p + M_q + FM_{pq} + LF_{ip} + LM_{iq} + LFM_{ipq} + \varepsilon_{ijpqk} \tag{5.21}$$

where y_{ijpqk} is the kth observation of the pqth cross the in the jth block in the ith site; μ is the overall mean; L_i is the ith fixed site (environment) effect, $i = 1$ to t; $B_{j(i)}$ is the fixed effect of the jth block nested within the ith site, $j = 1$ to b; F_p is the random general combining ability (GCA) effect of the pth female $\sim \mathrm{NID}\left(0, \sigma_F^2\right)$; M_q is the random general combining ability (GCA) effect of the qth male $\sim \mathrm{NID}\left(0, \sigma_M^2\right)$; FM_{pq} is the random cross effect of the pth and the qth parents $(p \neq q) \sim \mathrm{NID}\left(0, \sigma_{FM}^2\right)$; LF_{ip} is the random female by location Interaction $\sim \mathrm{NID}\left(0, \sigma_{LF}^2\right)$; LM_{iq} is the random male by location Interaction $\sim \mathrm{NID}\left(0, \sigma_{LM}^2\right)$; LFM_{ipq} is the random cross by location interaction effect $\sim \mathrm{NID}\left(0, \sigma_{LFM}^2\right)$; ε_{ijpqk} is the random error term $\sim \mathrm{NID}\left(0, \sigma_\varepsilon^2\right)$.

We can use ASReml models similar to diallel models for factorial mating designs. The difference is that male and female parents are distinct in the factorial mating design, so we would NOT use a field definition like '`male !AS female`' for the factorial mating because a male parent coded as '1' would not be the same parent as a female parent coded as '1'. Similarly, we can fit male, female, and male*female interaction effects in the model, but the factorial model coding is much simpler since we do not need to use the `and()` function to overlay the design matrices. On the contrary, we want to allow the male and female factors to be distinct and we will obtain two different estimates of GCA variance, one for male and another for female parents.

Similar to diallels, we can estimate causal genetic variances (additive, dominance) and derive linear combinations of variance components. Variance components observed for the female or male are 1/4 of additive genetic variance $\left(\sigma_F^2 = 1/4\sigma_A^2, \text{and}\, \sigma_M^2 = 1/4\sigma_A^2\right)$. These may be poor estimates since the number of female or males used in the mating are often small and the variance components associated with either the female and male effect may have large standard errors. A better estimate of the additive variance is the average of female and male variance components: $\sigma_A^2 = 2\left(\sigma_F^2 + \sigma_M^2\right)$. The phenotypic variance $\left(\sigma_p^2\right)$ is the sum of the all the variance components, except the variance component for the random block effect:

$$\sigma_{phen}^2 = \sigma_F^2 + \sigma_M^2 + \sigma_{FM}^2 + \sigma_{SF}^2 + \sigma_{SM}^2 + \sigma_{SFM}^2 + \sigma_\varepsilon^2 \tag{5.22}$$

The best estimate of narrow-sense heritability $\left(h_i^2\right)$ is:

$$h_i^2 = 2\left(\sigma_F^2 + \sigma_M^2\right)/\sigma_{phen}^2 \tag{5.23}$$

The mean of a half-sib family means with a common **female** parent is:

$$\bar{Y}_{..p..} = \mu + \bar{L}_. + \bar{B}_{..} + F_p + \frac{M_.}{m} + \frac{FM_{p.}}{m} + \frac{\overline{LF}_{.p}}{l} + \frac{\overline{LM}_{..}}{lm} + \frac{\overline{LFM}_{.p.}}{lm} + \frac{\bar{\varepsilon}_{..p..}}{lbmn} \tag{5.24}$$

The **variance of maternal half-sib family means** is:

$$Var\left(\bar{Y}_{..p..}\right) = \sigma^2_{phen\_f} = \sigma^2_F + \frac{\sigma^2_M}{m} + \frac{\sigma^2_{FM}}{m} + \frac{\sigma^2_{LF}}{l} + \frac{\sigma^2_{LM}}{lm} + \frac{\sigma^2_{LFM}}{lm} + \frac{\sigma^2_\varepsilon}{lbmn} \tag{5.25}$$

Notice that the contribution of the male effect to the phenotypic variance of female means is reduced by the number of males used to cross to each female parent.

The mean and variance for **paternal half-sib family means** are analogous:

$$\bar{Y}_{...q.} = \mu + \bar{L}_. + \bar{B}_{..} + \frac{F_.}{f} + M_q + \frac{FM_{.q}}{f} + \frac{\overline{LF}_{..}}{lf} + \frac{\overline{LM}_{.q}}{l} + \frac{\overline{LFM}_{..q}}{lf} + \frac{\bar{\varepsilon}_{...q.}}{lbfn} \tag{5.26}$$

$$Var\left(\bar{Y}_{...q.}\right) = \sigma^2_{phen\_f} = \frac{\sigma^2_F}{f} + \sigma^2_M + \frac{\sigma^2_{FM}}{f} + \frac{\sigma^2_{LF}}{lf} + \frac{\sigma^2_{LM}}{l} + \frac{\sigma^2_{LFM}}{lf} + \frac{\sigma^2_\varepsilon}{lbfn} \tag{5.27}$$

We can take the average of female and male parental mean phenotypic variances to get a pooled estimate of half-sib family phenotypic variance. Heritability estimates for maternal and paternal half-sib family means and a pooled average heritability can be obtained as:

$$
\begin{aligned}
h_f^2 &= \sigma^2_F / \sigma^2_{phen\_f} && \text{or} \\
h_m^2 &= \sigma^2_M / \sigma^2_{phen\_m} && \text{or} \\
h_{fm\_pooled}^2 &= 1/2 \frac{\left(\sigma^2_F + \sigma^2_M\right)}{\sigma^2_{phen\_pooled}}
\end{aligned}
\tag{5.28}
$$

Analysis of Cloned Progeny Test Data

In some plant and animal species, individuals can be cloned and their genetically identical copies can be tested in multiple environments. Cloning individuals for within family-selection is an especially common practice to develop varieties for deployment in eucalyptus species (Mullin et al. 2011) and many ornamental plants and vegetatively-propagated crops. Cloning allows breeders to capture hybrid vigor and non-additive genetic effects among and within crosses. Cloning individuals and testing in replicated field trials has the advantage of increasing the accuracy of genotypic value predictions compared to individual evaluations (Isik et al. 2004). With cloned progeny we can increase the heritability both by increasing the numerator (including all of the genotypic variance because broad-sense heritability is applicable to gain from selection) and by decreasing the denominator by averaging clonal means over more observations. If progenies from a mating design are cloned, we can also estimate total genotypic variance in addition to additive variance, providing a means to assess the relative importance of non-additive genetic variance.

Data

As an example of analysis of data from a clonal evaluation, we present a data set for loblolly pine (*Pinus taeda*) from the North Carolina State University Cooperative Tree Breeding Program. A single-pair mating design was used to produce eight full-sib families (crosses). Some of the parents used in crossing were related. Progeny from crosses were cloned via somatic embryogenesis. In total, 136 genetically unique progenies were cloned. Clones were field tested using an unbalanced incomplete block design at three locations. There were eight blocks at two of the locations and four blocks at the third. No blocks included all entries, and some entries appeared twice within some blocks. On average, genotypes were represented by about 15 clonal copies in the experiment. At age 8 years after planting, tree height diamater (dia8) and stem diameter were measured. Stem forking was recorded as a binary trait, 0 = no forking, 1 = forked stem. The objective was to estimate variance components and predict genetic values of clones for strictly within family selection. The first five rows of the data file (*Pine_clones*.csv) are shown below, *cloneid* refers to the code for the individual progeny genotype (not to the specific clonal copy within a genotype):

cloneid	parent1	parent2	cross	location	block	height	dia8	fork
18	14	16	J	Hill	2	35.9	7.0	0
18	14	16	J	Hill	4	38.3	6.6	0
18	14	16	J	Hill	6	27.7	5.8	0
18	14	16	J	Hill	8	37.0	6.4	0
18	14	16	J	Orchard	2	35.7	5.4	0

Statistical model

The following linear mixed model was fit to the data to estimate variance components and predict genetic values of cloned individuals for tree height. We have pedigree information that provides information on the additive relationships among each cloneid, so we use a form of the individual model and apply the additive relationship matrix to model the variance-covariance relationships among the clones:

$$y_{ijk} = \mu + L_i + B_{j(i)} + G_k + LG_{ik} + \varepsilon_{ijk} \tag{5.29}$$

where y_{ijk} is the kth genotype in the jth block of the ith location; μ is the overall mean; L_i is the ith fixed location (environment) effect, $i = 1$ to t; $B_{j(i)}$ is the random effect of the jth incomplete block nested within the ith location, $j = 1$ to b; G_k is the random individual progeny genotype effect $\sim N(0, A\sigma_A^2)$; LG_{ik} is the random individual by location interaction effect $\sim N(0, A\sigma_{AL}^2)$; and ε_{ijk} is the random error term $\sim NID(0, \sigma_\varepsilon^2)$. This model will provide estimates of additive and additive genetic – by – environment interaction variances.

We can extend this model to include any genetic variation among genotypes that is not captured by the genetic term with variance-covariance modeled according to the additive genetic relationship matrix. This will include non-additive genetic variance, although we cannot be more specific about which non-additive variances contribute to this additional term. It could include epistatic variances and dominance variance. The extended model includes two additional terms:

$$y_{ijk} = \mu + L_i + B_{j(i)} + G_k + Gr_k + LG_{ik} + LGr_{ik} + \varepsilon_{ijk} \tag{5.30}$$

The terms in this model are identical to the previous model, but we have added two additional factors: Gr_k, the residual genetic effects (non-additive) of the kth individual, $Gr_k \sim NID(0, \sigma_{Gr}^2)$; and LGr_{ik} the interaction between residual genetic effects (non-additive) of the kth individual and the ith environment, $LGr_k \sim NID(0, \sigma_{LGr}^2)$. So, this model includes two genetic effects for each individual k, one with a variance-covariance structure proportional to the A matrix, and the other with an independent effect distribution. Note that the sum of the variance components for additive variance and residual genetic variance does not necessarily equal the total genetic variance because the IID assumption of residual genetic effects is not correct if the progenies are related. Again, this is why we cannot attach a specific interpretation to the residual genetic variance other than to say it is non-additive variance.

The ASReml code for these models is given below.

Code example 5.5
Code example analyzing cloned progeny test data (*Code 5-5_clones.as*)

```
!OUTFOLDER    V:\Book\Book1_Examples\ch05_sca\outputfiles
Title: clonal data
#cloneid,parent1,parent2,cross,location,block,height,dia8,fork
#18,14,16,J,Hill,2,35.9,7,0
#18,14,16,J,Hill,4,38.3,6.6,0
#18,14,16,J,Hill,6,27.7,5.85,0
 cloneid !P
 female  !P
 male    !P
 cross   !A
 location    !A
 block *
 height8  dia8  fork

!FOLDER     V:\Book\Book1_Examples\data
Pine_clones_pedigree.txt  !SKIP 1  !ALPHA
Pine_clones.csv  !SKIP 1  !NOREORDER !DOPART 1

TABULATE    height8 ~ location block !COUNT
TABULATE    height8 ~ cloneid !COUNT
TABULATE    height8 ~ location !COUNT

!PART 1  # additive genetic variance model
height8   ~ mu location  ,
        !r location.block nrm(cloneid) nrm(cloneid).location
         residual units


!PART 2  # add ide term to catch non-additive genetic variance
height8   ~ mu location  ,
        !r  location.block nrm(cloneid) nrm(cloneid).location ,
            ide(cloneid) ide(cloneid).location
            residual units
```

- The term `nrm(cloneid)` fits the additive effects of individuals according to the pedigree information supplied in the pedigree file.
- Since the factor *cloneid* was defined as a pedigree factor in the field definitions, we need to use a special term to fit *cloneid* without the pedigree relationships. This is accomplished with the term `ide(cloneid)`, which takes a copy of a pedigree factor *cloneid* and fits it without the genetic relationships using an identity matrix for its variance-covariance structure.
- `!NOREORDER` qualifier keeps the variances in the order they are specified in the model.

The variance components estimates from above models are given below:

PART 1

```
  8 LogL=-3824.59      S2=  13.050       2068 df

          - - - Results from analysis of height8 - - -
  Akaike Information Criterion      7657.18 (assuming 4 parameters).
  Bayesian Information Criterion    7679.72
```

Model_Term			Gamma	Sigma	Sigma/SE	% C
location.block	IDV_V	24	0.108434	1.41509	2.67	0 P
nrm(cloneid)	NRM_V	153	0.438690	5.72501	6.15	0 P
units		2071 effects				
Residual	SCA_V	2071	1.00000	13.0502	30.94	0 P
cloneid	NRM	153				
nrm(cloneid).location		459 effects				
location	ID_V	1	0.101193E-06	0.132059E-05	0.00	0 B

- The variance associated with nrm(cloneid) is the additive genetic variance. The term nrm(cloneid).location refers to additive genetic – by – location interaction variance and has an estimated variance component of zero, so it could be dropped from the model.

PART 2

```
  7 LogL=-3824.52      S2=  13.047       2068 df

          - - - Results from analysis of height8 - - -
  Akaike Information Criterion      7661.03 (assuming 6 parameters).
  Bayesian Information Criterion    7694.84
```

Model_Term			Gamma	Sigma	Sigma/SE	% C
location.block	IDV_V	24	0.108445	1.41491	2.67	0 P
nrm(cloneid)	NRM_V	153	0.341199	4.45173	1.53	0 P
ide(cloneid)	IDV_V	153	0.528398E-01	0.689417	0.44	0 P
ide(cloneid).location	IDV_V	459	0.101193E-06	0.132030E-05	0.00	0 B
units		2071 effects				
Residual	SCA_V	2071	1.00000	13.0473	30.94	0 P
cloneid	NRM	153				
nrm(cloneid).location		459 effects				
location	ID_V	1	0.101193E-06	0.132030E-05	0.00	0 B

- The ide(cloneid) term explains very little variation (0.689), and the likelihood of this model is nearly identical to the previous model, so there is no evidence for significant non-additive genetic variance in this case.

Despite the lack of evidence for non-additive variance, for completeness we demonstrate here how to estimate heritability for individual tree observations from model 2 as:

$$h^2 = \frac{\sigma_A^2}{\sigma_A^2 + \sigma_{Gr}^2 + \sigma_{AL}^2 + \sigma_{LGr}^2 + \sigma_\varepsilon^2} \tag{5.31}$$

and the heritability for mean cloneid values averaged across replicated clone trees as:

$$h^2_{cm} = \frac{\sigma^2_A}{\sigma^2_A + \sigma^2_{Gr} + \frac{\sigma^2_{AL}}{l} + \frac{\sigma^2_{LGr}}{l} + \frac{\sigma^2_\varepsilon}{n}}$$

(5.32)

Where l is the number of locations, and n is the harmonic mean of the number of trees measured per *cloneid*.

The syntax in ASReml is given in the VPREDICT !DEFINE statement following model 2, note that we did not include the genotype – by – location variances in the phenotypic variance estimates, since they were zero in this example:

```
!PART 2    # add ide term to catch non-additive genetic variance
height8  ~ mu location ,
         !r  location.block nrm(cloneid) nrm(cloneid).location ,
             ide(cloneid) ide(cloneid).location
             residual units
VPREDICT !DEFINE
F Va       nrm(cloneid)    # Additive variance
F Vna      ide(cloneid)    # Non-additive variance
P Verr     units;Residual   # Residual variance
F Vp       Va + Vna + Verr  # Phenotypic variance for individual trees
F Vp_m     Va + Vna + Verr*0.082   # Phenotypic variance for clone
means, harmonic mean of trees per cloneid = 12.2
H h2       Va Vp # Heritability of individual tree values
H h2_cm    Va Vp_m # Heritability of clone means
```

In the estimate of phenotypic variance for clone means, the residual term is multiplied by the reciprocal of the harmonic mean of the number of trees measured per *cloneid* (1/12.2 = 0.082).

The output of the VPREDICT directive is in *Code 5-5_clones.pvc*:

```
           - - - Results from analysis of height8 - - -
 1  location.block        V    24       1.41491        0.52993
 2  nrm(cloneid)          V   153       4.45173        2.90963
 3  ide(cloneid)          V   153       0.689417       1.56686
 4  ide(cloneid).locatio  V   459       0.132030E-05   0.00000
 units                    2071 effects
 5  units;Residual               V 2071   13.0473      0.421697
 cloneid                  NRM    153
 nrm(cloneid).location          459 effects
 6  location                     V    1   0.132030E-05  0.0000
 7  Va   2          4.4517    2.9031
 8  Vna  3          0.68942   1.5512
 9  Verr 5          13.047    0.42170
10  Vp   7          18.188    1.5626
11  Vp_m 7          6.2110    1.5136
    h2           = Va  2      7/Vp  7    10=  0.2448 0.1406
    h2_cm        = Va  2      7/Vp_m 7   11=  0.7167 0.3074
```

- The heritability of individual tree measurements is only 0.24, but the heritability of clone mean values (averaged over clonal copies in different replications and environments) is much higher (0.72).

A portion of solution from *Code 5-5_clones.sln* file is shown here:

```
Model_Term              Level       Effect      seEffect
...
  nrm(cloneid)          1           1.495       1.666
  nrm(cloneid)          2          -0.2138      2.012
  nrm(cloneid)          3          -0.1706      1.945
...
  nrm(cloneid)          18          0.3146E-01  1.269
  nrm(cloneid)          19         -2.102       1.235
  nrm(cloneid)          20         -0.3655      1.147
...
  ide(cloneid)          1           0.000       0.8304
  ide(cloneid)          2           0.000       0.8304
  ide(cloneid)          3           0.000       0.8304
...
  ide(cloneid)          18         -0.1362      0.7665
  ide(cloneid)          19         -0.7971      0.7614
  ide(cloneid)          20         -0.2592      0.7492
```

- Solutions for *cloneid* levels 1–17 are for parents. The nrm(cloneid) solutions are the additive genetic effects, and these can be estimated for parents as well as progeny because of their known pedigree relationships.
- The solutions for *cloneid* levels 18 and higher are for the progenies that were directly measured in the experiment.
- The solutions for ide(cloneid) are the residual non-additive genetic effects. Notice that solutions for ide (cloneid) for the parents (levels 1, 2 and 3) are zero because the ide() effects are modeled as independent and we have no direct measurements on the parents.
- The ide(cloneid) values can be added to the nrm(cloneid) values for the progenies in order to estimate total genotypic value, which can be useful for selecting progenies to distribute clonally as new varieties for production.

For selecting progenies with best predicted breeding values to intermate to produce a new generation, however, the nrm (cloneid) values would be appropriate for making selections.

Electronic supplementary material: The online version of this chapter (doi:10.1007/978-3-319-55177-7_6) contains supplementary material, which is available to authorized users.

Abstract

Multivariate models are commonly used to estimate phenotypic, genetic and environmental variances, covariances, and correlations for multiple traits in plant and animal breeding programs. When traits are correlated, breeding value predictions from a multivariate model can be more accurate than univariate models. In this chapter we introduce multivariate models for two data sets: a maize inbred line multi-environment trial and pig data with pedigree information appropriate for an animal model.

Introduction

In breeding applications, multivariate models offer several advantages over their singe trait counterparts. For example, empirical breeding values (EBVs) estimated from multivariate models for low heritability traits (such as fitness and fertility) receive an increase in accuracy by exploiting their covariances with higher heritability traits (such as size and morphology characteristics). Even highly heritable traits can benefit from multivariate analyses, although the advantage might be limited. Genetic correlations can also be used to predict the effectiveness of indirect selections on one trait with the primary goal of improving a different trait. For example, if height is an easily observed trait that is highly correlated with yield, and it takes more time and effort to improve yield, then it may be possible to select on height to more efficiently make positive changes in yield in the population. Furthermore, multivariate models are able to accommodate the implicit structure of the selection process by which individuals may be culled sequentially based on a series of traits over their lifetime, so that data are not missing at random but are missing more frequently for later-recorded traits. In this and in other cases where individuals may have missing data for different traits, multivariate analysis is able to better account for the missing data. In contrast, the ordinary least squares (multivariate ANOVA) approach requires dropping records that have missing values for any one of the several traits being analyzed jointly. The more efficient use of missing or unbalanced data by mixed models analysis can have a substantial impact in cases where not all traits are measured on all experimental units, for example where one of the traits was not measured at one of the environments in a multi-environment plant breeding trial. Analysis of repeated measures of a single trait at multiple time points is also a specific type of multivariate analysis.

Quantitative genetics researchers are often interested in partitioning observed phenotypic variance into causal (genetics and environment) components, and multivariate analyses allow this type of approach to be extended to partitioning observed covariances and correlations into the underlying environmental and genetic covariances and correlations. In population and evolutionary genetics, genetic correlation estimates are useful to understand how different traits are genetically related to fitness and how natural selection for higher fitness affects or constrained by other traits.

While multivariate analyses offer several advantages, they come at a cost of increased computational demand. The memory requirements for REML analysis of multivariate models increase rather quickly especially when several random effects are included in the model and the number of parameters to be estimated increases. Also, convergence to globally optimal parameter values becomes more difficult with multivariate models, and the algorithm may converge to a local optimum solution that is not the best. Therefore the choice of starting values for variances and covariances in multivariate models becomes more important. As a rule of thumb, it is advisable to first perform initial single trait analysis models and then use the variance components estimates from the univariate models as starting values for subsequent multivariate models. The univariate estimates of variance components also serve as a useful check that the multivariate model produces reasonable parameter estimates. For some complicated models, it may help to run analyses with different starting values and evaluate the concordance of different runs.

Some Theory

Phenotypic values of different traits measured on the same subjects are often correlated. Both environmental factors and genetic effects contribute to observed correlations among phenotypic values for different traits. Environmental correlations arise when a common environmental variation affects multiple traits. For example, temperature can affect both the time to flowering and height of plants. Genetic correlations can arise by either of two causes: pleiotropy and linkage disequilibrium. Pleiotropy occurs when one or more genes influence more than one trait. Linkage disequilibrium can result in genotypic correlations even when loci affecting a pair of traits are distinct, because the alleles at different loci (affecting different traits) are not independent. This results in their effects being correlated, and the net effect of these trait correlations across all pairs of loci can generate a genetic correlation. In this book, we will not distinguish between the causes of genetic correlations, we will instead focus on estimating the value of the correlation coefficients.

Similar to how we partitioned the phenotypic variance for one trait in variance components, we can decompose the phenotypic covariance for two or more traits into covariance components associated with different genetic and non-genetic factors in the experiment. Then we can estimate the phenotypic correlation $r_{P(x, y)}$ between two traits x and y as the ratio of the phenotypic covariance to the square root of the product of the phenotypic variances for the two traits:

$$r_{P(x,y)} = \frac{\sigma_{G(x,y)} + \sigma_{\varepsilon(x,y)}}{\sqrt{\sigma_{Gx}^2 + \sigma_{\varepsilon x}^2}\sqrt{\sigma_{Gy}^2 + \sigma_{\varepsilon y}^2}} \tag{6.1}$$

where $\sigma_{G(x, y)}$ is genotypic covariance, $\sigma_{\varepsilon(x, y)}$ is residual environmental covariance, and $\sigma_{Gx}^2, \sigma_{\varepsilon x}^2, \sigma_{Gy}^2$ and $\sigma_{\varepsilon y}^2$ are genotypic or environmental variances for traits x and y, respectively.

The genotypic covariance is the sum of covariances between additive, dominance, and epistatic effects. Recall that genetic variances for traits x and y are due to additive, dominance, and epistatic interactions as follows:

$$\sigma_{Gx}^2 = \sigma_{Ax}^2 + \sigma_{Dx}^2 + \sigma_{AAx}^2 + \sigma_{ADx}^2 + \sigma_{DDx}^2 + \dots \tag{6.2}$$

$$\sigma_{Gy}^2 = \sigma_{Ay}^2 + \sigma_{Dy}^2 + \sigma_{AAy}^2 + \sigma_{ADy}^2 + \sigma_{DDy}^2 + \dots \tag{6.3}$$

In the same way, the genotypic covariance is the sum of covariances for each of these component effects across traits:

$$\sigma_{G(x,y)} = \sigma_{A(x,y)} + \sigma_{D(x,y)} + \sigma_{AA(x,y)} + \sigma_{AD(x,y)} + \sigma_{DD(x,y)} + \dots \tag{6.4}$$

where $\sigma_{A(x, y)}$ is the additive genetic covariance between traits x and y, and the other terms represent dominance and epistatic covariances between the traits. If additive and dominance genetic variances can be estimated for each trait using the experimental designs and analyses outlined in Chaps. 4 and 5, then additive genetic ($r_{A(x, y)}$), dominance genetic ($r_{D(x, y)}$) and residual environmental ($r_{\varepsilon(x, y)}$) correlations can also be estimated as ratios of covariance components to variance components:

$$r_{A(x,y)} = \frac{\sigma_{A(x,y)}}{\sqrt{\sigma_{Ax}^2}\sqrt{\sigma_{Ay}^2}} \tag{6.5}$$

$$r_{D(x,y)} = \frac{\sigma_{D(x,y)}}{\sqrt{\sigma_{Dx}^2}\sqrt{\sigma_{Dy}^2}} \tag{6.6}$$

$$r_{E(x,y)} = \frac{\sigma_{E(x,y)}}{\sqrt{\sigma_{Ex}^2}\sqrt{\sigma_{Ey}^2}} \tag{6.7}$$

In some cases, such as when we lack pedigree information and do not have an adequate mating design, experimental data do not allow us to partition the total genotypic variance into additive and other component variances. In these cases, the genotypic covariance and correlation also cannot be partitioned into any component pieces.

The Linear Mixed Model for Multivariate Models

As a toy example, suppose that we have measured height and yield on five individuals from two half-sib families. We would like to fit a bivariate model to height and yield to obtain variances, covariances and correlations of two traits. The data are given below:

Individual	Height	Yield
1	87	0.52
2	84	0.48
3	75	0.45
4	90	0.69
5	79	0.74

We will use the fact that individuals 1 and 2 are half-sibs and individuals 3, 4, and 5 are half-sibs later to model the variance-covariance structure of the genetic effects of individuals. The general form of the multivariate model is:

$$Y_{ij} = \mu_j + A_{ij} + \varepsilon_{ij} \tag{6.8}$$

where:

μ_j is the intercept for trait j, each trait has a separate intercept. We will use $j = 1$ and 2 as subscripts for the two traits. So, we switch from referring to traits as X and Y as Y_1 and Y_2.

A_{ij} is the effect of the additive genetic value trait j measured on individual i,

ε_{ij} is the residual deviation for on individual i.

Multivariate data and linear models can be organized in two distinct ways: the dependent observations can be organized as a matrix of data, where each column represents a different trait, or as a single vector. These forms are mathematically equivalent, but the matrix form may be easier to understand for first-time readers, whereas the vector form is required for some software (Holland 2006).

We first present the model for multivariate data organized as a matrix:

$$\mathbf{Y} = \mathbf{Xb} + \mathbf{Zu} + \mathbf{e}$$

$$\mathbf{Y}_{n \times d} = \mathbf{X}_{n \times (p+1)} \mathbf{b}_{(p+1) \times d} + \mathbf{Z}_{n \times r}\, \mathbf{u}_{r \times d} + \mathbf{e}_{n \times d} \tag{6.9}$$

Here, n is the number of rows (individuals or experimental units, in this example, five); and d is the number of dependent variables (in this example, two). The design matrix $\mathbf{X}$ has dimensions $n \times (p + 1)$, where p is the number of fixed predictors for one trait and an additional column is added for the intercept. $\mathbf{b}$ is the matrix of coefficients of fixed predictor effects to be estimated with dimensions $(p + 1) \times d$. The rows of $\mathbf{b}$ correspond to predictor variables and the columns are response variables. The design matrix $\mathbf{Z}$ has dimensions $n \times r$, where r is the number of random effects per trait, and $\mathbf{u}$ is an $r \times d$ matrix of random effects. The full set of matrices for this form is:

$$
\mathbf{Y} = \begin{bmatrix} Y_{11} & Y_{12} \\ Y_{21} & Y_{22} \\ Y_{31} & Y_{32} \\ Y_{41} & Y_{42} \\ Y_{51} & Y_{52} \end{bmatrix} = \begin{bmatrix} 87 & 0.52 \\ 84 & 0.48 \\ 75 & 0.45 \\ 90 & 0.69 \\ 79 & 0.74 \end{bmatrix} \tag{6.10}
$$

$$
= \begin{bmatrix} 1 \\ 1 \\ 1 \\ 1 \\ 1 \end{bmatrix} \begin{bmatrix} \mu_1 & \mu_2 \end{bmatrix} + \begin{bmatrix} 1 & 0 & 0 & 0 & 0 \\ 0 & 1 & 0 & 0 & 0 \\ 0 & 0 & 1 & 0 & 0 \\ 0 & 0 & 0 & 1 & 0 \\ 0 & 0 & 0 & 0 & 1 \end{bmatrix} \begin{bmatrix} A_{11} & A_{12} \\ A_{21} & A_{22} \\ A_{31} & A_{32} \\ A_{41} & A_{42} \\ A_{51} & A_{52} \end{bmatrix} + \begin{bmatrix} \varepsilon_{11} & \varepsilon_{12} \\ \varepsilon_{21} & \varepsilon_{22} \\ \varepsilon_{31} & \varepsilon_{32} \\ \varepsilon_{41} & \varepsilon_{42} \\ \varepsilon_{51} & \varepsilon_{52} \end{bmatrix}
$$

$$
\mathbf{X} = \begin{bmatrix} 1 \\ 1 \\ 1 \\ 1 \\ 1 \end{bmatrix} \quad \mathbf{b} = \begin{bmatrix} \mu_1 & \mu_2 \end{bmatrix} \quad \mathbf{Z} = \begin{bmatrix} 1 & 0 & 0 & 0 & 0 \\ 0 & 1 & 0 & 0 & 0 \\ 0 & 0 & 1 & 0 & 0 \\ 0 & 0 & 0 & 1 & 0 \\ 0 & 0 & 0 & 0 & 1 \end{bmatrix} \quad \mathbf{u} = \begin{bmatrix} A_{11} & A_{12} \\ A_{21} & A_{22} \\ A_{31} & A_{32} \\ A_{41} & A_{42} \\ A_{51} & A_{52} \end{bmatrix} \quad \mathbf{e} = \begin{bmatrix} \varepsilon_{11} & \varepsilon_{12} \\ \varepsilon_{21} & \varepsilon_{22} \\ \varepsilon_{31} & \varepsilon_{32} \\ \varepsilon_{41} & \varepsilon_{42} \\ \varepsilon_{51} & \varepsilon_{52} \end{bmatrix} \tag{6.11}
$$

In this very simple example, the $\mathbf{Z}$ matrix is an identity matrix, but this will not generally be true.

The alternate formulation that can make computations easier is to organize the observations in a single column vector of length nd:

$$\mathbf{Y}_{nd\times 1} = \mathbf{X}_{nd\times(p+1)d}\,\mathbf{b}_{(p+1)\,d\times 1} + \mathbf{Z}_{nd\times rd}\,\mathbf{u}_{rd\times 1} + \mathbf{e}_{nd\times 1} \tag{6.12}$$

$$\mathbf{Y} = \begin{bmatrix} Y_{11} \\ Y_{12} \\ Y_{21} \\ Y_{22} \\ Y_{31} \\ Y_{32} \\ Y_{41} \\ Y_{42} \\ Y_{51} \\ Y_{52} \end{bmatrix} = \begin{bmatrix} 87 \\ 0.52 \\ 84 \\ 0.48 \\ 75 \\ 0.45 \\ 90 \\ 0.69 \\ 79 \\ 0.74 \end{bmatrix} = \begin{bmatrix} 1 & 0 \\ 0 & 1 \\ 1 & 0 \\ 0 & 1 \\ 1 & 0 \\ 0 & 1 \\ 1 & 0 \\ 0 & 1 \\ 1 & 0 \\ 0 & 1 \end{bmatrix} \begin{bmatrix} \mu_1 \\ \mu_2 \end{bmatrix} + \mathbf{I_{10}} \begin{bmatrix} A_{11} \\ A_{12} \\ A_{21} \\ A_{22} \\ A_{31} \\ A_{32} \\ A_{41} \\ A_{42} \\ A_{51} \\ A_{52} \end{bmatrix} + \begin{bmatrix} e_{11} \\ e_{12} \\ e_{21} \\ e_{22} \\ e_{31} \\ e_{32} \\ e_{41} \\ e_{42} \\ e_{51} \\ e_{52} \end{bmatrix} \tag{6.13}$$

$$\mathbf{X} = \begin{bmatrix} 1 & 0 \\ 0 & 1 \\ 1 & 0 \\ 0 & 1 \\ 1 & 0 \\ 0 & 1 \\ 1 & 0 \\ 0 & 1 \\ 1 & 0 \\ 0 & 1 \end{bmatrix} \quad \mathbf{b} = \begin{bmatrix} \mu_1 \\ \mu_2 \end{bmatrix} \quad \mathbf{Z} = \mathbf{I}_{10} \quad \mathbf{u} = \begin{bmatrix} A_{11} \\ A_{12} \\ A_{21} \\ A_{22} \\ A_{31} \\ A_{32} \\ A_{41} \\ A_{42} \\ A_{51} \\ A_{52} \end{bmatrix} \quad \mathbf{e} = \begin{bmatrix} e_{11} \\ e_{12} \\ e_{21} \\ e_{22} \\ e_{31} \\ e_{32} \\ e_{41} \\ e_{42} \\ e_{51} \\ e_{52} \end{bmatrix} \tag{6.14}$$

In its most compact representation, the mixed model for the multivariate model appears the same as before: $\mathbf{Y} = \mathbf{Xb} + \mathbf{Zu} + \mathbf{e}$, and this is the same form we used for univariate analysis. However, the design matrices are different, and we can make the relationship between the structure of the multivariate and univariate mixed models clearer by rewriting the equation in a way that uses the $\mathbf{X}$ and $\mathbf{Z}$ matrices of the univariate form:

$$\mathbf{Y}_{nd} = \left(\mathbf{I}_d \otimes \mathbf{X}_{n\times(p+1)}\right)\mathbf{b}_{(p+1)\,d\times 1} + \left(\mathbf{I}_d \otimes \mathbf{Z}_{n\times r}\right)\mathbf{u}_{rd\times 1} + \mathbf{e}_{nd} \tag{6.15}$$

Again, $\mathbf{Y}$ is the vector of traits with nd rows (n individuals times d traits measured per individual), $\mathbf{I}_d$ is the identity matrix with dimension d, and $\mathbf{X}$ and $\mathbf{Z}$ are identical to the univariate model. The direct products of $\mathbf{I}_d$ with the univariate $\mathbf{X}$ and $\mathbf{Z}$ matrices generate the forms needed for the multivariate equations above:

$$\mathbf{I}_d \otimes \mathbf{X} = \begin{bmatrix} 1 & 0 \\ 0 & 1 \end{bmatrix} \otimes \begin{bmatrix} 1 \\ 1 \\ 1 \\ 1 \\ 1 \end{bmatrix} = \begin{bmatrix} 1 & 0 \\ 0 & 1 \\ 1 & 0 \\ 0 & 1 \\ 1 & 0 \\ 0 & 1 \\ 1 & 0 \\ 0 & 1 \\ 1 & 0 \\ 0 & 1 \end{bmatrix} \tag{6.16}$$

$$\mathbf{I}_d \otimes Z = \begin{bmatrix} 1 & 0 \\ 0 & 1 \end{bmatrix} \begin{bmatrix} 1 & 0 & 0 & 0 & 0 \\ 0 & 1 & 0 & 0 & 0 \\ 0 & 0 & 1 & 0 & 0 \\ 0 & 0 & 0 & 1 & 0 \\ 0 & 0 & 0 & 0 & 1 \end{bmatrix} = \mathbf{I}_{10} \tag{6.17}$$

We can solve Henderson's mixed model equations to obtain predictions for two traits, and the mixed model equations are the same as for univariate analysis (Chap. 2):

$$\begin{pmatrix} \mathbf{X'R^{-1}X} & \mathbf{X'R^{-1}Z} \\ \mathbf{Z'R^{-1}X} & \mathbf{Z'R^{-1}Z+G^{-1}} \end{pmatrix} \begin{pmatrix} \widehat{\mathbf{b}} \\ \widehat{\mathbf{u}} \end{pmatrix} = \begin{pmatrix} \mathbf{X'R^{-1}y} \\ \mathbf{Z'R^{-1}y} \end{pmatrix} \tag{6.18}$$

To use these equations to solve the multivariate models, we need to understand the form of the $\mathbf{R}$ and $\mathbf{G}$ matrices. In this example:

$$\mathbf{G} = \mathbf{A} \otimes \begin{bmatrix} \sigma_{A1}^2 & \sigma_{A12} \\ \sigma_{A12} & \sigma_{A2}^2 \end{bmatrix} \tag{6.19}$$

$$\mathbf{R} = \mathbf{I} \otimes \begin{bmatrix} \sigma_{\varepsilon1}^2 & \sigma_{\varepsilon12} \\ \sigma_{\varepsilon12} & \sigma_{\varepsilon2}^2 \end{bmatrix} \tag{6.20}$$

In each of these variance-covariance matrices for model effects, variances for the two traits are on the diagonal and the covariances between the traits are the off-diagonal elements. In our example, the first two individuals and the last three individuals represent half-sib families, so the $\mathbf{G}$ matrix is:

$$\mathbf{G} = \begin{bmatrix} 1 & 0.25 & 0 & 0 & 0 \\ 0.25 & 1 & 0 & 0 & 0 \\ 0 & 0 & 1 & 0.25 & 0.25 \\ 0 & 0 & 0.25 & 1 & 0.25 \\ 0 & 0 & 0.25 & 0.25 & 1 \end{bmatrix} \otimes \begin{bmatrix} \sigma_{A1}^2 & \sigma_{A12} \\ \sigma_{A12} & \sigma_{A2}^2 \end{bmatrix} =$$

$$\begin{bmatrix} \sigma_{A1}^2 & \sigma_{A12} & 0.25\sigma_{A1}^2 & 0.25\sigma_{A12} & 0 & 0 & 0 & 0 & 0 & 0 \\ \sigma_{A12} & \sigma_{A2}^2 & 0.25\sigma_{A12} & 0.25\sigma_{A2}^2 & 0 & 0 & 0 & 0 & 0 & 0 \\ 0.25\sigma_{A1}^2 & 0.25\sigma_{A12} & \sigma_{A1}^2 & \sigma_{A12} & 0 & 0 & 0 & 0 & 0 & 0 \\ 0.25\sigma_{A12} & 0.25\sigma_{A2}^2 & \sigma_{A12} & \sigma_{A2}^2 & 0 & 0 & 0 & 0 & 0 & 0 \\ 0 & 0 & 0 & 0 & \sigma_{A1}^2 & \sigma_{A12} & 0.25\sigma_{A1}^2 & 0.25\sigma_{A12} & 0.25\sigma_{A1}^2 & 0.25\sigma_{A12} \\ 0 & 0 & 0 & 0 & \sigma_{A12} & \sigma_{A2}^2 & 0.25\sigma_{A12} & 0.25\sigma_{A2}^2 & 0.25\sigma_{A12} & 0.25\sigma_{A2}^2 \\ 0 & 0 & 0 & 0 & 0.25\sigma_{A1}^2 & 0.25\sigma_{A12} & \sigma_{A1}^2 & \sigma_{A12} & 0.25\sigma_{A1}^2 & 0.25\sigma_{A12} \\ 0 & 0 & 0 & 0 & 0.25\sigma_{A12} & 0.25\sigma_{A2}^2 & \sigma_{A12} & \sigma_{A2}^2 & 0.25\sigma_{A12} & 0.25\sigma_{A2}^2 \\ 0 & 0 & 0 & 0 & 0.25\sigma_{A1}^2 & 0.25\sigma_{A12} & 0.25\sigma_{A1}^2 & 0.25\sigma_{A12} & \sigma_{A1}^2 & \sigma_{A12} \\ 0 & 0 & 0 & 0 & 0.25\sigma_{A12} & 0.25\sigma_{A2}^2 & 0.25\sigma_{A12} & 0.25\sigma_{A2}^2 & \sigma_{A12} & \sigma_{A2}^2 \end{bmatrix}$$

The additive effects for the two traits on a single individual have a covariance equal to the additive covariance. The additive effects for trait 1 on individual 1 and trait 2 on its half-sib have a covariance of 0.25 times the additive covariance. Measurements on unrelated individuals have zero covariance.

The residual effects for the two trait measurements on a single individual have a covariance equal to the residual trait covariance, but residual effects across individuals have zero covariance in the typical model where residuals for one trait are independent across individuals:

$$\mathbf{R} = \mathbf{I}_5 \otimes \begin{bmatrix} \sigma_{\varepsilon 1}^2 & \sigma_{\varepsilon 12} \\ \sigma_{\varepsilon 12} & \sigma_{\varepsilon 2}^2 \end{bmatrix} = \begin{bmatrix} \sigma_{\varepsilon 1}^2 & \sigma_{\varepsilon 12} & 0 & 0 & 0 & 0 & 0 & 0 & 0 & 0 \\ \sigma_{\varepsilon 12} & \sigma_{\varepsilon 2}^2 & 0 & 0 & 0 & 0 & 0 & 0 & 0 & 0 \\ 0 & 0 & \sigma_{\varepsilon 1}^2 & \sigma_{\varepsilon 12} & 0 & 0 & 0 & 0 & 0 & 0 \\ 0 & 0 & \sigma_{\varepsilon 12} & \sigma_{\varepsilon 2}^2 & 0 & 0 & 0 & 0 & 0 & 0 \\ 0 & 0 & 0 & 0 & \sigma_{\varepsilon 1}^2 & \sigma_{\varepsilon 12} & 0 & 0 & 0 & 0 \\ 0 & 0 & 0 & 0 & \sigma_{\varepsilon 12} & \sigma_{\varepsilon 2}^2 & 0 & 0 & 0 & 0 \\ 0 & 0 & 0 & 0 & 0 & 0 & \sigma_{\varepsilon 1}^2 & \sigma_{\varepsilon 12} & 0 & 0 \\ 0 & 0 & 0 & 0 & 0 & 0 & \sigma_{\varepsilon 12} & \sigma_{\varepsilon 2}^2 & 0 & 0 \\ 0 & 0 & 0 & 0 & 0 & 0 & 0 & 0 & \sigma_{\varepsilon 1}^2 & \sigma_{\varepsilon 12} \\ 0 & 0 & 0 & 0 & 0 & 0 & 0 & 0 & \sigma_{\varepsilon 12} & \sigma_{\varepsilon 2}^2 \end{bmatrix}$$

This form of the mixed model requires one row (equation) for each combination of individual and trait. If an individual is missing an observation for one trait, we would still represent that missing trait as a row in the mixed model equations so that we can maintain the structure of the univariate $\mathbf{X}$ and $\mathbf{Z}$ matrices, although the $\mathbf{Y}$ observation for that row would be missing.

Maize RILs Multivariate Model

We will use the balanced MaizeRILs.csv data presented in Chap. 2 as an example of a multivariate analysis. A set of 62 recombinant inbred lines (RILs) was tested using a incomplete block design with two complete replications at each of four locations. Each line had 20 plants in a plot. There are three traits for which we are interested in estimating variances and genetic correlations: days to pollen shed (pollen), days to silking (silking) and mean height of the five plants in each plot (height). A small subset of the data are given below:

location	rep	block	plot	RIL	pollen	silking	ASI	height
ARC	1	1	1	RIL-53	74	77	3	184.8
ARC	1	1	2	RIL-40	75	75	0	225.2
ARC	1	1	4	RIL-41	74	74	0	174.4
ARC	1	1	5	RIL-28	69	71	2	147.6
ARC	1	1	6	RIL-11	69	71	2	181.6

The linear model for this experiment is:

$$y_{ijk} = \mu + L_i + B(L)_{ij} + G_k + GL_{ik} + \varepsilon_{ijk} \tag{6.21}$$

Where μ = overall mean, L_i = effect of location i, $R(L)_{ij}$ = effect of the replication (complete block) j nested within location i, G_k = effect of genotype k (RIL effect), GL_{ik} = effect of interaction between genotype k and location i, ε_{ijk} = residual (experimental error) effect of the plot containing genotype k in complete block j of location i. We will assume that location is a fixed effect. Replication, RIL and location $\times$ RIL interaction effects are random. This model is an example where we can estimate the genotypic variances, covariances, and correlations, but it is not possible to partition the genotypic covariance into its additive and non-additive component pieces.

Multivariate models may have trouble converging; good starting values can save a lot of trouble. We recommend analyzing traits independently using univariate models first to obtain estimates of variance components for each. The variance estimates from univariate models can be used as starting values for multivariate analysis.

In the following code the same univariate model is fit to height, pollen and silking separately to estimate variance components for each.

Code example 6.1
Univariate analysis of maize RIL data (see code *Code 6-1_MultivariateMaize.as* for details).

```
!ARGS 1 !RENAME 1
Title: Maize RILs multivariate analysis
  location  !A   rep *   block *   plot !I     RIL  !A
  pollen    !M0
  silking   !M0
  ASI
  height    !M0

MaizeRILs.csv !SKIP 1   !DOPATH   $A

!PART 1  # univariate analysis
!CYCLE  height  pollen  silking
$I ~ mu location ,
      !r  rep.location RIL RIL.location
     residual idv(units)
```

As a reminder, we use the !ARGS qualifier (on top line) and the !CYCLE qualifier (immediately after the PATH 1 declaration) to control which parts of the program are executed (see Chap. 1 for more details):

- !ARGS 1 !RENAME 1: indicates that the string variable $A should be replaced by "1" when it appears later in the code and that the output files will include this string in their name.
- !DOPATH $A: indicates that the string listed after !ARGS on the top line will be substituted here, resulting in the qualifier '!DOPATH 1' being executed, directing the program to execute the code corresponding to '!PATH 1'.
- !CYCLE: qualifier replaces $I in the model with variables listed. So the same model is run for all three traits, one at a time, and the outputs are combined in a single output file.
- The variance structure for all random terms is the default IID structure. For example, the RIL effects are assumed to have the distribution $\sim \mathrm{N}\left(0, \sigma_{RIL}^2 \mathbf{I}_{62}\right)$. Since there are multiple random factors (replication, RIL, and RIL-by-location interaction), the variance matrix for all of the random effects is block-diagonal. $\mathbf{G} = \oplus_{j=1}^{L} \sigma_{uj}^2 = \sigma_{rep}^2 \mathbf{I}_8 \oplus \sigma_{RIL}^2 \mathbf{I}_{62} \oplus \sigma_{loc.RIL}^2 \mathbf{I}_{248}$.
- The residual errors have the IID variance structure of $\mathbf{R} = \sigma_e^2 \mathbf{I}_{474}$. This assumes that error variances are the same at all locations; we show in Chap. 8 how to relax this assumption and permit unique variances for each location.

A subset of the primary output file *Code 6-1_MultivariateMaize1.asr* is given below:

```
QUALIFIER: !DOPART    1 is active


... Cycle 1 value is height...
Univariate analysis of height

Model_Term                       Gamma      Sigma     Sigma/SE   % C
idv(location.rep)      IDV_V   8   0.207229   13.4463      1.31   0 P
idv(RIL)               IDV_V  62   4.67204   303.151       5.27   0 P
idv(location.RIL)      IDV_V 248   0.384246   24.9323      3.73   0 P
idv(units)                   496 effects
Residual               SCA_V 496   1.00000    64.8862     11.05   0 P


... Cycle 2 value is pollen ...
Univariate analysis of pollen
```

```
Model_Term                        Gamma      Sigma    Sigma/SE   % C
idv(location.rep)      IDV_V   8  0.381313   0.449696     1.36   0 P
idv(RIL)               IDV_V  62  4.14640    4.88999      5.21   0 P
idv(location.RIL)      IDV_V 248  0.486081   0.573253     4.32   0 P
idv(units)                   496  effects
Residual               SCA_V 496  1.00000    1.17934     11.05   0 P


...Cycle 3 value is silking...
Univariate analysis of silking

Model_Term                        Gamma      Sigma    Sigma/SE   % C
idv(location.rep)      IDV_V   8  0.229566   0.662877     1.32   0 P
idv(RIL)               IDV_V  62  1.72876    4.99185      4.95   0 P
idv(location.RIL)      IDV_V 248  0.289872   0.837013     3.08   0 P
idv(units)                   496  effects
Residual               SCA_V 496  1.00000    2.88753     11.05   0 P
```

Estimated variance components for all of the model terms for each trait are in bold font in the output above. Now that we have good estimates of the variance components, we are ready to move to a multivariate model, building the model in steps of increasing complexity. We start with a relatively simple multivariate model, in which we analyze all three traits simultaneously, but we use a diagonal variance-covariance structure for the random model effects, such that these effects have zero covariance across traits. In other words, the effects of a particular line on the three traits are independent. In PART 2 of the code we fit a diagonal variance `diag().id()` structure to all random terms and un-structured `id().us()` to residuals:

Code example 6.2
Multivariate analysis of maize RIL data, correlated residuals only (file *Code 6-1_MultivariateMaize.as* continued).

```
!ARGS 2 !RENAME 1
Title: Maize RILs multivariate analysis
...

!PART 2 # DIAGONAL structures, no starting values
height  pollen  silking ~ Trait Trait.location ,  # fixed effects
    !r  diag(Trait).id(rep.location),
        diag(Trait).id(RIL)  ,
        diag(Trait).id(RIL.location)
     residual    id(units).us(Trait)
```

The multivariate model requires some specific modifications:

- *Trait* is the multivariate version of intercept (mu) in the univariate model. It creates a vector holding the intercepts for each trait included in the analysis. '*Trait*' is a reserved term in ASReml and should not be used for any other purpose but for multivariate analysis.
- All fixed effects and random effects should involve an interaction with *Trait* to create appropriate design matrices for the effects. For example, it does not make sense to fit a common effect of a location on all three traits, whereas fitting the term *Trait.location* more sensibly estimates a separate effect for each location on each trait.

Variance structures (**G** matrix) for random effects in the multivariate model require more detailed specifications:

- The **G** variance structure has one sub-matrix for each random term: *Trait.RIL, Trait.location.RIL* and *Trait.location.rep*.
- `diag(Trait).id(RIL)` specifies that the variance structure for the random RIL effect is the product of two matrices; the block diagonal matrix **Σ** (`diag(Trait)`) and the identity matrix **I** (`id(RIL)`):

$$\mathbf{G}_{\text{Trait.RIL}} = \Sigma_{Trait.RIL} \otimes \mathbf{I}_{62} = \begin{bmatrix} \sigma^2_{RILh} & 0 & 0 \\ 0 & \sigma^2_{RILp} & 0 \\ 0 & 0 & \sigma^2_{RILs} \end{bmatrix} \otimes \mathbf{I}_{62} = \mathbf{G}_{RILh} \oplus \mathbf{G}_{RILp} \oplus \mathbf{G}_{RILs}$$

- Notice that the sub matrix $\Sigma_{\text{Trait.RIL}}$ has zero covariances in the off-diagonals because it is a block diagonal matrix. This models RIL effects across traits as uncorrelated.
- $\mathbf{G}_{RILh}$, $\mathbf{G}_{RILp}$, $\mathbf{G}_{RILs}$ are identical (IDV) sub-matrices for height, pollen and silking, respectively. Each sub matrix has the same 62×62 dimensions representing the variances of effects of the 62 RILs on each trait.
- The final dimension of $\mathbf{G}_{\text{Trait.RIL}}$ is 3 traits $\times$ 62 RILs = 186.
- *Trait.location.RIL* is a diagonal matrix with 744 dimensions (3 traits $\times$ 4 locations $\times$ 62 RILs).
- *Trait.location.rep* is a diagonal matrix with 24×24 dimensions (3 traits $\times$ 4 locations $\times$ 2 reps per location).

Residual variance structure (**R** matrix):

- `id(units).us(Trait)`: The **R** matrix is a direct product of the identity matrix **I** and an unstructured matrix **Σ**. Notice that the identity matrix **I** comes before **Σ** because traits are nested within experimental units.

$$\mathbf{R} = \mathbf{I}_{496} \otimes \Sigma = \mathbf{I}_{496} \otimes \begin{bmatrix} \sigma^2_{\varepsilon 1} & \sigma_{\varepsilon 12} & \sigma_{\varepsilon 13} \\ \sigma_{\varepsilon 12} & \sigma^2_{\varepsilon 2} & \sigma_{\varepsilon 23} \\ \sigma_{\varepsilon 13} & \sigma_{\varepsilon 23} & \sigma^2_{\varepsilon 3} \end{bmatrix}$$

$$\mathbf{R} = \begin{bmatrix} \sigma^2_{\varepsilon 1} & \sigma_{12} & \sigma_{13} & 0 & 0 & 0 & \cdots & 0 & 0 & 0 \\ \sigma_{21} & \sigma^2_{\varepsilon 2} & \sigma_{23} & 0 & 0 & 0 & \cdots & 0 & 0 & 0 \\ \sigma_{31} & \sigma_{32} & \sigma^2_{\varepsilon 3} & 0 & 0 & 0 & \cdots & 0 & 0 & 0 \\ 0 & 0 & 0 & \sigma^2_{\varepsilon 1} & \sigma_{12} & \sigma_{13} & \cdots & 0 & 0 & 0 \\ 0 & 0 & 0 & \sigma_{21} & \sigma^2_{\varepsilon 2} & \sigma_{23} & \cdots & 0 & 0 & 0 \\ 0 & 0 & 0 & \sigma_{31} & \sigma_{32} & \sigma^2_{\varepsilon 3} & \cdots & 0 & 0 & 0 \\ \cdots & \cdots & \cdots & \cdots & \cdots & \cdots & \cdots & \cdots & \cdots & \cdots \\ 0 & 0 & 0 & 0 & 0 & 0 & \cdots & \sigma^2_{\varepsilon 1} & \sigma_{12} & \sigma_{13} \\ 0 & 0 & 0 & 0 & 0 & 0 & \cdots & \sigma_{21} & \sigma^2_{\varepsilon 2} & \sigma_{23} \\ 0 & 0 & 0 & 0 & 0 & 0 & \cdots & \sigma_{31} & \sigma_{32} & \sigma^2_{\varepsilon 3} \end{bmatrix}$$

- The elements $\left(\sigma^2_{\varepsilon 1}, \sigma^2_{\varepsilon 2}, \sigma^2_{\varepsilon 3}\right)$ in the diagonal of the US matrix are residual variances for traits. Unlike the block diagonal structure used for other effects in the model, the residual effects for measurements of different traits on the same experimental unit are not independent, resulting in some non-zero off-diagonal elements. The dimensions of **R** are $496 \times 3 = 1488$. There are 496 sub matrices with dimensions of 3×3, one for each experimental unit.

A subset of the primary output file *Code 6-1_MultivariateMaize2.asr* is given below:

```
...

Multivariate analysis of height          pollen          silking
Summary of 496 records retained of 496 read


...
9 LogL=-2528.45     S2=  1.0000        1476 df

          - - - Results from analysis of height pollen silking - - -
Akaike Information Criterion      5086.90 (assuming 15 parameters).
Bayesian Information Criterion    5166.36

Model_Term                            Sigma      Sigma   Sigma/SE   % C
id(units).us(Trait)         1488 effects
Trait               US_V 1  1   65.7327      65.7327     11.03    0 P
Trait               US_C 2  1   0.514130     0.514130     0.94    0 P
Trait               US_V 2  2   1.39831      1.39831     11.29    0 P
Trait               US_C 3  1  -1.45659     -1.45659     -1.77    0 P
Trait               US_C 3  2   1.03521      1.03521      7.99    0 P
Trait               US_V 3  3   3.37118      3.37118     11.35    0 P
diag(Trait).id(location.rep)    24 effects
Trait               DIAG_V  1   13.3382      13.3382      1.31    0 P
Trait               DIAG_V  2   0.421237     0.421237     1.34    0 P
Trait               DIAG_V  3   0.607512     0.607512     1.30    0 P
diag(Trait).id(RIL)        186 effects
Trait               DIAG_V  1   304.254      304.254      5.28    0 P
Trait               DIAG_V  2   4.72115      4.72115      5.25    0 P
Trait               DIAG_V  3   4.87996      4.87996      5.02    0 P
diag(Trait).id(location.RIL)   744 effects
Trait               DIAG_V  1   23.4447      23.4447      3.58    0 P
Trait               DIAG_V  2   0.275315     0.275315     2.70    0 P
Trait               DIAG_V  3   0.248401     0.248401     1.16    0 P
Covariance/Variance/Correlation Matrix US Residual
  65.73      0.5363E-01 -0.9785E-01
  0.5141      1.398       0.4768
 -1.457       1.035       3.371
...
Finished: 30 Oct 2015 14:35:07.825   LogL Converged
```

- The model requires estimates of 15 parameters. The first six are variances and covariances between residual effects, followed by three parameters for *rep.location*, *RIL* and *RIL.location* terms. For the residuals, US_V refers to variance components in the unstructured matrix and US_C refers to the covariances between pairs of traits.
- At the bottom we see the 3×3 unstructured covariance/variance/correlation estimate matrix for the residuals. The diagonal elements are residual variances for height, pollen and silking. The above diagonal elements are correlations between residuals for different traits measured on the same plot. Pollen and silking traits have a moderate positive environmental correlation (0.4768) but the residual correlations of these two traits with height are much smaller. The correlation between residual effects on height and pollen is negative (-0.098). The elements below the diagonal are covariance components.

The model shown in !PATH 2 converged successfully without specifying initial values because we have fit a simple model to a balanced and relatively small data set. As models become more complex, however, good initial values may be helpful or even necessary to achieve convergence. We demonstrate two modifications of this model that are identical to model 2 but specify initial parameter values in two different ways. In model 3 we supply the initial parameter values using the !INIT qualifier directly inside the **G** and **R** structure definitions.

Code example 6.3
Multivariate analysis of maize RIL data, correlated residuals only, with initial parameter values (file *Code 6-1_MultivariateMaize.as* continued).

```
!ARGS 3 !RENAME 1 !OUTFOLDER  V:\Book\Book1_Examples\ch06_mvar\outputfiles
Title: Maize RILs multivariate analysis
...

!PART 3   # Diagonal structures with starting values
height   pollen  silking ~ Trait Trait.location ,
     !r  diag(Trait !INIT 13.44 0.45 0.663).id(rep.location),
         diag(Trait !INIT 303.15  4.9  4.992).id(RIL),
         diag(Trait !INIT 24.93 0.573 0.837).id(RIL.location)
     residual id(units).us(Trait 64.9 0.0  1.179 0.0  0.0 2.89)
```

In the **G** structure definitions, we need to supply three variance components values for each term. The variance components should be in the same order as the traits are listed as dependent variables on the left side of the model equation. Thus, we provide initial estimates of the rep.location variances for height (13.44), pollen (0.45), and silking (0.663) in that order with: diag(Trait !INIT 13.44 0.45 0.663).id(rep.location).

In the **R** structure definition Maize RILs multivariate model:, we need to provide three variance components plus three covariance components, since we are using an unstructured **R** matrix, which includes covariances as well as variances: residual id(units).us(Trait 64.9 0.0 1.179 0.0 0.0 2.89). The order of initial parameters follows the order of elements in the lower triangular representation of the symmetric matrix ($\sum$): $\sigma^2_{\varepsilon h}, \sigma_{\varepsilon h p}, \sigma^2_{\varepsilon p}, \sigma_{\varepsilon h s}, \sigma_{\varepsilon p s}, \sigma^2_{\varepsilon s}$. Notice we have zeros as starting values of covariances, since we have no idea what the covariances might be. We are not fixing the values to be zero, however, and the model solution will give us best estimates of the covariances.

Another way to provide initial values is to use the !ASSIGN qualifier to assign initial values as a string to a named variable, then include the named variable inside the structure definition for the appropriate term. For example, before the model definition we can write:

!ASSIGN repDIAG !INIT 13.44 0.45 0.663. This assigns the string value "!INIT 13.44 0.45 0.663" to a variable we named "repDIAG". Then in the model definition we can write: diag(Trait $repDIAG).id(rep.location), and the term acts like a macro variable that gets substituted by its assigned string value, resulting in a model term defined as: diag(Trait !INIT 13.44 0.45 0.663).id(rep.location). Note that the named variable cannot be longer than eight characters.

An advantage of this method is that we can define the string values over multiple lines, which can help us keep better track of the initial values when the model structures become more complex. For example, instead of defining the initial values for residual variances and covariances as we did in model 3 with: residual id(units).us(Trait 64.88 0.0 1.179 0.0 0.0 2.89), we can write the values in the form of the lower triangular matrix to more easily ensure that we put them in the correct order:

```
!ASSIGN RUS   !< !INIT
64.88
0.0  1.179
0.0  0.0    2.89  !>
```

Notice that for our convenience, this follows the form:

$$\begin{pmatrix} \sigma^2_{\varepsilon h} & & \\ \sigma_{\varepsilon h p} & \sigma^2_{\varepsilon p} & \\ \sigma_{\varepsilon h s} & \sigma_{\varepsilon p s} & \sigma^2_{\varepsilon s} \end{pmatrix}$$

The special characters !< and !> define a block of lines over which the string is defined. Since we assigned this string to a variable called 'RUS' (an arbitrary name chosen for our convenience to remind us that this is for an unstructured **R** matrix), then we simply include '$RUS' in the model where we want this string to be inserted.

Code example 6.4
Multivariate analysis of maize RIL data, correlated residuals only, initial values supplied using !ASSIGN (file *Code 6-1_MultivariateMaize.as* continued).

```
# DIAGONAL structure, with starting values using ASSIGN qualifier
!PART 4           .
!ASSIGN repDIAG !INIT 13.44    0.44969 0.66288
!ASSIGN RilDIAG !INIT 303.15   4.8899   4.9918
!ASSIGN locRilD !INIT 24.93    0.57325  0.8370
!ASSIGN RUS   !< !INIT
64.88
0.0  1.179
0.0  0.0     2.89  !>
height  pollen  silking ~ Trait Trait.location ,
    !r  diag(Trait $repDIAG).id(rep.location),
        diag(Trait $RilDIAG).id(RIL),
        diag(Trait $locRilD).id(RIL.location)
     residual   id(units).us(Trait $RUS)
```

This model is identical to Model 3 given before (PATH 3) and provides the same result as both Models 2 and 3, since the initial values provided did not affect the result in this case.

To model genetic correlations between traits, all we need to do is change the variance structure of the *Trait.RIL* term and make it either US (unstructured covariance matrix, Model 5) or CORGH (heterogeneous variances with correlations, Model 6).

Code example 6.5
Multivariate analysis of maize RIL data, correlated residuals and RIL effects with US structure (file *Code 6-1_MultivariateMaize.as* continued).

```
!PART  5      # US model for RILs and residuals, diagonal for other
factors
!ASSIGN RilUS !< !INIT
303.15
0.0  4.8899
0.0  0.0  4.9918    !>
!ASSIGN repDIAG !INIT 13.44    0.44969 0.66288
!ASSIGN locRilD !INIT 24.93    0.57325
0.8370
!ASSIGN RUS   !< !INIT
64.88
0.0  1.1793
0.0  0.0     2.8875  !>

height  pollen  silking ~ Trait Trait.location ,
    !r diag(Trait $repDIAG).id(rep.location),
       us(Trait $RilUS).id(RIL)  ,
        diag(Trait $locRilD).id(RIL.location)
     residual   id(units).us(Trait $RUS)
```

$$\begin{bmatrix} \sigma^2_{RILh} & \sigma_{RILhp} & \sigma_{RILhs} \\ \sigma_{RILhp} & \sigma^2_{RILp} & \sigma_{RILps} \\ \sigma_{RILhs} & \sigma_{RILps} & \sigma^2_{RILs} \end{bmatrix} \otimes \mathbf{I}_{62}$$

- In Model 5, we fit an unstructured matrix (us) and provide initial parameter values for the genotypic variances of three traits in the diagonal and genotypic covariances between trait pairs in the lower off-diagonal.
- The final $\mathbf{G}_{\text{Trait.RIL}}$ matrix (186 × 186 dimensions) is sparse and has a banded structure as shown below.

$$\mathbf{G}_{\text{Trait.RIL}} = \begin{bmatrix} \sigma^2_{h1} & \cdots & 0_{62} & \sigma_{hp} & \cdots & 0_{124} & \sigma_{hs} & \cdots & 0_{186} \\ \cdots & \cdots & \cdots & \cdots & \sigma_{hp} & \cdots & \cdots & \sigma_{hs} & \cdots \\ 0_{62} & \cdots & \sigma^2_{h62} & \cdots & \cdots & \sigma_{hp} & \cdots & \cdots & \sigma_{hs} \\ \sigma_{hp} & \cdots & \cdots & \sigma^2_{p1} & \cdots & \cdots & \sigma_{ps} & \cdots & \cdots \\ \cdots & \sigma_{hp} & \cdots & \cdots & \cdots & \cdots & \cdots & \sigma_{ps} & \cdots \\ 0_{124} & \cdots & \sigma_{hp} & \cdots & \cdots & \sigma^2_{p62} & \cdots & \cdots & \sigma_{ps} \\ \sigma_{hs} & \cdots & \cdots & \sigma_{ps} & \cdots & \cdots & \sigma^2_{s1} & \cdots & \cdots \\ \cdots & \sigma_{hs} & \cdots & \cdots & \sigma_{ps} & \cdots & \cdots & \cdots & \cdots \\ 0_{186} & \cdots & \sigma_{hs} & \cdots & \cdots & \sigma_{ps} & \cdots & \cdots & \sigma^2_{s62} \end{bmatrix}$$

Results of this model are in *Code 6-1_MultivariateMaize5.asr*:

```
  10 LogL=-2483.15    S2=   1.0000           1476 df

          - - - Results from analysis of height pollen silking - - -
Akaike Information Criterion      5002.31 (assuming 18 parameters).
Bayesian Information Criterion    5097.66

Model_Term                            Sigma         Sigma   Sigma/SE   %   C

id(units).us(Trait)         1488 effects
Trait            US_V 1 1    65.7139        65.7139      11.03   0   P

Trait            US_C 2 1    0.461181       0.461181      0.85   0   P

Trait            US_V 2 2    1.38395        1.38395      11.30   0   P

Trait            US_C 3 1   -1.54571       -1.54571      -1.88   0   P

Trait            US_C 3 2    1.01298        1.01298       7.91   0   P

Trait            US_V 3 3    3.34466        3.34466      11.35   0   P

...
Covariance/Variance/Correlation Matrix US Residual
   65.71       0.4836E-01  -0.1043
   0.4612      1.384        0.4708
  -1.546       1.013        3.345
Covariance/Variance/Correlation Matrix US us(Trait).id(RIL)
  303.4        0.5964       0.5493
  23.08        4.938        0.8547   <-----
  21.57        4.281        5.080
```

$$\begin{bmatrix} \sigma^2_{Gh} & r_{Ghp} & r_{Ghs} \\ \sigma_{Ghp} & \sigma^2_{Gp} & r_{Gps} \\ \sigma_{Ghs} & \sigma_{Gps} & \sigma^2_{Gs} \end{bmatrix}$$

- Genotypic variance and covariance components are associated with the term us(Trait).id(RIL), so we can write σ^2_{Gi} for the RIL variance component for trait i.

- Genotypic correlations between pairs of traits are given at the bottom of the output. The bold values in the diagonal of the Covariance/Variance/Correlation matrix are variance components associated with RIL effects for height, pollen and silking, respectively. The values above the diagonal are genotypic correlations, ranging from 0.55 and 0.86. The values below the diagonal are covariances between pairs of traits.

Model 6 is identical to model 5, but parameterized differently, using correlations instead of covariances for the off-diagonal terms of the variance-covariance matrix for the factor Trait.RIL. We fit a full correlation structure with heterogeneous variances (corgh) and provide initial values for the variances of three traits in the diagonal and correlations in the off diagonals. Although the result will be the same as for model 5, the corgh parameterization is computationally more efficient and may converge better in some circumstances.

Code example 6.6
Multivariate analysis of maize RIL data, correlated residuals and RIL effects with corgh structure (file *Code 6-1_MultivariateMaize.as* continued).

```
!PART 6 # Correlation model for RILs, diagonal for other terms
height  pollen  silking ~ Trait Trait.location ,
    !r diag(Trait).id(rep.location),
       corgh(Trait).id(RIL),
       diag(Trait).id(RIL.location)
     residual id(units).us(Trait)
```

Results of this model are in *Code 6-1_MultivariateMaize6.asr*:

```
 15 LogL=-2483.15    S2=  1.0000        1476 df
              - - - Results from analysis of height pollen silking - - -
Notice:  US structures were modified   2 times to make them positive definite.
         If ASReml has fixed the structure [flagged by B], it may not have
             converged to a maximum likelihood solution.
     Used !EMFLAG 0 Single standard EM update when AI update unacceptable
         You could try !GU (negative definite US) or use XFA instead.

Akaike Information Criterion     5002.31 (assuming 18 parameters).
Bayesian Information Criterion   5097.66
```

Model_Term				Sigma	Sigma	Sigma/SE	% C
id(units).us(Trait)		1488	effects				
Trait	US_V	1	1	65.7130	65.7130	11.03	0 P
Trait	US_C	2	1	0.460983	0.460983	0.85	0 P
Trait	US_V	2	2	1.38387	1.38387	11.30	0 P
Trait	US_C	3	1	-1.54575	-1.54575	-1.88	0 P
Trait	US_C	3	2	1.01289	1.01289	7.91	0 P
Trait	US_V	3	3	3.34453	3.34453	11.35	0 P
diag(Trait).id(rep.location)		24	effects				
Trait	DIAG_V	1		13.3226	13.3226	1.31	0 P
Trait	DIAG_V	2		0.421850	0.421850	1.34	0 P
Trait	DIAG_V	3		0.608367	0.608367	1.30	0 P
corgh(Trait).id(RIL)		186	effects				
Trait	COR_R	1		0.596378	0.596378	6.73	0 P
Trait	COR_R	2		0.549331	0.549331	5.55	0 P
Trait	COR_R	3		0.854742	0.854742	20.63	0 P
Trait	COR_V	1		303.397	303.397	5.28	0 P

```
Trait                         COR_V   2   4.93772    4.93772    5.26   0  P
Trait                         COR_V   3   5.07972    5.07972    5.04   0  P
diag(Trait).id(RIL.location)    744 effects
Trait                         DIAG_V  1   23.5349    23.5349    3.59   0  P
Trait                         DIAG_V  2   0.280093   0.280093   2.74   0  P
Trait                         DIAG_V  3   0.257035   0.257035   1.20   0  P
Covariance/Variance/Correlation Matrix US Residual
    65.71     0.4834E-01  -0.1043
    0.4610     1.384       0.4708
   -1.546      1.013       3.345
Covariance/Variance/Correlation Matrix COR corgh(Trait).id(RIL)
    303.4      0.5964      0.5493
    23.08      4.938       0.8548
    21.57      4.281       0.5080
```

- There is a notice in the output saying that the US structure was modified twice to make it positive definite. This is referring to the variance-covariance matrix of the residuals; twice during the convergence process, the estimated covariance parameters were going out of bounds of a consistent, positive definite matrix, but the program 'pushed them' back into bounds. The message also says *"If ASReml has fixed the structure [flagged by B], it may not have converged to a maximum likelihood solution."* This is a warning to check the labels in column 'C' for the estimates in the .asr file; in this example they are all 'P' meaning 'positive definite' so we don't have any problems. If you find some estimates labeled as 'B', it means they were forced in bounds. In this case, rerun the model with extra iteration, simplify the model or let some parameters go out of bounds using !GU.

- The warning message also suggests using a !GU (unbounded) qualifier for some parameters so that they can go out of bounds (variance components can be negative and covariance components can exceed ±1.0 times the square root of the product of the variances). This may be necessary in some cases to obtain convergence, but you may end up with an estimated genotypic correlation that exceeds 1.0, which may lead to some embarrassment.

- The warning message also suggests using an extended factor analytic structure (XFA), which can be a more parsimonious model with fewer parameters to help with model convergence. We will introduce factor analytic structures in Chap. 8 in the context of multi-environment models, but they can also be used for multivariate models. An example of the XFA model is included as the final model (model 9) in the file "Code06-1MultivariateMaize.as" for interested readers, but we will not pursue this any further in this chapter, since the XFA model only becomes more parsimonious when the number of traits exceeds three.

- The only difference from the model 5 output is that correlation estimates ('COR_R') are reported in the Sigma column rather than covariance estimates for Trait.RIL. The model likelihood, AIC, and parameter estimate values are all identical between models 5 and 6.

In model 7, we extend model 5 to include covariances between the effects of all model terms on different traits. We change the diag() structure for Trait.rep.location and Trait.RIL.location terms to us():

Code example 6.7
Multivariate analysis of maize RIL data, all effects random effects with US structure (file *Code 6-1_MultivariateMaize.as* continued).

```
!PART 7  #unstructured matrices for all model terms
height  pollen  silking ~ Trait Trait.location,
    !r  us(Trait).id(rep.location),
        us(Trait).id(RIL),
        us(Trait).id(RIL.location)
    residual id(units).us(Trait)
```

The results of this analysis are in *Code 6-1_MultivariateMaize7.asr*:

```
26 LogL=-2464.44      S2=  1.0000        1476 df
             - - - Results from analysis of height pollen silking - - -
Notice:  US structures were modified  32 times to make them positive definite.
         If ASReml has fixed the structure [flagged by B], it may not have
            converged to a maximum likelihood solution.
         Used !EMFLAG 0 Single standard EM update when AI update unacceptable
         You could try !GU (negative definite US) or use XFA instead.

Akaike Information Criterion     4976.88 (assuming 24 parameters).
Bayesian Information Criterion    5104.02

Model_Term                               Sigma        Sigma   Sigma/SE   % C
id(units).us(Trait)         1488 effects
Trait                 US_V  1  1    64.8862      64.8862      11.05    0 P
Trait                 US_C  2  1  -0.176168     -0.176168     -0.31    0 P
Trait                 US_V  2  2    1.17934       1.17934      11.05    0 P
Trait                 US_C  3  1   -1.78707      -1.78707      -2.02    0 P
Trait                 US_C  3  2    0.637361      0.637361      5.10    0 P
Trait                 US_V  3  3    2.88753       2.88753      11.05    0 P
us(Trait).id(rep.location)      24 effects
Trait                 US_V  1  1    13.4463       13.4463       1.31    0 P
Trait                 US_C  2  1   -1.72941      -1.72941      -1.11    0 P
Trait                 US_V  2  2    0.449696      0.449696      1.36    0 P
Trait                 US_C  3  1   -2.26286      -2.26286      -1.15    0 P
Trait                 US_C  3  2    0.542075      0.542075      1.36    0 P
Trait                 US_V  3  3    0.662875      0.662875      1.32    0 P
us(Trait).id(RIL)       186 effects
Trait                 US_V  1  1   303.151       303.151        5.27    0 P
Trait                 US_C  2  1    22.7931       22.7931       3.81    0 P
Trait                 US_V  2  2     4.88999       4.88999      5.21    0 P
Trait                 US_C  3  1    21.5031       21.5031       3.56    0 P
Trait                 US_C  3  2     4.18711       4.18711      4.70    0 P
Trait                 US_V  3  3     4.99185       4.99185      4.95    0 P
us(Trait).id(RIL.location)     744 effects
Trait                 US_V  1  1    24.9323       24.9323       3.73    0 P
Trait                 US_C  2  1     1.47768       1.47768      2.19    0 P
Trait                 US_V  2  2     0.573253      0.573253     4.32    0 P
Trait                 US_C  3  1     0.370188      0.370188     0.39    0 P
Trait                 US_C  3  2     0.562225      0.562225     3.74    0 P
Trait                 US_V  3  3     0.837013      0.837013     3.08    0 P
Covariance/Variance/Correlation Matrix US Residual
   64.89      -0.2014E-01  -0.1306
  -0.1762       1.179       0.3454
  -1.787        0.6374      2.888
Covariance/Variance/Correlation Matrix US us(Trait).id(rep.loc
   13.45      -0.7033     -0.7580
  -1.729        0.4497      0.9929
  -2.263        0.5421      0.6629
```

```
Covariance/Variance/Correlation Matrix US us(Trait).id(RIL)
   303.2      0.5920      0.5528
   22.79      4.890       0.8475
   21.50      4.187        4.992
Covariance/Variance/Correlation Matrix US us(Trait).id(RIL.loc
   24.93      0.3909      0.8104E-01
   1.478      0.5733      0.8117
   0.3702     0.5622      0.8370

                              Wald F statistics
      Source of Variation      NumDF              F-inc
   10 Trait                       3              14962.38
   11 Trait.location              9                224.35
   15 us(Trait).id(rep.location)      24 effects fitted
   17 us(Trait).id(RIL)             186 effects fitted
   20 us(Trait).id(RIL.location)    744 effects fitted
           6  possible outliers: see .res file
Finished: 28 Sep 2016 08:41:18.886   LogL Converged
```

Notice the large number of parameters that are estimated with this model. As model complexity increases, convergence on the REML solution can become more difficult (although in this particular case we did not have a problem). There are a number of approaches that users can take to attain convergence for complex models. One is to increase the number of iterations with the !MAXIT qualifier and see if convergence will happen eventually. Another possibility is to provide initial parameter estimates to get ASReml closer to the eventual solution at the first iteration. In some cases, re-parameterizing the model may help. As an example of reparameterization, model 8 uses CORGH() structures to produce a model identical to model 7 but is parameterized in terms of correlations, rather than the covariances of the US() structures. It turns out that convergence of model 8 in this example requires fitting missing values as sparse effects using the '!f mv' term. In general, this term is not required for multivariate analysis, but it sometimes helps convergence; on the other hand, sometimes it slows convergence (model 7 is an example). Users can try including it in cases where the model did not converge without it. In addition, '!f mv', along with the job qualifier 'ASUV', is required for models where the residual is something other than the id(units).us(Trait) structures used in the examples in this chapter. An example might be a repeated measures model where the residual effects of measurements taken at different time points are constrained to have uniform variances and covariances (id(units).coruv(Trait)).

Code example 6.8
Multivariate analysis of maize RIL data, all random effects correlated (file *Code 6-1_MultivariateMaize.as* continued).

```
!PART 8 #corgh matrices for all model terms
height  pollen  silking ~ Trait Trait.location ,
    !r  corgh(Trait).id(rep.location),
        corgh(Trait).id(RIL),
        corgh(Trait).id(RIL.location),
        !f mv
    residual id(units).us(Trait)
```

Results of this model are in *Code 6-1_MultivariateMaize8.asr*:

```
   14 LogL=-2464.44    S2=  1.0000        1476 df
            - - - Results from analysis of height pollen silking - - -
Notice:  US structures were modified   2 times to make them positive definite.
         If ASReml has fixed the structure [flagged by B], it may not have
             converged to a maximum likelihood solution.
      Used !EMFLAG 0 Single standard EM update when AI update unacceptable
         You could try !GU (negative definite US) or use XFA instead.

Akaike Information Criterion      4976.88 (assuming 24 parameters).
Bayesian Information Criterion    5104.01

Model_Term                              Sigma        Sigma    Sigma/SE   % C
id(units).us(Trait)          1488 effects
Trait                  US_V  1  1   64.8862       64.8862      11.05    0 P
Trait                  US_C  2  1  -0.176168     -0.176168     -0.31    0 P
Trait                  US_V  2  2    1.17934       1.17934      11.05    0 P
Trait                  US_C  3  1   -1.78707      -1.78707      -2.02    0 P
Trait                  US_C  3  2    0.637361      0.637361      5.10    0 P
Trait                  US_V  3  3    2.88753       2.88753      11.05    0 P
corgh(Trait).id(rep.location)   24 effects
Trait                  COR_R  1  -0.703295     -0.703295      -2.48    0 P
Trait                  COR_R  2  -0.757950     -0.757950      -3.07    0 P
Trait                  COR_R  3   0.992848      0.992848      31.00    0 P
Trait                  COR_V  1   13.4463       13.4463        1.31    0 P
Trait                  COR_V  2    0.449696      0.449696      1.36    0 P
Trait                  COR_V  3    0.662877      0.662877      1.32    0 P
corgh(Trait).id(RIL)         186 effects
Trait                  COR_R  1   0.591998      0.591998       6.62    0 P
Trait                  COR_R  2   0.552766      0.552766       5.54    0 P
Trait                  COR_R  3   0.847481      0.847481      19.86    0 P
Trait                  COR_V  1  303.151       303.151         5.27    0 P
Trait                  COR_V  2    4.88999       4.88999        5.21    0 P
Trait                  COR_V  3    4.99185       4.99185        4.95    0 P
corgh(Trait).id(RIL.location)  744 effects
Trait                  COR_R  1   0.390865      0.390865       2.26    0 P
Trait                  COR_R  2   0.810354E-01  0.810354E-01   0.38    0 P
Trait                  COR_R  3   0.811653      0.811653       6.09    0 P
Trait                  COR_V  1   24.9323       24.9323        3.73    0 P
Trait                  COR_V  2    0.573253      0.573253      4.32    0 P
Trait                  COR_V  3    0.837013      0.837013      3.08    0 P
Covariance/Variance/Correlation Matrix US Residual
  64.89      -0.2014E-01 -0.1306
 -0.1762      1.179       0.3454
 -1.787       0.6374      2.888
Covariance/Variance/Correlation Matrix COR corgh(Trait).id(rep.
  13.44      -0.7032     -0.7579
 -1.728       0.4496      0.9928
 -2.262       0.5419      0.6627
Covariance/Variance/Correlation Matrix COR corgh(Trait).id(RIL)
 303.2        0.5920      0.5528
  22.79       4.890       0.8475
  21.50       4.187       4.992
Covariance/Variance/Correlation Matrix COR corgh(Trait).id(RIL.
  24.93       0.3909      0.8104E-01
   1.478      0.5733      0.8117
   0.3702     0.5622      0.8370
```

Table 6.1 A sequence of variance structures fit to the maize multivariate model

Model term	Model 2–4	Model 5	Model 6	Model 7–8
rep.location	DIAG	DIAG	DIAG	CORGH
RIL	DIAG	US	CORGH	CORGH
RIL.location	DIAG	DIAG	DIAG	CORGH
Residual	US	US	US	US
# of parameters	15	18	18	24
LogL	−2528.45	−2483.15	−2483.15	−2464.44
AIC	5086.90	5002.31	5002.31	4976.88

Notice that the likelihoods and parameter estimates (for example, the highlighted genotypic correlations) are identical for models 7 and 8. Many of the trait correlations for non-genetic effects (rep and RIL – by – location) are far from zero, suggesting that this model might be better than the preceding models that fit those effects as independent across traits. We can compare the different models using their AIC values (Table 6.1):

Even after penalizing models 7 and 8 for the additional parameters introduced to account for the covariances between traits for *rep.location* and *RIL.location* effects, their AIC value is lower (better) than the other models, indicating that they are indeed best.

Linear Combinations of Variances and Covariances

We obtain the genotypic correlation estimates directly in the outputs from the multivariate models. However, we do not directly get standard errors of these estimates. There are several ways to obtain the standard errors of the correlation estimates from ASReml. In all cases, these are the 'delta method' approximate standard errors that are reasonable when sample sizes are sufficiently large (Holland 2006).

One way to obtain the standard error of the genotypic correlation estimate from the CORGH () model output is back-calculate from Sigma/SE. Notice that the parameter estimate is followed by the ratio of the estimate to its approximate standard error. For example, in *Code 6-1_MultivariateMaize8.asr*, we have the following correlation estimates for RILs:

```
Model_Term                      Sigma       Sigma   Sigma/SE    % C
...
corgh(Trait).id(RIL)         186 effects
  Trait         COR_R    1  0.591998    0.591998      6.62     0 P
  Trait         COR_R    2  0.552766    0.552766      5.54     0 P
  Trait         COR_R    3  0.847481    0.847481     19.86     0 P
...
```

From these values, we can back-calculate the standard errors (SE's) as the estimate (Sigma) divided by the ratio of the estimate to its standard error (Sigma/SE):

Parameter	Estimate	Estimate/SE	SE
r_{ghp}	0.591998	6.62	0.591998/6.62 = 0.089
r_{ghs}	0.552766	5.54	0.100
r_{gps}	0.847481	19.86	0.043

A better approach, however, is to get the standard errors of all the correlation coefficients estimated in the CORGH model by requesting a default .pvc output with the VPREDICT directive added following the model definition:

Code example 6.9
Multivariate analysis of maize RIL data, 'empty' VPREDICT !DEFINE statement used to generate a default .
pvc output (file *Code 6-1_MultivariateMaize.as* continued).

```
!PART 9 # corgh matrices for all model terms
height  pollen  silking ~ Trait Trait.location ,
   !r  corgh(Trait).id(rep.location),
       corgh(Trait).id(RIL),
       corgh(Trait).id(RIL.location),
       !f mv
     residual id(units).us(Trait)

VPREDICT !DEFINE
```

If we do not include any definitions of combinations of variance components, the .pvc output file will be generated with
the estimates of all of the parameters in the model, each followed by their standard error (*Code06-1MaizeMultivariateData8.pvc*):

```
            - - - Results from analysis of height pollen silking - - -
id(units).us(Trait)             1488 effects
 1 id(units).us(Trait);us(Trait)      V  1  1      64.8862      5.87205
 2 id(units).us(Trait);us(Trait)      C  2  1     -0.176168    -0.310000
 3 id(units).us(Trait);us(Trait)      V  2  2       1.17934     0.106728
 4 id(units).us(Trait);us(Trait)      C  3  1      -1.78707    -2.02000
 5 id(units).us(Trait);us(Trait)      C  3  2       0.637361    0.124973
 6 id(units).us(Trait);us(Trait)      V  3  3       2.88753     0.261315
corgh(Trait).id(rep.location)    24 effects
 7 corgh(Trait).id(rep.loc;corgh(Trait)   R   1 -  -0.703295   -2.48000
 8 corgh(Trait).id(rep.loc;corgh(Trait)   R   2 -  -0.757950   -3.07000
 9 corgh(Trait).id(rep.loc;corgh(Trait)   R   3      0.992848   0.320274E01
10 corgh(Trait).id(rep.loc;corgh(Trait)   V   1     13.4463    10.2644
11 corgh(Trait).id(rep.loc;corgh(Trait)   V   2      0.449696   0.330659
12 corgh(Trait).id(rep.loc;corgh(Trait)   V   3      0.662877   0.502180
corgh(Trait).id(RIL)             186 effects
13 corgh(Trait).id(RIL);corgh(Trait)      R   1      0.591998   0.894257E01
14 corgh(Trait).id(RIL);corgh(Trait)      R   2      0.552766   0.997773E01
15 corgh(Trait).id(RIL);corgh(Trait)      R   3      0.847481   0.426728E01
16 corgh(Trait).id(RIL);corgh(Trait)      V   1    303.151     57.5239
17 corgh(Trait).id(RIL);corgh(Trait)      V   2      4.88999    0.938578
18 corgh(Trait).id(RIL);corgh(Trait)      V   3      4.99185    1.00845
corgh(Trait).id(RIL.location)   744 effects
19 corgh(Trait).id(RIL.loc;corgh(Trait)   R   1      0.390865   0.172949
20 corgh(Trait).id(RIL.loc;corgh(Trait)   R   2      0.810354E-01 0.213251
21 corgh(Trait).id(RIL.loc;corgh(Trait)   R   3      0.811653   0.133276
22 corgh(Trait).id(RIL.loc;corgh(Trait)   V   1     24.9323     6.68426
23 corgh(Trait).id(RIL.loc;corgh(Trait)   V   2      0.573253   0.132697
24 corgh(Trait).id(RIL.loc;corgh(Trait)   V   3      0.837013   0.271757
Notice: The parameter estimates are followed by
          their approximate standard errors.
```

- The highlighted values are the estimate genotypic correlations and their approximate standard errors. Notice that the standard errors for the genotypic correlations match our computations above.

A third way to obtain the standard errors of the correlation estimates is to define them directly as functions of the covariance and variance components using the VPREDICT directive. There is no reason to use this approach if the correlations of interest are directly produced by the CORGH model, as shown above. Nevertheless, there may be more complex correlations or other functions of variance components of interest, such as the phenotypic correlation, that cannot be obtained from the CORGH model directly. The phenotypic correlation between a pair of traits x and y in this experiment is:

$$r_{p(x,y)} = \frac{Cov_{p(x,y)}}{\sqrt{V_{p(x)} V_{p(y)}}} = \frac{\sigma_{g(x,y)} + \sigma_{ge(x,y)} + \sigma_{\varepsilon(x,y)}}{\sqrt{\left(\sigma^2_{g(x)} + \sigma^2_{ge(x)} + \sigma^2_{\varepsilon(x)}\right)\left(\sigma^2_{g(y)} + \sigma^2_{ge(y)} + \sigma^2_{\varepsilon(y)}\right)}} \tag{6.22}$$

To estimate the phenotypic correlation, we need the covariances of genotype (RIL), genotype – by – environment (RIL – by – location), and residual effects between trait pairs. We will obtain these values by returning to the US() model (model 7) for these effects. We first run this model with no functions defined below the 'VPREDICT !DEFINE' statement. This will produce a default .pvc output (*Code 6-1_MultivariateMaize7.pvc*) that we need to check to ensure that we can reference the variance/covariance components correctly in our functions:

```
     - - - Results from analysis of height pollen silking - - -

id(units).us(Trait)          1488 effects
   1 id(units).us(Trait);us(Trait)     V  1  1     64.8862      5.87737
   2 id(units).us(Trait);us(Trait)     C  2  1    -0.176168    -0.310000
   3 id(units).us(Trait);us(Trait)     V  2  2      1.17934     0.106728
   4 id(units).us(Trait);us(Trait)     C  3  1     -1.78707    -2.02000
   5 id(units).us(Trait);us(Trait)     C  3  2      0.637361    0.124973
   6 id(units).us(Trait);us(Trait)     V  3  3      2.88753     0.261552
us(Trait).id(rep.location)     24 effects
   7 us(Trait).id(rep.locati;us(Trait)  V  1  1     13.4463     10.2644
   8 us(Trait).id(rep.locati;us(Trait)  C  2  1     -1.72941    -1.11000
   9 us(Trait).id(rep.locati;us(Trait)  V  2  2      0.449696    0.330659
  10 us(Trait).id(rep.locati;us(Trait)  C  3  1     -2.26286    -1.15000
  11 us(Trait).id(rep.locati;us(Trait)  C  3  2      0.542074    0.398584
  12 us(Trait).id(rep.locati;us(Trait)  V  3  3      0.662877    0.502180
us(Trait).id(RIL)              186 effects
  13 us(Trait).id(RIL);us(Trait)       V  1  1    303.151      57.5239
  14 us(Trait).id(RIL);us(Trait)       C  2  1     22.7931      5.98244
  15 us(Trait).id(RIL);us(Trait)       V  2  2      4.88999     0.938578
  16 us(Trait).id(RIL);us(Trait)       C  3  1     21.5031      6.04020
  17 us(Trait).id(RIL);us(Trait)       C  3  2      4.18711     0.890875
  18 us(Trait).id(RIL);us(Trait)       V  3  3      4.99185     1.00845
us(Trait).id(RIL.location)     744 effects
  19 us(Trait).id(RIL.locati;us(Trait)  V  1  1     24.9323      6.68426
  20 us(Trait).id(RIL.locati;us(Trait)  C  2  1      1.47768     0.674740
  21 us(Trait).id(RIL.locati;us(Trait)  V  2  2      0.573253    0.132697
  22 us(Trait).id(RIL.locati;us(Trait)  C  3  1      0.370188    0.949200
  23 us(Trait).id(RIL.locati;us(Trait)  C  3  2      0.562225    0.150328
  24 us(Trait).id(RIL.locati;us(Trait)  V  3  3      0.837013    0.271757
```

First, we can compute (again!) the correlations for RIL, RIL – by – location, and residual effects from these results to demonstrate some peculiarities of the function definitions with VPREDICT. To start with, we can compute all three genotypic correlations from the RIL variances and covariances with a single function definition starting with the code 'R', which will estimate all correlations if given a single variance-covariance matrix as the input. For this term, we can refer to the complete variance-covariance structure for RILs as 'us(Trait).id(RIL)', matching the label for this structure highlight in the .pvc output above.

Code example 6.10

Multivariate analysis of maize RIL data, genotypic correlation estimated from variance and covariance components (file *Code 6-1_MultivariateMaize.as* continued).

```
!PART 10    # unstructured matrices for all model terms
...
VPREDICT !DEFINE
R rg us(Trait).id(RIL)
```

This produces the following output below the default .pvc output lines (*Code 6-1_MultivariateMaize7.pvc*):

```
...
rg      2  1 = us(Tr 14/SQR[us(Tr 13*us(Tr 15]=  0.5920  0.0895
rg      3  1 = us(Tr 16/SQR[us(Tr 13*us(Tr 18]=  0.5528  0.0999
rg      3  2 = us(Tr 17/SQR[us(Tr 15*us(Tr 18]=  0.8475  0.0427
```

Again, the standard errors of the estimates match our previous computations.

Now, things get tricky when we try to estimate the correlations of RIL – by – location effects. Following the example above, it seems sensible to write:

Code example 6.11

Univariate analysis of maize RIL data, genotypic correlation of RIL-by-ocation

```
!PART 11    # unstructured matrices for all model terms
...
VPREDICT !DEFINE
R rg us(Trait).id(RIL)
R rge us(Trait).id(RIL.location)
```

But this produces an error:

```
...
ERROR: Failed to locate model term which matches us(Trait).id(RIL.location)
 Notice: VPREDICT line format is
         [C|D|F|S|V|X|R|H] in position 1, a label starting in position 3
         and an expression depending of line type. Skipping line:
         R rge us(Trait).id(RIL.location)
...
```

The problem is that, ASReml internally truncated the name of the term 'us(Trait).id(RIL.location)'. We can see the truncation in the default part of the .pvc output (*Code 6-1_MultivariateMaize7.pvc*):

```
...
us(Trait).id(RIL.location)        744 effects
  19 us(Trait).id(RIL.locati;us(Trait)      V   1   1   24.9323   6.68426
...
```

Model terms with long label names can be a problem. There are several different shortcuts to identify the appropriate term in the function definitions:

1. Sometimes, the truncated form of the parameter can be determined from the default *.pvc* output. See the second highlighted term in the output box above. We can substitute exactly this part of the parameter label before the ';' when referring to the term in the function definition:

```
...
R rge us(Trait).id(RIL.locati
...
```

2. Another way to refer to parameters derived from direct products is from the names of the component factors separated by ';' as the parameter label appears in the default .pvc output (see the highlighted part of the name below):

```
...
us(Trait).id(RIL.location)        744 effects
  19 us(Trait).id(RIL.locati;us(Trait)      V   1   1   24.9323   6.68426
...
```

So, we can write the function definition as:

```
...
R rge id(RIL.locati;us(Trait)
...
```

Or, in fact using any unique substring to refer to the consolidated term in the parameter name:

```
...
R rge RIL.loc;us(Trait)
...
```

3. Use the parameter number. From the default .pvc output, we can see that the composite term 'us(Trait).id(RIL.location)' is associated with six different parameters, numbered 19–24:

```
. . .
us(Trait).id(RIL.location)      744 effects
  19 us(Trait).id(RIL.locati;us(Trait)   V  1  1     24.9323      6.68426
  20 us(Trait).id(RIL.locati;us(Trait)   C  2  1      1.47768     0.674740
  21 us(Trait).id(RIL.locati;us(Trait)   V  2  2      0.573253    0.132697
  22 us(Trait).id(RIL.locati;us(Trait)   C  3  1      0.370188    0.949200
  23 us(Trait).id(RIL.locati;us(Trait)   C  3  2      0.562225    0.150328
  24 us(Trait).id(RIL.locati;us(Trait)   V  3  3      0.837013    0.271757
```

We can refer to such a consecutive sequence of parameter values using an index of numbers:

```
. . .
R rge 19:24
. . .
```

Similarly, we can request the correlations of the residual effects by referring to the error variance-covariance matrix as 'id (units).us(Trait)', 'units;us(Trait)', or '1:6':

```
. . .
R rerr id(units).us(Trait)
. . .
```

These three function definitions will produce nine correlation estimates (bold) and their standard errors in the .pvc output (*Code 6-1_MultivariateMaize7.pvc*):

```
rg     2  1 = us(Tr  14/SQR[us(Tr  13*us(Tr  15]=    0.5920    0.0895
rg     3  1 = us(Tr  16/SQR[us(Tr  13*us(Tr  18]=    0.5528    0.0999
rg     3  2 = us(Tr  17/SQR[us(Tr  15*us(Tr  18]=    0.8475    0.0427
rge    2  1 = us(Tr  20/SQR[us(Tr  19*us(Tr  21]=    0.3909    0.1732
rge    3  1 = us(Tr  22/SQR[us(Tr  19*us(Tr  24]=    0.0810    0.2119
rge    3  2 = us(Tr  23/SQR[us(Tr  21*us(Tr  24]=    0.8117    0.1332
rerr   2  1 = id(un   2/SQR[id(un   1*id(un   3]=   -0.0201    0.0640
rerr   3  1 = id(un   4/SQR[id(un   1*id(un   6]=   -0.1306    0.0629
rerr   3  2 = id(un   5/SQR[id(un   3*id(un   6]=    0.3454    0.0564
```

- Notice that the correlations are labeled using integers to identify the traits (1, 2, or 3). The user has to know the order that the traits are listed in the model statement to know which they refer to (height, pollen, and silk in this example).

Now to estimate the phenotypic correlation, we need a function to first estimate the phenotypic covariance, functions to extract the variances needed, and a final function to compute the correlation. First we demonstrate how to extract specific values from the variance-covariance results to compute the phenotypic correlation between height and pollen. Later we will demonstrate how to efficiently compute all of the phenotypic correlations at once.

To estimate the phenotypic covariance between height and pollen, we need to sum the RIL, RIL – by – location, and residual covariances for height and pollen. These are the second of the six values within each of the overall variance-covariance matrices for these three terms. We can refer to the ith value of such a composite term using '[i]' after the name of the overall term in the function definition. So the phenotypic covariance for height and pollen is computed using the following function:

Code example 6.12
Multivariate analysis of maize data. Phenotypic covariance

```
!PART 12
...
F Cov_p_hp   us(Trait).id(RIL)[2] +  us(Trait).id(RIL.locati[2] +
id(units).us(Trait)[2]
...
```

(note that in the program file, there is no line break in the function definition).

The phenotypic covariance provides us the numerator of the phenotypic correlation. To get the terms for the denominator, we need the phenotypic variances for height and pollen, which are each the sum of the first or third values in each of the variance-covariance matrices for RIL, RIL – by –location, and residual terms. Finally, we construct the correlation using 'R' to indicate this is a correlation, then providing the components of the correlation in the following order: (1) variance of one trait, (2) covariance, (3) variance of the second trait. This order must be used to define the correlation correctly. All together, this is:

Code example 6.13
Multivariate analysis of maize RIL data, phenotypic correlations (file *Code 6-1_MultivariateMaize.as* continued).

```
!PART 13
...
F Cov_p_hp   us(Trait).id(RIL)[2] +  us(Trait).id(RIL.locati[2] +
id(units).us(Trait)[2]
F Vp_h       us(Trait).id(RIL)[1] +  us(Trait).id(RIL.locati[1] +
id(units).us(Trait)[1]
F Vp_p       us(Trait).id(RIL)[3] +  us(Trait).id(RIL.locati[3] +
id(units).us(Trait)[3]
R rp_hp  Vp_h Cov_p_hp Vp_p  #   rpheno height pollen
```

This generates the following lines of output in the .pvc file (*Code 6-1_MultivariateMaize13.pvc*):

```
...
25 Cov_p_hp 14      24.095     6.0029
26 Vp_h 13          392.97     57.749
27 Vp_p 15          6.6426    0.94401
   rp_hp  = Cov_p_hp/SQR[Vp_h 13 *Vp_p 15 ]= 0.4716  0.0757
```

The phenotypic correlation coefficient is 0.47 with a standard error of 0.08. This approach of extracting the pieces of the phenotypic correlations can be cumbersome, but fortunately, ASReml provides a shortcut whereby we can compute linear combinations of whole matrices of composite terms. So, we can add the three covariance matrices (for RIL, RIL – by – genotype, and error effects) to create one matrix that contains all of the phenotypic variances and covariances. Then we provide that matrix to the 'R' function and it knows to estimate all of the correlations from the matrix. At the same time, this also provides a more efficient way to estimate the heritabilities for each trait, because the phenotypic variance-covariance matrix contains the denominators of the heritabilities (on a plot-basis) for the three traits:

Code example 6.14

Multivariate analysis of maize RIL data, heritabilities estimated from variance components (file *Code 6-1_MultivariateMaize.as* continued).

```
!PART 14
...
F Covp  us(Trait).id(RIL) +  RIL.loc;us(Trait) + id(units).us(Trait)
R rp Covp     #rpheno all three traits at once
F Vp_all us(Trait).id(RIL) +  RIL.loc;us(Trait) +
id(units).us(Trait)
H H_ht  us(Trait).id(RIL)[1] Covp[1] #heritability of height
H H_p   us(Trait).id(RIL)[3] Covp[3] #heritability of pollen
H H_s   us(Trait).id(RIL)[6] Covp[6] #heritability of silk
```

Now we get the three phenotypic correlations and the three heritabilities with their approximate standard errors in the .pvc output (*Code 6-1_MultivariateMaize14.pvc*):

```
   ...
   28 Covp 13              392.97         57.749
   29 Covp 14              24.095         6.0029
   30 Covp 15              6.6426        0.94401
   31 Covp 16              20.086         6.0860
   32 Covp 17              5.3867        0.89736
   33 Covp 18              8.7164         1.0312
       rp     2  1 = Covp  29/SQR[Covp  28*Covp  30]=  0.4716   0.0757
       rp     3  1 = Covp  31/SQR[Covp  28*Covp  33]=  0.3432   0.0787
       rp     3  2 = Covp  32/SQR[Covp  30*Covp  33]=  0.7079   0.0403
       H_ht         = us(Trait  13/Covp 13    28=  0.7714   0.0362
       H_p          = us(Trait  15/Covp 15    30=  0.7362   0.0405
       H_s          = us(Trait  18/Covp 18    33=  0.5727   0.0535
   Notice: The parameter estimates are followed by
           their approximate standard errors.
```

Predictions from Multivariate Models

The predicted breeding values of height will be different from the multivariate model than from the univariate model, because information on pollen and silk for a particular RIL influence is predicted value for height. If the traits are correlated, we expect lower standard errors of predictions from the multivariate model because of the additional information used in computing the prediction. For example, we can obtain the predicted values for RILs from the univariate model by adding the statement after model 1:

```
predict RIL
```

The predictions are found in output file *Code06-1MaizeMultivariateData1.pvs*:

```
...
---- ---- ---- ----      1 height  ---- ---- ---- ----
Predicted values of height
The SIMPLE averaging set:  location
The ignored set: rep
RIL              Predicted_Value Standard_Error Ecode
RIL-1               182.1005          3.0964 E
RIL-11              182.8557          3.0964 E
RIL-12              185.1213          3.0964 E
...
---- ---- ---- ----      1 pollen  ---- ---- ---- ----
Predicted values of pollen
The SIMPLE averaging set:  location
The ignored set: rep
RIL              Predicted_Value Standard_Error Ecode
RIL-1                78.3525          0.4467 E
RIL-11               74.4663          0.4467 E
RIL-12               77.8667          0.4467 E
...
---- ---- ---- ----      1 silking ---- ---- ---- ----
Predicted values of silking
The SIMPLE averaging set:  location
The ignored set: rep
RIL              Predicted_Value Standard_Error Ecode
RIL-1                82.7819          0.6487 E
RIL-11               76.8206          0.6487 E
RIL-12               78.8077          0.6487 E
...
```

We can compare the univariate model predictions to the multivariate model predictions by adding the following statement after model 8:

```
predict Trait.RIL
```

The predictions are found in output file *Code06-1MaizeMultivariateData8.pvs*:

```
---- ---- ---- ---- ---- ----      1  ---- ---- ---- ---- ---- ----
The SIMPLE averaging set:  location
The ignored set: rep
RIL           Trait          Predicted_Value Standard_Error Ecode
RIL-1         height            182.6460          3.0728 E
RIL-1         pollen             78.4022          0.4433 E
RIL-1         silking            82.5288          0.6266 E
RIL-11        height            182.2226          3.0728 E
RIL-11        pollen             74.5053          0.4433 E
RIL-11        silking            76.7818          0.6266 E
RIL-12        height            184.9607          3.0728 E
RIL-12        pollen             77.8466          0.4433 E
RIL-12        silking            78.9407          0.6266 E
...
```

The predictions are slightly different for the two models, for example RIL-1 has a predicted height of 182.1 cm from the univariate model and 182.6 cm from the multivariate model. This occurs because the predicted value of height in the multivariate model is influenced by the observations of silk and pollen observations on that line, since the traits have genetic (and other) correlations. As expected, the standard errors of predictions are lower from the multivariate model than the univariate model, since more information enters into the multivariate predictions. The largest decrease in standard errors occurs with silking (SE of predictions decrease from 0.65 to 0.63), since it has the lowest heritability but moderate to high genetic correlation with the other, higher heritability, traits. Hence silking predictions have the most to gain from information on the related traits.

The Animal Model in a Multivariate Re-visitation

In the previous section, we partitioned phenotypic covariance into components due to genotype, genotype-by-environment, and residual effects, but we were not able to partition the genotypic covariance or correlation into additive and non-additive components. In this section we will extend the pedigree-based 'animal model' first introduced in Chap. 4 to incorporate correlations between traits, and use the additive genetic relationships among the individuals to estimate the additive genetic covariance and correlation between traits. We will again use the same example data set, found in "*pig_data.txt*":

pig	sire	dam	year	sex	pen	weanage	weanwt	adg	weight	loinarea
133	2	1	2004	1	52	21	13.25	2.0	264	5.34
654	2	1	2004	1	58	21	12.45	2.0	266	6.62
655	2	1	2004	1	54	21	13.55	1.6	215	5.50
153	2	1	2004	1	56	21	15.60	1.9	267	7.04
656	4	3	2004	2	59	21	10.40	1.5	210	3.94
657	4	3	2004	2	57	21	10.10	0.9	153	3.74

We will conduct a bivariate analysis of the traits '*weanwt*' (animal weight at weaning) and '*weight*' (weight at the end of the growing test period). Recall from Chap. 4 that it is possible to fit a model that includes the breeding values of the pigs themselves, the breeding values of the dam maternal effects, the covariance between the direct and maternal breeding values, plus a term for maternal 'environmental' effects, but for *weanwt*, the complete model did not converge in theoretical bounds. Therefore, in this chapter we will not attempt to fit the maternal environmental effect term ide(dam) and focus instead on fitting the direct and maternal additive genetic values, showing how we can incorporate covariances across traits for all of these effects. Univariate analyses for the two traits can be performed with the following ASReml model. Previous tests indicated that we should include *year*, *sex*, and *weanage* as fixed covariates for *weanwt* and *year*, *sex*, and *pen* as fixed covariates for *weight*.

Code example 6.15
Univariate animal models with maternal effects (part of file **Code 6-2_BivariateAnimal.as**).

```
!RENAME 1   !ARGS 1
Title: Bivariate Animal model
  pig    !P
  sire   !P
  dam    !P
  year !A    sex !A    pen !A
  weanage   weanwt   adg   weight   loinarea
pig_pedigree.txt   !SKIP 1   !ALPHA    # Pedigree file
pig_data.txt   !SKIP 1 !DOPATH $1 !Workspace 1600 !MAXIT 300 !EXTRA
14
```

```
### Univariate models
!PART 1
weanwt ~  year  sex  weanage,
        !r str(pig dam corgh(2).nrm(pig))

!PART 2
weight ~   year sex pen,
        !r  str(pig dam corgh(2).nrm(pig))
```

- Recall that `str()` causes the terms inside the parentheses to share a common variance-covariance coefficient matrix. Therefore the term `str(pig dam corgh(2).nrm(pig))` means that the model includes the effect of the pig being measured and the effect of its mother (dam) with a common variance-covariance structure so that both are proportional to the additive 'numerator' relationship matrix based on the pedigree `nrm(pig)`. We allow direct genetic and maternal genetic effects to each have a unique variance and a covariance between them with the `corgh(2)` structure.

These models produce the following univariate outputs:

Output for PART 1 (*Code 6-2_BivariateAnimal1.asr* file).

```
          - - - Results from analysis of weanwt - - -
    Akaike Information Criterion      5717.47 (assuming 4 parameters).
    Bayesian Information Criterion    5740.84

Model_Term                      Gamma      Sigma    Sigma/SE    % C
Residual         SCA_V 2556   1.00000    2.15065      12.27    0 P
corgh(2).nrm(pig)          5414 effects
2               COR_R    1 -0.742342  -0.742342      -4.41    0 P
2               COR_V    1  0.246000   0.529062       1.63    0 P
2               COR_V    2  1.91557    4.11973       11.09    0 P
pig             NRM    2707
Covariance/Variance/Correlation Matrix COR pig
  0.5291      -0.7423
 -1.096        4.120
```

- Recall that this output indicates that the direct effect of a pig's breeding value on its own weaning weight phenotype has a variance of 0.591, whereas the breeding values of the dam effects on their progeny is much larger (4.120), and the two effects have a correlation of −0.74.

Output for PART 2 (*Code 6-2_BivariateAnimal2.asr*).

```
          - - - Results from analysis of weight - - -
    Akaike Information Criterion      18611.96 (assuming 4 parameters).
    Bayesian Information Criterion    18635.24
Model_Term                      Gamma       Sigma     Sigma/SE    % C
Residual         SCA_V 2556   1.00000     387.074       13.01    0 P
corgh(2).nrm(pig)          5414 effects
2               COR_R    1 -0.423929E-01 -0.423929E-01  -0.22    0 P
2               COR_V    1  0.465325     180.115        3.38    0 P
2               COR_V    2  0.438955     169.908        5.37    0 P
pig             NRM    2707
```

```
Covariance/Variance/Correlation Matrix COR pig
  180.1      -0.4240E-01
 -7.417       169.9
```

- This output indicates that the direct effect of a pig's breeding value on its own weight phenotype has a variance of 180.1, whereas the breeding values of the dam effects on their progeny is 169.9, and the two effects are not correlated (−0.042).

Breeding values for direct effects on weight at the end of the study period are much more important (variance 180.1) than they were for weaning weight. This makes sense because the effect of a pig's genotype on its phenotype is expected to be stronger later in life than at weaning, at which point its weight is largely dependent on its mother's care.

Now we fit a simple multivariate model that includes only the direct genetic effects (no maternal genetic effects) with separate variances for the traits but no covariances across traits using the DIAG() structure. We are allowing traits to have separate residual variances and covariances across traits using the unstructured US() matrix for residual effects. In other words, the model assumes that traits are not correlated at the level of genetic effects but the residuals are correlated.

Code example 6.16
Bivariate animal model with diagonal structure for pig effects (part of file **Code 6-2_BivariateAnimal.as**).

```
### Bivariate models
!PART 3   # Diagonal structure for pig, us for residual
weanwt weight ~ Trait.year ,
          Trait.sex ,
          at(Trait,1).weanage, # at(Trait,weanwt).weanage also
works
          at(Trait,2).pen, # at(Trait,weight).pen also works
    !r  diag(Trait).pig  !f mv  # Diagonal structure
  residual idv(units).us(Trait) # Unstructured residuals
```

Structure for pig effect

$$\begin{bmatrix} \sigma_{A1}^2 & 0 \\ 0 & \sigma_{A2}^2 \end{bmatrix} \otimes \mathbf{A}_{5414}$$

Notice that we can use at() to fit *weanage* as a fixed covariate only for *weanwt* and *pen* as a fixed effect only for *weight*. The **G** structure for the animal effect is block diagonal formed by the Kronecker product of two matrices; a block-diagonal matrix of trait additive genetic variances and the **A** matrix derived from the pedigree. We could also write the term diag(Trait).pig as diag(Trait).nrm(pig) to clarify that the pig effects have variance-covariance structure proportional to the numerator relationship matrix (**A**). It is not necessary, however, since we defined pig as a pedigree-associated effect in the field definitions, so by default it is associated with the **A** matrix, not an identity matrix. This produces the following output:

Output for PART 3 (*Code 6-2_BivariateAnimal3.asr*).

```
 14 LogL=-2389.06    S2=  1.0000       5037 df
              - - - Results from analysis of weanwt weight - - -
 Akaike Information Criterion   24788.13 (assuming 5 parameters).
 Bayesian Information Criterion  24820.75
```

```
Model_Term                           Sigma      Sigma    Sigma/SE   % C
idv(units).us(Trait)        5112 effects
Trait             US_V  1  1    2.89147    2.89147      20.06   0 P
Trait             US_C  2  1   12.4938    12.4938       12.83   0 P
Trait             US_V  2  2   385.789    385.789       16.57   0 P
diag(Trait).pig             5414 effects
Trait             DIAG_V   1   1.98813    1.98813        9.49   0 P
Trait             DIAG_V   2   315.811    315.811        8.65   0 P
pig               NRM   2707
Covariance/Variance/Correlation Matrix US Residual
    2.891        0.3741
    12.49        385.8
```

- *Weanwt* has additive genetic variance of 1.98813 whereas *weight* has additive genetic variance of 315.811. The residual environmental correlation between two traits is 0.3741.

If the `!nodisplay` argument is used in the ASReml job file, a scatter plot of residuals is produced by default. The plot is split into two subplots, one for each trait in the model (Fig. 6.1). The first plot is for the trait listed first (weanwt).

For model 4, we extend the multivariate animal model to include correlations between direct genetic effects of pig breeding values using a CORGH structure:

Code example 6.17
Bivariate animal model with CORGH structure for pig effects (part of file **Code 6-2_BivariateAnimal.as**).

```
!PART 4  # US structure for pig and residual
weanwt weight ~ Trait.year ,
          Trait.sex ,
          at(Trait,1).weanage,
          at(Trait,2).pen,
!r  corgh(Trait).pig  !f mv  # CORGH structure
   residual idv(units).us(Trait) # Unstructured residuals
```

Structure for pig effect
$$\begin{bmatrix} \sigma^2_{A1} & \sigma_{A12} \\ \sigma_{A12} & \sigma^2_{A2} \end{bmatrix} \otimes \mathbf{A}_{5414}$$

Fig. 6.1 Residual plots of two traits from the bivariate model

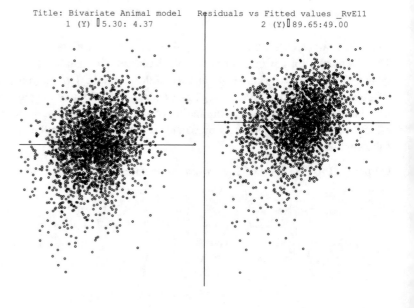

```
Title: Bivariate Animal model   Residuals vs Fitted values _RvE11
       1 (Y) 5.30: 4.37                2 (Y) 89.65:49.00
```

Output for PART 4 (*Code 6-2_BivariateAnimal4.asr*).

```
          - - - Results from analysis of weanwt weight - - -
Akaike Information Criterion    24788.44 (assuming 6 parameters).
Bayesian Information Criterion  24827.59

Model_Term                           Sigma         Sigma    Sigma/SE   % C
idv(units).us(Trait)         5112 effects
Trait            US_V  1  1   2.85247       2.85247      19.36    0 P
Trait            US_C  2  1   11.3732       11.3732       8.35    0 P
Trait            US_V  2  2   378.932       378.932      15.85    0 P
corgh(Trait).pig             5414 effects
Trait            COR_R  1  0.956054E-01  0.956054E-01    1.24    0 P
Trait            COR_V  1  2.05446       2.05446         9.35    0 P
Trait            COR_V  2  328.723       328.723         8.54    0 P
pig              NRM   2707
Covariance/Variance/Correlation Matrix US Residual
   2.852        0.3459
   11.37        378.9
Covariance/Variance/Correlation Matrix COR corgh(Trait).pig
   2.054        0.9561E-01
   2.485        328.7
```

- The additive genetic correlation between traits is low in this model (~ 0.10), and its inclusion does not appear warranted, since the AIC of this model increased compared to model 3. We will maintain this term in subsequent models for completeness.

For model 5, we introduce the maternal effects that can be explained as due to additive genetics effects. We introduce this as a separate term from the direct genetic effects, which means they are modeled as independent (zero covariance) from the direct genetic effects. However, we allow covariances within direct and maternal effects across traits with CORGH structures on each.

Code example 6.18
Bivariate animal model with pig and dam effects, independent of each other (part of file 'Code 6-2_BivariateAnimal.as').

```
!PART 5  # add maternal additive effects, but no correlation with
direct effects
weanwt weight ~ Trait.year ,
          Trait.sex ,
          at(Trait,1).weanage,
          at(Trait,2).pen,
 !r  corgh(Trait).pig, #direct genetic effects CORGH
     corgh(Trait).dam  !f mv  # maternal genetic effects CORGH
   residual idv(units).us(Trait) # Unstructured residuals
```

Structure for pig and dam effects -

$$\begin{bmatrix} \sigma^2_{A1} & & & \\ \sigma_{A12} & \sigma^2_{A2} & & \\ 0 & 0 & \sigma^2_{Am1} & \\ 0 & 0 & \sigma_{Am12} & \sigma^2_{Am2} \end{bmatrix} \otimes \mathbf{A}$$

This produces the following output:

Output for PART 5 (*Code 6-2_BivariateAnimal5.asr*)

```
       - - - Results from analysis of weanwt weight - - -
    Akaike Information Criterion    24133.59 (assuming 9 parameters).
    Bayesian Information Criterion  24192.31

    Model_Term                         Sigma      Sigma   Sigma/SE  % C
    idv(units).us(Trait)         5112 effects
    Trait              US_V 1  1   2.39593    2.39593     18.42   0 P
    Trait              US_C 2  1  11.2124    11.2124       8.40   0 P
    Trait              US_V 2  2  381.957    381.957      14.79   0 P
    corgh(Trait).pig             5414 effects
    Trait              COR_R  1 -0.994588   -0.994588     -0.10   0 P
    Trait              COR_V  1  0.108348E-01 0.108348E-01  0.05   0 P
    Trait              COR_V  2  187.036    187.036        4.18   0 P
    pig              NRM   2707
    corgh(Trait).dam             5414 effects
    Trait              COR_R  1  0.193259   0.193259       2.41   0 P
    Trait              COR_V  1  3.61617    3.61617       11.30   0 P
    Trait              COR_V  2  174.592    174.592        6.39   0 P
    dam              NRM   2707
    Covariance/Variance/Correlation Matrix US Residual
      2.396       0.3706
     11.21        382.0
    Covariance/Variance/Correlation Matrix COR corgh(Trait).pig
     0.1083E-01 -0.9946
     -1.416       187.0
    Covariance/Variance/Correlation Matrix COR corgh(Trait).dam
      3.616       0.1933
      4.856       174.6
```

- Fitting direct additive and maternal genetic effects as independent terms suggests that the direct additive genetic effect is almost zero (0.0108) and the trait is pretty much under control of maternal genetic effect (187.036). The correlation of -0.994588 is misleading in this case because of almost zero additive direct genetic variance.

Based on AIC, this model is superior to any of the previous multivariate models. Now we can extend this model by allowing covariances between the pig and dam effects within and across traits. This is done by extending the univariate model term `str(pig dam corgh(2).nrm(pig))` to its multivariate analog: `str(Trait.pig Trait.dam corgh(4).nrm(pig)`. In the multivariate case, the dimension of the CORGH matrix is four because there are two direct effects (one for each trait) for each pig and two direct effects for each dam. This will produce a matrix with four variances and six covariances between all pairs of effects:

Code example 6.19
Bivariate animal model with correlated pig and dam effects (part of file **Code 6-2_BivariateAnimal.as**).

```
!PART 6  # CORGH structure for pig + maternal genetic effect
weanwt weight ~ Trait.year ,
           Trait.sex ,
           at(Trait,1).weanage,
           at(Trait,2).pen,
 !r str(Trait.pig Trait.dam corgh(4).nrm(pig)) !f mv # CORGH
structure
   residual idv(units).us(Trait) # Unstructured residuals
```

Structure for pig and dam effects

$$
\begin{bmatrix}
\sigma^2_{A1} & & & \\
\sigma_{A12} & \sigma^2_{A2} & & \\
\sigma_{A1\,Am1} & \sigma_{A2\,Am1} & \sigma^2_{Am1} & \\
\sigma_{A1\,Am2} & \sigma_{A2\,Am2} & \sigma_{Am12} & \sigma^2_{Am2}
\end{bmatrix} \otimes \mathbf{A}
$$

Output for PART 6 (*Code 6-2_BivariateAnimal6.asr*).

```
            - - - Results from analysis of weanwt weight - - -
Akaike Information Criterion    24121.12 (assuming 13 parameters).
Bayesian Information Criterion   24205.94

Model_Term                      Sigma        Sigma   Sigma/SE  % C
idv(units).us(Trait)            5112 effects
Trait          US_V 1 1    2.12811         2.12811      12.03   0 P
Trait          US_C 2 1    10.1967         10.1967       6.20   0 P
Trait          US_V 2 2    383.131         383.131      12.94   0 P
corgh(4).nrm(pig)               10828 effects
4              COR_R  1  0.772225E-01  0.772225E-01     0.27   0 P
4              COR_R  2 -0.713218      -0.713218       -4.71   0 P
4              COR_R  3 -0.206750      -0.206750       -1.61   0 P
4              COR_R  4 -0.302079      -0.302079       -1.37   0 P
4              COR_R  5 -0.134150E-01 -0.134150E-01    -0.07   0 P
4              COR_R  6  0.256988       0.256988        2.99   0 P
4              COR_V  1  0.577114       0.577114        1.75   0 P
4              COR_V  2  185.100        185.100         3.48   0 P
4              COR_V  3  4.10084        4.10084        11.11   0 P
4              COR_V  4  176.907        176.907         5.53   0 P
pig            NRM    2707
Covariance/Variance/Correlation Matrix US Residual
   2.128        0.3571
  10.20         383.1
Covariance/Variance/Correlation Matrix COR Trait.pig
  0.5771    0.7722E-01  -0.7132      -0.3021
  0.7981    185.1       -0.2067      -0.1342E-01
 -1.097    -5.696        4.101        0.2570
 -3.052    -2.429        6.922       176.9
```

This model improves the AIC a bit more, but the variance component for direct genetic effects on *weanwt* (0.5771) appears small here, so maybe it is not needed in the model. A way to test that is to fit a reduced model that does not include the direct genetic effect on *weanwt* but does include it for *weight*. To demonstrate how to fit a complex variance-covariance structure that includes effects on only some of the trait, we will use `at()` to fit the direct pig effects only for *weight*, but include this term inside the `str()` function so that those direct effects have a covariance with the maternal genetic effects. Since we are excluding the direct effects on *weanwt*, the dimension of the CORGH matrix in this case is reduced from four to three and we will get estimates for three variance components and their three pairwise correlations:

Code example 6.20
Reduced bivariate animal model with correlated pig and dam effects but no direct genetic effects on *weanwt* (part of file '**Code 6-2_BivariateAnimal.as**').

```
!PART 7  # drop direct genetic effect on weanwt
weanwt weight ~ Trait.year ,
            Trait.sex ,
            at(Trait,1).weanage,
            at(Trait,2).pen,
!r str(at(Trait,2).pig Trait.dam us(3).nrm(pig)) !f mv #CORGH
structure
   residual idv(units).us(Trait) #
Unstructured residuals
```

Structure for pig and dam effects

$$-\begin{bmatrix} \sigma^2_{A2} & & \\ \sigma_{A2\,Am1} & \sigma^2_{Am1} & \\ \sigma_{A2\,Am2} & \sigma_{Am12} & \sigma^2_{Am2} \end{bmatrix} \otimes A$$

Output for PART 7 (*Code 6-2_BivariateAnimal7.asr*).

```
        - - - Results from analysis of weanwt weight - - -
Akaike Information Criterion    24133.44 (assuming 9 parameters).
Bayesian Information Criterion  24192.16

Model_Term                      Sigma      Sigma    Sigma/SE   % C
idv(units).us(Trait)            5112 effects
Trait           US_V  1  1     2.40114    2.40114    31.93    0 P
Trait           US_C  2  1     10.5130    10.5130    13.24    0 P
Trait           US_V  2  2     384.584    384.584    13.25    0 P
us(3).nrm(pig)           8121 effects
3               US_V  1  1     182.037    182.037     3.52    0 P
3               US_C  2  1    -2.56877   -2.56877    -0.80    0 P
3               US_V  2  2     3.62664    3.62664    12.29    0 P
3               US_C  3  1     7.01333    7.01333     0.21    0 P
3               US_C  3  2     5.11434    5.11434     2.30    0 P
3               US_V  3  3     169.590    169.590     5.48    0 P
pig             NRM    2707
Covariance/Variance/Correlation Matrix US Residual
   2.401       0.3460
   10.51       384.6
```

```
Covariance/Variance/Correlation Matrix US at(Trait,2).pig
  182.0      -0.1000      0.3992E-01
 -2.569       3.627       0.2062
  7.013       5.114        169.6
```

Notice that the AIC value for this model (24133.44) is worse (larger) than the AIC value for the more complete model, model 6 (24121.12) suggesting that we should not drop the direct genetic effects on *weanwt* and that model 6 is better for estimating variances and correlations and for predicting breeding values.

Electronic supplementary material: The online version of this chapter (doi:10.1007/978-3-319-55177-7_7) contains supplementary material, which is available to authorized users.

Abstract

In this chapter we describe spatial analyses that deal with heterogeneity of errors in a post-hoc analysis, rather than in the experimental design. These spatial analyses can improve the estimation of genetic effects by modeling more accurately the spatial distribution of error effects in a field trial. Selecting an optimal model can be complicated when models differ for both fixed and random effects. We demonstrate a process by which models can be compared to choose a model that is likely to provide the best predictions of genetic values.

Background

An assumption of the ordinary least squares analysis of variance is that error effects are independently and identically distributed (IID). This assumption is probably reasonable in many animal breeding and genetics studies, where an animal itself is an experimental unit, and is able to move about, limiting the effects of spatial trends in its environment. This assumption may not hold for many field experiments in plant or tree breeding, however. When possible, plant breeders select homogeneous field sites for experiments, but some fields may exhibit a high level of heterogeneity due to differences in soil type, fertility, water holding capacity, and so forth. As the number of experimental entries becomes larger, the ability to partition experimental fields into blocks of homogeneous experimental units becomes more difficult. Randomized complete blocks are an excellent experimental design when a single plot of each experimental entry can be fit into each of several homogeneous blocks, but this becomes increasingly difficult as the number of test genotypes becomes large.

Incomplete block designs, such as lattices (Cochran and Cox 1957) and alpha designs (Patterson and Williams 1976) were developed to handle situations where experimental units are not uniform within complete blocks. Incomplete block designs introduce unbalance into the experimental design, which can be well handled by mixed models analysis that simultaneously estimates the effects of random complete blocks and incomplete blocks along with the (random or fixed) treatment effects. Incomplete block designs will be very efficient when the experimental units within an incomplete block are homogeneous, but there is no guarantee that this will be the case in field experiments. In general, it is unlikely that an incomplete block design will capture the heterogeneity of experimental units in a field with unknown spatial trends (although some trends may be obvious before planting, many others will not; still other spatial trends will only appear after planting due to trends in management and data collection). Row-column alpha designs (John and Eccleston 1986) permit blocking in both the row and column directions while optimizing balance in the precision of pairwise entry comparisons as part of the design.

In this chapter we consider both randomized complete block and incomplete block field designs, and also describe spatial analyses that deal with heterogeneity of errors in a post-hoc analysis, rather than in the experimental design. These spatial analyses can improve the estimation of genetic effects by modelling more accurately the spatial distribution of error effects in a field. The potential improvement in analysis of field data by spatial analysis should not be taken as a reason to abandon the principles of good experimental design. Indeed, we recommend using incomplete block and row-column designs when possible for large experiments, and to compare traditional incomplete block design analyses with spatial analyses, and perhaps combine aspects of both analyses to optimally model the non-genetic variation in field experiments as a way to improve the estimation of breeding values of the test materials.

Modeling Spatial Effects

The randomized complete block design is one approach to handling heterogeneity in the experimental units by fitting the main effects of blocks in the analysis. The extent to which plots in different replications are different on average is attributed to the block effects, and the variation explained by the block effects is accounted for in the model itself, thus reducing the residual error variance compared to an analysis that ignored block effects. Incomplete block and row-column designs extend this concept further, absorbing additional variation due to the effects of incomplete blocks within complete blocks, at a cost of more parameters to estimate, introduction of imbalance to the experimental design, and different levels of precision for pairwise entry comparisons. When possible, we recommend the use of these incomplete block and row-column designs for large experiments designed to evaluate breeding values or genotypic values of crops or trees.

In some cases, for practical reasons, it is not possible to use a row-column design. For example, the North Carolina State University maize breeding program conducts research at several experimental research stations that manage experiments for a wide range of crops and research projects. Fields are rotated among crops and projects over years, and the identity of the fields to be used for planting are typically not known until after the time required to design experiments to allow seeds to be counted and packaged for planting. Thus, alpha lattices are typically used for experimental designs, with the knowledge that the incomplete blocks will likely not represent rectangular subsections of the field. Nevertheless, they may still capture some portion of field gradients, so are useful. The use of repeated check varieties (as in various forms of the augmented design, (Federer and Raghavarao 1975) in either a systematic or random fashion (Müller et al. 2010) helps to provide information on spatial trends and allows them to be modeled in this way.

In addition to designing experiments *a priori* to account for variability within complete blocks, there are post-hoc analysis approaches that attempt to model field position effects. For example, row and column effects can be included in the analysis by using trend analysis (which fits orthogonal polynomial variables to row and column positions; (Brownie et al. 1993), splines (Gilmour et al. 1997), penalized splines (Rodríguez-Álvarez et al. 2016), or even by simply fitting row and column as random effects if there is sufficient overlap of entries among rows and columns. Alternatively, one can model the error effects as being correlated due to spatial proximity as a way to account for spatial heterogeneity (see next section). Finally, these approaches are not exclusive, one can fit row and column factors in the model as fixed or random effects, and also model correlations among the residuals. A combination of incomplete block or row-column designs, spatial analysis, and model selection can improve accuracy of variety selection trials considerably (Qiao et al. 2000, 2004). A key challenge when these approaches are used is to avoid over-fitting the model and to choose an optimal model among the many various possibilities.

Variance-Covariance Matrix of Residuals

The typical ANOVA model for randomized complete blocks assumes that residuals are independent and identically distributed ($\varepsilon_i \sim iidN(0, \sigma_e^2)$). This means that the structure of the variance-covariance matrix of residuals, **R**, for plots 1 to n is:

$$\mathbf{R} = \begin{matrix} & \begin{matrix} 1 & 2 & 3 & \cdots & n \end{matrix} \\ \begin{matrix} 1 \\ 2 \\ 3 \\ \vdots \\ n \end{matrix} & \begin{bmatrix} \sigma_\varepsilon^2 & 0 & 0 & \cdots & 0 \\ 0 & \sigma_\varepsilon^2 & 0 & 0 & 0 \\ 0 & 0 & \sigma_\varepsilon^2 & 0 & 0 \\ \vdots & 0 & 0 & \ddots & 0 \\ 0 & 0 & 0 & 0 & \sigma_\varepsilon^2 \end{bmatrix} \end{matrix} = \mathbf{I}_n \sigma_\varepsilon^2 \tag{7.1}$$

In order to write the model to permit spatial correlations we need to understand the formation of variance structures via direct products ($\otimes$). Recall from Chap. 3 that the direct product of two matrices is

$$\mathbf{A}_{mxp} \otimes \mathbf{B}_{nxq} = \begin{bmatrix} a_{11}B & \cdots & a_{1p}B \\ \vdots & \ddots & \vdots \\ a_{m1}B & \cdots & a_{mp}B \end{bmatrix} \tag{7.2}$$

For example, for matrices A and B:

$$\mathbf{A} = \begin{bmatrix} x & z \\ y & w \end{bmatrix}, \quad \mathbf{B} = \begin{bmatrix} D & G & K \\ E & H & L \\ F & I & N \end{bmatrix} \tag{7.3}$$

Their direct product is:

$$\mathbf{A} \otimes \mathbf{B} = \begin{bmatrix} xD & xG & xK & zD & zG & zK \\ xE & xH & xL & zE & zH & zL \\ xF & xI & xN & zF & zI & zN \\ yD & yG & yK & wD & wG & wK \\ yE & yH & yL & wE & wH & wL \\ yF & yI & yN & wF & wI & wN \end{bmatrix} \tag{7.4}$$

A typical model for spatially correlated errors for field trials is the separable first-order autoregressive model in two dimensions (AR1 × AR1) (Cullis et al. 1998). This models the residuals as correlated based on the distance between plots in the row and column directions (with the correlation decreasing as increasing powers of the correlation with distance), and with different values of the spatial correlation in the two directions. If the field is arranged as a rectangular grid or rows and columns, we can use the direct product to efficiently write the complex variance-covariance matrix of residuals, $\mathbf{R}$.

By sorting the experimental units rows within columns, the variance-covariance matrix of the residuals can be written as a direct product of two sub matrices:

$$\sigma_e^2 \, \Sigma_c(\rho_c) \otimes \Sigma_r(\rho_r) \tag{7.5}$$

where $\Sigma_r(\rho_r)$ is the correlation matrix for the row model with dimension $r \times r$ and ρ_r is the auto-correlation parameter in the row direction; $\Sigma_c(\rho_c)$ is the correlation matrix for the column model with dimension $c \times c$ and ρ_c is the auto-correlation parameter in the column direction. Both are symmetrical matrices, so we show only the lower triangle elements:

$$\Sigma_r = \begin{bmatrix} 1 & & & & \\ \rho_r & 1 & & & \\ \rho_r^2 & \rho_r & 1 & & \\ . & . & . & . & \\ \rho_r^{r-1} & \rho_r^{r-2} & \rho_r^{r-3} & \cdots & 1 \end{bmatrix} \quad \Sigma_c = \begin{bmatrix} 1 & & & & \\ \rho_c & 1 & & & \\ \rho_c^2 & \rho_c & 1 & & \\ . & . & . & . & \\ \rho_c^{c-1} & \rho_c^{c-2} & \rho_c^{c-3} & \cdots & 1 \end{bmatrix} \tag{7.6}$$

In this model, correlations will be greater between pairs of residuals on adjacent plots than between residuals that are separated by more than one row or column, and the correlations will tend toward zero as the distance between plots grow. The residual correlation in the row direction is raised to the power of the distance between the two plots in the row dimension (where the distance is measured either as geographic distance or simply as the number of rows apart). Similarly, the residual correlation in the column direction is raised to a power equal to the distance between the columns that the two plots are located in. Since the correlation coefficients ρ_c and ρ_r are less than or equal to one, the pairwise correlation between plots will decrease as the distance between the plots increases. The two-dimensional pattern permits the correlation to change differently as distance increases in the row direction compared to in the column direction.

As an example, consider a field layout with 12 plots arranged in a grid of 3 rows and 4 columns:

	Column 1	Column 2	Column 3	Column 4
Row 3	9	10	11	12
Row 2	5	6	7	8
Row 1	1	2	3	4

An AR1 × AR1 residual structure would require three residual parameter estimates: σ_e^2, ρ_c, and ρ_r. The $\mathbf{R}$ matrix would be given by the product of the error variance and the direct product of the row and column AR1 matrices:

$$\Sigma_{\mathbf{r}}(\rho_r) = \begin{bmatrix} 1 & \rho_r & \rho_r^2 \\ \rho_r & 1 & \rho_r \\ \rho_r^2 & \rho_r & 1 \end{bmatrix} \tag{7.7}$$

$$\Sigma_c(\rho_c) = \begin{bmatrix} 1 & \rho_c & \rho_c^2 & \rho_c^3 \\ \rho_c & 1 & \rho_c & \rho_c^2 \\ \rho_c^2 & \rho_c & 1 & \rho_c \\ \rho_c^3 & \rho_c^2 & \rho_c & 1 \end{bmatrix} \tag{7.8}$$

$$\mathbf{R} = \sigma_e^2\, \Sigma_{\mathbf{r}}(\rho_r) \otimes \Sigma_c(\rho_c) = \sigma_e^2 \begin{bmatrix}
1 \\
\rho_c & 1 \\
\rho_c^2 & \rho_c & 1 \\
\rho_c^3 & \rho_c^2 & \rho_c & 1 \\
\rho_r & \rho_r\rho_c & \rho_r\rho_c^2 & \rho_r\rho_c^3 & 1 \\
\rho_r\rho_c & \rho_r & \rho_r\rho_c & \rho_r\rho_c^2 & \rho_c & 1 \\
\rho_r\rho_c^2 & \rho_r\rho_c & \rho_r & \rho_r\rho_c & \rho_c^2 & \rho_c & 1 \\
\rho_r\rho_c^3 & \rho_r\rho_c^2 & \rho_r\rho_c & \rho_r & \rho_c^3 & \rho_c^2 & \rho_c & 1 \\
\rho_r^2 & \rho_r^2\rho_c & \rho_r^2\rho_c^2 & \rho_r^2\rho_c^3 & \rho_r & \rho_r\rho_c & \rho_r\rho_c^2 & \rho_r\rho_c^3 & 1 \\
\rho_r^2\rho_c & \rho_r^2 & \rho_r^2\rho_c & \rho_r^2\rho_c^2 & \rho_r\rho_c & \rho_r & \rho_r\rho_c & \rho_r\rho_c^2 & \rho_c & 1 \\
\rho_r^2\rho_c^2 & \rho_r^2\rho_c & \rho_r^2 & \rho_r^2\rho_c & \rho_r\rho_c^2 & \rho_r\rho_c & \rho_r & \rho_r\rho_c & \rho_c^2 & \rho_c & 1 \\
\rho_r^2\rho_c^3 & \rho_r^2\rho_c^2 & \rho_r^2\rho_c & \rho_r^2 & \rho_r\rho_c^3 & \rho_r\rho_c^2 & \rho_r\rho_c & \rho_r & \rho_c^3 & \rho_c^2 & \rho_c & 1
\end{bmatrix} \tag{7.9}$$

In this symmetric matrix, element R[5, 4] is the correlation between residuals on plots row 5 and column 4, which are separated by one row and three columns, so the correlation is $\rho_r\rho_c^3$.

As a numerical example, if the correlation in the row direction were 0.9 and the correlation in the column direction were 0.5, the following **R** matrix would result:

$$\mathbf{R} = \sigma_e^2 \begin{bmatrix}
1 \\
0.5 & 1 \\
0.25 & 0.5 & 1 \\
0.125 & 0.25 & 0.5 & 1 \\
0.9 & 0.45 & 0.225 & 0.1125 & 1 \\
0.45 & 0.9 & 0.45 & 0.225 & 0.5 & 1 \\
0.225 & 0.45 & 0.9 & 0.45 & 0.25 & 0.5 & 1 \\
0.1125 & 0.225 & 0.45 & 0.9 & 0.125 & 0.25 & 0.5 & 1 \\
0.81 & 0.405 & 0.2025 & 0.10125 & 0.9 & 0.45 & 0.225 & 0.1125 & 1 \\
0.405 & 0.81 & 0.405 & 0.2025 & 0.45 & 0.9 & 0.45 & 0.225 & 0.5 & 1 \\
0.2025 & 0.405 & 0.81 & 0.405 & 0.225 & 0.45 & 0.9 & 0.45 & 0.25 & 0.5 & 1 \\
0.10125 & 0.2025 & 0.405 & 0.81 & 0.1125 & 0.225 & 0.45 & 0.9 & 0.125 & 0.25 & 0.5 & 1
\end{bmatrix}$$

In this example, you can see that the pairwise error correlation is greatest between plots in the same column and adjacent rows, for example plot pairs [row,col] [1,5] and [8, 4], with $r = 0.9$. It is lowest between the most distant plots, for example plot pairs [12,1] and [9,4], with $r = 0.10125$.

Model Selection

The objective of analysis of field trial data is to select a model that most accurately and parsimoniously characterizes the extraneous spatial effects in the field, and consequently provides the most accurate and precise estimates of variety effects. Recommendations for appropriate methods for model selection vary among authors. Brownie et al. (1993) and Brownie and Gumpertz (1997) gave highest priority to modelling fixed trends, and suggest selecting an appropriate residual variance-covariance (**R** matrix) structure only after accounting for fixed field trends. In contrast, Gilmour et al. (1997) recommend modelling the residual structure first, then fitting fixed trend effects after selecting an appropriate **R** structure. These authors also differ as to the relative importance of maximizing precision of variety predictions vs. model fitting criteria when selecting models.

Model selection can be conducted with likelihood ratio tests if two models being compared are nested models. That is, the two models share the same fixed effects, and one model (the "reduced model") has a subset of the random effects in the other model (the "full model"). If the two models share the same fixed effects but are not nested, they can be compared with information criteria such as Akaike's (AIC) or Schwarz's Bayesian Information Criteria (BIC)(Lynch and Walsh 1998; Welham et al. 2010). A technical issue complicates model selection procedures with mixed models, however: models that have different fixed effects and different random effects cannot be compared on the basis of their residual likelihoods or likelihood-based information criteria (AIC or BIC). REML proceeds by first absorbing fixed effects, then estimating the maximum likelihood random part of the model given the residual deviations from the fixed part of the model (Lynch and Walsh 1998). Thus, the likelihood spaces of models with different fixed effects estimated with REML are not equivalent, and the models cannot be compared on the basis of likelihoods. This means that we cannot compare a model with fixed trend effects and IID **R** structure ($\mathbf{R} = \sigma^2_e \mathbf{I}$) to a model with no fixed trend effects and spatially correlated **R** structure on the basis of their likelihoods.

We suggest the following procedure to conduct model selection for spatial analyses, to allow consideration of models with and without fixed trend effects, random block terms, and complex **R** structures:

1. Fit a base model with variety effects only; inspect graphical plots of residuals against rows and columns of the field grid to get a sense of the potential importance of field trends.
2. Fit a model with random blocks. These can include complete and incomplete blocks.
3. Fit a model with random row and column terms. Models 2 and 3 are not nested but can be compared by AIC or BIC (Kehel et al. 2010; Lynch and Walsh 1998). Users should be cautious about fitting this model if the experiment has limited replication in the absence of repeated check plots. This step involves fitting n_r and n_c effects, one for each row and column in the field grid, and this may result in an over fitted model.
4. Fit a model with fixed polynomial trend effects (up to 4th order in both directions (Brownie et al. 1993). This model is typically more parsimonious than Model 3 because it fits a maximum of eight effects for the trends, and perhaps can be used in some cases where Model 3 cannot be fit due to limited replication. (Alternatively, fit smoothing splines as fixed effects to capture extraneous variation in the row and column directions (Gilmour et al. 1997). Model 4 cannot be compared to models 2 or 3 based on likelihood criteria because their fixed effects differ. Nevertheless, the F-tests for the fixed trend terms can be inspected for significance, and the models can be compared on the basis of residual error variance, F-test for variety effects, and average standard error for pairwise variety comparisons. Choose the model that has the highest significance for the variety F-test and the lowest pairwise variety mean comparison standard error.
5. Add AR1 × AR1 **R** structure to best model selected in previous step and test the significance of this structure using a likelihood ratio test. It is possible that the spatial correlations are important in only one direction, which can be judged roughly by the magnitude of the correlation coefficient and the ratio of the correlation coefficient to its standard error. In such a case, a nested reduced model with AR1 × ID structure can be fit and tested with a likelihood ratio test.
6. After fitting the **R** structure, the fixed trend effects can be rechecked based on their F tests. Or if random row, column, or block structures were selected in step 3, they can be re-tested with a likelihood ratio test comparing models in steps 3 and 5. If not significant, they can be dropped to obtain a final parsimonious model.

Example of Spatial Analyses of Field Trial Data

We take as an example data from a barley field variety trial, composed of a randomized complete block with six replications of 25 varieties. These data are provided as one of the example data sets distributed with ASReml ("barley. asd", included with the data sets for this book) and their analysis is discussed in the ASReml user's guide example section (Gilmour et al. 2009), and in greater detail in Gilmour et al. (1997). The field arrangement was a regular grid of 10 rows and 15 columns. Complete blocks (reps) are composed of 25 plots representing the intersection of 5 rows and 5 columns. To demonstrate some of the features of ASReml Version 4, we created a new data file "barley_missing_data.csv" which is identical to the original data set, but we deleted the record for the last plot in the experiment (positioned in row 10, column 15 of the spatial plot grid).

A sequence of 10 models following the procedure outlined above is given in the program file "Code 7-1_SpatialAnalysis.as". A sample R script for drawing residual heat maps is provided in "Code 7-2_Heatmaps.R". Some key statistics for each model are presented in Table 7.1. An equivalent code for ASReml-R based on code provided by Miraslov Zoric (Institute of Field and Vegetable Crops, Serbia) is also included in the file 'Code 7-3_Spatial_asremlR.R'.

Code example 7.1
Spatial analysis of barley data (see example Code 7-1_SpatialAnalysis.as file for more details).

```
!ARG 1 !RENAME 1 !DOPART $1
# change part number to run different models
Title: Spatial analysis of barley data
  Rep        6   # Six replicates of 5x5 plots in 2x3 arrangement
  RowBlk    30   # Rows within replicates numbered across replicates
  ColBlk    30   # Columns within reps numbered across replicates
  row       10   # Field row
  column    15   # Field column
  variety   25
  yield

barley_missing_data.csv !skip 1 !ROWFACTOR row !COLFACTOR column !BRIEF -1

!PART 1   # no rep effects specified, simple residual
yield ~ mu variety !f mv
      residual units
predict variety
```

Model 1: No block or spatial effects
The base model is very simple, the only random term is the residual which has the default IID variance-covariance structure. We highlight a few new qualifiers and terms introduced in this model. First, we introduce !ROWFACTOR and !COLFACTOR as data file qualifiers. These terms are interpreted by ASReml as indicating the two factors that define the two dimensions of the spatial grid of the experimental unit. In this example, the dimension factors were conveniently names 'row' and 'column', but they could have had any other name. This qualifier is important because in combination with the term '!f mv' in the model, they will check that all combinations of the two factors exist in the data set (e.g., all possible combinations of row and column), and if some are missing, they will be created internally to complete the grid. The complete grid will be required in later models where we fit spatial correlations among the residuals. '!f mv' indicates that ASReml should estimate missing values fit as 'sparse fixed' effects. We include another data file qualifier, '!BRIEF -1' to tell ASReml that we want to obtain estimates of the fixed effects in the .asr file directly (otherwise we have to find them in the .sln file). We use this for convenience because later models will include some fixed effect covariates and we want to be able to quickly judge how important their effects are. For model 1, we do not need any of these additional qualifiers and terms, but they will

Table 7.1 A sequence of ten models used to analyze barley data set: their random and fixed effects, R structure, log likelihood, residual variance, F-statistic for H$_0$: no variety differences, and the average standard error of a difference (SED) between two variety means

Model: R structure	Random effects	Fixed effects	LogL	AIC	Residual variance	F	SED
1: IID	–	variety	−747	1496.72	44108	2.41	122
2: IID	rep	variety	−738	1479.98	34936	3.04	108
3: IID	row, col	variety	−732	1469.11	25523	3.79	99
4: IID	–	variety, pol(row,4), pol(col, 4)	−696[a]	1394.11	31604	3.31	106
5: IID	–	variety, pol(row,4), pol(col, 3)	−700	1401.09	31334	3.36	105
6: AR1×AR1	row, col	variety	−695	1400.25	35663	13.08	59
7: AR1×AR1	–	variety	−695	1396.74	39002	12.91	60
8: AR1×AR1+units	–	variety	−692	1391.72	46143+4927	10.11	61
9: ID×AR1+units	–	variety	−705	1416.06	42615+0	8.68	68

[a]Models 4 and 5 have different fixed effects than other models, they cannot be compared to each other or to other models using likelihood-based criteria, such as AIC

become important later. (However, if `!ROWFACTOR` and `!COLFACTOR` qualifiers are included without a complete spatial grid we must also use `!f mv` in the model, even if we are not fitting a spatial model. Furthermore, we also need to specify 'residual units', otherwise the default residual structure becomes a spatially correlated model).

In this example there are 149 plots, but by augmenting the data set to complete the grid, we end up with estimated effects for 150, so the residual variance structure is $\mathbf{R} = \sigma_e^2 \mathbf{I}_{150}$. Key diagnostic statistics for this and subsequent models are given in Table 7.1. "Average SED" (Table 7.1) refers to the average standard error of a difference between two variety means, and it is given at the bottom of the .pvs output file from ASReml if the "predict variety" statement is included after the model.

Model 2: RCBD model
We can fit the complete block effects of the RCBD design by adding a random rep effect to the model:

```
!PART 2   # Random rep effect
yield ~ mu variety !f mv !r rep
     residual units
predict variety
```

The model has two random terms; replications (compete blocks; $\mathbf{u}_r \sim N(0, \sigma_r^2 \mathbf{I}_6)$, and a residual component (IID residuals, or 'unit' effects, $\mathbf{R} = \sigma_e^2 \mathbf{I}_{150}$). We can compare diagnostic statistics between the base and RCBD models (Table 7.1).

The statistics in Table 7.1 demonstrate that the RCBD model is better than the base model. The spatial distribution of residuals can be seen in heat maps of the residuals plotted against row and column position for each model (Fig. 7.1). In each heat map, each box represents the residual effect on a plot, with darker boxes indicating plots with larger positive residuals, whereas lighter boxes indicate plots with larger negative residuals, and gray boxes have near zero residual (Fig. 7.1). The blocking structure of the randomized complete block design is superimposed on the heat map with black lines. The heat map for model 1 shows obvious spatial trends in the data, with generally small groups of adjacent plots sharing residuals of the same sign. This is a very patchy spatial distribution of residuals. The RCBD model (model 2) captures the generally negative

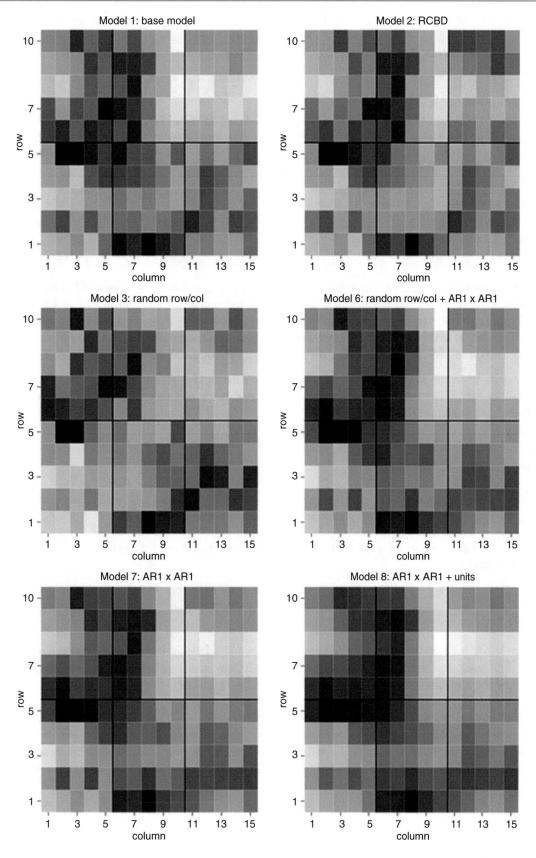

Fig. 7.1 Heat map of residuals from analysis of barley yield trial data: Model 1 (base model fitting only variety effects), Model 2 (RCBD model), Model 3 (random row and column effects), Model 6 (row and column effects plus AR1 × AR1 residual), Model 7 (AR1 × AR1 residual), and Model 8 (AR1 × AR1 plus units residual)

effect of the block in the upper right hand of the field grid, so that the residuals from this block are not uniformly negative (Fig. 7.1). Nevertheless, it is obvious that much of the spatial heterogeneity occurs in patches that are not delimited by the borders of the complete blocks. Thus, more complex spatial analysis models might better capture the heterogeneity of this field.

Model 3: Random row and column effects
Even though this experiment was not designed as a row-column incomplete block designs, we can nevertheless fit the effects of rows and columns in the analysis at the cost of imbalance in the analysis and the fitting of 10 row and 15 column effects.

```
!PART 3 # random row and column
yield ~ mu variety !f mv !r row column
      residual units
predict variety
```

A substantial reduction in error variance and average SED, along with increases in the variety F-test and log likelihood of model 3 compared to the RCBD model suggests that the random row and column effects do a better job of modelling and controlling the spatial variability than the complete blocks (Table 7.1). We cannot use a likelihood ratio test to compare these two models, because they are not nested models, but we can compare the models with AIC and BIC as both have the same set of fixed effects. AIC and BIC are reported in the .asr outputs, but here we demonstrate how they are computed:

AIC $= -2\ln L + 2k$, where $\ln L$ is the log likelihood of a model, k is the number of random covariance parameters. So, in this example $k = 2$ for Model 2 (one variance component each for replication and residual), and $k = 3$ for Model 3 (variance components for row, column, and residual).

AIC (model 2) $= -2(-743.340) + 4 = 1480$

AIC (model 3) $= -2(-736.828) + 6 = 1470$

The model with lower AIC value (in this case model 3) is considered the better model. The BIC has a stronger penalty for fitting additional model terms than the AIC, so we can also compare the models using BIC:

$BIG = -2\ln L + k\ln(n)$, where $\ln L$ and k are the same as in the AIC definition, and n is the total number of observations. $n = 149$ for both models,

BIC (model 2) $= -2(-738) + 2(\ln(149)) = 1486$

BIC (model 3) $= -2(-732) + 3(\ln(149)) = 1479$

The difference in BIC between the models is a little smaller, because the improvement in likelihood in model 3 is penalized more for the additional parameter used. The results of AIC and BIC are consistent in this case, the random row and column terms do a better job than the complete blocks of the RCBD design in controlling residual variance.

Model 4: Fixed row and column orthogonal polynomial trends
Fitting random row and column effects in model 3 provided an improved model fit, but required estimating 25 additional effects. We can attempt to capture the important linear and non-linear trends in the field more parsimoniously by fitting up to 4th order orthogonal polynomials in the row and column directions instead. Typically these are fit as fixed covariates.

```
!PART 4
yield ~ mu variety pol(row,-4) pol(column,-4) !f mv
      residual units
predict variety pol(row,-4) 0 pol(column,-4) 0
```

The term pol(row,−4) creates 4 orthogonal polynomials from first to 4th order based on the row factor. The row and column polynomials are significant based on their F-tests in the output:

```
                            Wald F statistics
       Source of Variation  NumDF   DenDF_con F-inc     F-con M P-con
   8 mu                         1    116.0 10208.11   9908.85 . <.001
   6 variety                   24    116.0     3.36      3.31 A <.001
   9 pol(row,-4)                4    116.0     4.06      4.03 A 0.004
  10 pol(column,-4)            4    116.0    10.20     10.20 A <.001
```

Notice that each trend term has four degrees of freedom. We get a combined F-test for all four row polynomials together, for example. We can check if we really need to fit up to 4th order polynomials by examining the individual regression coefficient estimates and their standard errors in the .asr output file:

```
                  Solution           Standard Error   T-value   T-prev
 10 pol(column,-4)
              1    -103.240             24.5588         -4.20
              2    -74.5544             30.4151         -2.45      0.75
              3    150.740              34.4541          4.38      5.10
              4    -0.136026E-01        32.8986         -0.00     -3.28

  9 pol(row,-4)

              1    -63.3770             23.5678         -2.69
              2    -5.70253             28.7630         -0.20      1.56
              3    50.1794              27.1655          1.85      1.42
              4    63.9285              25.2398          2.53      0.37
```

This indicates that the linear trend effect in the column direction is -103.2 (where the units are scaled to the change that would occur by moving halfway across the field), and its standard error is 24.6. A rough guide to significance for these regression coefficients is that their absolute value should be more than about twice their standard error to be significant (T-value with absolute value greater than two). In this example, it appears that the 4th order column polynomial is not significant and the 2nd and perhaps 3rd order row polynomials are not significant. We should therefore consider dropping the 4th order column polynomial and the 2nd and 3rd order row polynomials. ASReml does not have a simple function to fit, for example, the 1st and 4th order row polynomials only, however. Thus, if we want to fit the 4th order row polynomial, we must fit 1st through 4th order terms.

We cannot compare Models 4 and 3 on the basis of likelihoods, because they have different fixed effects. Model 3 has a smaller error variance and SE for variety comparisons and a larger variety F statistic than Model 4, however, suggesting it is a better model. But before abandoning the fixed polynomial trend effects, we can attempt to improve the model by dropping the 4th order column term.

Model 5: Fixed 1st to 4th order row and 1st to 3rd order column trends
Model 5 is a small modification to Model 4, differing only in that we now fit up to 3rd order column trends.

```
!PART 5 # Fixed 1st-4th order row and 1st-3rd order column trends
yield  ~ mu variety pol(row,-4) pol(column,-3) !f mv
       residual units
```

Model 5 seems better than Model 4 in every aspect, except for log likelihood, but we know that we cannot compare the models on that basis. The overall F-test for column trend polynomial is improved by fitting only up to 3rd order:

```
                              Wald F statistics

Source of Variation       NumDF      DenDF_con   F-inc      F-con M   P-con

8 mu                          1       117.0    10296.12   10030.47 .   <.001
6 variety                    24       117.0        3.39       3.36 A   <.001
9 pol(row,-4)                 4       117.0        4.10       4.07 A   0.004
10 pol(column,-3)             3       117.0       13.72      13.72 A   <.001
```

But based on the F-test for variety effects and the precision of variety differences, we still prefer Model 3, so we will not attempt to fit fixed spatial covariates any further.

As an aside, if we attempt to predict variety effects from either Model 4 or Model 5 with the usual statement:

```
predict variety
```

We do not get variety predictions; instead we get the following error message in the .pvs file:

```
Predict statement 1 aborted: cannot average over pol(column,-4)

Nominate a particular level of column to predict.
```

We need to explicitly include values of the covariates at which we want to make the predictions. A sensible approach is to make predictions where the **scaled row and column variables are zero, which is the center of the field**:

```
predict variety pol(row,-4) 0 pol(column,-3) 0
```

Model 6: Random row and column effects plus spatially correlated residuals in row and column direction (AR1 × AR1)
Now we augment Model 3 by allowing the residual effects to be correlated following the autoregressive spatial correlation pattern in both the row and column directions. We now specify the **R** structure as the direct product of two sub-matrices that model the correlations in column and row directions.

```
!PART 6   # random row and column + AR1 x AR1 residual
yield ~ mu variety !f mv,
     !r row column
   residual ar1v(row).ar1(column)
predict variety
```

In the functional specifications of **R**, residual ar1v(row).ar1(column) tells ASReml that **R** is a direct product of two correlation matrices $\left(\mathbf{R} = \sigma_e^2 \sum_r(\rho_r) \otimes \sum_c(\rho_c)\right)$, giving a two-dimensional (row and column) first order separable autoregressive spatial structure for error. Note that $\sum_c(\rho_c)$ is not a summation, but a correlation matrix structure, as shown previously in this chapter. We allow different correlations along the column and row directions.

The variance components estimates for this model are shown in the .asr output file as:

Model_Term			Gamma	Sigma	Sigma/SE	%	C
row	IDV_V	10	0.833561E-01	2972.74	0.64	0	P
column	IDV_V	15	0.122790E-06	0.437907E-02	0.00	0	B
ar1v(row).ar1(column)			150 effects				
Residual	**SCA_V**	**150**	**1.00000**	**35663.2**	**4.40**	**0**	**P**
row	**AR_R**	**1**	**0.482957**	**0.482957**	**5.60**	**0**	**P**
column	**AR_R**	**1**	**0.644831**	**0.644831**	**7.36**	**0**	**P**

The term labelled with 'Residual' and 'SCA_V' is the residual error variance component; the following terms labelled 'row' or 'column' and 'AR_R' are the autoregressive spatial correlations in the row and column directions.

Models 3 and 6 can be compared with a likelihood ratio test, as Model 3 is a nested model relative to Model 6. The difference in log likelihoods is considerable (about 37 units) with the addition of only two new parameters to the model (the two spatial error correlations), so it is obvious that Model 6 is preferred. The F-test for variety has increased dramatically from 3.79 to 13.08, and the average SED has been cut nearly in half (Table 7.1). Clearly, we have made a significant advance in modelling the spatial variation with this model. It may be strange to note, therefore, that the error variance of this model is actually quite a bit higher than that of Model 3. This is actually typical of AR1 × AR1 models. They achieve an improvement in model fit not by necessarily reducing the estimated residual error variance, but by modelling residuals more accurately through the spatial correlation structure. Recall that the variance of a comparison between two values is:

$$Var(X - Y) = Var(X) + Var(Y) - 2Cov(X, Y) \tag{7.10}$$

When we assume independent errors, then we assume that the covariance between the residuals of different plots is zero, so that there is also no covariance between the means of two varieties, and we can ignore the last term in this variance. If we model errors with spatial covariances, however, then the precision of each pairwise entry comparison will be different and it will depend on how close the plots of each entry are to each other. In this example, with only 25 entries, the two variety plots in the same rep, and even those in different reps, will not be too far apart; with a strong spatial correlation of residuals, the residuals on these plots will have a substantial positive covariance. Thus, if X and Y in the equation above represent the means of varieties X and Y from this experiment, the terms $Var(X)$ and $Var(Y)$ will be larger in Model 6 than in Model 3, but the $Cov(X,Y)$ term will also be fairly large and positive in Model 6 (and zero in Model 3), **reducing the variance of the variety comparison** to an extent that it more than offsets the increase in the error variance.

The effect of the AR1 × AR1 model on the spatial structure of residuals is, again, not to reduce the magnitude of the residuals, but instead to smooth out the edges of the patches so that negative residuals tend to group together (Fig. 7.1).

When a spatially correlated **R** structure is specified in ASReml, the program by default will generate two graphical displays of the residuals. One is the semivariogram, which displays the expected variance of a difference between residuals plotted against plot distance in row and column directions. For Model 6 of this example, the semivariogram is shown in Fig. 7.2.

Interpretation of semivariograms can be difficult especially for large trials, but sharp undulations in the pattern are an indication of poor modelling of the spatial trends (Gilmour et al. 1997). In this example, the semivariogram does not smoothly plateau as is ideal, but it does not indicate major deficiencies in the model.

Fig. 7.2 Semi-variogram of
residuals from Model 6 (random
row and column effects plus AR1
× AR1 residual structure)

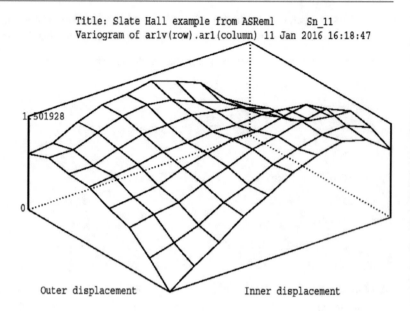

Fig. 7.3 Plot of residual trends
from Model 6 (random row and
column effects plus AR1 × AR1
residual structure)

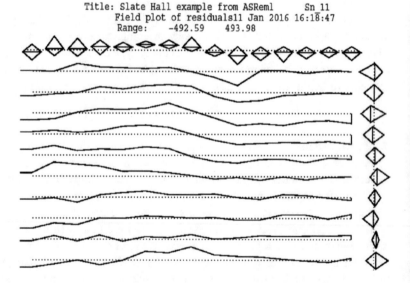

Another default graphical output of spatial models in ASReml is a plot of residual trends across columns for each row, which is an alternative display of the same information shown in the heat map, but also includes the row and column marginal summaries of residuals (Fig. 7.3).

While Model 6 is an improvement over previous models, it is no longer clear that the random main effects for row and column are needed. The variance for column effects is zero, so it can be dropped, but the variance component for row effects is also smaller than its standard error. It is not surprising that modelling the spatial correlations in row and column directions absorbs much of the variation that was previously associated with the row and column main effects and vice-versa. Hence, the next step is to drop the row and column random effects.

Model 7: Spatially correlated residuals in row and column directions (AR1 × AR1) only

```
!PART 7   # AR1 x AR1 residual only
yield ~ mu variety  !f mv
    residual ar1v(row).ar1(column)
predict variety
```

Comparing the results of Models 6 and 7 demonstrates a discrepancy among model selection criteria that can sometimes occur. According to the likelihood criteria, Model 7 is better. Since Model 7 is a nested model relative to Model 6, we can test the null hypothesis of no row and no column main effects using a likelihood ratio test (with two degrees of freedom), and there is almost no difference in likelihood between the models, so we would not reject the null hypothesis, and dropping row and column main effects seems justified. The AIC is also improved by dropping row and column main effects. However, the F-test for variety differences, the residual error variance, and the standard error for variety mean comparisons all are worse for Model 7, suggesting we should keep the row and column main effects (or, alternatively, that they only appear better for Model 6 due to overfitting). Our suggestion is to use likelihood criteria to compare models and to favor parsimony when possible, to avoid overfitting models. In this example, Model 7 has the better AIC and fewer parameters, so we suggest this model should be preferred.

A modification of the spatial residual model is to partition the residual variance into two separate pieces, one of which has spatial correlation, and the other of which has an independent distribution with no correlations. The independent part of the residual can be fit in the model by specifying 'units' in the random part of the model rather than as part of the residual. Recall that 'units' is a reserved term in ASReml that refers to the effects of individual observations, so the variance associated with the term is a residual variance.

Model 8: Spatially correlated residuals in row and column directions (AR1 × AR1) with units variance

```
!PART 8   # AR1 x AR1 residual + units
yield ~ mu variety  !f mv,
      !r units
      residual ar1v(row).ar1(column)
predict variety
```

The term *units* is added to the random part of the model, but it is really a residual variance term. The *units* term is sometimes called the *nugget effect* or the *measurement error*. It tells ASReml to construct an additional random term with one level for each experimental unit (e.g., tree or plot) so that a second (independent) error term can be fitted in addition to the (non-independent) residual effect modelled in the **R** structure. The unit or nugget variance is identically and independently distributed. So, this separates the residual effects into one portion that is independent among plots, and another part that is correlated between plots according to the spatial distances. If there is some part of the error that may be affected by spatial positioning (e.g, soil fertility and water holding capacity) and another part that is not affected by field position (e.g., the variation due to measurement error that is constant no matter where in the field the data are collected), this model makes sense.

We can also use this form of the model, where we fit *units* as the IID residual term and include the spatial variance structure due to row and column in the random part of the model:

```
yield ~ mu variety  !f mv,
      !r ar1v(row).ar1(column)
      residual units
```

The results are identical, although the labelling of terms in the output will differ. In both cases the actual residual variance structure is:

$$\mathbf{R} = \sigma_e^2 \left(\gamma_n I_{150} + \sum_r (\rho_r) \otimes \right) \sum_c (\rho_c)$$

The variance components estimates for this model are:

Model_Term			Gamma	Sigma	Sigma/SE	%	C
units	IDV_V	150	0.106788	4927.46	2.72	0	P
ar1(row).ar1(column)			150 effects				
Residual	SCA_V	150	1.00000	46142.7	2.73	0	P
row	AR_R	1	0.684346	0.684346	6.68	0	P
column	AR_R	1	0.844095	0.844095	12.31	0	P

The first term is the unit or nugget variance that has independent distribution, the second term is the residual variance component, which follows the AR1 × AR1 distribution, and the spatial correlation coefficients follow. Note that the unit variance is an order of magnitude smaller than the spatial error variance, indicating that most of the residual effects are spatially correlated. Also notice that the spatial correlation coefficients in both directions increased in Model 8 compared to Model 7. This occurs because we have partitioned the residuals into a part that is affected by spatial correlations and a part that is not, so the residual effects that remain in the spatially correlated part of the model have a stronger correlation.

Comparing the error variance between models 8 and 7, we see that the residual variance has increased dramatically, although the average SED has not changed much. Again, this occurs because the precision of comparisons of varieties means takes into account the correlations among their residuals, and as those correlations become more strongly positive, then they offset the inflation of the residual variance component. The variety F-test has decreased a bit, which brings into question whether this is really a better model or not. We can make a formal test of the significance of the units variance by noting that Model 7 is a reduced nested model compared to Model 8 and that they differ by one parameter, so the likelihood ratio test is about 6 with one degree of freedom, so the p-value is less than 0.05. The addition of the unit variance significantly improves the model, even though it reduces the precision of the variety comparisons by a small amount. The semivariogram does not change noticeably, but the heat map shows that the trend toward smoothing the patchiness of the residuals observed in Model 7 is enhanced (Fig. 7.1). Again, this occurs, because Model 8 estimates even higher levels of the spatial correlations, meaning that it models strong positive correlations among adjacent plots.

Model 9: Spatial correlation in one direction
Results from Model 7 suggest that the spatial correlations are strong and significant in both row and column directions. Therefore, there is no reason to drop one or the other correlation. For demonstration purposes only, however, we show a reduced model whereby we replace the AR1 × AR1 **R** structure with an ID × AR1 structure, which has residuals correlated based solely on their distance in the column direction only.

```
!PART 9 # ID x AR1 + units
yield ~ mu variety   !f mv,
        !r units
        residual idv(row).ar1(column)
predict variety
```

This code specifies a two-dimensional spatial structure for error but with spatial correlation in the column direction only, and independence in the row direction (ID); that is $\sim \sigma_e^2 \mathbf{I}_{10} \otimes \sum_r(\rho_r)$. Variance components estimates from this model are:

Model_Term			Gamma	Sigma	Sigma/SE	%	C
units	IDV_V	150	0.101193E-06	0.431237E-02	0.00	0	B
idv(row).ar1(column)			150 effects				
Residual	SCA_V	150	1.00000	42615.3	4.66	0	P
column	AR_R	1	0.731845	0.731845	11.78	0	P

Comparing Models 8 and 9, it is obvious that dropping the residual correlation in the row direction significantly reduces the fit of the model, so we should not consider the ID × AR1 **R** structure further. We select Model 8 as our final model and estimate the variety means from this model using the predict statement.

Heritability Estimate from Spatial Model

In the barley example we assumed that the variety effect is fixed because we were interested in comparing the variety means and selecting the desired ones. If the analyst is interested in estimation of some population genetic parameters, such as variance components, heritability and genetic correlations, then the variety effect can be treated as random. However, the spatially correlated residual model introduces some complications for estimating heritability, since we saw previously that correlated errors model may provide a better fit to the data but with a larger residual variance. The typical heritability estimate uses only the variance components and ignores the correlations among residual effects, so if the residual error variance increases, the heritability estimate will decrease. Intuitively, breeders will not prefer a model with lower heritability. We show here that an alternative estimator of heritability is more appropriate in this case and will better reflect the effectiveness of selection with correlated errors models.

For demonstration, we will compare two alternative models where we fit variety as a random effect. In the first case (Model 10) we fit replications as a random effect and use an IID error structure. In the second case (Model 11), we drop the replication main effect and fit the residuals using an AR1 × AR1 structure with IID units. The ASReml code is given below.

***Models 10 and 11: variety as random effect, with reps and IID residuals or with AR1 × AR1 + units residual
structure***

```
# treat variety as random to estimate heritability
!PART 10 # Random rep and variety, IID residual
yield ~ mu  !f mv,
       !r rep variety
       residual units
predict variety #marginal prediction
predict variety !AVERAGE rep #conditional prediction
VPREDICT !DEFINE
F pheno_plt variety + units
F pheno_lm  variety + units*0.167
H H_plt variety pheno_plt
H H_lm variety pheno_lm


!PART 11 # Random variety and AR1xAR1 row/col, IID nugget residual
yield ~ mu  !f mv,
        !r variety ar1v(row).ar1(column)
        residual units
predict variety #marginal prediction
predict variety row 0 column 0 #marginal prediction
predict variety !AVERAGE row column #conditional prediction
VPREDICT !DEFINE
F pheno_plt variety + units + ar1v(row).ar1(column)[2]
F pheno_lm variety + units*0.167  + ar1v(row).ar1(column)[2]*0.167
H H_plt variety pheno_plt
H H_lm variety pheno_lm
```

For both models, we want to estimate heritability on a plot-basis (H_plt) and on a line mean-basis (H_lm) using the
VPREDICT directive. For Model 10, the phenotypic variance among plots is the sum of the variety and residual variances;
for Model 11, we include both the nuggets (units) variance and the residual variances. The summary results from the two
models are:

Results of Model 10

```
 7 LogL=-869.009           S2= 34921.     148 df     0.2678       0.3415

Final parameter values           0.2678          0.3415

              - - - Results from analysis of yield - - -

Akaike Information Criterion      1744.02 (assuming 3 parameters).

Bayesian Information Criterion    1753.01


          Approximate stratum variance decomposition
```

```
Stratum       Degrees-Freedom Variance Component Coefficients

Model_Term                   Gamma      Sigma      Sigma/SE    %  C

rep           IDV_V    6   0.267832    9352.85        1.37     0  P

variety       IDV_V   25   0.341482    11924.8        2.30     0  P

units               150 effects

Residual      SCA_V   150  1.00000     34920.6        7.71     0  P
```

Results of Model 11

```
  11 LogL=-822.656          S2= 5194.8 148 df
                    - - - Results from analysis of yield - - -
Akaike Information Criterion      1655.31 (assuming 5 parameters).
Bayesian Information Criterion    1670.30

Model_Term                   Gamma        Sigma      Sigma/SE    %  C

variety       IDV_V    25    3.00380      15604.2       3.09     0  P

units 150 effects

Residual      SCA_V   150    1.00000      5194.83       2.90     0  P

ar1v(row).ar1(column)    150 effects

row           AR_R     1     0.694413     0.694413      6.91     0  P

row           AR_V     1     8.76838      45550.3       2.68     0  P

column        AR_R     1     0.845972     0.845972     12.37     0  P
```

Since there are no fixed factors in either model, we can compare the model fits using likelihoods and AIC/BIC. It is clear that Model 11 is a better fitting model by all criteria. However, when we estimate the heritabilities using functions of variance components, we find that the heritability estimate from Model 11 is lower because the residual variances are larger from the correlated errors model:

Model 10 .pvc

```
                 - - - Results from analysis of yield - - -
1 rep                     V    6    9352.85    6826.90
2 variety                 V   25    11924.8    5184.70

units                   150 effects
3 units;Residual          V  150    34920.6    4529.26
4 pheno_plt 2      46845.      6366.5
5 pheno_lm 2       17757.      5129.1

H_plt        = variety 2/pheno_pl 4= 0.2546      0.0895

H_lm         = variety 2/pheno_lm 5= 0.6716      0.1040
```

Model 11 pvc

```
                              - - - Results from analysis of yield - - -
    1 variety                           V    25      15604.2      5049.90

units                         150 effects

    2 units;Residual                    V   150      5194.83      1791.32

ar1v(row).ar1(column)         150 effects

    3 ar1v(row).ar1(column);ar1v(row)   R    1       0.694413     0.100494

    4 ar1v(row).ar1(column);ar1v(row)   V    1       45550.3      16996.4

    5 ar1v(row).ar1(column);ar1(column) R    1       0.845972     0.683890E-01

    6 pheno_plt 1                66349.     18115.

    7 pheno_lm 1                 24079.     5823.5

      H_plt    = variety    1/pheno_pl    6=          0.2352       0.0848

      H_lm     = variety    1/pheno_lm    7=          0.6481       0.1075
```

As discussed previously, the explanation of this apparent paradox is that the variance of variety mean differences is actually smaller in Model 11 because the correlated residuals introduce a covariance term that is subtracted from the variance components (Eq. 7.9). The standard error of a difference between two variety means involves the nugget variance component plus the spatial variance component *minus the average covariance between plot residuals for the two varieties*. If we ignore the spatial variance component and use only the units variance, we underestimate the phenotypic variance. If we include the spatial variance component without accounting for the covariances of the residual effects, we overestimate the phenotypic variance. To properly account for the correlated residuals, we should use an alternative approach for estimating the heritability that is appropriate for complex residual structures and unbalanced experimental designs introduced by Cullis et al. (2006) and discussed by Piepho and Möhring (2007):

$$H_C = 1 - \frac{\bar{V}_{BLUP\_difference}}{2\widehat{\sigma}_G^2} \tag{7.11}$$

Where $\bar{V}_{BLUP\,difference}$ is the average variance of a difference between a pair of variety BLUPs.

Notice that this definition of heritability is related to reliability of breeding value predictions introduced in Chaps. 2 and 4 and its square root of accuracy (Eq. 4.25). The reliability of a random effect prediction is the squared correlation between predicted and true values of the effect u and can be estimated as a function of the prediction error variance (PEV) and the genetic variance component (Mrode 2014):

$$\widehat{r}^2_{g,\widehat{g}} = 1 - \frac{PEV}{\widehat{\sigma}_g^2} \tag{7.12}$$

In Chap. 4, we used a specific form of this equation to obtain the accuracy of a breeding value prediction. In our current example, the variety effects correspond to inbred genotypes and are not interpretable as breeding values, but we can use this same general formula. The expectation of average reliability across all tested genotypes is the generalized heritability applicable to predicting response to selection based on genotype BLUPs (Piepho and Möhring 2007):

$$E\left(\widehat{r}^2_{g,\widehat{g}}\right) = \frac{\sigma_g^2}{\sigma_P^2} = h^2 \tag{7.13}$$

To understand the relationship between Eqs. 7.10 and 7.11, notice the relationship between the average standard error of prediction differences (which we obtain from ASReml using the prediction directive) in the case of uncorrelated errors and PEV:

$$SE \text{ of prediction differences} = \sqrt{Var(BLUP \text{ differences})} = \sqrt{Var\left(\widehat{Y}_i - \widehat{Y}_{i'}\right)} = \sqrt{2Var\left(\widehat{Y}\right)_i} = \sqrt{2PEV}$$

Substituting the average standard error of prediction differences into the reliability formula gives:

$$1 - \frac{PEV}{\widehat{\sigma}_g^2} = 1 - \frac{\left(SE \text{ of prediction differences}/\sqrt{2}\right)^2}{\widehat{\sigma}_g^2} = 1 - \frac{\widehat{V}_{BLUP \text{ difference}}}{2\widehat{\sigma}_g^2}$$

ASReml provides the average standard error of a difference between variety predictions (the square root of $\bar{V}_{BLUP \text{ difference}}$ in Eq. 7.10) in the last line of the prediction output file (.pvs):

Average standard error of difference between variety BLUPs from Model 10:

```
SED: Overall Standard Error of Difference 88.68
```

Average standard error of difference between variety BLUPs from Model 11:

```
SED: Overall Standard Error of Difference 57.92
```

These values are the square roots of the average variances for BLUP differences that we need for the Cullis estimators of heritability. For model 10, we have:

$$H_C = 1 - \frac{\bar{V}_{BLUP\_difference}}{2\widehat{\sigma}_G^2} = 1 - \frac{88.68^2}{2(11924.8)} = 1 - 0.33 = 0.67 \tag{7.14}$$

Notice that in the case of balanced data and IID residuals, this estimator of heritability is identical to the classical estimate of heritability on a line mean-basis based on variance components. However, for Model 11, the estimate of heritability on a line mean-basis is:

$$H_C = 1 - \frac{\bar{V}_{BLUP\_difference}}{2\widehat{\sigma}_G^2} = 1 - \frac{57.92^2}{2(15604.2)} = 1 - 0.11 = 0.89 \tag{7.15}$$

The generalized heritability for line means from the AR1 × AR1 model is substantially higher, and this is congruent with the result that it is a better fitting model. The estimate on line basis using V_BLUP_difference (0.89) is also higher than the estimate (0.648) we obtained from Model 11 (.pvc file). The estimator of heritability for spatial analysis proposed by Cullis et al. (2006) is preferred over the variance components-based estimator for models with non-IID residual variance structures, because the variance components-based estimators are biased downward when the spatial error correlations are positive.

Notice that we can obtain the correct heritability estimates also by computing the average reliabilities of the varieties if we request conditional predictions averaged across the observed levels of replications (in Model 10) or row and column (in Model 11). The marginal predictions themselves are fine, but their standard errors are too large, since they include variation due to the variation across replications (or across rows and columns). The marginal predictions are obtained with the following predict statements:

```
!PART 10 # Random rep and variety, IID residual
yield ~ mu  !f mv,
        !r rep variety
        residual units
predict variety #marginal prediction
...
!PART 11 # Random variety and AR1xAR1 row/col, IID nugget residual
yield ~ mu  !f mv,
        !r variety ar1v(row).ar1(column)
        residual units
predict variety #marginal prediction
predict variety row 0 column 0 #marginal prediction
```

For Model 10, the marginal predictions for the first three varieties are:

variety	Predicted_Value	Standard_Error	Ecode
1	1291.0009	74.4751	E
2	1494.2845	74.4751	E
3	1461.4679	74.4751	E
...			

SED: Overall Standard Error of Difference 88.68

For Model 11, the marginal predictions are:

variety	Predicted_Value	Standard_Error	Ecode
1	1262.2420	98.4421	E
2	1510.4295	98.4350	E
3	1403.2221	98.7673	E
...			

SED: Overall Standard Error of Difference 57.92

Notice that we obtain the same marginal predictions and standard errors if we ignore the row and column effects in the predict statement ('predict variety') or if we predict them at level '0' of row and column ('predict variety row 0 column 0').

The conditional predictions are obtained by explicitly requesting their values at the average levels of the other random effects:

```
!PART 10 # Random rep and variety, IID residual
predict variety !AVERAGE rep # conditional prediction
...
!PART 11 # Random variety and AR1xAR1 row/col, IID nugget residual
predict variety !AVERAGE row column # conditional prediction
```

Notice that we obtain the same predictions for Model 10 and the same average standard error of a variety difference, but the individual conditional predictions have lower standard errors.

Model 10 conditional predictions:

```
variety      Predicted_Value          Standard_Error      Ecode

1                  1291.0009                  63.1485      E

2                  1494.2845                  63.1485      E

3                  1461.4679                  63.1485      E

...

SED: Overall Standard Error of Difference 88.68
```

Notice that almost all of the standard errors of the individual variety predictions are the same; their average value (63.31) times the square root of 2 gives about the overall standard error of variety prediction differences (88.68). Similarly, if we compute reliability for each variety i as:

$$\hat{r}^2_{gi,gi} = 1 - \frac{PEV_i}{\hat{\sigma}^2_g} = 1 - \frac{(SE_i)^2}{11924.8} \tag{7.16}$$

the average of the reliabilities from Model 10 is 0.66, almost equal to the estimate obtained in Eq. 7.11.

The conditional predictions obtained from Model 11 are different from the values predicted by ignoring the row and column factors, since now we are averaging over the levels of row and column:

```
variety      Predicted_Value          Standard_Error    Ecode
1                  1288.4823                  40.9721    E

2                  1536.6699                  42.1394    E

3                  1429.4624                  40.7044    E

...

SED: Overall Standard Error of Difference 57.92
```

However, the relative differences among variety predictions remain identical (they are all increased by 26.24 compared to the marginal predictions), and the overall standard error of difference has not changed. The individual prediction standard errors, however, are much lower for the conditional predictions than the marginal predictions from the same model, and their average value is about $\sqrt{2}$ times greater than the overall standard error of differences. The average reliability of variety predictions for Model 11 is equal to the Cullis estimator of heritability obtained previously:

$$\hat{r}^2_{gi,gi} = 1 - \frac{(SE_i)^2}{15604.2} = 0.89.$$

where SE_i is the average standard error of variety prediction (41.4278).

Electronic supplementary material: The online version of this chapter (doi:10.1007/978-3-319-55177-7_8) contains supplementary material, which is available to authorized users.

© Springer International Publishing AG 2017

F. Isik et al., *Genetic Data Analysis for Plant and Animal Breeding*, DOI 10.1007/978-3-319-55177-7_8

Abstract

Most field tests for plant breeding are replicated across different environments to measure the performance of breeding stocks across a range of environmental conditions to which a cultivar might be exposed. Multi-environment trials provide information about the adaptability of genotypes to specific environments or to sets of environments. The variance-covariance structures introduced in preceding chapters can be used to model genotype-by-environment interactions in multi-environmental trials. The number of parameters required to fit fully specified multi-environment trial models increases faster than the number of environments, so more parsimonious models are preferred when the number of environments is large. In this chapter we compare an unstructured matrix that involves separate parameters for genetic variance within each environment and for genetic correlations between each pair of environments to more parsimonious models, such as factor analytic structures, which require fewer parameters. Factor analytical structures can often efficiently capture the genotype-by-environment patterns without requiring extraordinary model complexity.

Introduction

In multi-environmental trials (MET) a set of genotypes or families are raised in a number of environments. The objectives are to compare genotypes across a range of environments and identify those that are generally adaptable across the testing environments, or to identify superior genotypes for subsets of the testing environments. If broad adaptation is not possible, then the breeder may instead prioritize selecting different genotypes with good performance in subsets of the environments. Proper analysis of MET data can reveal not only which genotypes are 'best' overall or in subsets of environments, but also can reveal the relationships among environments in terms of the genotype by environment (GxE) interaction patterns. This information can be used to improve the efficiency of breeding programs by identifying highly correlated clusters of environments that may represent oversampling of similar environments.

In addition to using METs to estimate GxE interactions, METs can serve a practical purpose in reducing the risk of losing the genetic materials due to environmental catastrophes. In many cases, breeders test a subset of material that is available in a given year and establish new field trails as new material becomes available. A subset of the genetic material (such as 'check' varieties) is used across multiple years to establish connections between testing series (years). A series of field trials established over time are also METs.

For yield and growth traits, large differences are often observed among environments. This occurs because of variation in soil fertility, precipitation, temperatures, and pathogen pressure. In perennial species, additional variation may be introduced because the ages of tests may differ among environments. For example, different growth rates may cause significant GxE interaction due to differences in the magnitude of genotypic variances across sites, even if genotype ranks do not change across environments (Cockerham 1963; Cooper and DeLacy 1994). This is a form of GxE interaction that does not hinder breeding gains, but is simply caused by the scale effect.

Ignoring heterogeneity in the variances can introduce bias in the predictions of breeding values and estimates of genetic variances, particularly if the breeding units are not replicated across all environments (Hill 1984). Accounting for heterogeneity in the data improves the accuracy of evaluations. In crop trials and in forest tree field tests, which may be balanced across sites within a year, accounting for heterogeneity of error variances in the mixed model can improve genotype predictions by giving more weight to information from environments with lower error variances.

Historically, such problems were not easy to handle with ordinary least squares ANOVA, but the flexibility of mixed models permits fitting complex multi-environment models that account for differences between residual as well as genotypic variances among sites. Further, mixed models approaches allow modelling of the pairwise genetic correlations among environments that provide a more realistic treatment than assuming that all pairs of environments have a common correlation, as was done in traditional ANOVA.

MET: General Approach and Considerations

Depending on the number of sites, we can perform one-stage or two-stage MET analysis. One stage is preferable but may not be feasible if there are large numbers of trials. Two-stage analysis proceeds as follows:

- Analyse data for each environment separately to check the data quality and estimate means and variances. We recommend this step even for one-stage analysis.
- If field position coordinates of plots are available, select the optimal spatial model for each site and predict site-specific genotype values for varieties
- Save predictions and their standard errors in a file
- Conduct a combined analysis across the sites based on the site-specific predictions. In combined analysis, some or most of the variances can be fixed to help with model convergence.
- The second stage can be weighted by the inverse of the variances of predictions of values from the first stage (Welham et al. 2010).

With increased computing power, one-stage analysis has become more feasible for large data sets. ASReml facilitates fitting different models for within-environment non-genetic effects and variation for different sites in the multi-environment single stage analysis. Differences among sites can include: (1) different field designs and covariates, (2) different spatial models within sites, and (3) heterogeneous variances across sites.

Modelling genetic correlations between each pair of sites using an unstructured (US) covariance matrix is feasible if there are a few sites and many entries. For s environments, the US covariance model requires s within-environment genetic variances and $s(s-1)/2$ pairwise environment covariances. If there are many sites, the number of parameters to estimate becomes very large. In such cases, factor analytic (FA or XFA) models are often more appropriate for modelling complex GxE patterns because they are more parsimonious, involving fewer parameter estimates. ASReml also allows fixing some variances and correlations that are at the boundary of theoretically allowable values to help models converge. For a given MET data set, we can consider a hierarchy of models of increasingly complex variance-covariance structures for both residuals (the **R** matrix) and genotype-environment effects (the **G** matrix). Like spatial analysis of field trials, a major focus of MET analysis is on selecting the best fitting model (while avoiding over fitting) to account for heterogeneity and predict the breeding values of genetic entries with high confidence.

Typical **R** structures that can be tested in MET analyses include:

- **IDV** *structure*: one common error variance for all environments
- **DIAG** *structure*: heterogeneous error variances across sites
- **AR1** *structure*: heterogeneous spatially correlated R structure: each environment has unique error variance and two-dimensional spatial error correlation pattern.

Commonly used **G** structures for MET include:

- **DIAG** *structure*: each environment has a unique genetic variance, but there are no correlations between environments (1 parameter for **G**)
- **CORUV** *structure*: constant genetic correlation between environments and genetic variance within environments (2 parameters for **G**). We show below that this is the traditional ANOVA structure for multi-environment models. This structure is also called "compound symmetry."
- **CORUH** *structure*: constant genetic correlations between pairs of environments but heterogeneous genetic variances within environments (with $s + 1$ variance parameters). If there are $s = 10$ environments, then $s + 1 = 11$ variance parameters are needed.
- **US** *structure*: unstructured covariance and heterogeneous variance. Each environment has a unique genetic variance and each pair of environments has a unique covariance, with $s(s + 1)/2$ variance parameters. For 10 environments, $10(11)/2 = 55$ variance parameters are needed.
- **CORGH** = **US** *structure*: This is also a fully heterogeneous genetic correlation and variance structure, so is equivalent to the US structure, but it is parameterized in terms of correlations instead of covariances between environments.
- **FA***n* and **XFA***n* *structures*: Factor analytic and extended factor analytic models that model heterogeneous within-environment genetic variances and unique pairwise correlations between environments, but the correlations are constrained to capture only the first n multivariate factors in the data. This requires $s(k+1)-k(k-1)/2$ parameters, where k is the number of factors modelled. For ten environments, an FA1 model requires **20** parameter estimates. This is a large reduction compared to US and CORGH structures.

Statistical Models

The classical ANOVA model for a cross-classified design of m genotypes evaluated at s environments with b complete blocks at each site is:

$$Y_{ijkl} = \mu + E_i + G_j + GE_{ij} + B(E)_{ik} + \varepsilon_{ijkl} \qquad (8.1)$$

where μ is the overall mean, E_i is the fixed effect of environment i; G_j is the random effect of genotype j, $G_j \sim N\left(0, \sigma_G^2\right)$; GE_{ij} is the random interaction between genotype j and environment i, $GE_{ij} \sim N\left(0, \sigma_{GE}^2\right)$; $B(E)_{ik}$ is the random effect of block k nested in environment i, $B(E)_{ik} \sim N\left(0, \sigma_B^2\right)$; ε_{ijkl} is the residual error associated with the experimental unit l of genotype j in k-th block of environment i, $\varepsilon_{ijkl} \sim N\left(0, \sigma_e^2\right)$.

From the analysis of variance, we can estimate the variance components and compute the means of genotypes at specific sites and across all environments. Importantly, this model assumes that there are no correlations between different factors in the model. Based on that assumption, the covariance between the values of a genotype at two environments i and i' is:

$$Cov\left(Y_{ij.}, Y_{i'j.}\right) = Cov\left(G_j, G_j\right) + Cov\left(GE_{ij}, GE_{i'j}\right) = \sigma_G^2 + 0 \qquad (8.2)$$

Where $Y_{ij.}$ and $Y_{i'j.}$ are values of a genotype j at two environments; G_j is the genetic value of genotype j, which is the same at the two environments; GE_{ij} and $GE_{i'j}$ are interaction effects of genotype j with the environments. By definition, the covariance of the genotype j effect with itself is the variance of genotype effects. The covariance between interaction effects of genotype j with two different environments is zero. So, the covariance of genotype j at two environments is the variance of genotypes (σ_G^2). The variance of true genotypic values within an environment (measured without error) is:

$$Var\left(Y_{ij.}\right) = Var\left(G_j\right) + Var\left(GE_{ij}\right) = \sigma_G^2 + \sigma_{GE}^2 \qquad (8.3)$$

So, the correlation between true values of one genotype at any two sites is

$$r\left(Y_{ij}, Y_{i'j}\right) = \frac{Cov\left(Y_{ij}, Y_{i'j}\right)}{\sqrt{Var\left(Y_{ij}\right)Var\left(Y_{i'j}\right)}} = \frac{\sigma_G^2}{\sqrt{\left(\sigma_G^2 + \sigma_{GE}^2\right)\left(\sigma_G^2 + \sigma_{GE}^2\right)}} = \frac{\sigma_G^2}{\sigma_G^2 + \sigma_{GE}^2} \qquad (8.4)$$

The ratio $\sigma_G^2 / \left(\sigma_G^2 + \sigma_{GE}^2\right)$ is sometimes called a 'type B genetic correlation'. Typically, type B genetic correlations refer to correspondence in performance of family means at different environments (Yamada 1962). The ratio is bounded as $0 \leq r_B \leq 1$. A value of $r_B = 0$ indicates no correspondence between performance of a genotype in different environments, whereas $r_B = 1$ suggests perfect correspondence between performance of genotypes in different environments (Burdon 1977). If we analyze two sites separately and estimate breeding values of genotypes tested at these two sites, the product-moment correlation between breeding values would be similar to r_B.

Thus, this model assumes that the genotypic variances expressed within all environments are equal: $\sigma_{G1}^2 = \sigma_{G2}^2 = \sigma_{G3}^2, \ldots, = \sigma_{Gs}^2$ and that the correlation of genotypic values between environments is the same for all pairs of environments, $r_{12} = r_{13}, \ldots, = r_{(s-1)s}$. The mixed model approach will allow us to relax these assumptions, but the way to do this may not be immediately obvious, as it combines the genotype and genotype-by-environment factors into a single compound model factor of genotype nested within environment: $G(E)_{ij}$. This formulation then allows us to specify the pattern of genotypic variances within environments and also the correlation structure for the effects of a common genotype across environments. We start by specifying the nested model as

$$Y_{ijkl} = \mu + E_i + G(E)_{ij} + B(E)_{ik} + \varepsilon_{ijkl} \qquad (8.5)$$

The effects in the model are the same as in the cross-classified model given in Eq. 8.1, but we have combined G_j and GE_{ij} into a single factor, $G(E)_{ij}$. We can start with the assumption that the distribution of $G(E)_{ij}$ is identical and independently distributed (*iid*): $G(E)_{ij} \sim N\left(0, \sigma^2_{G(E)}\right)$. Under this assumption, the covariance between values of a common genotype at different environments is zero:

$$Cov\left(Y_{ij}, Y_{i'j}\right) = Cov\left[G(E)_{ij}, G(E)_{i'j}\right] = 0 \tag{8.6}$$

and the variance of true genotypic values within environments is due solely to genotype-by-environment interaction variances as they were defined in the cross-classified model:

$$Var\left(Y_{ij}\right) = Var\left[G(E)_{ij}\right] = \sigma^2_{GE} \tag{8.7}$$

Of course, the *independent, identical distribution* assumption is usually worse than the original cross-classified model we started with, but writing the model in this form and using mixed models analysis gives great flexibility to specify a range of alternate assumptions and model forms. For example, we can make the model equivalent to the cross-classified analysis by changing the variance-covariance structure of the compound $G(E)_{ij}$ effects so that they have a common variance within environments and a common covariance across environment pairs (three environments in this example):

$$Cov\left(Y_{ij}, Y_{i'j}\right) = Cov\left[G(E)_{ij} + G(E)_{i'j}\right] = \sigma^2_G$$

$$Var\left[G(E)_{ij}\right] = \begin{bmatrix} \sigma^2_G + \sigma^2_{G(E)} & \sigma^2_G & \sigma^2_G \\ \sigma^2_G & \sigma^2_G + \sigma^2_{G(E)} & \sigma^2_G \\ \sigma^2_G & \sigma^2_G & \sigma^2_G + \sigma^2_{G(E)} \end{bmatrix} \otimes \mathbf{I}_m = \sigma^2_G \tag{8.8}$$

where $\mathbf{I}_m$ is the identity matrix with $m \times m$ dimensions for m genotypes. For example, with three environments, the variance-covariance matrix in Eq. 8.8 has dimension 3×3. By changing the structure of the 3×3 matrix in this Kronecker product, we can then allow for genotypic variances and covariances to vary among environments and pairs of environments, respectively. For example, at the other extreme of model complexity, we can allow each environment to have its own genetic variance and each pair of environments to have their own covariance. This is the unstructured (US) covariance model for genotype within environment effects and it involves six unique parameters:

$$Var\left[G(E)_{ij}\right] = \begin{bmatrix} \sigma^2_{G(E1)} & \sigma^2_{G21} & \sigma^2_{G31} \\ \sigma^2_{G12} & \sigma^2_{G(E2)} & \sigma^2_{G32} \\ \sigma^2_{G13} & \sigma^2_{G23} & \sigma^2_{G(E3)} \end{bmatrix} \otimes \mathbf{I}_m = \sigma^2_G \tag{8.9}$$

The US covariance formulation of the **G** matrix for METs involving large numbers of genotypes and environments may often fail to converge. For example, the unstructured **G** matrix in an experiment involving 50 environments requires estimation of 1275 parameters. Clearly, estimation of such a large number of parameters can be computationally prohibitive.

Factor analytic (FA) covariance structures for METs offer a more parsimonious approach to capture the complexity of covariances among many environments while limiting the number of parameters that require estimation (Smith et al. 2001, 2005; Thompson et al. 2003). For s trials, the number of parameters to be estimated for the US model is $p = s(s + 1)/2$, whereas for FA models it is $s(k + 1) - k(k - 1)/2$, where k is the number of factors (Thompson et al. 2003). The reduction in parameters requiring estimation can be noted for the case of 50 environments and $k = 1$ factor, for which only 100 parameters are estimated compared to the 1275 required for the unstructured model.

The US and compound symmetry models can be formulated as specific cases of the FA model. For example, if we fit the maximum of $k = s - 1$ factors, we recapitulate the US model with $s(s + 1)/2$ parameters. At the other extreme, we can create the compound symmetry model in this framework by fitting $k = 1$ factor and forcing the site loadings (explained below) to be equal, requiring only two parameters, one factor to generate the correlation between environments and one variance component (Cullis et al. 2014; Meyer 2009).

If the vector of genetic effects nested within sites is written as $\mathbf{u_g}$, we can conceive of these effects being arranged as a matrix of effects with m rows (for m genotypes) and s columns (for s environments). Conceptually, then, this matrix of effects can be subjected to factor analysis, in which the patterns of genotype response across environments are modelled as interactions between genotype effects and one or a small number of factors that underlie the environmental influences on genotype-within-environment phenotypes. FA models can be interpreted as random regression models of genotype and GE effects on k unknown environmental covariates, in which each genotype has its own slope (genotypic scores) but a common intercept (Crossa et al. 2006). The slopes measure the sensitivity of genotypes to hypothetical environmental factors represented in the model by the numerical 'loadings' for each site in each factor (Piepho et al. 2007; Smith et al. 2005). In this model, the genotypic effect for genotype j in site i (u_{gij}) is a sum of k multiplicative terms (Cullis et al. 2014; Smith et al. 2002):

$$u_{gij} = \lambda_{1i}f_{1j} + \lambda_{2i}f_{2j} + \ldots + \lambda_{ki}f_{kj} + \delta_{ij} \tag{8.10}$$

The terms in the multiplicative model include λ_{1i}, the **loading** for environment i on the first factor; f_{1j}, the genetic effect (**score**) of genotype j on the first factor; λ_{ki}, the loading for environment i on factor k; f_{kj}, the score of genotype j on factor k, and δ_{ij} is the deviation of the observed genetic effect of genotype j in environment i from its predicted value based on the multiplicative factor model fit. Factor analysis is related to principal components analysis but whereas principal components decomposition of the matrix of GE effects would identify eigenvectors based on their ability to account for the variation within and covariance between environments, the FA model identifies factors that maximally explain the covariance among environments and introduces an additional unique variance to capture any additional variation within each environment.

The FA models are named based on the number of the k factors (multiplicative terms) included in the model, e.g., FA1, FA2, and FAk. Our hope is to identify a model that can accurately describe the observed variance-covariance relationships among and within environments with as few factors as possible.

For a given number of factors k selected, the covariance between a genotype's performance in different environments is estimated as (Smith et al. 2002):

$$Cov(Y_{ij}, Y_{i'j}) = \sum_{f=1}^{k} \lambda_{fi}\lambda_{fi'} \tag{8.11}$$

Notice that this generates a unique covariance for each pair of environments if loadings differ among the environments. The variance of genotypic effects within an environment is estimated as:

$$Var(Y_{ij}) = \sum_{f=1}^{k} \lambda_{fi}^2 + \sum_{j=1}^{m} \frac{Var(\delta_{ij}^2)}{m} \tag{8.12}$$

The second piece of this expected variance is the average site-specific variance over all m genotypes within environment i. This is the within-site variance that is not accounted for by the factor loadings, and will be designated Ψ_{gi} (Smith et al. 2002).

$$Var(Y_{ij}) = \sum_{m=1}^{k} \lambda_{mi}^2 + \Psi_{gi} \tag{8.13}$$

Writing the vector of genotypic effects within environments in the $m \times s$ matrix form, we have:

$$\mathbf{u_g} = (\mathbf{\Lambda} \otimes \mathbf{I}_m)\boldsymbol{f} + \boldsymbol{\delta} \tag{8.14}$$

Where $\mathbf{I}_m$ is the identity matrix with dimensions $m \times m$, $\mathbf{\Lambda}$ is the matrix of environment loadings (with dimension $s \times k$), $\boldsymbol{f}$ is the vector of genotypic scores with dimensions $mk \times 1$, and $\boldsymbol{\delta}$ is a vector of residual genetic effects (with dimensions $ms \times 1$). If the genotypic effects are additive breeding values with an additive relationship matrix $\mathbf{A}$, then the variances of $\boldsymbol{f}$ and $\boldsymbol{\delta}$ are:

$$Var(\boldsymbol{f}) = \mathbf{A} \otimes \mathbf{I}_k \tag{8.15}$$

$$Var(\boldsymbol{\delta}) = \mathbf{A} \otimes \mathbf{\Psi} \tag{8.16}$$

where $\mathbf{I}_k$ is a $k \times k$ identity matrix and $\mathbf{\Psi}$ is an $s \times s$ diagonal matrix with site-specific genetic variances (ψ_s) on the diagonal and zero covariance between sites. $\mathbf{A}$ can be replaced with $\mathbf{I}_m$ if relationships are unknown and families are assumed independent, or with some other relationship matrix, such as the realized relationship matrices described in Chap. 11.

The variance of additive genotypic effects across all trials is:

$$Var\left(\mathbf{u}_g\right) = \mathbf{G}_g = \left(\mathbf{\Lambda}\mathbf{\Lambda}^T + \mathbf{\Psi}\right) \otimes \mathbf{A} \tag{8.17}$$

Typically, the model fitting process starts by fitting an FA1 model and proceeds to fit more complex ($k > 1$) models. Since the models are nested we can use likelihood ratio tests (LRT), Akaike Information Criterion (AIC), or Bayesian Information Criterion (BIC) to select models, although at some point model convergence may hinder fitting more complex models and we can stop. Smith et al. (2014) suggested that AIC and LRT might select models that are too complex (overfit), whereas BIC which penalizes model complexity more, might select underfit models that miss some important signal in the data. They suggest measuring goodness-of-fit for each model based on both the percent variance explained by k factors at within each individual environment (V_i) and averaged across environments ($\bar{V}$) as follows

$$V_i = 100 \frac{\sum_{r=1}^{k} \lambda_{ri}^2}{\sum_{r=1}^{k} \lambda_{ri}^2 + \psi_i^2} \tag{8.18}$$

$$\bar{V} = 100 \frac{tr\left(\Lambda\Lambda^T\right)}{tr\left(\Lambda\Lambda^T + \mathbf{\Psi}\right)} \tag{8.19}$$

where $tr()$ is the trace of the matrix (sum of diagonal elements) (Smith et al. 2014). The first factor accounts for as much of the covariances of genotype performances among environments as possible; subsequent factors are independent of previous factors and explain consecutively less covariance. Smith et al. (2015) recommend a model where the proportion of variation within most environments is high and few environments have low variance explained. These metrics are useful diagnostics, but unfortunately, they do not provide a model selection criterion. The choice of the number of factors to fit remains complicated; ideally a few factors can capture most of the patterns in the observed data, which is ideal for reducing the number of parameters.

Formulation of FA models in ASReml

In ASReml, FA models are specified in a covariance form, correlation form, or in an extended factor analytic (XFAk) form (Gilmour et al. 2014). In the **covariance** formulation of FA models, the variance is given as the direct product of an FA covariance matrix for sites (environments) and a genotype effect correlation matrix (which could be IDV or a numerator or other relationship matrix for genotype effects). The FA covariance structure for sites is parameterized as $\mathbf{\Lambda}\mathbf{\Lambda}^T + \mathbf{\Psi}$, where $\mathbf{\Lambda}$ is the matrix of loadings on the covariance scale. As an example, the covariance matrices for FA1 model with m unrelated genotypes tested at four sites would be:

$k = 1$ factor

$$\mathbf{\Lambda} = \begin{bmatrix} \lambda_{11} \\ \lambda_{12} \\ \lambda_{13} \\ \lambda_{14} \end{bmatrix}, \mathbf{\Psi} = \begin{bmatrix} \Psi_1 & 0 & 0 & 0 \\ 0 & \Psi_2 & 0 & 0 \\ 0 & 0 & \Psi_3 & 0 \\ 0 & 0 & 0 & \Psi_4 \end{bmatrix},$$

$$\mathbf{G}_g = \left[\mathbf{\Lambda}\mathbf{\Lambda}^T + \mathbf{\Psi}\right] \otimes \mathbf{I}_m = \begin{bmatrix} \lambda_{11}^2 + \Psi_1 & \lambda_{11}\lambda_{12} & \lambda_{11}\lambda_{13} & \lambda_{11}\lambda_{14} \\ \lambda_{11}\lambda_{12} & \lambda_{12}^2 + \Psi_2 & \lambda_{12}\lambda_{13} & \lambda_{12}\lambda_{14} \\ \lambda_{11}\lambda_{13} & \lambda_{12}\lambda_{13} & \lambda_{13}^2 + \Psi_3 & \lambda_{13}\lambda_{14} \\ \lambda_{11}\lambda_{14} & \lambda_{12}\lambda_{14} & \lambda_{13}\lambda_{14} & \lambda_{14}^2 + \Psi_4 \end{bmatrix} \otimes \mathbf{I}_m$$

The covariance matrices for FA2 model with m unrelated genotypes tested at four sites would be:

$k = 2\ factors$

$$\Lambda = \begin{bmatrix} \lambda_{11} & \lambda_{21} \\ \lambda_{12} & \lambda_{22} \\ \lambda_{13} & \lambda_{23} \\ \lambda_{14} & \lambda_{24} \end{bmatrix},$$

$$\mathbf{G}_g = \begin{bmatrix} \lambda_{11}^2 + \lambda_{21}^2 + \Psi_1 & \lambda_{11}\lambda_{12} + \lambda_{21}\lambda_{22} & \lambda_{11}\lambda_{13} + \lambda_{21}\lambda_{23} & \lambda_{11}\lambda_{14} + \lambda_{21}\lambda_{24} \\ \lambda_{11}\lambda_{12} + \lambda_{21}\lambda_{22} & \lambda_{12}^2 + \lambda_{22}^2 + \Psi_2 & \lambda_{12}\lambda_{13} + \lambda_{22}\lambda_{23} & \lambda_{12}\lambda_{14} + \lambda_{22}\lambda_{24} \\ \lambda_{11}\lambda_{13} + \lambda_{21}\lambda_{23} & \lambda_{12}\lambda_{13} + \lambda_{22}\lambda_{23} & \lambda_{13}^2 + \lambda_{23}^2 + \Psi_3 & \lambda_{13}\lambda_{14} + \lambda_{23}\lambda_{24} \\ \lambda_{11}\lambda_{14} + \lambda_{21}\lambda_{24} & \lambda_{12}\lambda_{14} + \lambda_{22}\lambda_{24} & \lambda_{13}\lambda_{14} + +\lambda_{23}\lambda_{24} & \lambda_{14}^2 + \lambda_{24}^2 + \Psi_4 \end{bmatrix} \otimes \mathbf{I}_m$$

In the **correlation** parameterization of FA models, the factor loadings are scaled by the genetic variances within sites. The matrix of loadings is now referred to as $\mathbf{F}$ and is analogous to the Λ matrix in the covariance form. For example, for an FA1 model on the correlation scale:

$$F = \begin{bmatrix} f_{11} \\ f_{12} \\ f_{13} \\ f_{14} \end{bmatrix} = \begin{bmatrix} \dfrac{\lambda_{11}}{\sqrt{\lambda_{11}^2 + \Psi_1}} \\ \dfrac{\lambda_{12}}{\sqrt{\lambda_{12}^2 + \Psi_2}} \\ \dfrac{\lambda_{13}}{\sqrt{\lambda_{13}^2 + \Psi_3}} \\ \dfrac{\lambda_{12}}{\sqrt{\lambda_{14}^2 + \Psi_4}} \end{bmatrix}$$

The off-diagonal elements of the product $\mathbf{FF}^T$ are the correlations between environments. However, $\mathbf{FF}^T$ is not a correlation matrix because its diagonal elements are not equal to 1. Therefore, we create a correlation matrix $\mathbf{C}$ by adding to the diagonal elements of $\mathbf{FF}^T$ to make them 1:

$\mathbf{C} = \mathbf{FF}^T + \mathbf{E}$, where $\mathbf{E}$ is a diagonal matrix defined as $\mathbf{E} = \mathrm{diag}(1 - \mathbf{F}^2)$.

We can then generate the covariance matrix for sites as $\mathbf{DCD}$, where D is an $s \times s$ diagonal matrix whose elements are square roots of the genetic variance within each site, i.e. $D_{11} = \sqrt{\lambda_{11}^2 + \Psi_1}$ for an FA1 model. Then the covariance structure for lines within sites is:

$$\mathbf{G}_g = [DCD] \otimes \mathbf{I}_m$$

Notice that $\mathbf{DF}$ in the FA correlation model is equal to Λ in the FA covariance formulation. Similarly, $\mathbf{DED}$ in the FA correlation formulation is equal to Ψ in the FA covariance formulation (Gilmour et al. 2014).

The *covariance* and *correlation* formulations of model parameterizations can have convergence problems and produce zero or even negative site-specific variances. This can occur when the factors alone ($\Lambda\Lambda^T$ or $\mathbf{FF}^T$) explain all of the variance within a site or predict more variance than is actually observed, such that one or more elements of Ψ or E are zero or negative. This situation is referred to as a *Heywood* case (Smith et al. 2001). **Extended factor analytical** (XFAk) models were developed to avoid convergence problems related to *Heywood* cases and also to increase computational efficiency (Meyer 2009; Thompson et al. 2003). XFA models have the same parameterization as FA covariance models $\Lambda\Lambda^T + \Psi$, but the algorithm used to fit the model is different. The common factors ($\lambda_{ri}f_{rj}$) are fit separately from the specific factors ($\delta_{g_{ij}}$), which leads to greater sparsity in the mixed model equations; furthermore, if a site-specific variance is zero, the $\delta_{g_{ij}}$ effects at that site are set to zero without hindering convergence for estimating the other model effects (Meyer 2009; Thompson et al.

2003). In ASReml syntax, the parameters in the *covariance* and *correlation* models are specified in the order of loadings Λ or F followed by specific variances, Ψ or E. In contrast, in XFAk models, the specific variances, Ψ, are specified first, followed by the loadings, Λ.

Example: Analysis of Pine Polymix MET Data

Polymix mating involves pollinating a set of individuals using bulked pollen from another set of individuals to reduce the cost of breeding. The goal is to predict breeding values of females for half-sib family selection. The Cooperative Tree Improvement Program at North Carolina State University used polymix breeding in the third cycle of loblolly pine (*Pinus taeda* L.) selection to predict the general performance of female parents (McKeand and Bridgwater 1998). In one of the test series, 70 individuals were mated with bulked pollen collected from another set of 40 individuals. Progeny from crossing were considered half-sibs with known mother and different fathers. A randomized complete block design was used with 20 blocks. Each female parent had one progeny in each block, for a total of 20 progeny at a site at the time of the planting. The experiment was replicated at 12 sites in the southeastern US. Height of tree, stem volume, fusiform rust disease incidence (present $= 1$, absent $= 0$) and stem straightness (1–6, 1 being the most strait) were assessed at age 6 years. A subset of the data is given below (*polymix.csv*).

female	male	site	block	height	volume	rust	stemform
16	0	101	18	25.5	0.76	0	1
16	0	101	5	21.5	1.09	0	4
16	0	101	6	24.5	1.19	0	4
16	0	101	3	29.0	2.15	0	5
16	0	101	17	32.0	2.85	0	2
16	0	101	16	27.0	1.64	0	3

Summarize Data for Each Site

This section allows us to check that our program is reading the data correctly and also provides summary statistics for each site. The only terms included in the linear model are fixed intercept and site effects and a random female effect.

Code example 8.1
Analysis of pine polymix data (see Code 8-1_MET.as for more details)

```
!ARGS 1 height !RENAME 2 !OUTFOLDER V:\Book\Chapter8_met\outfiles
Title: Pine polymix progeny tests
 female !I   male    !I
 site   !I   block   !I
 height  volume   rust   stemform

!FOLDER V:\Book\Chapter8_met
polymix.csv  !SKIP 1 !DOPART $A !FCON !DDF 2

!PART 1
! Obtain data summary for sites # this is a comment
TABULATE height volume ~ site !count
TABULATE height volume ~ female !count
# Model 1
$B ~ mu site !r female
```

- The line starting with the exclamation point and space writes the text that follows to the primary output file (.asr). This is a way to include comments preceding the output.

- We requested tabulation (TABULATE) of height for each site. This will generate a file with a *.tab* extension including the mean, standard deviation, minimum, and maximum of plant height measures as well as the total number of observations for each site. The second TABULATE statement generates summary statistics for female parents.

The output from TABULATE (*Code 8-1_MET1_height.tab*) includes descriptive statistics for sites. The range of site means for height growth is 21.8–29.7. The number of observations per site ranged from 1125 to 1372. The approximate F-tests in the primary output file *.asr* are given below.

```
Wald F statistics
   Source of Variation    NumDF    DenDF_con F-inc     F-con M P-con
9 mu                          1       69.4 79076.11   79076.11 . <.001
3 site                       11    15313.2  1205.49    1205.49 A <.001
```

The large and significant F-value for site effect indicates that the variation among sites is significant. A common observation in multi environmental trials (MET) data is that when the mean values for a trait vary significantly among environments, often the error variances may also differ significantly among site. This is often simply a matter of scale, with larger variances associated with larger observed measurements. In such cases, the default assumption of homogeneous residuals at all sites ($\varepsilon \sim N(0, \sigma^2_e \mathbf{I}_n)$) may not hold.

Analyze Each Site Separately to Obtain Variances

The second step is to analyse each site separately to obtain site-specific error variances and genetic variances. The model for each site is: $Y_{ijk} = \mu + B_i + G_j + \varepsilon_{ijk}$, where B_i is the random block effect, G_j is the random female effect and ε_{ijk} is the random residual. Variance components from individual sites can be used as starting parameters when we run the combined MET analysis and attempt to fit heterogeneous error variances. One way to run the same model for different sites is to use '! FILTER site !SELECT n' in combination with !ARGS:

```
!ARGS 2 height 1 2 3 4 5 6 7 8 9 10 11 12 !RENAME 3 !OUTFOLDER V:\Book
Title: Multi Environmental Trials
  female  !I    male  !I
  site    !I    block !I
  height  volume  rust  stemform

!FOLDER V:\Book\Book1_Examples\data
polymix.csv  !SKIP 1  !DOPART $A
...
!PART 2
!FILTER site .!SELECT $C
! Individual site analysis
! Model: y = mu + rep + GCA + e
$B ~ mu !r  female block
```

- The argument **$A** after naming the data file indicates the point at which the first argument ('2') will be substituted (the PART to analyse).
- The argument **$B** in the models indicates the point at which the second argument ('height') will be substituted (the trait to analyse).
- **$C** indicates the point at which the program will iteratively substitute the remaining arguments, one at a time ('1' through '12'). Here $C indicates the level of site to select when filtering the data set in the current iteration. !FILTER v !

SELECT n together are used to select data from a single site for analysis. The v is the number or name of a data field ('site' in this case) and n is the value of the field to be selected. It can be an integer (as in this example) or a character string in quotes. This is similar to using the BY statement in SAS procedures.

Different output files will be created for each site (file names will include three variable suffix values corresponding to PART, TRAIT, and SITE). In the output files we see large differences between sites for both genetic (range 0.45–1.48) and residual variances (3.18–11.41). Block differences at each site also explain considerable variation and should remain in subsequent multi-environment models. Heritability estimates had a range of 0.28–0.61. Now that we have a sense of the heterogeneity in the data, we will keep in mind that our final model should reflect this. Before we include such complexity in the combined model, however, we will start with the simplest model, a cross-classified ANOVA.

Model 3: Cross-Classified ANOVA

We can perform the combined analysis across environments using the traditional cross-classified genotype-environment model. The variance structures for random effects, including the residual, are scaled identical and independent (IID) variances. The linear model is $Y_{ijk} = \mu + S_i + SB_{ij} + F_k + SF_{ik} + \varepsilon_{ijk}$, where Y_{ijk} is the observation on a progeny of female k in block j at site i, S_i is the i-th site effect, SB_j is the random block effect nested within site, F_k is the random female effect, SF_{ik} is the random female by site interaction effect and ε_{ijkl} is the random residual associated with the data point. We can fit site as fixed effect since we have a balanced design in this case and we are not interested in making predictions or inferences about site effects or variances. Female is a random factor. Therefore the site-by-female interaction is random, even if site effect is considered fixed effect.

```
!PART 3
! Traditional cross-classified model
$B ~ mu site  ,
    !r  female ,
       site.female ,
       site.block
```

A small subset of the output from .asr file is given below.

OUTPUT 3

```
         7 LogL=-3323.03 S2= 7.1215 15379 df

- - - Results from analysis of height - - -
Akaike Information Criterion 46654.05 (assuming 4 parameters).
Bayesian Information Criterion 46684.61

Model_Term                    Sigma  Sigma/SE  %  C
female       IDV_V    70  0.563185     5.42   0  P
site.block   IDV_V   240  0.931300     9.53   0  P
site.female  IDV_V   840  0.174454     5.95   0  P
Residual     SCA_V 15391  7.12148     84.68   0  P
```

- All variance components seem to be significant since they are at least two times their standard errors (Sigma/SE column).

Model 4: Compound Symmetry

We can modify the factorial family-environment model to be a family nested within environment model as: $Y_{ijk} = \mu + S_i + SB_{ij} + SF_{ik} + \varepsilon_{ijk}$, where the terms are the same as above, except that by removing the family main effect, the family effects become nested within sites. We can recover a model equivalent to the cross-classified ANOVA model by fitting a *compound symmetry model* (`coruv` **G** structure) to the nested *site.female* effect. This fits a common genetic variance within sites and a common pairwise correlation between sites. We assume a uniform variance (`coruv`) for female within site effects and a uniform correlation (`coruv`) or covariance between pairs of sites. The model in ASReml is:

PART 4

$$
\begin{bmatrix}
\sigma^2_{g(e)} & \rho & \rho & \rho \\
\rho & \sigma^2_{g(e)} & \rho & \rho \\
\rho & \rho & \sigma^2_{g(e)} & \rho \\
\rho & \rho & \rho & \sigma^2_{g(e)}
\end{bmatrix} \otimes \mathbf{I}_m
$$

```
!PART 4
! IID R and G structure
$B ~ mu site  !r
    coruv(site).id(female) ,   ←
    site.block
```

- The covariance structure shown to the right is for four sites only as an example.
- In the model there is no female main effect. It appears with site as a consolidated term (*site.female*).

OUTPUT 4

```
     8  LogL=-3323.03       S2=   7.1215        15379 df

            - - - Results from analysis of height - - -
  Akaike Information Criterion      46589.43 (assuming 15 parameters).
  Bayesian Information Criterion   46704.04

  Model_Term                           Sigma    Sigma/SE   % C
  site.id(block)           IDV_V  240  0.931300       9.53   0 P
  Residual                 SCA_V 15391  7.12148      84.68   0 P
  coruv(site).id(female) 840 effects
  site                     COR_R    1  0.763497      16.76   0 P
  site                     COR_V    1  0.737639       6.88   0 P
```

- The parameters related to *site.female* effect are labeled with "*site* COR_R" for pairwise correlation between sites (identical for pairs of sites) or with "*site* COR_V" for the female within site variance component.

The relationship of COR_R and COR_V estimates from model 4 to the variance components from the cross-classified model (PART 3) may not be immediately obvious but they are indeed the same model. We have just changed how the model is parameterized. Notice that the residual LogL of models 3 and 4 are identical. Recall that the cross-classified ANOVA model produced a variance component estimate of 0.5632 for *female* effect and a variance component of 0.1744 for the *female.site* interaction effect. The sum of these two variance components from the ANOVA model (0.5632 + 0.1744) is equal to the variance component for the compound term *site.female* in the nested model, $Var(Y_{ij}) = Var\left(G(E)_{ij}\right) = \sigma^2_G + \sigma^2_{GE} = 0.737$.

Further, the ratio of *female* to the sum of *female* and *site.female* variance components estimated from the ANOVA cross-classified model, $0.5632 / (0.5632 + 0.1744) = 0.76$ is equal to the pairwise site correlation estimate from the nested CORUV model (COR_R = 0.76).

Model 5: Heterogeneous Residuals and Block Effects

In the models above we assumed that residuals and blocks nested within sites have scaled identity variance structures. However, we saw in part 2 that the models fit within each site separately resulted in widely different residual variances. Checking for heterogeneity in the residual variances across sites is a recommended practice. We can perform a formal test of the null hypothesis that the residual variances are uniform among sites by fitting the block diagonal residual structure `residual sat(site).id(units)`, which fits a separate residual variance for each site, and comparing the resulting log likelihood to model 3 or 4. The LogL for the heterogeneous R structure model was -2909 while it was $LogL = -3323$ for the homogeneous residual model (OUTPUT 4). The likelihood ratio test statistic would be $2(-2909 - (-3323)) = 828$ with 11 degrees of freedom (1 residual variance versus 12). Clearly a chi-square value of 828 with 11 df is significant (the critical value of chi-square for 11 df is 19.67 at $p = 0.05$), so we can safely reject the null hypothesis of equal residual variances among sites. We can also test the assumption that the block within site variances are equal among environments by fitting a heterogenous block within site variance structure with the model term `idh(site).block` (or, equivalently, `at(site).block`). The heterogeneous block within site variance model was also significantly better, so for the remaining examples in this chapter, we will use both heterogeneous residual and block variances across sites. Next, we will focus on fitting different **G** structures to model the variance-covariance relationships among family-within-site effects.

Model 6: CORUH G Structure

The compound symmetry structure, `coruv()` of genetic effects in models 4 and 5 assumes that the random genotype and genotype by environment interaction effects are constant. It involved only two genetic parameters; a variance and a correlation. This is an underfit model, as we shall see in the following models. We can relax a uniform G structure by allowing different genetic (*female*) variances at each site. This makes sense since there appeared to be large differences between sites for female variance components, with a range of 0.45 (site 12) to 1.48 (site 1) observed among the individual site models in part 2. Part 6 of our ASReml program fits a CORUH model:

PART 6

```
########### CORUH G structure
!PART 6
$B ~ mu site  !r ,
    coruh(site).id(female)  ,←
    idh(site).block

  residual sat(site).id(units)
```

$$\begin{bmatrix} \sigma^2_{g(e)1} & \rho & \rho & \rho \\ \rho & \sigma^2_{g(e)2} & \rho & \rho \\ \rho & \rho & \sigma^2_{g(e)3} & \rho \\ \rho & \rho & \rho & \sigma^2_{g(e)4} \end{bmatrix} \otimes I_m$$

- The G structure for $F(E)_{ij}$ effects is a direct product of the $s \times s$ variance-covariance matrix for a common female's effects within and across sites (although we only show a matrix for four sites in the example above) and the identity matrix (assuming females are unrelated). The variance function `coruh()` fits a heterogeneous variance structure to female effects. The `coru` stands for uniform correlation, and **h** indicates heterogeneous variances.

A subset of the results is given below:

OUTPUT 6

```
   11 LogL=-2850.48      S2=  1.0000        15379 df


           - - - Results from analysis of height - - -
     Akaike Information Criterion    45774.97 (assuming 37 parameters)
     Bayesian Information Criterion  46057.67

     Model_Term                              Sigma   Sigma/SE   % C
     sat(site,01).id(units)         1359
     Residual_1              SCA_V 1359     10.1027     25.24    0 P
     ...
     sat(site,12).id(units)         1333
     Residual_12             SCA_V 1333      5.43416    25.31    0 P
     idh(site).block                 240
     site                    DIAG_V   1      0.225986    1.86    0 P
     ...
     site                    DIAG_V  12      0.275270    2.37    0 P
     coruh(site).id(female)          840
     site                    COR_R    1      0.854844   22.52    0 P
     site                    COR_V    1      1.18904     4.28    0 P
     site                    COR_V    2      0.540631    4.44    0 P
     ...
     site                    COR_V   12      0.486392    3.79    0 P
```

- The logL of model 6 is -2850.48. This is a substantial improvement over the modified model 5 that included heterogeneous block and residual variances (results not shown).

Models 7 and 8: US and CORGH Structures

In model 6 we relaxed the constant variance assumption and fit a heterogeneous variance structure for the female within site effect. However, the coruh() model may still be too restrictive because it assumes a constant genetic correlation between pairs of environments. The general form of the variance structure for female effects would have different variances at each environment and different correlations (or covariances) between pairs of environments. In other words, fitting the us() and corgh() structures with $p = \frac{s(s+1)}{2}$ parameters. As the number of environments increases, model convergence and reliability of parameters become an issue. Therefore these structures are not recommended for multi environmental models with large numbers of environments (Smith et al. 2005). In this example, the number of parameter estimates for the female within site effect for these models is 78. The ASReml code to fit these models are included as models 7 and 8 in the example code file "Code 8-1_MET.as"; we were able to attain convergence only for model 8 (the CORGH model), which had a log likelihood of -2804.06, Akaike Information Criterion of 45812.12, and Bayesian Information Criterion of 46591.48.

Model 9: FA1 Covariance Structure

In PART 9 we fit the FA1 ($k = 1$) model to the data using the covariance parameterization.

```
!PART 9
$B ~ mu site !r    ,
facv(site).id(female) ;
  idh(site).block
       residual sat(site).id(units)
```

$$\begin{bmatrix} \lambda_{11}^2 + \Psi_1 & \lambda_{11}\lambda_{12} & \lambda_{11}\lambda_{13} & \lambda_{11}\lambda_{14} \\ \lambda_{11}\lambda_{12} & \lambda_{12}^2 + \Psi_2 & \lambda_{12}\lambda_{13} & \lambda_{12}\lambda_{14} \\ \lambda_{11}\lambda_{13} & \lambda_{12}\lambda_{13} & \lambda_{13}^2 + \Psi_3 & \lambda_{13}\lambda_{14} \\ \lambda_{11}\lambda_{14} & \lambda_{12}\lambda_{14} & \lambda_{13}\lambda_{14} & \lambda_{14}^2 + \Psi_4 \end{bmatrix} \otimes \mathbf{I}_m$$

- The variance-covariance structure for the compound term site.female is the direct product of an FA1 matrix for site effects and an identity matrix for female effects. If pedigree information were available on the females, we could use nrm (female) to account for genetic relationships among females.

A subset of the *.asr* output file is given here:

OUTPUT 9: FA1 covariance model

```
  21 LogL=-2829.63      S2=   1.0000        15379 df


           - - - Results from analysis of height - - -
  Akaike Information Criterion     45755.27 (assuming 48 parameters).
  Bayesian Information Criterion   46122.03


  Model_Term              Sigma        Sigma/SE   % C
  ...
  facv(site).id(female)          840 effects
  site        FACV_L  1   1  0.824275          5.86    0 P
  site        FACV_L  1   2  0.678831          9.57    0 P
  site        FACV_L  1   3  0.720423          8.44    0 P
  site        FACV_L  1   4  0.743635          9.22    0 P

  ...
  site        FACV_L  1  12  0.675770          9.18    0 P
  site        FACV_V  0   1  0.769548          3.40    0 P
  site        FACV_V  0   2  0.856239E-01      1.73    0 P
  site        FACV_V  0   3  0.669189E-01      0.84    0 P
  site        FACV_V  0   4  0.153460E-06      0.00    0 B

  ...
  site        FACV_V  0  12  0.126198E-01      0.24    0 P
```

$$\mathbf{\Lambda} = \begin{bmatrix} \lambda_{11} \\ \lambda_{12} \\ \lambda_{13} \\ \lambda_{14} \\ \vdots \\ \lambda_{112} \end{bmatrix}$$

$$\mathrm{diag}(\mathbf{\Psi}) = \begin{bmatrix} \Psi_1 \\ \Psi_2 \\ \Psi_3 \\ \Psi_4 \\ \vdots \\ \Psi_{12} \end{bmatrix}$$

```
  Covariance/Variance/Correlation Matrix FACV facv(site).id(female
  1.45  0.63  0.64  0.68  0.38  0.55  0.68  0.68  0.63  0.68  0.64  0.68
  0.56  0.55  0.86  0.92  0.51  0.74  0.92  0.92  0.84  0.92  0.86  0.91
  0.59  0.49  0.59  0.94  0.52  0.76  0.94  0.94  0.86  0.94  0.89  0.93
  0.61  0.50  0.54  0.55  0.55  0.80  1.00  1.00  0.91  1.00  0.94  0.99
  0.36  0.30  0.32  0.33  0.63  0.44  0.55  0.55  0.51  0.55  0.52  0.55
  0.58  0.48  0.51  0.52  0.31  0.77  0.80  0.80  0.73  0.80  0.76  0.79
  0.77  0.63  0.67  0.69  0.41  0.66  0.87  1.00  0.91  1.00  0.94  0.99
  0.62  0.51  0.54  0.56  0.33  0.53  0.71  0.57  0.91  1.00  0.94  0.99
  0.67  0.55  0.58  0.60  0.35  0.57  0.76  0.61  0.78  0.91  0.86  0.90
  0.61  0.51  0.54  0.55  0.33  0.53  0.70  0.56  0.60  0.56  0.94  0.99
  0.81  0.67  0.71  0.73  0.43  0.70  0.92  0.75  0.80  0.74  1.10  0.93
  0.56  0.46  0.49  0.50  0.30  0.48  0.63  0.51  0.55  0.50  0.67  0.47
```

- The log likelihood of the FA1 model is -2829.63, which is similar to the `corgh()` model (model 8 log likelihood $= -2804.06$), although the FA1 model requires only 24 parameters for the genotype within environment covariance matrix compared to 78 for the `us()`/`corgh()` models. By capturing the variance/covariance structure well with many fewer parameters, the FA1 model has much better (lower) Akaike and Bayesian Information Criteria than the `corgh()` model.
- In the *.asr* output file, site loadings on the correlation scale are labeled 'FACV_L'. Values with label 'FACV_V' are the site-specific genetic variances (the diagonal elements of Ψ).
- The within-site genetic variances and between site covariances and correlation estimates are given in the 'covariance/variance/correlation matrix' at the bottom of the output. In the example output above, we highlighted in bold the estimates for the first four environments.
- The diagonal elements of the FACV covariance matrix are obtained as the squared loadings plus the site-specific variances. For example, for site 1, the variance (element [1,1] in the covariance/variance/correlation matrix) is:

$$\sigma^2_{g(e)1} = \lambda^2_{11} + \Psi_1 = (0.824275)^2 + 0.769548 = 1.45$$

- Notice that a relatively large additional site-specific variance must be added to the squared loading for site 1 to obtain a good estimate of the within-site genetic variance. In contrast, for site 4, its within-site variance is estimated accurately by the square of its loading, so its site-specific variance is close to zero.
- The estimated genetic covariance between a family's performance at sites 1 and 2 (element [2,1] in the covariance/variance/correlation matrix) is simply the product of their loadings:

$$\sigma_{g12} = \lambda_{11}\lambda_{12} = (0.824275)(0.678831) = 0.56$$

- The estimated genetic correlation between a family's performance at sites 1 and 2 (element [2,1] in the covariance/variance/correlation matrix) is the covariance divided by the square root of the product of the within-site genetic variances:

$$r_{g12} = \frac{\lambda_{11}\lambda_{12}}{\sqrt{\left(\lambda^2_{11} + \Psi_1\right)\left(\lambda^2_{12} + \Psi_2\right)}} = \frac{0.56}{\sqrt{(1.45)(0.55)}} = 0.62$$

- In this example, female effects represent half-sib family means, so the genetic variance and covariance estimates are a quarter of the additive genetic variances/covariances. The correlation estimates are additive genetic correlations.

Model 10: FA1 Correlation Structure

In PART 10 we fit the FA1 model using the correlation parameterization.

```
!PART 10
$B ~ mu site !r   fa(site).id(female)  ,
                idh(site).block
     residual sat(site).id(units)
```

A subset of the *.asr* output file is given here:

OUTPUT 10: FA1 correlation model

```
    3 LogL=-2829.63       S2=  1.0000        15379 df


            - - - Results from analysis of height - - -
    Akaike Information Criterion     45754.09 (assuming 48 parameters)
    Bayesian Information Criterion   46120.84


    Model_Term                    Sigma        Sigma/SE   % C
    fa(site).id(female)                 840 effects
    fa(site)      FA_R  1  1  0.684752         7.55    0 U
    fa(site)      FA_R  1  2  0.918306        19.67    0 U
    fa(site)      FA_R  1  3  0.941158        14.25    0 U
    fa(site)      FA_R  1  4  0.999995         0.00    0 B

    ...
    fa(site)      FA_R  1 12  0.986463        17.51    0 U
    fa(site)      FA_V  0  1  1.44893          4.34    0 U
    fa(site)      FA_V  0  2  0.546377         4.46    0 U
    fa(site)      FA_V  0  3  0.585880         3.60    0 U
    fa(site)      FA_V  0  4  0.552736         3.84    0 U
    ...
    fa(site)      FA_V  0 12  0.469227         3.76    0 U
```

$$\mathbf{F} = \begin{bmatrix} f_{11} \\ f_{12} \\ f_{13} \\ f_{14} \\ \vdots \\ f_{112} \end{bmatrix}$$

$$\mathrm{diag}(\mathbf{D})^2 = \begin{bmatrix} D_{11}^2 \\ D_{22}^2 \\ D_{33}^2 \\ D_{33}^2 \\ \vdots \\ D_{1212}^2 \end{bmatrix}$$

```
Covariance/Variance/Correlation Matrix FA fa(site).id(femal
1.45  0.63  0.64  0.68  0.38  0.55  0.68  0.68  0.63  0.68  0.64  0.68
0.56  0.55  0.86  0.92  0.51  0.74  0.92  0.92  0.84  0.92  0.86  0.91
0.59  0.49  0.59  0.94  0.52  0.76  0.94  0.94  0.86  0.94  0.89  0.93
0.61  0.50  0.54  0.55  0.55  0.80  1.00  1.00  0.91  1.00  0.94  0.99
0.36  0.30  0.32  0.33  0.63  0.44  0.55  0.55  0.51  0.55  0.52  0.55
0.58  0.48  0.51  0.52  0.31  0.77  0.80  0.80  0.73  0.80  0.76  0.79
0.77  0.63  0.67  0.69  0.41  0.66  0.87  1.00  0.91  1.00  0.94  0.99
0.62  0.51  0.54  0.56  0.33  0.53  0.71  0.57  0.91  1.00  0.94  0.99
0.67  0.55  0.58  0.60  0.35  0.57  0.76  0.61  0.78  0.91  0.86  0.90
0.61  0.51  0.54  0.55  0.33  0.53  0.70  0.56  0.60  0.56  0.94  0.99
0.81  0.67  0.71  0.73  0.43  0.70  0.92  0.75  0.80  0.74  1.10  0.93
0.56  0.46  0.49  0.50  0.30  0.48  0.63  0.51  0.55  0.50  0.67  0.47
```

- In the *.asr* output file, residual variances and genetic correlations for pairs of sites and genetic variances are reported (under the column heading 'sigma'). Site loadings on the correlation scale are labeled 'FA_R' in the output. Values with label 'FA_V' are genetic variances within each site (which are the sum of the squared site loadings and the site-specific variance). Several site loadings on the correlation scale are very close to one (FA_R = 0.9995) and are constrained at the boundary flagged by 'B'.
- The genetic variance and correlation estimates are also given in the covariance/variance/correlation matrix at the bottom of the output. Notice that the model likelihood and the covariance/variance/correlation estimates are identical for models 9 and 10. The only difference is in how the parameter estimates are reported.
- The loadings on the correlation scale are equal to the covariance model loadings divided by the square root of the within-site genetic variance. For example, for site 1, the correlation loading is equated to the covariance model parameters as:

$$f_{11} = \frac{\lambda_{11}}{\sqrt{\lambda_{11}^2 + \Psi_1}} = \frac{0.824275}{\sqrt{1.45}} = 0.68$$

- The estimates labelled as 'FA_V' are the squared diagonal elements of the **D** matrix, equal to the within-site variances estimated from the covariance parameterization. For example, for site 1: $D_{11}^2 = \lambda_{11}^2 + \Psi_1 = 1.45$

The between-site genetic correlations are obtained directly as products of the correlation loadings (the 'FA_R' values in the output), which are the elements of the $\mathbf{F}$ vector. As an example, consider the loadings for only the first four environments:

$$\mathbf{F} = \begin{bmatrix} 0.6847 \\ 0.9183 \\ 0.9412 \\ 0.9999 \end{bmatrix}$$

We can construct something close to the correlation matrix from the product $\mathbf{FF}^T$.

$$\mathbf{FF}^T = \begin{bmatrix} 0.47 & 0.63 & 0.64 & 0.68 \\ 0.63 & 0.84 & 0.86 & 0.92 \\ 0.64 & 0.86 & 0.89 & 0.94 \\ 0.68 & 0.92 & 0.94 & 0.99 \end{bmatrix}$$

The off-diagonal elements of the product are the correlations between pairs of environments, e.g.

$r_{12} = 0.6847 * 0.9183 = 0.63$.

However, the diagonal elements are not equal to one, so $\mathbf{FF}^T$ is not a proper correlation matrix. For example, the element $(1,1)$ of $\mathbf{FF}^T$ is $(0.6847)^2 = 0.47$. Therefore we construct a matrix $\mathbf{E} = \mathrm{diag}(\mathbf{1} - \mathbf{F}^2)$ and add it to $\mathbf{FF}^T$ to make the correlation matrix $\mathbf{C}$, which now has diagonal elements equal to exactly one:

$$\mathbf{E} = \begin{bmatrix} 1 - (0.6847)^2 & 0 & 0 & 0 \\ 0 & 1 - (0.9183)^2 & 0 & 0 \\ 0 & 0 & 1 - (0.9412)^2 & 0 \\ 0 & 0 & 0 & 1 - (0.9999)^2 \end{bmatrix}$$

$$\mathbf{C} = \mathbf{FF}^T + \mathbf{E} = \begin{bmatrix} 1 & 0.63 & 0.64 & 0.68 \\ 0.63 & 1 & 0.86 & 0.92 \\ 0.64 & 0.86 & 1 & 0.94 \\ 0.68 & 0.92 & 0.94 & 1 \end{bmatrix}$$

The $\mathbf{D}$ matrix has square roots of the genetic variances within each site on the diagonal:

$$\mathbf{D} = \begin{bmatrix} \sqrt{1.45} & 0 & 0 & 0 \\ 0 & \sqrt{0.55} & 0 & 0 \\ 0 & 0 & \sqrt{0.59} & 0 \\ 0 & 0 & 0 & \sqrt{0.55} \end{bmatrix}$$

The correlation matrix for family within site effects is obtained as:

$$\mathbf{G} = \mathbf{DCD}^T \otimes \mathbf{I}\sigma_F^2 = \begin{bmatrix} \mathbf{1.45} & 0.56 & 0.60 & 0.61 \\ 0.56 & \mathbf{0.55} & 0.49 & 0.51 \\ 0.60 & 0.49 & \mathbf{0.59} & 0.54 \\ 0.61 & 0.51 & 0.54 & \mathbf{0.55} \end{bmatrix} \otimes \mathbf{I}\sigma_F^2$$

Now, consider how this model can be reformulated in terms of a covariance matrix. The loadings on the covariance scale ($\mathbf{\Lambda}$) are equal to the product $\mathbf{DF}$ from the correlation parameterization:

$$\mathbf{\Lambda} = \mathbf{DF} = \begin{bmatrix} \sqrt{1.45} & 0 & 0 & 0 \\ 0 & \sqrt{0.55} & 0 & 0 \\ 0 & 0 & \sqrt{0.59} & 0 \\ 0 & 0 & 0 & \sqrt{0.55} \end{bmatrix} \begin{bmatrix} 0.6847 \\ 0.9183 \\ 0.9412 \\ 0.9999 \end{bmatrix} = \begin{bmatrix} 0.8243 \\ 0.6788 \\ 0.7204 \\ 0.7436 \end{bmatrix}$$

This is the set of loadings we obtained with the covariance forms of the FA1 model (model 9).

Model 11: XFA*1* Structure

The XFA1 model is a third model equivalent to FACV1 and FA1, but has a different parameterization that improves computational efficiency.

```
!PART 11
! XFA extended factor analytical G
!CONTINUE 3
$B ~ mu site  !r  xfa1(site).id(female) ,
                  idh(site ).block
     residual sat(site).id(units)
```

OUTPUT 10: A subset of the output from XFA1 model

```
   3 LogL=-2829.63      S2=  1.0000       15379 df


          - - - Results from analysis of height - - -
   Akaike Information Criterion      45747.27 (assuming 44 parameters)
   Bayesian Information Criterion   46083.46


   Model_Term                   Sigma    Sigma/SE    % C
   ...
   xfa1(site).id(female)  910 effects
     site     XFA_V  0  1    0.769547        3.40     0 P
     site     XFA_V  0  2    0.856240E-01    1.73     0 P
     site     XFA_V  0  3    0.669197E-01    0.84     0 P
     site     XFA_V  0  4    0.00000         0.00     0 B

   ...
     site     XFA_V  0 12    0.126183E-01    0.24     0 P
     site     XFA_L  1  1    0.824247        5.53     0 P
     site     XFA_L  1  2    0.678790        8.77     0 P
   ...
     site     XFA_L  1 11    0.988628        8.24     0 P
     site     XFA_L  1 12    0.675730        8.41     0 P
```

$$\mathrm{diag}(\boldsymbol{\Psi}) = \begin{bmatrix} \Psi_1 \\ \Psi_2 \\ \Psi_3 \\ \Psi_4 \\ \vdots \\ \Psi_{112} \end{bmatrix}$$

$$\boldsymbol{\Lambda} = \begin{bmatrix} \lambda_{11} \\ \lambda_{12} \\ \lambda_{13} \\ \lambda_{14} \\ \vdots \\ \lambda_{112} \end{bmatrix} \qquad \mathbf{F} = \begin{bmatrix} f_{11} \\ f_{12} \\ f_{13} \\ f_{14} \\ \vdots \\ f_{112} \end{bmatrix}$$

```
   Covariance/Variance/Correlation Matrix XFA xfa1(site).id(female
   1.45  0.63  0.64  0.68  0.38  0.55  0.68  0.68  0.63  0.68  0.64 0.68  0.68
   0.56  0.55  0.86  0.92  0.51  0.74  0.92  0.92  0.84  0.92  0.86 0.91  0.92
   0.59  0.49  0.59  0.94  0.52  0.76  0.94  0.94  0.86  0.94  0.89 0.93  0.94
   0.61  0.50  0.54  0.55  0.55  0.80  1.00  1.00  0.91  1.00  0.94 0.99  1.00
   0.36  0.30  0.32  0.33  0.63  0.44  0.55  0.55  0.51  0.55  0.52 0.55  0.55
   0.58  0.48  0.51  0.52  0.31  0.77  0.80  0.80  0.73  0.80  0.76 0.79  0.80
   0.77  0.63  0.67  0.70  0.41  0.66  0.87  1.00  0.91  1.00  0.94 0.99  1.00
   0.62  0.51  0.54  0.56  0.33  0.53  0.71  0.57  0.91  1.00  0.94 0.99  1.00
   0.67  0.55  0.58  0.60  0.36  0.57  0.76  0.61  0.78  0.91  0.86 0.90  0.91
   0.62  0.51  0.54  0.55  0.33  0.53  0.70  0.56  0.60  0.56  0.94 0.99  1.00
   0.81  0.67  0.71  0.74  0.43  0.70  0.92  0.75  0.80  0.74  1.10 0.93  0.94
   0.56  0.46  0.49  0.50  0.30  0.48  0.63  0.51  0.55  0.50  0.67 0.47  0.99

   0.82  0.68  0.72  0.74  0.44  0.71  0.93  0.76  0.81  0.75  0.99 0.68  1.00
```

- The residual LogL is the same as it was for models 9 and 10. The AIC/BIC values of model 11 are different because ASReml does not count any site-specific variances in Ψ that are fixed at zero as parameters in the XFA model, whereas these parameters are set to very small values close to zero in the FACV and FA models. This artificially makes the XFA1 appear to be a better fitting model, but effectively the models are all the same.
- The parameter estimates in the output for the XFA model are identical to the FACV model, except they appear in different order.
- Parameter estimates labeled 'XFA_V' are the site-specific variances (the diagonal elements of Ψ), four of which are fixed at 0 in this example.
- The values labeled 'XFA_L' are site loadings on the covariance scale (Λ).
- The covariance/variance/correlation matrix for sites is given at the bottom of the output. This matrix is identical to the matrices estimated by models 9 and 10, except that one extra row and one extra column are added to the matrix.
- The extra column added to the right side of the matrix contains the factor loadings on the correlation scale (equal to the $\mathbf{F}$ vector in the FA model).
- The additional row at the bottom of the matrix has the factor loadings on the covariance scale (Λ).

To aid with model diagnosis and selection, a plot of the proportion of within-site variances estimated by the factor part of the model appears in the *.res* file ("Code 8-1_MET11_height.res"). The column labeled "%expl" corresponds to the within-site genetic variances described in Eqs. 8.12 and 8.13:

```
DISPLAY of variance partitioning for XFA structure in xfa1(site).id(female)
  Lvl  |----+----+----+----+----+----+----+----+----+----|  TotalVar %expl   PsiVar  Loadings
    1  |                        1                       |    1.4489  46.9    0.7695    0.8242
    2  |                                     1          |    0.5464  84.3    0.0856    0.6788
    3  |                                       1        |    0.5859  88.6    0.0669    0.7204
    4  |                                          1     |    0.5527 100.0    0.0000    0.7435
    5  |            1                                   |    0.6266  30.7    0.4341    0.4387
    6  |                          1                     |    0.7724  64.4    0.2751    0.7051
    7  |                                          1     |    0.8738 100.0    0.0000    0.9348
    8  |                                          1     |    0.5706 100.0    0.0000    0.7554
    9  |                               1                |    0.7832  83.6    0.1286    0.8091
   10  |                                          1     |    0.5568 100.0    0.0000    0.7462
   11  |                                       1        |    1.1033  88.6    0.1259    0.9886
   12  |                                          1|       0.4692  97.3    0.0126    0.6757
    0  |----+----+----+----+----+----+----+----+-- Average  0.7408  82.0    0.1582    0.7517
```

In above output less than half of the variation among females within sites 1 and 5 (highlighted in the output) is explained by the factor part of the model, so fairly large site-specific variances ("PsiVar") are needed to explain the observed variation at those environments.

Model 12: XFA2 Structure

In PART 12 the *XFA with 2 factors* is fitted. The XFA2 model assumes that two factors explain the correlation structure between pairs of sites:

```
!PART 12
! XFA2 extended factor analytical G
$B ~ mu site  !r  xfa2(site).id(female) ,
                  idh(site).block
     residual sat(site).id(units)
```

The output of the XFA2 model follows (result may differ slightly due to different starting values).

```
     12 LogL=-2820.25     S2=  1.0000        15379 df


          - - - Results from analysis of height - - -
     Akaike Information Criterion    45744.51 (assuming 52 parameters)
     Bayesian Information Criterion   46141.83


     Model_Term                 Sigma      Sigma/SE   % C
     ...
     xfa2(site).id(female)           980 effects
     site        XFA_V  0  1  0.719280          3.22   0 P
     site        XFA_V  0  2  0.694826E-01      1.44   0 P
     site        XFA_V  0  3  0.599184E-01      0.77   0 P
     site        XFA_V  0  4 -0.100965         -2.05   0 P
     site        XFA_V  0  5  0.00000           0.00   0 F
     site        XFA_V  0  6  0.203934          1.27   0 P
     site        XFA_V  0  7 -0.163581E-01     -0.22   0 P
     site        XFA_V  0  8 -0.227097E-01     -0.51   0 P
     site        XFA_V  0  9  0.100274          1.13   0 P
     site        XFA_V  0 10 -0.506460E-02     -0.08   0 P
     site        XFA_V  0 11  0.689859E-01      0.53   0 P
     site        XFA_V  0 12  0.00000           0.00   0 F
     site        XFA_L  1  1  0.835118          0.00   0 F
     site        XFA_L  1  2  0.677372          8.61   0 P
     site        XFA_L  1  3  0.718671          7.84   0 P
     site        XFA_L  1  4  0.743573          8.60   0 P
     ...
     site        XFA_L  1 12  0.678952          8.58   0 P
     site        XFA_L  2  1 -0.179462         -0.83   0 P
     site        XFA_L  2  2  0.140464          1.08   0 P
     site        XFA_L  2  3  0.108155          0.73   0 P
     site        XFA_L  2  4  0.232243          1.68   0 P
     ...
     site        XFA_L  2 12 -0.107097         -0.81   0 P


     Covariance/Variance/Correlation Matrix XFA xfa2(site).id(female
      1.45  0.61  0.63  0.68  0.25  0.60  0.70  0.70  0.66  0.69  0.69  0.71  0.69 -0.15
      0.54  0.55  0.88  1.02  0.65  0.69  0.92  0.94  0.81  0.93  0.84  0.87  0.92  0.19
      0.58  0.50  0.59  1.03  0.63  0.72  0.94  0.96  0.84  0.95  0.87  0.90  0.94  0.14
      0.58  0.54  0.56  0.51  0.84  0.76  1.05  1.08  0.91  1.06  0.94  0.98  1.05  0.33
      0.25  0.40  0.40  0.50  0.69  0.21  0.52  0.59  0.37  0.58  0.37  0.40  0.54  0.84
      0.64  0.45  0.48  0.47  0.15  0.77  0.83  0.81  0.79  0.80  0.82  0.84  0.81 -0.28
      0.78  0.63  0.67  0.69  0.40  0.67  0.86  1.03  0.93  1.01  0.97  1.00  1.01 -0.03
      0.63  0.52  0.55  0.58  0.37  0.53  0.71  0.56  0.93  1.03  0.96  1.00  1.02  0.05
      0.70  0.53  0.57  0.57  0.27  0.61  0.76  0.62  0.78  0.92  0.90  0.93  0.92 -0.15
      0.61  0.51  0.54  0.56  0.36  0.52  0.69  0.57  0.60  0.55  0.95  0.98  1.00  0.05
      0.87  0.65  0.70  0.70  0.32  0.75  0.93  0.75  0.84  0.73  1.10  0.97  0.95 -0.18
      0.59  0.44  0.48  0.48  0.23  0.51  0.64  0.51  0.57  0.50  0.70  0.47  0.99 -0.16
      0.84  0.68  0.72  0.74  0.45  0.71  0.93  0.76  0.81  0.74  1.00  0.68  1.00  0.00
     -0.18  0.14  0.11  0.23  0.70 -0.24 -0.03  0.04 -0.14  0.03 -0.19 -0.11  0.00  1.00
```

- The XFA2 model has a better log likelihood than the XFA1 model (-2820.25 vs. -2829.63) but it uses 12 additional parameters to capture additional variation. Depending on the penalty used for adding parameters to the model, the XFA2 model could be considered better or worse than the XFA1 model. The XFA2 model has better Akaike Information Criterion than the XFA1 model (45744.51 for XFA2 vs. 45747.27 for XFA1) but worse Bayesian Information Criterion

(46141.83 for XFA2 vs. 46083.46 for XFA1). Therefore, choice of XFA1 vs XFA2 model in this case is not clear cut and is up to the judgement of the researcher.

- Parameter estimates labeled 'XFA_V' are the site-specific variances and 'XFA_L' are loadings. For the XFA2 model, the loadings are indexed by the factor number (1 or 2) and the site number (1 through 12):
  - XFA_L 1 1 refers to the loading on the first factor for the first site,
  - XFA_L 1 2 refers to the loading on the first factor for the second site,
  - XFA_L 2 1 refers to the loading on the second factor for the first site and so forth.
- For the XFA2 model, Λ_g has s rows for sites and two columns for two factors. The loadings for the first four environments on the two factors are:

$$\Lambda = \begin{bmatrix} 0.84 & -0.18 \\ 0.68 & -0.14 \\ 0.72 & 0.11 \\ 0.74 & 0.23 \end{bmatrix}$$

- Notice that the loadings on the first factor are different than the loadings in XFA1 model. For the second factor, some sites had negative loadings. The factor loadings are not unique solutions, and other solutions can be produced.
- The last two columns in the XFA output (orange color vectors) are site loadings on the correlation scale. Notice that correlations can go out of theoretical bounds (>1) in the XFA2 model.
- Also notice that some of the site-specific variances (for example, site 4) are negative. The genetic variance predicted at site 4 based on the two factors is the sum of the squared loadings for site 4: $\sum_{r=1}^{2} \lambda_{i4}^2 = (0.74)^2 + (0.23)^2 = 0.60$. However, this is an overestimate of the genetic variance within site 4. So, a negative site-specific variance needs to be added to the sum of squared loadings to get a better estimate of the within-site variance: $Var\big(G(E)_4\big) = \sum_{r=1}^{2} \lambda_{i4}^2 + \Psi_4 = 0.60 - 0.10 = 0.50$. This is within rounding error of element [4,4] of the covariance/variance/correlation matrix in the output above.

Model 13: XFA3 Structure

In PART 13 of the example code, the *XFA with 3 factors* model is fitted. We do not show the output from this model, as its AIC and BIC values are worse than the XFA2 model. A summary of the models fit to pine polymix data is given in Table 8.1:

Model LogL values decrease as model complexity (the number of parameters) increases. AIC value follows the same trend until the FA2 model, after which the penalty for additional parameters outweighs the improvement in likelihood. The BIC penalizes additional parameters more stringently, such that the simple CORUV model (equivalent to the classical factorial model) has the best BIC value. In such situations, model choice is not clear cut, but we note that the FA1/XFA1 model provides a good compromise between model fit and number of parameters, such that it has second best AIC and third best BIC.

MET Models with ASReml-R

For interested readers, an R markdown file (*Code 8-2_pine_met.Rmd*) and its knitted output (*Code 8-2_pine_met.html*) are provided to show the sequence of analyses using ASReml-R. In ASReml-R, the US and FA3 models did not converge despite using initial values and update.asreml() function of ASReml-R. Another detail that ASReml-R users should be aware of is that for FA models with $k > 1$, the factor loading solutions are not unique and ASReml-R produces different

Table 8.1 Model fit statistics (log likelihood, Akaike Information Criterion (AIC), Bayesian Information Criterion (BIC), standard error of differences of family mean predictions, and number of parameters for the site.female term) for pine polymix data

Model G Structure	LogL	AIC	BIC	SED	G(E) Parameters
CORUV (Model 5)	−2865.30	45782.59	**45981.25**	0.2418	2
CORUH (Model 6)	−2850.48	45774.97	46057.67	0.2425	13
CORGH (Model 8)	−2804.06	45812.12	46591.48	0.2395	78
XFA1 (Models 9/10/11)	−2829.04	45747.27	46083.46	0.2438	24
XFA2 (Model 12)	−2818.66	**45744.51**	46141.83	0.2413	56
XFA3 (Model 13)	−2814.42	45754.63	46236.20	0.2412	63

Models differ for site.female compound term only

solutions than ASReml standalone (Cullis et al. 2010). We show in the code how to perform an orthogonal rotation of the ASReml-R loading solutions to match the ASReml standalone solutions.

Genetic Prediction with FA Models

From MET data, we can predict family values within sites or averaged across sites. Typically, the family means across sites are most useful for selection, but there are cases in which one site or group of sites may be distinct (with a low or negative correlation with other sites) and we may want to predict performance specifically in different groups of sites. Here, we demonstrate how to predict family values within and across sites with ASReml and how the predictions relate to the model effect estimates. We start with the simplest model (cross-classified and compound symmetry) and continue to the XFA1 model.

Model 3 – Cross-classified predictions

For the cross-classified model, we can obtain the across-site and within-site predictions using:

```
predict female !PLOT female
predict site.female
```

- The first predict statement is for prediction of female effect across all the sites. The second predict statement is site-specific predictions for females.
- The !PLOT qualifier produces a postscript graphic of predictions for females with one standard error.

The across-site predictions appear in the *.pvs* file above the site-specific predictions:

```
---- ---- ---- ---- ---- ----   1   ---- ---- ---- ---- ----
  Predicted values of height
  The SIMPLE averaging set:  site
  The ignored set: block

female    Predicted_Value   Standard_Error Ecode
    16         27.9609            0.1803      E
    18         26.4248            0.1860      E
   580         26.0100            0.2595      E
...
SED: Overall Standard Error of Difference     0.2518

---- ---- ---- ---- ---- ----   2   ---- ---- ---- ---- ----
  Predicted values of height
  The ignored set: block

site    female   Predicted_Value Standard_Error Ecode
 101      16       25.7900           0.4339          E
 101      18       23.9636           0.4317          E
 101     580       23.1160           0.4708          E
...
 102      16       31.1890           0.4308          E
 102      18       29.5329           0.4318          E
 102     580       29.3640           0.4666          E
...
 113      16       29.0128           0.4308          E
 113      18       27.4434           0.4350          E
 113     580       21.4658           0.5513          E
...
```

- There are two separate predictions in the *.pvs* file. The first output at the top is predicted breeding values of females across the sites.
- In the second part of the prediction we included the predictions for three families (16, 18, 580) at sites 101, 102, and 103. Female 580 did not have data at site 103 but its value is predicted. Notice that the standard error of this site-specific prediction for female 580 is 0.5513, higher than standard error of its predictions in sites 101 and 102. This is because in the absence of data for the particular site-family combination, the prediction is based on the main effects of the site and the family, with the predicted interaction effect exactly zero. Users may want to exclude such predictions from the *.pvs* file. This can be done by requesting the site-specific predictions with the qualifier **!present site female**.

```
predict site.female !present site female
```

Model 4 – CORUV predictions

Although we do not have family main effects in the CORUV model, we can nevertheless predict family values across sites as well as within sites using the same prediction statements as for the cross-classified model:

```
predict female
predict site.female !present site female
```

This produces predictions identical to the cross-classified model:

```
female    Predicted_Value    Standard_Error  Ecode
    16          27.9609             0.1803     E
    18          26.4248             0.1860     E
   580          26.0100             0.2595     E
...
SED: Overall Standard Error of Difference 0.2518

---- ---- ---- ---- ---- ---- 2 ---- ---- ---- ---- ---- ----
 Predicted values of height
 The ignored set: block
 Warning: 6 non-estimable [empty] cell(s) may be omitted from the table.

site  female  Predicted_Value Standard_Error Ecode
101     16    25.7900              0.4339        E
101     18    23.9636              0.4317        E
101    580    23.1160              0.4708        E

site  female  Predicted_Value Standard_Error Ecode
102     16    31.1890              0.4308        E
102     18    29.5329              0.4318        E
102    580    29.3640              0.4666        E
...
site  female  Predicted_Value Standard_Error Ecode
113     16    23.4069              0.4373        E
113     18    21.8089              0.4383        E
...
```

- Since we used "!present site female" there is no prediction for female 580 at site 113.

Model 11 – XFA1 predictions

The FA and XFA models partition the family within environment effects into a part due to the multiplicative interactions between factor loadings and family scores, and a second part due to site-specific genetic deviations for the family. This permits some flexibility in the across-sites and within-site family predictions, as the predictions can account for or ignore the site-specific genetic deviations. Recall that the predicted effect for genotype j at environment i, accounting for both the factor loadings and the site-specific effects for an FAk model is:

$$\widehat{u}_{gij} = \widehat{\lambda}_{1i}\widehat{f}_{1j} + \widehat{\lambda}_{2i}\widehat{f}_{2j} + \ldots + \widehat{\lambda}_{ki}\widehat{f}_{kj} + \widehat{\delta}_{ij} \tag{8.20}$$

This is a prediction with narrow inference: it is the family's effect in the specific environment i included in the experiment. The predicted value of the family also includes the intercept and site effect:

$$\widehat{Y}_{ij} = \mu + \widehat{S}_i + \widehat{u}_{gij} \tag{8.21}$$

We can also make a prediction of the family's site-specific value based only on the factors, ignoring the site-specific deviations:

$$\widehat{u}^*_{gij} = \widehat{\lambda}_{1i}\widehat{f}_{1j} + \widehat{\lambda}_{2i}\widehat{f}_{2j} + \ldots + \widehat{\lambda}_{ki}\widehat{f}_{kj} \tag{8.22}$$

$$\widehat{Y}^*_{ij} = \mu + \overline{\overline{S}}_. + \widehat{u}^*_{gij} \tag{8.23}$$

This type of prediction has a wider inference: it refers to the family's predicted effect in a future environment that is perfectly correlated with environment i.

Similarly, the predictions across sites can refer to the average performance across the set of environments actually included in the study:

$$\widehat{Y}_j = \mu + \overline{\overline{S}}_. + \overline{\widehat{u}}_{g.j} \tag{8.24}$$

This is equal to averaging the site-specific predictions including the site-specific genetic deviations. A prediction with wider inference would ignore the site-specific deviations and refer to performance at a hypothetical 'average' environment by predicting at the mean values of the factors:

$$\widehat{Y}^*_j = \mu + \overline{\overline{S}}_. + \overline{\widehat{u}}^*_{gij} = \mu + \overline{\overline{S}}_. + \frac{1}{r}\sum_{k=1}^r \widehat{\lambda}_k \widehat{f}_{kj} \tag{8.25}$$

Here we demonstrate how to obtain these various predictions from ASReml, using the XFA1 model. In this case, we need only account for loadings and scores for a single factor; for models with $k > 1$, the sum of the products of loadings and scores over factors are needed.

The usual marginal predictions of family values across sites ($\widehat{Y}_{.j}$, the narrow-scope inference that includes the site-specific genetic deviations) are obtained as:

```
$B ~ mu site   !r  xfa1(site).id(female) ,
                   diag(site).id(block)
                   residual sat(site).id(units)
   predict female   !PLOT female !AVE block site
```

Here we use `!AVE block site` to get the conditional predictions with appropriate standard errors for computing reliability (which we will show in the next section). The predictions for females across sites produce values similar to the other models (with differences due to allowing the within-site variances and between-site correlations to vary):

```
female           Predicted_Value  Standard_Error Ecode
  16             27.9630          0.1752 E
  18             26.4088          0.1815 E
...
  580            25.9551          0.2453 E
...
SED: Overall Standard Error of Difference    0.2439
```

The standard errors of the predictions are a bit smaller than in the previous model because of a better model fit. For female 580 at site 1, the SE of prediction is 0.2453 from XFA1 model compared to 0.2595 in CORUV model.

Prediction of family effects at a hypothetical 'average' environment can be accomplished with:

```
predict female !AVE site 12*0 0.752 !ONLY xfa1(site).id(female)
```

Here, the qualifier !ONLY xfa1(site).id(female) tells ASReml to make the prediction only using the parameter estimates of the XFA1 part of the model. The qualifier !AVE site 12*0 0.752 refers to coefficients for the XFA1 model parameters: we set the coefficients for the first 12 parameters (the site-specific genetic variances) to zero to exclude them, and then we specify 0.752 as the average value of the site loadings on the first (and only) factor ($\bar{\lambda}_{1.}$).

The output from this predict statement in the .pvs file is an effect prediction, one could add it to the overall mean to get a predicted value:

```
female   Predicted_Value Standard_Error Ecode
  580    -0.4136         0.2583         E
```

The standard error of this predicted effect is a bit larger than the standard error for the average of site-specific predictions because it is predicted for a new, untested environment.

The predicted values of females at individual sites including the site-specific genetic deviation effects are easily obtained with:

```
predict site.female !present site female !AVE block
```

For example, the predicted value of family 580 at the first site is:

```
site    female   Predicted_Value  Standard_Error Ecode
101     580      22.7646          0.6318         E
```

We can also obtain this predicted value as the sum of the intercept, the site 101 effect, and the predicted effect of family 580 at site 101:

$$\widehat{Y}_{1.580} = \mu + \widehat{S}_1 + \widehat{\lambda}_{1.1}\widehat{f}_{1.580} + \widehat{\delta}_{1.580}$$

The values needed to compute this prediction are found in the .sln file:

```
    Model_Term               Level     Effect    seEffect
    site                        101      0.000       0.000
...
    site                        113     -1.942      0.2195
    mu                            1      23.78       0.1989
 diag(site).block          101.018     -0.1346      0.3062
...
 xfa1(site).id(female      101.580     -1.016       0.6309
...
```

The term labelled 'xfa1(site).id(female' is the predicted genetic effect of family 580 at site 101, including the site specific genetic deviation:

$$\widehat{u}_{g1.580} = \widehat{\lambda}_{1.1}\widehat{f}_{1.580} + \widehat{\delta}_{1.580} = -1.016.$$

So the predicted value is: $\widehat{Y}_{1.580} = 23.78 + 0 - 1.016 = 22.764$, matching the prediction given directly in the *.pvs* file. We can also obtain the predicted effect of family 580 in site 101 based on only the FA1 part of the model as:

$$\widehat{u}^{*}_{g1.580} = \widehat{\lambda}_{1.1}\widehat{f}_{1.580}$$

We have already shown that the loading for site on the first factor is obtained in the *.asr* file:

```
 Model_Term               Sigma     Sigma/SE  % C
...
 site          XFA_L  1  1  0.824247    5.75  0 P
...
```

The factor score for family 580 is found in the last set of effect estimates in the *.sln* file. Note that the genotype factor scores are not printed out for the FACV or FA formulations of the model, only for XFA forms:

```
 xfa1(site).id(female    1.580    -0.5501    0.3436
```

The predicted effect for this combination of family and site based only on the factor is:

$\widehat{u}^{*}_{1.580} = 0.824247^{*}(-0.5501) = -0.4533$, and the predicted value is:

$$\widehat{Y}^{*}_{1.580} = \mu + \widehat{S}_1 + \widehat{u}^{*}_{g1.580} = 23.78 + 0 + -0.4533 = 23.327$$

One can also obtain the factor-based family within site effects with a predict statement that excludes the site-specific genetic deviations:

```
 predict female !AVE site 12*0 0.824 !ONLY xfa1(site).id(female)
```

This is very similar to the predict statement used previously to get the family effect prediction within a hypothetical 'average' environment, but in this case we use the loading for the first environment (0.824) instead of the average loading, resulting in the following prediction in the *.pvs* file:

```
 female Predicted_Value Standard_Error Ecode
 580      -0.4533            0.2831       E
```

Estimating Heritability and Reliability from FA Models

Estimating heritability as a function of observed variance components can be tricky when there are consolidated (compound) terms and complex covariance structures in the model, as in FA or US models. One difficulty is defining the appropriate function of variance components, for example if we have a model in which the genotypic variance is different for every environment. Understanding the labelling of parameter estimates in the function definitions in ASReml adds some additional complexity. Another difficulty can be having different mating designs such as half-sib families and full-sib families in the same data. In this case calculation of causal genetic variances (e.g. additive genetic variance) may not be obvious.

Before considering how to extend heritability estimates to complex MET models, it helps to consider the concepts of heritability, genetic variance, and environmental variance in the context of replicated family evaluation trials that often occur in tree and crop breeding experiments. Conceptually, the simplest assumption is that we have a reference population of genotypes from which the parents of the families are sampled, and, similarly, we have sampled the testing environments at random from the target population of environments, usually production environments within a defined geographic range (Cooper and DeLacy 1994). The variance components estimates for genotype main effects, environment main effects, and genotype-by-environment interaction effects refer to the variability in these conceptual reference populations (Dudley and Moll 1969).

In this context, the expected response to selection based on an individual's phenotype when its progenies are evaluated in an independent environment depends on the narrow-sense heritability, $h_i^2 = \sigma_A^2 / \sigma_P^2$. We can estimate the pieces (additive genetic variance $\widehat{\sigma}_A^2$, and phenotypic variance $\widehat{\sigma}_P^2$) of this heritability estimator from a half-sib family evaluation like the pine polymix example using the traditional cross-classified analysis model as $\widehat{\sigma}_A^2 = 4\widehat{\sigma}_F^2$ and $\widehat{\sigma}_P^2 = \widehat{\sigma}_F^2 + \widehat{\sigma}_{FE}^2 + \widehat{\sigma}_\epsilon^2$. Thus, the narrow-sense heritability that is appropriate to predict response to selection among individual trees is:

$$h^2 = \frac{4\sigma_F^2}{\sigma_F^2 + \sigma_{FE}^2 + \sigma_\epsilon^2} \tag{8.26}$$

where σ_F^2 is the variance component due to family main effects, σ_{FE}^2 is the variance component due to family-by-environment interaction, and σ_ϵ^2 is the experimental error variance. Below, we will describe how to generalize this heritability estimator to more complex models such as the US and FA models with heterogeneous residual variances. Here, we will consider the appropriate heritability estimator to predict response to selection among family means. If we select superior families based on their means across environments and measure the response observed by growing remnant half-sib progenies in an independent environment sampled from the same reference population of environments, response to selection is a function of the selection differential and the heritability of family means defined using the cross-classified model structure as:

$$h_f^2 = \frac{\sigma_F^2}{\sigma_F^2 + \frac{\sigma_{FE}^2}{s} + \frac{\sigma_\epsilon^2}{sr}} \tag{8.27}$$

where s is the number of environments and r is the number of blocks per environment from which the means were calculated (Holland et al. 2003).

We can begin to generalize the estimator of family means-basis heritability by first considering the case where we have unbalanced data, with different numbers of plot measurements and environmental replications among families. One modification for unbalanced data is to use harmonic means of numbers of environments (s_h) and total plots (n_h) in which each family is measured (Holland et al. 2003):

$$h_f^2 = \frac{\sigma_F^2}{\sigma_F^2 + \frac{\sigma_{FE}^2}{s_h} + \frac{\sigma_\epsilon^2}{n_h}} \tag{8.28}$$

Another modification is the Cullis heritability estimator we introduced in Chap. 7 (Cullis et al. 2006):

$$h_{fC}^2 = 1 - \frac{\bar{V}_{BLUP\_difference}}{2\widehat{\sigma}_f^2} \tag{8.29}$$

The variance of the BLUP differences can be obtained from ASReml by squaring the average standard error of differences provided in the *.pvs* file when across-site family predictions are requested. Related to this estimator is the average of the prediction reliabilities, as introduced in Chap. 7.

A third modification is the bootstrapping method (Piepho and Möhring 2007). Note that no modification of the individual-basis narrow-sense heritability estimator is required when data are unbalanced because the selection units are individuals rather than family mean values.

To continue generalizing, when we have heterogeneous residual error variances, such that there are s distinct residual variances, the denominator of the narrow-sense heritability involves an average of the within-environment error variances:

$$h^2 = \frac{4\sigma_F^2}{\sigma_F^2 + \sigma_{FE}^2 + \bar{\sigma}_\epsilon^2} \tag{8.30}$$

Where $\bar{\sigma}_\epsilon^2$ is the average within-environment error variance. The *family mean-basis heritability estimate with heterogeneous error variances* includes a weighted average of within-environment variances:

$$h_f^2 = \frac{\sigma_F^2}{\sigma_F^2 + \frac{\sigma_{FE}^2}{s_h} + \frac{1}{s}\sum_{i=1}^{s}\frac{\sigma_{\epsilon i}^2}{r_{hi}}} \tag{8.31}$$

Where $\sigma_{\epsilon i}^2$ is the error variance within the ith environment and r_{hi} is the harmonic mean of number of plots per family in the ith environment. The Cullis estimator can also be used in this situation.

Finally, we generalize to the situation where the model has no genotype main effects, but rather *genotype effects nested in environments*. The response to selection among individual phenotypes as measured by their half-sib relatives grown in an independent environment is a function of a heritability estimator equal to the covariance of the selection and response individuals divided by the variance of individuals under selection (Nyquist 1991; Holland et al. 2003):

$$h^2 = \frac{E\left[Cov\left(f_{ij}, f_{i'j}\right)\right]}{V\left(f_{ij}\right)} \tag{8.32}$$

This is easily constructed from the estimated common genetic covariance between environments ($\widehat{\sigma}_{gii'}$) and common within-environment genetic variance component ($\widehat{\sigma}_{gi}^2$) from the CORUV model:

$$h^2 = \frac{\widehat{\sigma}_{gii'}}{\widehat{\sigma}_{gi}^2 + \bar{\sigma}_\epsilon^2} \tag{8.33}$$

For family-based selection, the predicted value of a family across sites is the average of its within-site predictions. To predict the response to selection based on this mean value as measured in an independent environment from the same reference population of environments used for the evaluation experiment, we want the expected covariance of the family mean to its value in an independent environment divided by the phenotypic variance of the family means:

$$h_f^2 = \frac{E\left[Cov\left(\bar{f}_{.j}, f_{i'j}\right)\right]}{V\left(\bar{f}_{.j}\right)} \tag{8.34}$$

Note that in the simple case of a model with family main effects and a common genotype-by-environment variance, the expected covariance of a family mean value with the family's value in an independent environment is estimated by the family variance component, and we have the usual heritability estimator for this model.

Considering the CORUV or compound symmetry model, we can use $\widehat{\sigma}_{gii'}$ and $\widehat{\sigma}^2_{gi}$ to estimate heritability:

$$E\left[\widehat{\sigma}_{gii'}\right] = E\left[Cov\left(\bar{f}_{.j}, f_{i'j}\right)\right] = \sigma^2_f$$

$$E\left[\widehat{\sigma}^2_{gi}\right] = E\left[V\left(ge_{ij}\right)\right] = \sigma^2_f + \sigma^2_{fe}$$

$$V\left(\bar{f}_{.j}\right) = \sigma^2_f + \frac{\sigma^2_{fe}}{s_h} + \frac{1}{s}\sum_{i=1}^{s}\frac{\sigma^2_{\varepsilon i}}{r_{hi}} = \frac{(s_h-1)\widehat{\sigma}_{gii'} + \widehat{\sigma}^2_{gi}}{s_h} + \frac{1}{s}\sum_{i=1}^{s}\frac{\sigma^2_{\varepsilon i}}{r_{hi}}$$

$$\widehat{h}^2_f = \frac{Cov\left(\bar{f}_{.j}, f_{i'j}\right)}{V\left(\bar{f}_{.j}\right)} = \frac{\widehat{\sigma}_{gii'}}{\frac{(s_h-1)\widehat{\sigma}_{gii'} + \widehat{\sigma}^2_{gi}}{s_h} + \frac{1}{s}\sum_{i=1}^{s}\frac{\sigma^2_{\varepsilon i}}{r_{hi}}} \tag{8.35}$$

If we have the more complex case of unequal pairwise variances among environments, our best estimate of the expected value of the covariance between the family mean value and its value in an independent environment is the average of the observed pairwise genotypic covariances between environments:

$$\widehat{Cov}\left(\bar{f}_{.j}, f_{i'j}\right) = \overline{Cov}\left(\bar{f}_{.j}, f_{i'j}\right) = \frac{1}{s(s-1)/2}\sum_{i=1}^{s-1}\sum_{i'=i+1}^{s}\widehat{\sigma}_{gii'} = \overline{\widehat{\sigma}_{gii'}} \tag{8.36}$$

The variance among family mean predictions is complicated if we have unbalanced data; it is the average over families of the variance of average family-by-environment effects:

$$\widehat{V}\left(\bar{f}_{.j}\right) = \bar{V}\left(\bar{f}_{.j}\right) = \frac{1}{n_f}\sum_{j=1}^{f}V\left(\frac{\sum_{i=1}^{S}ge_{ij}}{s_j}\right)^2 + \frac{1}{s^2}\sum_{i=1}^{s_j}\frac{\sigma^2_{\varepsilon i}}{r_{hi}} \tag{8.37}$$

Here, the value s_j refers to the number of environments in which family j was tested. The effects of a common family at different environments are not independent, so we need to include the covariances among these terms as well as their variances in this case:

$$\bar{V}\left(\bar{f}_{.j}\right) = \frac{1}{f}\left[\sum_{j=1}^{f}\frac{1}{s_j^2}\sum_{i=1}^{s_j}V\left(ge_{ij}\right) + \sum_{j=1}^{f}\frac{1}{s_j^2}\sum_{i=1}^{s_j}\sum_{i'\neq i}^{s_j}C\left(ge_{ij}, ge_{i'j}\right)\right] + \frac{1}{s^2}\sum_{i=1}^{s}\frac{\sigma^2_{\varepsilon i}}{r_{hi}}$$

$$= \frac{1}{f}\left[\sum_{j=1}^{f}\frac{1}{s_j^2}\sum_{i=1}^{s_j}\widehat{\sigma}^2_{gi} + \sum_{j=1}^{f}\frac{1}{s_j^2}\sum_{i=1}^{s_j}\sum_{i'\neq i}^{s_j}\widehat{\sigma}_{gii'}\right] + \frac{1}{s^2}\sum_{i=1}^{s}\frac{\sigma^2_{\varepsilon i}}{r_{hi}} \tag{8.38}$$

If data are balanced, this simplifies to:

$$\bar{V}\left(\bar{f}_{.j}\right) = \frac{\overline{\widehat{\sigma}^2_{gi}}}{s} + \frac{(s-1)\overline{\widehat{\sigma}_{gii'}}}{s} + \frac{1}{s^2}\sum_{i=1}^{s}\frac{\sigma^2_{\varepsilon i}}{r} = \frac{\overline{\widehat{\sigma}^2_{gi}}}{s} + \frac{(s-1)\overline{\widehat{\sigma}_{gii'}}}{s} + \frac{\overline{\sigma^2_{\varepsilon i}}}{sr} \tag{8.39}$$

Translating this to the model with family main effects, the variance of family mean values is:

$$\bar{V}\left(\bar{f}_{.j}\right) = \frac{\sigma^2_F + \sigma^2_{FE}}{s} + \frac{(s-1)\sigma^2_F}{s} + \frac{\overline{\sigma^2_{\varepsilon i}}}{sr}$$

$$= \sigma^2_F + \frac{\sigma^2_{FE}}{s} + \frac{\overline{\sigma^2_{\varepsilon i}}}{sr} \tag{8.40}$$

This simplifies further in the case of homogenous error variances to the standard estimator of heritability from multi-environment trials with balanced data:

$$\bar{V}\left(\bar{f}_{.j}\right) = \sigma_F^2 + \frac{\sigma_{FE}^2}{s} + \frac{\sigma_\epsilon^2}{sr} \tag{8.41}$$

Putting the average covariance between families across environments as the numerator and the average variance of family means across environments as the denominator as the heritability estimate, we get for the case of unbalanced data and heterogeneous genetic variances and covariances across sites:

$$h_f^2 = \frac{\overline{\hat{\sigma}_{gii'}}}{\frac{1}{f}\left[\sum_{j=1}^f \frac{1}{s_j^2}\sum_{i=1}^{s_j}\hat{\sigma}_{gi}^2 + \sum_{j=1}^f\frac{1}{s_j^2}\sum_{i=1}^{s_j}\sum_{i'\neq i}^{s_j}\hat{\sigma}_{gii'}\right] + \frac{1}{s^2}\sum_{i=1}^s\frac{\sigma_{ei}^2}{r_{hi}}} \tag{8.42}$$

In the case of balanced data but heterogeneous error variances, this simplifies to:

$$h_f^2 = \frac{\overline{\hat{\sigma}_{gii'}}}{\frac{\overline{\hat{\sigma}_{gi}^2}}{s} + \frac{(s-1)\overline{\hat{\sigma}_{gii'}}}{s} + \frac{\overline{\hat{\sigma}_{ei}^2}}{sr}} \tag{8.43}$$

Now we will use the parameter estimates from different models for the pine polymix data to estimate heritability of family means across environments. About 3% of plots are missing in this data set, so we should use Eq. 8.42, which involves the mean within-site variances weighted by the harmonic mean of replications per family at each site, but for simplicity, and because the level of imbalance is low, we will use the balanced data formula (Eq. 8.43), substituting the harmonic mean of the number replications per family and site (17.8) for the value r.

The harmonic mean of trees per family per site can be computed easily in R from a data frame (called "ds" in this example) holding our data:

```
trees_per_site <- aggregate(height ~ female + site, data=ds, length)
(nh = 1 / mean(1/trees_per_site$height))
```

First, we estimate narrow-sense and family mean-basis heritabilities for the *cross-classified* MET model with homogeneous error variances:

$$h^2 = \frac{4\sigma_F^2}{\sigma_F^2 + \sigma_{FE}^2 + \sigma_\epsilon^2} = \frac{4(0.563)}{0.563 + 0.174 + 7.12} = 0.29$$

$$h_f^2 = \frac{\sigma_F^2}{\sigma_F^2 + \frac{\sigma_{FE}^2}{s} + \frac{\sigma_\epsilon^2}{rs}} = \frac{0.563}{0.563 + \frac{0.174}{12} + \frac{7.12}{17.8*12}} = 0.92$$

Using the parameter estimates (rounded to the third decimal) from the compound symmetry (CORUV) model, we get the same results:

$$h^2 = \frac{4r_g\hat{\sigma}_{gii'}}{\hat{\sigma}_{gi}^2 + \sigma_\epsilon^2} = \frac{4(0.763)(0.738)}{0.738 + 7.12} = 0.29$$

$$h_f^2 = \frac{\overline{\hat{\sigma}_{gii'}}}{\frac{\hat{\sigma}_{gi}^2}{s} + \frac{(s-1)\overline{\hat{\sigma}_{gii'}}}{s} + \frac{\sigma_\epsilon^2}{rs}} = \frac{r_g\hat{\sigma}_{gi}^2}{\frac{\hat{\sigma}_{gi}^2}{s} + \frac{(s-1)r_g\hat{\sigma}_{gi}^2}{s} + \frac{\sigma_\epsilon^2}{rs}}$$

$$= \frac{(0.763)(0.738)}{\frac{0.738}{12} + \frac{11(0.763)(0.738)}{12} + \frac{7.12}{17.8*12}} = 0.92$$

The estimates for the CORUV model with heterogeneous error variances are:

$$h^2 = \frac{4r_g\widehat{\sigma}_{gii'}}{\widehat{\sigma}_{gi}^2 + \overline{\sigma_{ei}^2}} = \frac{4(0.8428)(0.6607)}{0.6607 + 7.21} = 0.28$$

$$h_f^2 = \frac{r_g\widehat{\sigma}_{gi}^2}{\dfrac{\widehat{\sigma}_{gi}^2}{s} + \dfrac{(s-1)r_g\widehat{\sigma}_{gi}^2}{s} + \dfrac{1}{s^2}\sum_{i=1}^{s}\dfrac{\sigma_{ei}^2}{r}}$$

$$= \frac{(0.843)(0.661)}{\dfrac{0.661}{12} + \dfrac{11(0.843)(0.661)}{12} + \dfrac{10.23 + 3.18 + \ldots + 9.40}{12^2 * 17.8}} = \frac{0.557}{0.598} = 0.93$$

Finally, for the XFA1 model with heterogeneous error variances, recall that the lower diagonal of the variance-covariance matrix of family within environment effects is (for 4 sites out of 12):

```
1.449 0.6288 0.6445 0.6848....

0.6260 0.6848 0.6445 0.6755....

0.5595 0.5464 0.8643 0.9183....

0.8395 0.9183 0.8643 0.9059....

...
```

The heritability estimate is based on the estimated variance-covariance matrix of family within environment effects. For this model, the average of the diagonal elements (0.7408) is the mean within-site family variance, and the average of the off-diagonal elements (0.563) is the average covariance between sites. The mean of the 12 site-specific residual variances is 7.15:

$$h^2 = \frac{4\overline{\widehat{\sigma}}_{gii'}}{\widehat{\sigma}_{gi}^2 + \sigma_{ei}^2} = \frac{4(0.563)}{0.741 + 7.15} = 0.28$$

In the heritability for individual measurements, the mean within-site family variance in the numerator is multiplied by 4 because the mean within-site female variance is 1/4 of the additive genetic variance due to the half-sib family structure in the data. In contrast, for the estimate of heritability on a family mean basis, we use the family variance component directly in the numerator, since our inference is to selection among the family mean predictions:

$$h_f^2 = \frac{\overline{\widehat{\sigma}}_{gii'}}{\dfrac{\overline{\widehat{\sigma}}_{gi}^2}{s} + \dfrac{(s-1)\overline{\widehat{\sigma}}_{gii'}}{s} + \dfrac{1}{s^2}\sum_{i=1}^{s}\dfrac{\sigma_{ci}^2}{r_{hi}}} = \frac{0.563}{\dfrac{0.741}{12} + \dfrac{11(0.563)}{12} + \dfrac{7.15}{12^2 * 17.8}} = 0.92$$

Again, for the XFA1 model we obtained similar narrow-sense and family mean heritabilities.

We can also estimate the family mean-basis heritability using the Cullis estimator by taking the average of the standard error of across-site family differences (SED) from the *.pvs* file:

```
SED: Overall Standard Error of Difference 0.2438
```

$$h_{fC}^2 = 1 - \frac{\bar{V}_{BLUP\_difference}}{2\widehat{\sigma}_f^2} = 1 - \frac{(0.2438)^2}{2\left(\overline{\widehat{\sigma}}_{gii'}\right)} = 1 - \frac{(0.2438)^2}{2(0.563)} = 0.947$$

Table 8.2 Family predictions across sites from model 11 (XFA1) and their reliabilities

Female	Predicted_Value	Standard_Error	REL
16	27.9630	0.1638	0.952
18	26.4088	0.1705	0.948
414	27.5957	0.1711	0.948
...			
580	25.9551	0.2372	0.900
...			
Average			0.947

This is close to the family mean-basis heritability based on variance components.

One more way to estimate the family mean-basis heritability is as the average of the prediction reliabilities, using the formula:

$$REL = 1 - \frac{PEV}{\sqrt{\sigma_f^2}} = 1 - \frac{PEV}{\sqrt{\overline{\sigma}_{ii'}}} \tag{8.44}$$

From the XFA model, we will use 0.563 in the denominator of the reliability equation. From the *.pvs* output of the statement 'predict female !AVE block site' in model 11, we can compute reliabilities (Table 8.2):

The average of the reliabilities (0.947) is identical to the Cullis estimator of family mean heritability.

We can obtain the estimates based on functions of variance components using the VPREDICT !DEFINE option in ASReml. As shown in Chap. 6, the easiest way to get the correct labels of parameter estimates from a complex ASReml model is to use VPREDICT !DEFINE at the end of the model and leave a blank line after it to generate a *.pvc* file with names and numbers of parameters identified. In this example we will estimate heritability from the XFA1 structure in model 11.

```
!PART 10
! XFA extended factor analytical G
$B ~ mu site  !r  xfa1(site).id(female) ,
                  diag(site).block
    residual sat(site).id(units)


predict female !present female site

!PART 10
VPREDICT !DEFINE
V female xfa1(site)   # Letter, label, coefficient
# OR use the following
X female xfa1(site)
```

- The components labeled 'female' were created using **V female xfa1(site)**.
- **V** is the function to convert components from XFA to unstructured (US) model parameters (i.e., to provide the within-environment variances and each of the pairwise environment covariances), 'female' is the label we assign and 'xfa1(site)' is the identifier of the variance component.

The output found in "Code 8-1_MET11_height.pvc" is given below:

```
          - - - Results from analysis of height - - -

  sat(site,01).id(units)              1359 effects
      1 sat(site,01).id(units);Residual_1              9.81099    0.389325
  ...
     12 sat(site,12).id(units);Residual_12             5.46531    0.218263
  ...
  xfa1(site).id(female)              910 effects
     25 xfa1(site).id(female);xfa1(site)  V  0  1  0.769546      0.226337
  ...
     36 xfa1(site).id(female);xfa1(site)  V  0 12  0.126178E-01  0.525742E-01
     37 xfa1(site).id(female);xfa1(site)  L  1  1  0.824246      0.149050
  ...
     48 xfa1(site).id(female);xfa1(site)  L  1 12  0.675730      0.803484E-01
     49 female         1.4489    0.32321    (variance site 1)
     50 female         0.55949   0.13227    (cov 1,2)
     51 female         0.54638   0.11257     (variance site 2)
     52 female         0.59378   0.14588     (cov 1,3)
     53 female         0.48899   0.97253E-01 (cov 2,3)
     54 female         0.58588   0.15189     (variance site 3)
     55 female         0.61280   0.14262     (cov 1,4)
  ...
    124 female         0.50422   0.97664E-01 (cov_9,12)
    125 female         0.66804   0.13021     (cov_10,12)
    126 female         0.46923   0.11384     (cov_11,12)

  Notice: The parameter estimates are followed by
          their approximate standard errors.
```

- Coefficients are identified by the numbers in the first field and by labels. For example, residual variances for sites are numbered from 1 to 12, and labeled as

```
      1 sat(site,01).id(units);Residual_1
      2 sat(site,02).id(units);Residual_2
```

- The fields named `female` (numbered from 49 to 126) are female within-site variance components (bold) and covariances between pairs of site. If we rearrange them in matrix format it will be more obvious how they relate to the US parameterization (for the first 4 sites):

```
          site1    site2    site3    site4
  site1   1.449
  site2   0.5595   0.5464
  site3   0.5938   0.4890   0.5859
  site4   0.6128   0.5047   0.5356   0.5527
```

In the following example, we compute phenotypic variances, additive genetic variances, and heritabilities for selection among individual trees or family means.

PART 10

```
!PART 10
! XFA1 extended factor analytical G
...
VPREDICT !DEFINE
V female xfa1(site)    # defines 14:23
# sum of error variances
F err        1+2+3+4+5+6+7+8+9+10+11+12 #
# mean error variance
F err.m    err*.08333    # 1/12=0.08333
# sum of within-site female variances
F fem.site      49+51+54+58+63+69+76+84+93+103+114+126
# mean within-site female variance
F fem.sitem    129*0.08333    # 1/12=0.08333
# sum of between-site pairwise covariances (there are 66 of them...)
F cov        50+52+53+54+56+...+125 # cov12+cov13+...+covii'
# mean covariance, 1/66
F covm        cov*.01515
# Additive genetic variance (numerator for h2i)
F Additive    covm*4.0
# phenotypic var
F phen        fem.sitem + err.m
# phenotypic variance of family means = covm
F phen_f    fem.sitem*0.0833 + covm*0.9167 + err.m*0.00468
# narrrow-sense heritability
H h2i        Additive  phen
# family-mean heritability
H h2f        covm  phen_f
```

A subset of the output (*Code 8-1_MET11_height.pvc*) is given below:

```
        - - - Results from analysis of height - - -
...
 127 err  1          85.820          1.0644
 128 err.m127         7.1514          0.88693E-01
 129 fem.site 49      8.8899          1.1220
 130 fem.sitem129     0.74079         0.93495E-01
 131 cov 50           37.190          5.6923
 132 covm131          0.56348         0.86247E-01
 133 Additive132      2.2539          0.34499
 134 phen128          7.8922          0.12610
 135 phen_f130        0.58104         0.86409E-01
     h2i        = Additive132 133/phen128  134=  0.2856 0.0407
     h2f        = covm131   132/phen_f13 135=  0.9211 0.0105
 Notice: The parameter estimates are followed by
         their approximate standard errors.
```

These estimates agree with our computations above. As the number of environments increases, the number of covariance for pairs of environment increases (66 in this example). This makes heritability calculations cumbersome in ASReml. Care is needed to make sure variances and covariances are selected correctly.

Biplots from FA Models

Biplots can be useful visualizations of GxE interactions from FA models. The responses of genotypes to environments on a two-dimensional surface are frequently reported in the plant breeding literature. A biplot displays site loadings and genotype scores simultaneously. R code to read in results from an XFA2 model produced by ASReml standalone and to generate a biplot is provided in "*Code 8-3_biplot.R*". Another form of the biplot using output of ASReml-R is provided in "*Code 8-2_pine_met.Rmd*". Figure 8.1 was produced by the first set of code and displays site loadings as vectors in blue on the two factors and family scores in black. This figure shows a typical problem with biplots: if the number of genotypes or families is large, the plot becomes very busy and hard to read. Nevertheless, even from this plot, it is clear that site 105 affected genotype performance differently than other sites. This is congruent with the result observed in the correlation estimates from the XFA models that indicate that this site had the lowest average correlation with other sites. A large number of genotypes are at the center of the plot; genotypes that are closer to the end of a particular site vector have scores with the same sign and similar magnitude of that site's loading compared to the rest of the population. This indicates that those families have their most positive effect at that environment. For example, family 421's score is near the loading for site 105, indicating that it has the most favorable effect at that site. Indeed, family 421 has the highest predicted value at site 105 (30.7, compared to a population mean of 28.9 at that site). Biplots are descriptive, however, and should be interpreted cautiously as they may not depict all aspects of the GxE interactions, including crossover interactions (Yang et al. 2009).

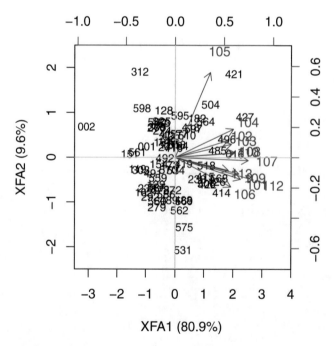

Fig. 8.1 Biplot for site loadings (*blue vectors*) and female scores (*black text labels*) on two factors from the XFA2 model. Site 105 has high within-site variation but has the smallest correlation with other sites. Genotypes 421, 504 and 427 have large positive scores for both factors

Electronic supplementary material: The online version of this chapter (doi:10.1007/978-3-319-55177-7_9) contains supplementary material, which is available to authorized users.

© Springer International Publishing AG 2017

F. Isik et al., *Genetic Data Analysis for Plant and Animal Breeding*, DOI 10.1007/978-3-319-55177-7_9

Abstract

Genetic data sets available to breeders are increasing in size, both in numbers of markers and in numbers of breeding individuals or lines genotyped. The scale of the data sets requires breeders to use software to perform quality control checks, visualize, and manipulate data. Breeders will often want to combine genetic marker data with physical or linkage map information, phenotypic data, and pedigree information. In this chapter we demonstrate the use of base R and the *synbreed* package of R to process an example data set from a maritime pine breeding population. The *synbreed* package defines the `gpData` object class, which can hold phenotypes, genotypes, pedigree, and genetic map information. This package is particularly useful to streamline data manipulation and analyses that combine genotype and phenotype data. Readers should be aware that algorithm and software developments for genomic data are areas of active research, more efficient and powerful methods are constantly being developed.

Marker Data and Some Definitions

Due to advances in next generation DNA sequencing and other genomics technologies, large amounts of data can be produced efficiently, even at the scale of animal and plant breeding populations. Compared to traditional phenotypic data collected from field or greenhouse trials, genetic marker data sets can be very large. For example, genotyping 2,000 individuals for 10,000 SNP markers is feasible for many breeding programs. The data set resulting from this genotyping will contain 20 million data points.

Raw marker data are usually not ready for analyses. When genotype data are obtained, often the first task is to organize, summarize, and reformat the data for subsequent analysis by available software packages. Genotype data come in different formats from different labs or companies, and usually require reformatting before they can be merged or used as input data for analysis. Also, some software may require removal of **monomorphic loci** (at which all individuals have the same genotype) before analysis. Finally, missing data and heterozygotes may need to be handled in different ways for different analysis software.

With large data sets, manual curation and formatting is a bad approach in general. It becomes very difficult to consistently make reformatting changes, and is also likely to introduce new errors. In this section, we present some approaches to reformatting, summarizing, and visualizing marker data using the *synbreed* R package (Wimmer et al. 2012). We focus on this package because in addition to providing useful functions for manipulating data, it interfaces with other packages in R, such as the *BLR*, the previous version of BGLR package (Pérez et al. 2010). The *synbreed* package provides estimates of important population genetic parameters, such as linkage disequilibrium (LD), and to fit statistical models for genomic estimated breeding values and model validation (Wimmer et al. 2012). Conveniently, it also can output relationship matrices formatted for use in ASReml.

Before describing code for reformatting, we review a typical raw data format and describe a modified format that is often more desirable for subsequent analysis. A common situation is to receive SNP genotype results as **base pair calls**, such as GG or CC for the homozygous genotypes and GC for the heterozygous genotype (e.g. from Illumina GenomeStudio). In some cases the marker genotypes are **pairs of observed nucleotides**, e.g. G/G, C/C, and G/C. For most quantitative analyses of marker data, however, it is preferable to recode genotypes from base pair calls to counts of minor alleles carried by an individual at the locus.

In general, most analysts use several software packages to accomplish various data analysis tasks, because no single package can handle all the different types of data manipulations, summary, and data visualizations. Similarly, algorithms and software developed to handle large numbers of markers are numerous, and they are constantly being revised to meet the challenges of large and complex data generated by constantly evolving methods of sequencing and genotyping. Readers should be aware that algorithm and software developments for genomic data are active areas of research. The tools covered here and subsequent chapters may be out of date in a few years and more efficient versions of the packages may become available.

As an example, Table 9.1 shows a subset of marker data in character format, reflecting the **SNP base pair calls** for a sample of diploid individuals:

We may prefer to transform this data set by first transposing columns and rows so that individuals are in rows instead of columns. Then the minor and major alleles at each locus can be easily identified by their frequencies in each column. By definition, **a minor allele has a frequency less than 0.5**; different bases could be the major allele at different loci. In Table 9.2 we have transposed the previous data matrix and identified the minor allele call above the column headers ("MA = ...");

Table 9.1 A matrix of SNP base pair calls (in rows) for individuals

MarkerID	Ind1	Ind2	Ind3	Ind4	Ind5	Ind6	...
01-256	GG	GG	GG	GG	GG	GG	GG
01-71	AC	AA	AA	CC	AA	AA	AA
01-559	CC	CC	CC	CC	CC	CC	CC
01-431	GG	AG	GG	0	GG	AG	AG
...	AA	GG	GG	0	GA	GG	GG

- The rows are marker names; usually strings of text and numbers.
- The columns are subject or sample names (Ind1, Ind2, ...).
- Ind1 is homozygous (GG) at locus 01-256 but heterozygous (AC) at locus 01-71.
- The 0 values are missing genotypes due to a failed or low quality SNP assay. Missing genotypes may be coded in a variety of formats, including "NA" or with just a blank space.
- Sometimes, there is no segregation for a marker in the genotyped population. For example, all the individuals are homozygous CC genotypes for marker 01-559. This marker is "fixed" (monomorphic) in the sample of individuals. It is possible that this locus may be segregating in a different population sample, however.

Table 9.2 A transposed matrix of SNP base pair calls. Individuals are in rows; base pair calls (markers) are in columns. Minor allele of each locus is given at the top of the columns

	MA=A	MA=C	MA=.	MA=G	MA=A
MarkerID	01-256	01-71	01-559	01-431	...
Ind1	GG	AC	CC	GG	AA
Ind2	GG	AA	CC	AG	GG
Ind3	GG	AA	CC	GG	GG
Ind4	GG	CC	CC	NA	NA
Ind5	GG	AA	CC	GG	GA
Ind6	GG	AA	CC	AG	GG
...	GG	AA	CC	AG	GG

Notice that there is no minor allele for locus 01-559. Although we do not see allele A for locus 01-256 in the small sample of six individuals in the table, the minor allele at this locus, A, is found in some individuals not included in the table.

Now we wish to recode the base pair calls numerically as gene content of the minor allele, such that an individual that is homozygous for the minor allele carries two copies of the minor allele and is coded as 2, a heterozygous individual has one copy of the minor allele and is coded as 1, and a homozygote for the major allele carries no copies of the minor allele and is coded as 0. One could also recode the data as gene content for the major allele, all subsequent analyses will still work, but their interpretation has to be adjusted to reflect how the data were coded. As a specific example, locus 01-71 in Table 9.2 is segregating for nucleotides A and C, where C is the minor allele, so genotypes at this locus can be recoded for gene content of the minor allele as AA = 0, AC or CA = 1 and CC = 2. Ind1 has one minor allele (C) for locus 01-71 but Ind2 does not have any minor alleles at this locus.

In Table 9.3 we have the minor allele count for each combination of locus and individual instead of letters representing base pair calls. Further, the "0" calls for missing scores in the original data set must be changed now to "NA" (or some other non-numeric code) so that they are distinguished from homozygous major allele calls. The sample of this data set would appear as follows after re-coding:

Table 9.3 Minor allele count for each locus and individual

MarkerID	01-256	01-71	01-559	01-431	...
Ind1	0	1	0	2	2
Ind2	0	0	0	1	0
Ind3	0	0	0	1	0
Ind4	0	2	0	NA	NA
Ind5	0	0	0	2	1
Ind6	0	0	0	1	0
...	0	0	0	1	0

Let's look at a few definitions of genetic terms, which may be encountered during the data analyses.

Allele Frequencies

By definition the **minor allele** is *the allele at a locus with frequency less than 0.5 in the population*. With biallelic SNP markers, a marker locus can have only two alleles. Sequence data on populations may reveal more than two bases segregating at a position, and the researcher will have to make some choices on how to handle such data. One possibility is to simply drop such loci from the analysis. Another option is to create two or more different biallelic markers from a multiallelic locus. One marker could have scores representing the counts of one rare allele in each individual, while the second marker represents the counts of a different rare allele. For this book, however, we will assume that SNP data are strictly biallelic or have been recoded to follow a biallelic format. Given this assumption, a sample of n individuals can have three different genotypes (e.g. AA, AT, TT) at the locus. The allele frequency (p_j) of marker j can be calculated from genotypes in 0, 1, 2 format as

$$p_j = \frac{\sum_{i=1}^{n} w_{ij}}{2n} \quad j = 1, \ldots, p \tag{9.1}$$

Where w_{ij} is the marker genotype for individual i and marker j and n is the number of individuals (Wimmer et al. 2012).

To demonstrate the computation of allele frequencies in a simple example, the following is a small sample data set of genotype calls at two sequence positions, M1 and M2, in five samples:

```
Sample   M1    M2
1        AA    GA
2        AT    GG
3        TA    AA
4        TT    AA
5        TT    AA
```

The frequency of allele A at position M1 is p_A = *sum of copies of A over all individuals/2 times of number of individuals* = 4/ 2 × 5 = 0.4. The frequency of the T allele is $q_T = (1 - p_A) = 1 - 0.4 = 0.6$. If the data were recoded in 0, 1, 2 format, the minor allele frequency at M1 would be calculated as (2 + 1 + 1 + 0 + 0)/(2 × 5) = 4/10 = 0.4, and the major allele frequency is still $(1 - p_A) = 1 - 0.4 = 0.6$. Although A is the minor allele at position M1, it is the major allele at position M2.

Hardy-Weinberg Equilibrium (HWE)

The simple definition of HWE is '*genotype and allele frequencies in a random mating population remain constant from one generation to another if there are no disturbing factors (mutation, migration, genetic drift and selection)*'. In such an ideal population the two alleles an individual receives, one from each parent are independent, such that genotype frequencies can be predicted from allele frequencies.

An example for a biallelic locus, let the allele frequencies be p for A allele, and q for T allele ($p + q = 1$). In order to predict the genotype frequencies, all we need are allele frequencies p and q:

The frequency of AA $= p^2$
The frequency of AT $= pq + qp = 2pq$
The frequency of TT $= q^2$
Genotype frequencies sum to $(p + q)^2 = p^2 + 2pq + q^2 = 1$

In the absence of selection, mutation, migration and genetic drift, allele frequencies and genotype frequencies remain constant from one generation to the next (HWE). Comparing the predicted genotype frequencies to observed genotype frequencies in a population can provide insight to the importance of inbreeding or selection in the population.

Polymorphism Information Content

The polymorphism information content (PIC) measures the probability of differentiating the allele transmitted by a given parent to its offspring given the marker genotype of father, mother, and offspring (Botstein et al. 1980):

$$PIC = 1 - \sum_{i=1}^{n} p_i^2 - \sum_{i=1}^{n} \sum_{j=i+1}^{n-1} 2p_i^2 p_j^2 \tag{9.2}$$

where p_i and p_j are frequencies of alleles, n is the number of alleles ($n = 2$ for SNPs). Loci with values close to 1 are more desirable, but can only occur with many alleles. The maximum PIC for a biallelic locus occurs when the allele frequencies are equal, and PIC $= 0.375$.

Heterozygosity

A diploid locus has two alleles, each inherited from a different parent. If an individual has the same alleles e.g. AA at a given locus then the individual is homozygous. If the individual has two different alleles at the given locus, e.g. AT or TA, then the individual is heterozygous. **Heterozygosity** *is simply the proportion of heterozygous individuals in the population*. This parameter is used to define the genetic variation in a population. The **allelic diversity**, sometimes called the **expected heterozygosity**, is the expected proportion of heterozygous individuals in the data set when HWE holds. Allelic diversity is a better measure of diversity when there is inbreeding because the frequency of heterozygotes does not indicate the actual segregation at the locus.

Linkage Disequilibrium (LD)

Gametic phase disequilibrium, more commonly referred to as **Linkage Disequilibrium** (LD) is *the non-random association of alleles at different loci*. Its converse, **Linkage Equilibrium** (LE), is *random association between alleles at different loci*, such that the frequency of allelic combinations (genotype frequency) is predicted accurately by the products of the individual allele frequencies. Alleles at loci located close together on the same homologous chromosome tend to retain their linkage relationship after meiosis, because recombination events are relatively infrequent (often only one or two per chromosome), and are therefore unlikely to occur between two loci that are physically near each other. Recombination is more likely to occur between two loci that are far apart on a chromosome. Over many generations, most alleles at many pairs of loci reach linkage equilibrium because of the accumulated effects of recombination between homologous chromosomes. The exception to this is loci that are extremely close together, new mutations (which have not undergone sufficient meiotic generations to reach linkage equilibrium with neighboring loci), or pairs of loci that have an interaction that confers a selective advantage or disadvantage on individuals that inherit specific combinations of alleles – such loci can remain in linkage disequilibrium for many generations.

LD impacts the accuracy of predictions from genomic data (Habier et al. 2007) and the resolution of association analyses (Flint-Garcia et al. 2003). Breeders have little ability to change it at the level of the entire species (Hayes et al. 2009), although LD can be increased by creation of sub-populations with limited effective population size (Grattapaglia and Resende 2011).

Commonly used statistics to measure LD in a population are the coefficient of disequilibrium D, the scaled coefficient of disequilibrium D', and the coefficient of determination, r^2. For example, consider two loci A and B, each with two alleles Aa and Bb (Fig. 9.1).

Locus A

Locus B		A	a	Total
	B	h_{AB}	h_{aB}	p_B
	b	h_{Ab}	h_{ab}	p_b
	Total	p_A	p_a	

Fig. 9.1 Two loci A and B each with two alleles. $h_{AB}, h_{Ab}, h_{aB}, h_{ab}$ are probability of observing gamete frequencies (allele combinations). p_A, p_a, p_B, p_b are allele frequencies

If gamete frequencies are equal to the product of their allele frequencies, this implies that the alleles at different loci are independent, and therefore in linkage equilibrium (LE):

$$h_{AB} = p_A p_B$$
$$h_{Ab} = p_A p_b$$
$$h_{aB} = p_a p_B$$
$$h_{ab} = p_a p_b$$

As an example, suppose allele A has a frequency of $p_A = 0.6$ and allele B has a frequency of $p_B = 0.4$. If the two loci are in LE then the expected gamete frequency would be $h_{AB} = p_A p_B = 0.6^*0.4 = 0.24$.

Linkage disequilibrium can be measured as D, the difference between the observed gamete frequencies and the gamete frequencies expected under linkage equilibrium:

$$D_{AB} = h_{AB} - p_A p_B$$
$$D_{Ab} = h_{Ab} - p_A p_b$$
$$\ldots$$

If two loci are independent then D would be zero and we say the two loci are in LE in the population. A value greater or smaller than zero suggests LD. LD can also be calculated as

$$D = h_{AB} h_{ab} - h_{Ab} h_{aB}$$

D is dependent on allele frequency, which makes it difficult to compare LD among pairs of loci. The maximum value of D is:

$$D_{max} = \min\{p_A p_B, (1 - p_A)(1 - p_B)\} \quad \text{when } D < 0$$
$$D_{max} = \min\{p_A(1 - p_B), (1 - p_A)p_B\} \quad \text{when } D > 0$$

D_{max} is the smaller of $p_A p_B$ and $(1 - p_A)(1 - p_B)$ when D is negative.

D_{max} is the smaller of $p_A(1 - p_B)$ and $(1 - p_A)p_B$ when D is positive.

D can be standardized (scaled) following (Lewontin 1964):

$$D' = \frac{D}{D_{max}} \tag{9.3}$$

D' has the range of -1 to 1, with the extreme values of -1 and $+1$ indicating no evidence of recombination between markers. D' is not a desirable statistic for small samples or loci with rare alleles, as it tends to be inflated under those conditions, so is also not entirely comparable across loci with different allele frequencies. A more commonly used statistic to measure LD is the coefficient of determination (r^2 or Δ^2).

$$r^2 = \frac{D^2}{p_A(1-p_A)p_B(1-p_B)} \tag{9.4}$$

The statistic r^2 has a range of 0–1 with 0 indicating linkage equilibrium and 1 indicating complete disequilibrium. LD for each pair of loci tested can be plotted against the physical distance between the loci as a way to understand the relationship between LD and physical distance (Reich et al. 2001) in the population. We will give an example later in the chapter.

Software and Tools for Processing Marker Data

We will demonstrate the use of the base R supplemented with packages from the Bioconductor project (http://www. bioconductor.org) and from CRAN for processing and analysing genome-wide marker data. The Bioconductor project is focused on producing tools for R users involved in high-throughput genomic research, and provides a wide array of packages for analysis of microarray, high-throughput sequencing, and SNP genotyping experimental data.

Introduction to the *Synbreed* Package

There are several R packages available, such as *GenABEL* (Aulchenko et al. 2007) and *genetics* (Warnes et al. 2013) to manage and visualize marker data, and new tools are being developed constantly. *Synbreed* is a comprehensive package composed of functions to manipulate and analyze marker, phenotype, genetic map and pedigree data (Wimmer et al. 2012). The package is available through CRAN at http://cran.r-project.org/web/packages/synbreed. In this chapter we will give examples using *synbreed* for exploratory marker data analysis, visualization and data processing. In subsequent chapters we will demonstrate with examples using *synbreed* to impute missing genotypes, calculate genomic relationships from markers, and fit statistical models to estimate genomic-based breeding values.

Synbreed creates an object of data class called `gpData` for analysis. The components of this object include phenotypes, the genetic map, genotype scores, covariates and pedigree information for a set of individuals. An object of class `gpData` is stored in a sparse binary format, which efficiently stores large amounts of data. The *synbreed* package can be downloaded and installed following Code Example 9.1:

Code example 9.1
Exploratory marker data analysis using *synbreed* package (Code 9-1_Data steps.R). The code reads different data sets (pedigree, phenotype, genotype and genetic map) and creates a unified object

```
##################################################
# Marker data analysis - Code09-1_Data steps.R
# This script is to read different types of data and organize
##################################################

# Remove everything in the working environment
  rm(list=ls())

##################################################
### Download the packages IF NOT installed already
install.packages('synbreed', dependencies=T)

# Upload the package to R environment
library(synbreed)

### Working directory to read files
path <- "/Users/.../data"
# set the path as working directory
setwd(path)

# Directory to save figures
gpath <- "/Users/.../Images"

# See the list of files in working directory
list.files()
```

Maritime Pine Data Example

We will use data adapted from Isik et al. (2016) to demonstrate marker data processing and visualization in this chapter and will return to this data set later in Chaps. 11 and 12 to demonstrate genomic BLUP and genomic selection methods. A SNP chip was used to genotype 654 maritime pine (*Pinus pinaster*) trees with 2,600 SNP markers. The population was composed of two generations of breeding: The G0 generation (founders) had 184 individuals. The G1 generation (descended from G0) had 470 individuals. Some samples were dropped because of errors in identification or because of lack of phenotype. Many of the 2,600 markers were placed on 12 linkage groups (Chancerel et al. 2011, 2013). We created four simulated phenotypic traits for each individual to demonstrate how to incorporate phenotype data into a common `gpData` object with the pedigree, marker, and map information. First, let's look at the elements used to create the `gpData` object.

Genotypic Data

The pedigree and genotype data are combined in a file (*maritime pine genoped.csv*) where individuals are in rows and markers in columns. The first four fields are unique individual IDs, code for parent 1, code for parent 2, and generation number. Genotypes are base pair calls (e.g., "AA", "AT", "TT"). Missing values were coded as NA. In the following code we read the data using the `read.csv` function of R and assign the unique tree IDs as row names to the resulting data frame.

```
### Load GENOTYPES and PEDIGREE file
# The first 4 columns are ID, Parent1, Parent2, Generation
# The rest of columns are genotypes "AA", "AB", "BB"
genoped <- read.csv(" maritime pine genoped.csv ", header= T,
          stringsAsFactors = F, colClasses = c(Par1 = "character",
          Par2 = "character"), row.names =1,check.names=F)

# Show first six entries and six columns of genoped
head(genoped[, 1:6])
```

The input format is a little tricky, we force the Par1 and Par2 columns containing the parental codes in the pedigrees to be read as character type, this is required in later analyses when we create a pedigree-based relationship matrix, so that offspring will be matched correctly to their parents, which are coded as character type in the row names of the data frame. The output from `head(genoped[, 1:6])` is:

	Par1	Par2	gener	384_PP2C-246	AJ300737-208	BX666025-170
0001	0	0	0	GG	GG	GG
0003	0	0	0	AG	GG	GG
0003-3	0003	0	1	AG	GG	GG
0004	0	0	0	GG	GG	GG
0005	0	0	0	GG	GG	GG
0006	0	0	0	GG	GG	GG

Phenotypic Data

For input, either a *data frame* or an *array* can be used. We read in the simulated phenotype trait data from the file '*maritime pine simul pheno.csv*':

```
### Load PEHNOTYPE data
pheno <- read.csv("maritime pine simul pheno.csv", header=TRUE,
        strip.white = TRUE, na.strings = c("NA",""),
        stringsAsFactors = F, row.names = 1, check.names=F)

# print the first 6 lines
head(pheno)
```

The output from head(pheno) shows the individual IDs as row names and the trait values for the four simulated phenotypes.

	Sim.Pheno	Sim.Pheno.2	Sim.Pheno.3	Sim.Pheno.4
0001	8.75	52.76	87.39	75.37
0003	10.36	51.18	84.54	74.86
0003-3	9.60	53.23	86.09	73.66
0004	9.43	48.83	86.11	75.70
0005	9.80	51.90	84.43	73.14
0006	9.17	52.98	85.19	75.01

It is important that the rownames of the objects to be combined in the *gpData* object match, so that marker data and phenotype data can be combined appropriately. We test that they match by using the identical function in R. When we run it we should see 'TRUE' in the R console.

```
# check that the row names are identical for geno and pheno
# (although order may be different)
identical(sort(rownames(genoped)), sort(rownames(pheno)))
```

Pedigree

The pedigree file can be an independent file. In this example, the pedigree information was already combined with the genotype scores. Here, we extract the pedigree information from the genoped data frame. We then use the create.pedigree function in *synbreed* to create a pedigree object from the data, this function sorts the pedigree to make sure that an individual is listed in the ID field first before it is listed as a Parent1 or Parent2:

```
# Create PEDIGREE object. The variables must be factors.
ped <- create.pedigree(rownames(genoped), genoped$Par1, genoped$Par2,
 genoped$gener)
# Use synbreed to summarize pedigree file
summary(ped)
# Print the first 6 lines
head(ped)
```

The first three columns in the ped object are character strings, while the last field ('gener') is an integer type. Zeroes in the parental codes indicate that the parents are unknown. The "ID" variable in the ped object is used to match pedigrees to the rownames of the geno and pheno components of the gpData object we will create.

```
> summary(ped)
Number of
      individuals  654
      Par 1        189
      Par 2         90
      generations 2
> head(ped)
     ID Par1 Par2 gener
1 0001    0    0     0
2 0003    0    0     0
4 0004    0    0     0
5 0005    0    0     0
6 0006    0    0     0
8 0008    0    0     0
```

Convert Marker Data to a Matrix and Match with Pedigree

The *synbreed* package requires marker data in a matrix format with row names taken from the pedigree file. In the following code we extract the marker score columns from the `genoped` data frame and use the `as.matrix` function of R to create the matrix of markers scores. We then check that the row names of the resulting matrix and the values of the pedigree `ped $ID` variable are identical.

```
### Create matrix of MARKERS and match with ped$ID
# remove some columns from the marker scores in genoped
gmat <- as.matrix(subset(genoped, select = -c(Par1, Par2, gener)))
# Check if row names identical?
identical(sort(rownames(gmat), sort(ped$ID))
# Show first six entries and 5 columns
head(gmat[,1:5])
```

The output shows a small part of the matrix of genotypes. The pedigree fields are dropped. Notice that the elements have double quotes because `gmat` is a matrix of character strings.

```
> head(gmat[,1:5])

       384_PP2C.246  AJ300737.208  BX666025.170  F51TW9001ARTXE.215  F51TW9001B0OJP.1031
0001      "GG"          "GG"          "GG"             "AA"                "AA"
0003      "AG"          "GG"          "GG"             "AA"                "AA"
0003-3    "AG"          "GG"          "GG"             "AA"                "AA"
0004      "GG"          "GG"          "GG"             "AA"                "AA"
0005      "GG"          "GG"          "GG"             "AA"                "AA"
0006      "GG"          "GG"          "GG"             "AA"                "AA"
```

Marker Map

In the following code we read "*maritime pine genetic map.csv*". The file contains the genetic position in **centiMorgans** (cM) of markers on 12 chromosomes of maritime pine. The **centiMorgan** (cM) is a unit of genetic distance or crossover frequency on a chromosome. Markers separated by one cM have about a 1% chance of recombination in a single generation. The data frame resulting from reading in the file has marker locus IDs as rownames and columns *chr* (chromosome indicator from 1 to 12) and *pos* (cM position on the chromosome). Variables *chr* and *pos* must be numeric. There are 2,600 loci listed in the file but only 2,258 of them are mapped (*chr* and *pos* of 342 loci are missing).

```
### Read the genetic map
map <- read.csv("maritime pine genetic map.csv", header = T,
            stringsAsFactors = F, sep = ",", dec = ".",
            na.string = "NA", row.names = 1, check.names = F )

# The Map file must have the same order as Marker file
# rownames of *markpos* must match *gmat* matrix (marker ID)
rownames(map) <- colnames(gmat)
# Show first six entries
head(map)
```

The output shows the positions of markers on chromosomes. Notice that the first two markers are not mapped on any linkage group and they have missing values for *chr* and *pos*.

```
> head(map)

                      chr      pos
384_PP2C-246          NA       NA
AJ300737-208          NA       NA
BX666025-170           7    92.60
F51TW9001ARTXE-215     4     7.66
F51TW9001B0OJP-1031   10   125.19
F51TW9001BF8Q1-922     1    83.05
```

The variable *pos* can be in cM units (as in this example) or in base pairs, if the physical position of the markers are known. Not all species and populations have genetic maps, and the map object is not required.

Putting All the Elements Together

The function create.gpData merges the individual data sources (genotype, phenotype, genetic map, pedigree) into a single object. The rownames of the genotype and phenotype inputs should match the "ID" column of the pedigree object. The return value is a gpData object, which is a modified list with elements pheno, geno, map, pedigree, covar, and info. The str function returns a compact display of the internal **str**ucture of the gp data object, and is especially well suited to display the (abbreviated) contents of lists of the gp data object.

```
# Object of class *gpData* includes all data required for analysis.
  gp <- create.gpData(geno=gmat, pheno=pheno, ped=ped, map=map)
# Use str function to examine the gp object
  str(gp)
# Summarize the object
  summary(gp)
```

The result of str(gp) is a view of the elements contained inside the gp object. Element covar is a data frame with 654 individuals and 4 variables. The id field is character, phenotyped, genotyped and family fields are logical. This element indicates which individuals are included in the genotype and phenotype data. The elements named 'pheno', 'geno', 'map,' and 'pedigree' hold phenotypic, genotypic, genetic map, and pedigree information, respectively. We used a subset of the original pheno object, including only the first phenotype (subsetting not shown).

```
> str(gp)
List of 7
$ covar :'data.frame':     654 obs. of 4 variables:
..$ id : chr [1:654] "0001" "0003" "0004" "0005" ...
..$ phenotyped: logi [1:654] TRUE TRUE TRUE TRUE TRUE TRUE ...
..$ genotyped : logi [1:654] TRUE TRUE TRUE TRUE TRUE TRUE ...
..$ family : logi [1:654] NA NA NA NA NA NA ...
$ pheno : num [1:654, 1, 1] 8.75 10.36 9.6 9.43 9.8 ...
..- attr(*, "dimnames")=List of 3
.. ..$ : chr [1:654] "0001" "0003" "0003-3" "0004" ...
.. ..$ : chr "Sim.Pheno"
.. ..$ : chr "1"
$ geno : chr [1:654, 1:2600] "AG" "AG" "GG" "AG" ...
..- attr(*, "dimnames")=List of 2
.. ..$ : chr [1:654] "0001" "0003" "0003-3" "0004" ...
.. ..$ : chr [1:2600] "F51TW9001A34NZ-1129" "F51TW9001A34NZ-1367" "F51TW9001A34NZ-175"
"BX679184-252" ...
$ map :Classes 'GenMap' and 'data.frame':     2600 obs. of 2 variables:
..$ chr: num [1:2600] 1 1 1 1 1 1 1 1 1 1 ...
..$ pos: num [1:2600] 0 0 0 3.24 3.24 3.24 3.24 7.65 7.65 9.22 ...
$ pedigree :Classes 'pedigree' and 'data.frame':     654 obs. of 4 variables:
..$ ID : chr [1:654] "0001" "0003" "0004" "0005" ...
..$ Par1 : chr [1:654] "0" "0" "0" "0" ...
..$ Par2 : chr [1:654] "0" "0" "0" "0" ...
..$ gener: int [1:654] 0 0 0 0 0 0 0 0 0 0 ...
$ phenoCovars: NULL
$ info :List of 3
..$ map.unit: chr "cM"
..$ codeGeno: logi FALSE
..$ version : chr "gpData object was created by synbreed version 0.11-29"
- attr(*, "class")= chr "gpData" - attr(*, "class")= chr "gpData"
```

The output of the summary function also helps to understand the contents of a gpData object, including the number of individuals phenotyped and genotyped, descriptive statistics of response variables, the number of markers, the proportion of missing marker scores, and the number of markers per chromosome:

```
> summary(gp)
object of class 'gpData'
covar
     No. of individuals 654
     phenotyped          654
     genotyped           654
pheno
     No. of traits:    1

Sim.Pheno
Min.   : 7.950
1st Qu.: 9.472
Median : 9.994
Mean   : 9.988
3rd Qu.:10.488
Max.   :12.112

geno
     No. of markers 2600
     genotypes AA AC AG AT CC CG GA GC GG TA TC TG TT
     frequencies 0.1427341 0.03877382 0.1259845 0.01232004 0.1483598
     0.01716655 0.004114326 0.01691778 0.1732586 0.01316455 0.1194337
     0.04364679 0.1422689
     NA's 0.186 %
map
     No. of mapped markers 2258
     No. of chromosomes 12

     markers per chromosome

  1   2   3   4   5   6   7   8   9  10  11  12
172 182 201 188 172 205 194 154 206 175 208 201

pedigree
Number of
     individuals 654
     Par 1        189
     Par 2        90
     generations 2 2
```

Recoding Loci and Imputing Missing Genotypes

The *synbreed* `codeGeno` function is used to perform the following data processing steps on a gpData object (Wimmer et al. 2012):

1. Remove markers with too many missing values
2. Recode marker genotypes from base pairs (or allele pairs) to counts of minor allele
3. Impute missing genotype scores
4. Recode minor and major alleles if imputation results in the 'less common' allele in the raw data becoming the 'more common' allele
5. Remove markers with low minor allele frequency (MAF)
6. Remove redundant markers

In the following example, we use `codeGeno` function to recode the genotypes from base pair calls to numeric minor allele frequency counts. At the same time, we impute the missing genotypes (`impute.type="random"`) as random samples from the expected genotype frequency distribution, given the observed allele frequencies and assuming HWE. The random imputation method is not generally recommended, we use it in this example because it is fast; more details about alternative methods for imputation of missing genotypes are described in Chap. 10. We also discard markers with MAF of 0.01 or less (`maf=0.01`), or with 10% or more missing genotypes (`nmiss=0.1`).

If there are heterozygotes in the original base pair call genotypes, we need to include the argument '`label.heter =`'. The value of this argument should be either a vector of strings identifying all possible heterozygous codes in the original data or a function that returns TRUE given any genotype code string that refers to a heterozygote. If the original data are coded as base pair calls like "A/T", "AT", "A:T", or "A|T" to indicate heterozygotes, as in this example, we can just set the argument to: `label.heter = "alleleCoding"`.

In the example, `gp` is the object of class `gpData`, and the option `verbose=TRUE` prints out a report in the R console about details of what is being done. Finally, the `save` function of R saves the object `gp.num` in the working directory for future use. For further data analysis, the object can be loaded into the R environment with the command '`load(file ="maritime pine codeGeno data.rda")`' instead of running all the data steps to re-create the `gp.num` object.

```
### Recoding genotype calls to gene content,
gp.num <- codeGeno(gp, label.heter ="alleleCoding", maf =0.01,
 nmiss=0.1, impute = TRUE, impute.type = "random", verbose = TRUE)
# Data summary. This may take a while depending on the data
summary(gp.num)

# Save the object gp.num for future use.
save(gp.num, file = "maritime pine codeGeno.rda")
```

The following output will be printed to the terminal because we requested the option `verbose=TRUE`:

```
step 1   : 2 marker(s) removed with > 10 % missing values
step 2   : Recoding alleles
step 4   : 31 marker(s) removed with maf < 0.01
step 7   : Imputing of missing values
step 7d  : Random imputing of missing values
step 8   : No recoding of alleles necessary after imputation
step 9   : 0 marker(s) removed with maf < 0.01
```

```
step 10 : No duplicated markers removed
End     : 2567 marker(s) remain after the check

Summary of imputation
total number of missing values            : 3016
number of random imputations              : 3016
```

After processing to remove markers with high missing data rates or low MAF, 2,567 markers remain. Imputation replaced 3,016 NA scores with random draws from expected genotype distribution given the observed allele frequencies for each marker. The resulting genotype data are coded 0, 1 and 2. A subset of the summary of gp.num object is given below. The only differences compared to the gp object is that the genotypes are now allele contents, and no genotype scores are missing.

```
        > summary(gp.num)
object of class 'gpData'
covar
    No. of individuals 654
            phenotyped 654
            genotyped  654
...
geno
    No. of markers 2567
    genotypes 0 1 2
    frequencies 0.4924888 0.396962 0.1105492
    NA's 0.000 %
```

Genetics Package for Estimating Population Parameters

An example below demonstrates the calculation of population genetic parameter estimates, such as HWE, polymorphism information content and heterozygosity using the *genetics* package.

Code example 9.2
Using genetics package to calculate some population parameters (Code 9-2_Population parameters.R)

```
### Using genetics package for population genetics parameters
# It requires pairs of observed nucleotides (A/B) as genotypes
#~~~~~~~~~~~~~~~~~~~~~~~~~~~~~~~~~~~~~~~~~~~~~~~~~~~~~~#
library(genetics)

### Working directory to read files
path <- "/.../Book/Book1_Examples/data"
setwd(path)
```

The first part of the script reads the original marker and pedigree data file with base call encoding (e.g., AA, AT, TT), extracts just the marker data and creates a matrix of marker genotypes.

```
### Load GENOTYPES and PEDIGREE file
genoped <- read.csv("maritime pine genoped.csv ", header= T,
         stringsAsFactors = F, colClasses = c(Par1 = "character",
         Par2 = "character"), row.names =1,check.names=F)

# Subset to only the marker data
marker <- as.matrix(subset(genoped, select = -c(Par1, Par2, gener)))

# Show first six entries and 5 columns of object genoped
head(marker[, 1:5])
```

To conduct analysis, the *genetics* package requires genotype calls to be pairs of observed nucleotides separated by a slash (e.g., "A/T") as genotypes. In this example, our data are formatted without the slash ("AT") so we need to do a little work to add the slash between the two alleles of each individual's genotype. Genotype conversion is not done for missing values (NA). Then we use the *genetics* package function makeGenotypes to convert it to a special data frame format that is used by other analysis functions of the *genetics* package.

```
# Function to convert the base pairs to allelic pairs separated by /
reformat <- function(str) {
  # List of non NA values and only modifies them
  sel = (!is.na(str))
  str[sel] = paste(substr(str[sel], 1, 1), substr(str[sel], 2, 2), sep
= "/")
  return(str)
}


genosDF = reformat(marker)

# Check that all NA locations are the same in both data sets
identical(is.na(genosDF), is.na(marker))
length(which(is.na(genosDF))) #3157
length(which(is.na(marker))) #3157

# Create a data set with genotypes of the form "A/T" and NA
# Convert columns in a dataframe to genotypes or haplotypes
gpine <- makeGenotypes(genosDF) # Very slow !
head(gpine[, 1:4])
```

Some of the calculations such as HWE can take a long time. For demonstration purposes we create a factor named gpine1locus with one locus and use the functions summary(), HWE.exact(), and HWE.chis() to check the segregation at the locus and test it for deviation from Hardy-Weinberg equilibrium expectations.

```
# Create a genotype object with only one locus
gpine1locus <- gpine[, 1]
#Genetic parameters can be displayed using the summary function
summary(gpine1locus)
#HW
HWE.exact(gpine1locus) #Exact test of H-W equilibrium
HWE.chisq(gpine1locus) #Perform Chi-square test for H-W equilibrium
```

The output of these commands is shown below. The summary function prints out the allele and genotype frequencies, heterozygosity, and polymorphic information content of the first locus. After the summary, we performed exact and chi-squared tests of Hardy-Weinberg Equilibrium at this locus; both tests indicate no evidence to reject the null hypothesis that the population genotype frequencies are in HWE.

```
> head(gpine[,1:4])
         384_PP2C-246 AJ300737-208 BX666025-170 F51TW9001ARTXE-215
0001            G/G          G/G          G/G                 A/A
0003            G/A          G/G          G/G                 A/A
0003-3          G/A          G/G          G/G                 A/A
0004            G/G          G/G          G/G                 A/A
0005            G/G          G/G          G/G                 A/A
0006            G/G          G/G          G/G                 A/A
> summary(gpine1locus)
Number of samples typed: 653 (99.8%)

Allele Frequency: (2 alleles)
    Count Proportion
G    1286      0.98
A      20      0.02
NA      2        NA

Genotype Frequency:
     Count Proportion
G/G    633      0.97
G/A     20      0.03
NA       1        NA

Heterozygosity (Hu)  = 0.03018195
Poly. Inf. Content   = 0.02970406

> HWE.exact(gpine1locus)
    Exact Test for Hardy-Weinberg Equilibrium

data:  gpine1locus
N11 = 633, N12 = 20, N22 = 0, N1 = 1286, N2 = 20, p-value = 1

> HWE.chisq(gpine1locus)
    Pearson's Chi-squared test with simulated p-value (based
    on 10000 replicates)
data:  tab
X-squared = 0.15794, df = NA, p-value = 1
```

It is awkward to conduct this test for each locus one at a time. Instead, we can use the apply() function of R to apply the HWE.chisq function to all of the loci at once. This function runs a bit slow, but returns a list of results for all the markers. Then we can use the sapply() function to extract the *p*-values from all of the tests and simplify the return object to a named vector. It is a simple matter to find the smallest *p*-values corresponding to the loci with most extreme deviations from HWE, and go back and inspect those loci.

```
#Perform HWE chi-square tests on all loci
#This will be a bit slow!
all.HWE = apply(gpine[,], 2, FUN = function(x) HWE.chisq(genotype(x)))

#all.HWE is a long list,
#each component is the result from a single marker test
#here we extract with [[]] the p-value for the test of each marker
all.HWE.p = sapply(all.HWE, "[[", "p.value")
all.HWE.p = sort(all.HWE.p)

#show the most extreme deviations from HWE with lowest p-values
head(all.HWE.p)

#check the data for the most extreme marker
summary(gpine[[names(all.HWE.p[1])]])

#re-check the HWE test with the exact test
HWE.exact(gpine[[names(all.HWE.p[1])]])
```

The results demonstrate that some markers, such as 'BX253931-1759', deviate quite substantially from HWE expectations, due to a deficiency of heterozygotes. Users should consider whether such markers should be removed from further analysis or if this result is biologically significant.

```
> head(all.HWE.p)
    BX253931-1759       BX253126-1425 F51TW9001AQDJL-451
       0.00009999          0.00049995         0.00069993
    FN695545-1347       BX252003-1657       BX680071-376
       0.00079992          0.00079992         0.00099990

> summary(gpine[[names(all.HWE.p[1])]])

Number of samples typed: 649 (99.2%)

Allele Frequency: (2 alleles)
    Count Proportion
C    737        0.57
G    561        0.43
NA    10          NA

Genotype Frequency:
     Count Proportion
C/C    240        0.37
C/G    257        0.40
G/G    152        0.23
NA       5          NA

Heterozygosity (Hu)  = 0.4911857
Poly. Inf. Content   = 0.3703614

> HWE.exact(gpine[[names(all.HWE.p[1])]])

    Exact Test for Hardy-Weinberg Equilibrium

data:  gpine[[names(all.HWE.p[1])]]
N11 = 240, N12 = 257, N22 = 152, N1 = 737, N2 = 561,
p-value = 9.81e-07
```

Data Summary and Visualization

Data on breeding populations may be complex and involve information from markers, phenotypes, genetic maps, and pedigrees. Graphical summaries ('data visualization') are increasingly useful to understand large and complex multi-dimensional data sets. Although regular R graphic functions, such as `hist` and `plot`, can be used to visualize data, more advanced functions or more coding may be needed to produce higher quality images. The *ggplot2* package of R provides an excellent framework for producing a wide array of customizable data visualizations (Wickham 2010). Some of the specialized packages, such as *synbreed*, also implement visualizations from their specifically formatted internal objects.

In the following example we use functions from various packages to visualize and summarize the maritime pine data, following Isik et al. (2016). Before running the script *Code 9-3_Visualize.R*, users must first load the R data set 'maritime pine gpData.rda' previously created by *Code 9-1_Data steps.R* into the working environment. We show the results of visualizations here, leaving the details in the supplementary code.

Genetic Map

Low-density marker maps can be obtained by using the `plotGenMap` function of *synbreed* package.

Code example 9.3
Visualizing genetic map using the plotGenMap function (see details in R code *Code 9-3_Visualize.R*)

```
###################################################
# Marker data analysis - Code09-3_Visualize.R
###################################################
### Using a function to check if packages are installed.
 # If they are not, the function download them from CRAN and
 # load them into the R session.
 # Source: https://gist.github.com/stevenworthington/3178163
 # function
 ipak <- function(pkg){
   new.pkg <- pkg[!(pkg %in% installed.packages()[, "Package"])]
   if (length(new.pkg))
     install.packages(new.pkg, dependencies = TRUE)
   sapply(pkg, require, character.only = TRUE)   }

 # usage
 packages <- c("plyr", "dplyr",  "ggplot2", "RColorBrewer",
 "LDheatmap", "Hmisc", "GGally", "synbreed")
 ipak(packages)
### Load the R data set
 load("maritime pine gpData.rda")
 ...
 ### Visualize Genetic MAP
 plotGenMap(gp.num, dense=F, nMarker=T, ylab="pos [cM]")
```

Changing the argument `dense=F` to `dense=T` generates high density maps. If there are many markers, it may take some time to generate the high density map. The `nMarker` option controls whether to show the number of markers on the linkage group or not. If true, the numbers of markers are printed below each linkage group in a genetic map (Fig. 9.2).

Fig. 9.2 Genetic map of maritime pine. The *horizontal lines* show the position of markers (in centimorgans) on chromosomes. The numbers of markers on each chromosome are printed at the *bottom*

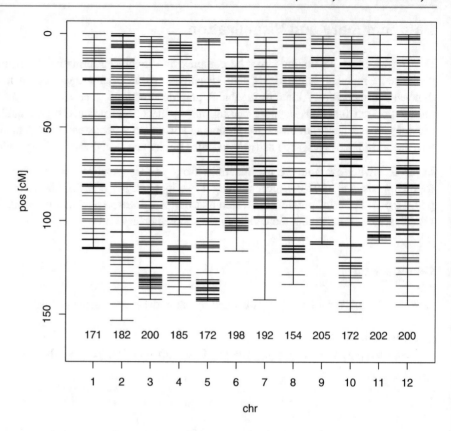

Pairwise Linkage Disequilibrium

The `pairwiseLD` function in the *synbreed* package estimates pairwise Linkage Disequilibrium (LD) between markers, expressed as r^2, using an object of class `gpData`. The option `type=` controls if the output of pairwise LD estimates is a matrix or a data frame. Some functions or packages require the matrix format. Pairwise LD calculation is carried out for all chromosomes by option 'chr= ' but it can be limited to a single chromosome by specifying a chromosome number, e.g. 'chr=1' for chromosome 1.

```
# For chromosome 1 use chr=1, otherwise leave it as chr= for all chr
# Save results as data frame.
LD.df <- pairwiseLD(gp.num, chr= , type="data.frame")
# Extract LD for chromosome 1
chr1 <- LD.df$chr_1
str(chr1)
```

We subset the `LD.df` to have a data frame for chromosome 1 only. There are five variables in the `chr1` data frame, as shown in the output below. The first two variables are `marker1` and `marker2`. They are followed by `r` and r^2 and `dist`. The "dist" is the distance between marker positions, which could be in cM or in base pairs, depending on the input object.

```
'data.frame':    14535 obs. of 5 variables:
$   marker1:   chr    "F51TW9001A34NZ-1129"    "F51TW9001A34NZ-1129"    "F51TW9001A34NZ-1129"
"F51TW9001A34NZ-1129" ...
```

```
$ marker2  : chr "F51TW9001A34NZ-1367" "F51TW9001A34NZ-175" "BX679184-252" "CR394043-2000" ...
$ r     : num 1 0.3577 -0.1219 0.0295 -0.0317 ...
$ r2    : num 1 0.127972 0.014866 0.000868 0.001007 ...
$ dist  : num 0 0 3.24 3.24 3.24 3.24 7.65 7.65 9.22 9.22 ...
```

In the code below we use `ggplot2` to produce mean LD for chromosome 1. We first use a function named "meanLD" to calculate the mean of a given vector (r^2 values in chr1) and call it in the `ggplot` code.

```
#~~~      Figure 9.3      ~~~
meanLD <- function(x, digits = 2) {
  meanLD <- round(mean(x), digits = digits)
  paste("r ==", meanLD)   }

par(mar=c(4,4,1,1)+0.1) #sets margins of plotting area
labels = data.frame(x = 30, y = 1, label = meanLD(chr1$r2))
LDplot <- ggplot(LD.df$chr_1, aes(x=dist, y=r2) ) +
  labs(x='Distance',  y='LD') + theme_bw() +
  geom_point(shape=1, colour="gray") + geom_smooth() +
  geom_text(data=labels, aes(x=x, y=y, label=label), parse=TRUE)
LDplot
```

The mean LD on chromosome 1 is very low ($r^2 = 0.011$) and decays rapidly as the distance between markers increases (Fig. 9.3). The LD values between some pairs of markers is very high (>0.5). By definition LD is non-random association between markers. If a population has genetically related individuals, as in a breeding population or if there is a genetic structure (two different founder groups) in the population, LD estimates might be biased (Mangin et al. 2012). Very high LD values can also be artifact of having physically linked markers belonging to the same contig or bias in composite linkage genetic maps (Plomion et al. 2014). Some bias in LD estimation can be reduced by using the genomic relationships of the individuals (Isik et al. 2016).

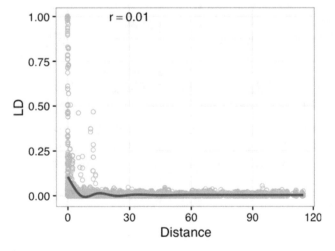

Fig. 9.3 LD between pairs of markers as a function of distance (cM) on chromosome 1. LD decays rapidly as the distance increases

Pairwise LD on chromosome 1

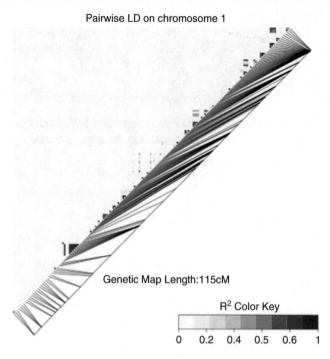

Fig. 9.4 The pairwise LD for 171 markers on chromosome 1 as the upper diagonal of the matrix. The *lines* on the diagonal show the location of markers (genetic distance)

The pairwise LD can also be visualized by using the LDMap function as shown below. The function requires a matrix of LD values. We create a matrix of LD values for chromosome 1 by using the pairwiseLD function with option chr=1 and save it as LD.mat.chr1. We then use the LDMap function to generate a map of LD on the chromosome (Fig. 9.4).

```
#~~~      Figure 9.4      ~~~
# Save LD results as matrix
LD.matrix <- pairwiseLD(gp.num, chr=, type="matrix")
# Subset for chromosome 1 only
LD.mat.chr1 <- pairwiseLD(gp.num, chr=1, type="matrix")
# Figure
LDMap(LD.mat.chr1, gp.num, onefile=T, flip=F)
```

Each colored rectangle represents the squared correlation (r^2) measure of LD between a pair of SNPs. Since LD is near zero for most pairs of loci, the area is mostly empty. We see high LD depicted by blocks of dark red color, which only occur among SNPs at relatively close map positions. The map positions are illustrated on the diagonal line (Shin et al. 2006).

The LD (r^2) between adjacent markers along each chromosome can also be visualized as a heat map (Fig. 9.5) using the function plotNeighbourLD of the *LDheatmap* package (Shin et al. 2006). The function uses the genetic map in gp.num object created by synbreed and LD.matrix, the matrix of pairwise LD calculated previously.

```
#~~~      Figure 9.5      ~~~
# LD between adjacent markers on chromosomes
par(mar=c(1,1,1,1))
    plotNeighbourLD(LD.matrix, gp.num, nMarker=T, dense=T,)
```

Fig. 9.5 Pairwise LD between adjacent markers on each chromosome as a heat map. The *dark* bands suggest very high LD between adjacent markers distributed on a chromosome

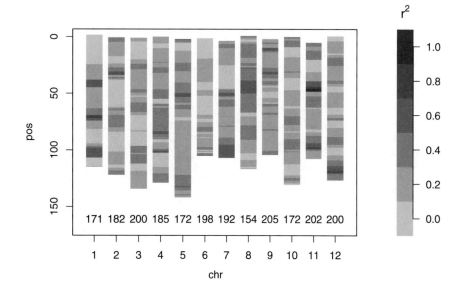

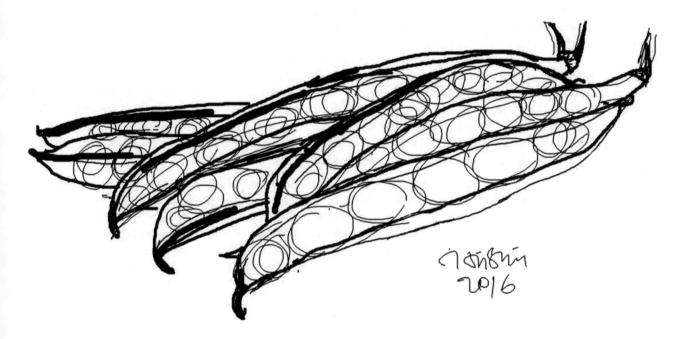

Electronic supplementary material: The online version of this chapter (doi:10.1007/978-3-319-55177-7_10) contains supplementary material, which is available to authorized users.

F. Isik et al., *Genetic Data Analysis for Plant and Animal Breeding*, DOI 10.1007/978-3-319-55177-7_10

Abstract

The term 'imputation' refers to the estimation of values to replace missing observations in a data set. In this chapter, we are interested in imputing missing genotype values. The chapter is not intended as a rigorous treatment of the use and implications of imputation in breeding values prediction, but rather as a brief overview of some common strategies currently employed. For the reader interested in a more in-depth treatment, references are provided within the chapter.

Introduction

The imputation process can be carried out either using only the information contained within the dataset of interest, or by incorporating additional data from other sources (such as a reference panel of haplotypes from a haplotype discovery project) if available. In both cases, the availability of at least one reference sequence for the species is needed to locate markers on chromosomes and place them in correct order with respect to the physical sequence. A larger number of sequenced individuals from the species is required to use reference haplotype panels, and these are not available for many crop and tree species. The availability of reference genomes is likely to improve within the next few years for many organisms. Within this chapter, we will discuss the two basic scenarios for imputation: (1) imputation from a densely genotyped reference panel to an experimental population genotyped at lower marker density, and (2) imputation of more or less randomly missing genotype values using only the experimental population itself. In both cases, we assume that markers are ordered correctly based on a common reference sequence. Recently, a few methods have been proposed to perform imputation of unordered marker sets that do not require a reference genome (Money et al. 2015; Rutkoski et al. 2013). As these methods are relatively new and untested in large populations, we will not cover these in this edition of the book.

With the advent of the genomic era there is an unprecedented amount of genotypic information being generated routinely. Within this context there are two main reasons for imputing data. The first is that no assay is perfect, and so some fraction of genotype data will be missing from virtually every genotype dataset. The second reason is that high-density genotyping (or sequencing) is expensive, and imputing genotype data can be much less expensive than genotyping. Dropping the individual or locus records in which these missing values occur is an option for single-marker methods such as association genetics, provided that the overall level of missing data is relatively low for each locus. Analytical methods that consider more than one locus at a time are often much more sensitive to missing data, because it is not cost-effective to drop individuals that are missing data at any one of the loci being analyzed. In the extreme case in which all marker loci in the genome are analyzed at once, the approach of dropping individuals with missing genotype data can mean discarding hundreds or thousands (or even hundreds of thousands) of correct data values in order to remove a small proportion of missing values. Furthermore, predictions of genetic merit of individuals using linear models and the realized genomic relationship matrix often require that there be no individual markers missing values. Some software do accept missing values, but those programs must carry out some imputation internally before the data can be analyzed, and often the imputation used is ad hoc and not necessarily accurate or optimal.

Imputing missing genotypic data can be less expensive and almost as accurate as experimentally determining the genotypes. Imputation methods that exploit linkage disequilibrium (LD) among SNP loci can often impute missing genotypes with 90% or greater accuracy (Huang et al. 2012). Breeders frequently work with populations derived from structured mating designs in which many individuals share one or both parents in common, and can use the LD created by such mating designs to impute genotypes in progeny based on known genotypes in parents. High-density SNP genotyping in parents of a population allows accurate imputation of missing genotypes in progeny of those parents that are genotyped at a much lower density. This approach provides the benefits of high-density SNP genotyping with much of the cost-effectiveness of low-density genotyping, and can work very well in populations designed specifically to take advantage of within-family LD for imputation (Habier et al. 2009; Yu et al. 2008). As noted above, linkage disequilibrium can be a powerful asset in imputing missing values with high accuracy. To exploit LD, however, it is essential to know which SNPs are in LD with any particular SNP that has missing values. LD across a population of unrelated individuals is typically a function of population history and genetic or physical distance between loci, so finding SNPs that are physically or genetically tightly linked is valuable. In species for which a reference genome sequence is available, the chromosomal positions of SNP loci are often available from the genome sequence, but for species that lack a reference genome sequence, alternative approaches need be used to identify loci in LD with a target SNP.

The Idea Behind Imputation

The following diagram shows a three-generation pedigree of diploid individuals, with a segment of one homologous pair of chromosomes shown as haplotypes below the symbol for each individual. This figure outlines one common scenario for imputation: genotype values are not missing randomly, rather, some part of the population has been genotyped at higher density than other members of the population, and we wish to impute all the SNPs identified in the high density genotyping platform to the whole sample.

The SNP loci in bold are those genotyped in all individuals; the loci in gray are genotyped only in the ancestors, and the genotypes of progeny at those loci are to be inferred by comparing the haplotypes at the bold SNP loci from related individuals. In the first generation we have 8 distinct haplotypes (four per cross) that can be passed to the progeny of the crosses, **a1**, **a2**, **b1**, **b2**, **c1**, **c2**, and **d1**, **d2**. In the second generation we can resolve which haplotype has been passed to the two progeny: **a1** and **b1** to the first individual and **c1** and **d1**. In the third generation only bold markers were genotyped. In this case the haplotypes of *individual 1* are ambiguous: it is not possible to know whether it inherited the **a1** and the **d1** haplotypes, or the **b1** and the **c1** haplotypes. *Individual 2* inherited the **a1** and the **c1** haplotypes. *Individuals 3 and 4* inherited the **b1** haplotype from one parent and **d1** haplotype from the other. You should notice that imputation essentially relies on knowing/or inferring the haplotype phase of individuals and then use linkage disequilibrium (LD) information from nearby markers to replace missing genotypes.

In most cases the methods used for imputation rely on a known genome assembly (so that the order of markers is known before analysis) and a combination of maximum likelihood and Bayesian methods to predict missing genotypes. We will shortly describe one of these methods but Figs. 10.1 and 10.2 shows in a conceptual way how these software work, without concern for the details of the algorithms involved. The diplotype (genotype) of an individual consists of a pair of haploid gametes (haplotypes) encoded as unphased pairs of alleles at a certain position.

In the most common situation arising with genomic information in breeding we observe diplotypes as opposed to haplotypes. The reasons for this are manifold. Diplotypes can be obtained easily through high-throughput platforms, while there are not equivalent platforms for haplotypes. In addition, methods for estimating genomic relationships and predicting genetic values from SNP genotype data are for the most part geared toward the use of diplotypes, rather than haplotypes, although the use of haplotypic values has been proposed. As a result, one common way of encoding genotypes numerically is as minor allele content, or the number of copies of the minor allele, as discussed in Chap. 9.

For example, assuming that A and C are the major alleles, and T and G are the minor alleles, at loci 1 and 2, respectively, then the minor allele content for the three possible genotypes is as fol7lows:

locus1	Content	locus2	content
AA	0	CC	0
AT/TA	1	CG/GC	1
TT	2	GG	2

This method of encoding SNP genotypes is readily used in subsequent analyses that relate genetic information to phenotype information, to estimate allele additive and dominance substitution effects on a phenotype; genetic merit of individuals across all markers or genetic merit based on additive, dominance and total genetic marker effects.

Pedigree Free Imputation

Several pedigree-free methods have been used in livestock and plant science to perform genotype imputations. These methods were mostly developed for application in human populations where pedigree information is for the most part not available or relatively incomplete. Other methods that effectively make use of deep pedigree information have been proposed, particularly for livestock (Hickey et al. 2012; Sargolzaei et al. 2014; VanRaden et al. 2013). While some of these pedigree-based methods are extremely effective, pedigree-free methods remain very popular due to their overall versatility.

The concept of phasing, as outlined in previous sections is closely related to that of imputing missing genotypes. Genotype phasing tries to determine the non-observed haplotypes that underlie the observed genotypes. Genotype imputation is a

Fig. 10.1 Imputation of missing loci using haplotypes. The SNP loci in bold are those genotyped in all individuals; the loci in gray are genotyped only in the ancestors

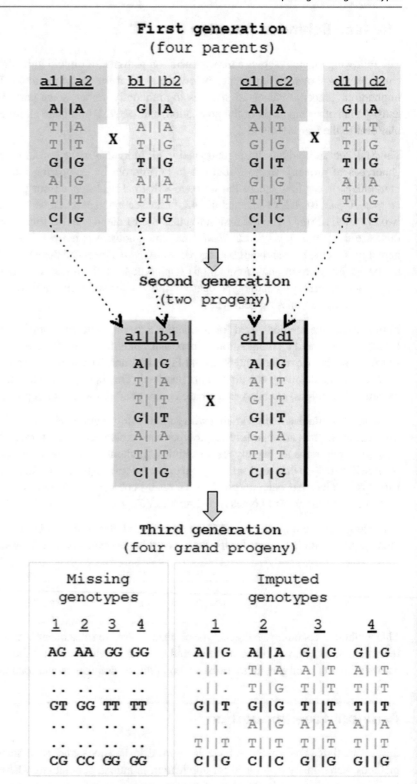

process that is very similar to phasing. It attempts to model the non-observed haplotypes using a reduced proportion of observed genotypes. All pedigree-free imputation methods rely on the same "intuition". In a genomic region there are a few haplotypes that are common and all the remaining ones are likely to be derivations of existing haplotypes through either recombination or mutation. So the majority of methods effectively try to recognize "haplotype clusters" as a common consensus haplotype and a set of related (derived) haplotypes. Effectively, the possible haplotypes originate from a finite

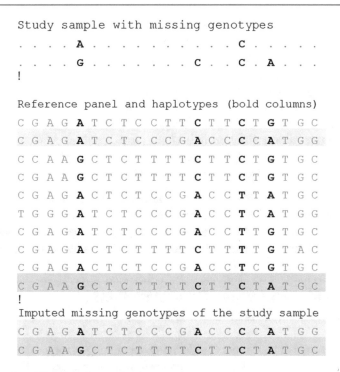

```
Study sample with missing genotypes

. . . . . A . . . . . . . . . . . C . . . . .

. . . . . G . . . . . . . C . . C . A . . .
!

Reference panel and haplotypes (bold columns)
C G A G A T C T C C T T C T T C T G T G C
C G A G A T C T C C C G A C C C A T G G
C C A A G C T C T T T T C T T C T G T G C
C G A A G C T C T T T T C T T C T G T G C
C G A G A C T C T C C G A C C T T A T G C
T G G G A T C T C C C G A C C T C A T G G
C G A G A T C T C C C G A C C T T G T G C
C G A G A C T C T T T T C T T T T G T A C
C G A G A C T C T C C G A C C T C G T G C
C G A A G C T C T T T T C T T C T A T G C
!
Imputed missing genotypes of the study sample
C G A G A T C T C C C G A C C C A T G G
C G A A G C T C T T T T C T T C T A T G C
```

Fig. 10.2 Imputation concept based on LD and a reference haplotypes

number of clusters. So the methods attempt to compute the probability of a particular haplotype given a cluster *p(haplotype/ cluster)*. Different methods differ in the way they recognize and organize these clusters but the general intuition remains the same. After having inferred the cluster membership based on the observed genotypes, the algorithms then impute missing genotypes based on their cluster membership.

Pedigree-free imputation algorithms are for the most part hidden Markov Model (HMM)-based approaches that try to approximate the demographic history of a sample and capture linkage disequilibrium information. Hidden Markov Models have a basic structure that make them suitable for both of these tasks. HMM have underlying hidden states that cannot be observed and they use other observed data that is controlled by the non-observed hidden states when attempting to resolve these hidden states. When using HMM for genotype phasing and/or imputation, the observed (non-missing) genotypes form the observed states and the non-observed haplotypes are represented by the hidden states. Transition probabilities, which are related to recombination fractions and haplotype frequencies, are used to determine the ways in which hidden states can change from one to another. Currently the most common HMM used for genotype phasing/imputation are implemented in four software packages: fastPHASE (Scheet and Stephens 2006), Beagle (Browning and Browning 2007), IMPUTE2 (Howie et al. 2009), and MaCH (Li et al. 2010). Within this chapter we will provide an example of imputation using Beagle, since this software has established itself as a "de facto" standard for most of crop/livestock applications. Nonetheless all proposed software have pros and cons. Marchini and Howie (2010) and Browning and Browning (2011) have compared these algorithms theoretically and from the point of view of application to human data sets. The reader is referred to these publications for an in-depth treatment of the subject.

Beagle is a mature software that is currently at version *4.1*. Each new version has changes and improvements of both the algorithm used and the formats and options available to the user, so we will not focus on the specifics of the software beyond what is strictly necessary. The reader can refer to software webpage for a complete overview of the software options. (https:// faculty.washington.edu/browning/beagle/beagle.html). Furthermore we will not cover in depth the details of the Beagle algorithm. Details on the inception and evolution of the method can be found in related papers (Browning and Browning 2011, 2013; Browning and Yu 2009; Browning 2006).

We have mentioned before how imputation software rely on the concept of clustering haplotypes into blocks. A simple solution to find haplotype blocks would be to use sliding windows of markers to cover the entire genome. While this approach is simple, fixed window size would not be able to effectively capture LD patterns. Alternatively, windows of all possible sizes could be used to account for linkage structure, but this would result in extremely high computation costs.

Browning (2006) proposed an heuristic method for building localized haplotype clusters by taking into account linkage disequilibrium. The algorithm uses a Markov Chain approach of variable length. As a result, haplotypes are clustered at each marker and the number of clusters is not fixed *a priori*. The algorithm is an extension of the cursive writing recognition algorithm proposed by Ron and colleagues (Ron et al. 1995). The algorithm represents haplotypes as directed acyclic graphs and it builds trees representing each distinct haplotype. For each edge it counts the number of haplotypes "passing through" that particular edge, essentially weighting each edge. After that it iteratively merges nodes for which the transition probability to downstream nodes are sufficiently similar. Beagle builds on this algorithm by adding an iterative procedure (similar to the expectation-maximization algorithm) to perform phasing and imputation. In a nutshell, the algorithm is initialized making a predetermined number of copies of each available genotype. Missing alleles are then imputed at random using allele frequencies and haplotypes are likewise created by randomly phasing.

An iterative approach follows by which localized haplotype clusters are built following the scheme described above and haplotypes pairs are sampled using an HMM for each genotype. This step is repeated several times and in the end the most likely haplotype for each genotype is imputed using the Viterbi algorithm, a particular HMM algorithm that is better at reconstructing sequences, such as haplotypes, than algorithms for imputing individual sites separately.

Imputation from Densely Genotyped Reference Panel to Individuals Genotyped at Lower Density

The following is a simplified example of how to use Beagle to impute missing genotypes from a reference panel to a related set of individuals from the same population, but with genotyped at lower marker density. This approach relies on the assumption that the reference panel accurately represents the haplotype frequencies of the set to be imputed, so the experimental set of individuals should be sampled from the same population as the reference set of genotypes. We use simulated data for 500 individuals with SNPs called on two autosomes. On the first chromosome we have 3000 SNP markers while on the second we have 2000. It should be noted that in this example we have simulated markers with a relatively low linkage disequilibrium (average r^2 between markers <0.05). This is a typical situation of populations with a large effective population size. The data are contained in two files. The first file holds the diplotypes (*ImputeGenotype*):

```
844  0 2 0 0 0 0 0 0 0 2 0 0 2 0 0 0 0 2 0 0 0 2 4 0 0 0 4 4 3 2 0 4 4 0 0 3 0
     0 2 3 0 0 0 3 0 0 0 4 2 2 0 3 2 2 0 0 4 4 3 0 0 0 4 3 0 0 4 4 0 0 4 0 2 2 0 0
     0 0 0 3 0 0 0 3 0 3 0 0 4 0 0 0 0 0 0 0 0 2 0 2 0 0 0 2 2 0 0 0 0 0 3 2 3 0 5
854  0 0 4 3 0 3 2 0 2 0 0 0 4 3 0 0 3 0 0 3 0 4 4 0 0 0 0 0 0 4 0 0 0 0 4 4 0
     0 0 0 0 0 0 2 0 3 0 0 2 3 0 0 2 2 0 0 0 0 2 0 2 4 3 2 3 3 0 0 0 3 3 3 2 3 4 0
     0 0 0 0 0 0 3 3 3 0 0 0 3 0 0 0 0 0 0 0 0 0 0 0 0 0 2 0 0 3 0 0 0 4 3 0 0
```

The bold numbers are individual IDs. The subsequent numbers are the numerical allele content diplotype codes for each SNP marker, with the marker genotypes given in order of their physical position. The genotype codes 3 and 4 represent the two phased heterozygous (0/1 and 1/0, respectively). Finally, the code 5 represents unknown (./.) genotypes.

You should notice that in this case the diplotypes are already phased (e.g., we already know which chromosome individuals have inherited from each parent). This is done in this example to keep calculations to a minimum, yet it is not a requirement of the software the phased '3' and '4' heterozygous codes could be replaced with the non-phased '1' diplotype code for heterozygotes.

The second file (*ImputeMarkerMap*) is a map file containing the chromosome and position of each marker.

```
1    1_78635      0    78635
1    1_153228     0    153228
1    1_243172     0    243172
1    1_334883     0    334883
1    1_417617     0    417617
1    1_489702     0    4897021
```

In this case the first field corresponds to the chromosome number. The second column contains the marker name. The third column is reserved for the genetic linkage map. The map is assumed to be in centiMorgans (cM). The genetic map is used to account for heterogeneity of recombination events during imputation. If the genetic map is not provided Beagle will assume a constant recombination rate of 1cM/Mb. In the example provided the genetic map is omitted (every value in the columns is given as 0). Finally, the fourth column contains the physical position of the markers on the chromosome (in Mb).

We are going to start by reading the two files in R and prepare them in the correct Beagle format.

Code example 10.1
Creating VCF genotype format for Beagle v4 for imputation with a reference panel (see 'Code 10-1_ GenotypeSetUpBeagle4.R' for details).

```
########################################################################
### To create VCF File Genotype format for BEAGLE 4.0 or greater  ###
########################################################################
rm(list=ls()); gc();

## change to working directory
setwd"/Users/.../data")

## read in Marker File
map <- read.table(file="ImputeMarkerMap",sep=" ",header=FALSE)

## read in true phased genotype matrix
# column 1 is ID and rest of columns are genotypes
# 0 - 1 1; 2 - 2 2; 3 - 1 2; 4 - 2 1
geno <- read.table(file="ImputeGenotype",sep=" ",header=FALSE)

## take off animal IDs
ID <- geno[,c(1)]

## remove IDs from animal
geno <- geno[ ,-c(1)]
## Loop over markers and create a beagle file for each genotype
```

Beagle from version 4.0 and later requires a Variant Call Format (VCF) 4.2 as input. A VCF file is a text file commonly used to store genotype data. The VCF file has three sections: a meta-information section (lines beginning with ##), a header section (one line beginning with #LG or #CHROM) and a section where data lines are accumulated. See an example below.

```
##fileformat=VCFv4.2
##FORMAT=<ID=GT, Number=1,Type=Integer, Description="Genotype">
##FORMAT=<ID=GP, Number=G,Type=Float, Description="Genotype Probabilities">
#CHROM POS    ID     REF ALT QUAL FILTER INFO FORMAT 844      854   855
2      294726 2_2947 C   T   .    PASS   .    GT     0/0      0/0   0/1
2      254517 2_2940 G   A   .    PASS   .    GT:GP  0/1:0.02 0/0   0/1:.
...
```

A detailed explanation of the file format can be found at http://faculty.washington.edu/browning/beagle/intro-to-vcf.html.

The next section of the R code accomplishes a few tasks. First it converts minor allele content to the genotype format (0/0, 0/1, 1/0, 1/1) required. Furthermore it splits the data in order to create a couple of sets. A first set contains a reduced panel of 100 individuals for which we want to perform imputation. For these individuals the script deletes marker information sampling only every fifth marker. The remaining 400 individuals are placed in a reference set in which all markers are genotyped. The scripts attach the required meta-data and header lines to each of the two sets and write them to four files. Two reference files, one per chromosome: *ref_data1.inp*, and *ref_data2.inp*, respectively, and two reduced files (the ones that will be imputed): *Geno_data1.inp*, *Geno_data2.inp*, respectively. As BEAGLE works with compressed files the script then proceed to gzip all files.

```
for(c in 1:2)
{
  cat("Starting Chromosome",c,"\n")
  ## VCF file requires 9 fields Chromosome pos ID Ref Alt Qual Filter
Info Format Sample1 Sample2......
  ## Grab correct chromosome index
  X <- which(map$V1 == c)
  ## grab SNP in order
  snpfile <- geno[ ,X]
  f<-seq(from= 1, to=dim(snpfile)[2],by=5)
  ## Loop Through SNP and Animals to set up right genotypes
  beagle <- matrix(NA, nrow =(ncol(snpfile)), ncol =(nrow(snpfile)))
  dim(beagle)
  for(i in 1:ncol(snpfile))
  {
    for(j in 1:nrow(snpfile))
    {
##########################################################
### This can be changed to 0 1 2 part and just assume ###
### all 1 are 1/2 and then let phasing figure it out  ###
##########################################################
      genot <- snpfile[,i]
```

```
      ## Genotype is a 0 ##
      beagle[i,j] <- ifelse(genot[j] == 0, "0/0", beagle[i,j])
      ## Genotype is a 2 ##
      beagle[i,j] <- ifelse(genot[j] == 2, "1/1", beagle[i,j])
      ## Genotype is a 3 ##
      beagle[i,j] <- ifelse(genot[j] == 3, "0/1", beagle[i,j])
      ## Genotype is a 4 ##
      beagle[i,j] <- ifelse(genot[j] == 4, "1/0", beagle[i,j])
      ## Genotype is missing ##
      beagle[i,j] <- ifelse(genot[j] == 5, "./.", beagle[i,j])
    }
  if((i - 1000) %% 1000 == 0){cat("Chromosome",c," snp ",i,"\n")}
  if(i == ncol(snpfile)){ cat("Total SNP ",i,"\n")}
  rm(genot)
}

 ## Set up columns prior to genotype data ##
 first9 <- cbind(map[X,1],
                 map[X,4],
                 paste(map[X,1],map[X,4],sep="_"),
                 rep("C",length(X)),
                 rep("T",length(X)),
                 rep(".",length(X)),
                 rep("PASS",length(X)),
                 rep(".",length(X)),
                 rep("GT",length(X)))
 # combine first 9 with genotypes
 beagle <- cbind(first9,beagle)

 ## add in column
 row <-
c("#CHROM","POS","ID","REF","ALT","QUAL","FILTER","INFO","FORMAT",ID)
 ## put animal ID on top of unphased genotypes
 beagle <- rbind(row,beagle)
 b1<-beagle[f,c(1:9,10:110)]
 b2<-beagle[,c(1:9,111:509)]
 b2[c(2:nrow(b2)),c(10:ncol(b2))] <-
gsub("/","|",b2[c(2:nrow(b2)),c(10:ncol(b2))])
   ## Output with appropriate header files

cat("##fileformat=VCFv4.2",file=paste("Geno_data",c,".inp",sep=""),sep
="\n")

cat("##FORMAT=<ID=GT,Number=1,Type=Integer,Description='Genotype'>",fi
le=paste("Geno_data",c,".inp",sep=""),append=TRUE,sep="\n")
```

```
  write.table(b1,file=paste("Geno_data",c,".inp",sep=""),sep=
"\t",append=TRUE, row.names=FALSE,col.names=FALSE, quote=FALSE)

  ## Output with appropriate header files

cat("##fileformat=VCFv4.2",file=paste("ref_data",c,".inp",sep=""),sep=
"\n")

cat("##FORMAT=<ID=GT,Number=1,Type=Integer,Description='Genotype'>",fi
le=paste("ref_data",c,".inp",sep=""),append=TRUE,sep="\n")
  write.table(b2,file=paste("ref_data",c,".inp",sep=""),sep=
"\t",append=TRUE, row.names=FALSE,col.names=FALSE, quote=FALSE)

  rm(beagle,snpfile, b1, b2)

  ## Note - the remaining section of this loop will fail on a
  ## windows OS. Windows uses shell() instead of system()
  ## further, windows does not have a native gzip application
  ## So, Windows users can flag out the commands in this block
  ## And instead use Cygwin shell to execute:
  ## 'gzip Geno_data1.inp' and 'gzip Geno_data2.inp'Zip File
  ## First make sure it is deleted so will zip it
  string = paste("rm Geno_data",c,".inp.gz",sep="")
  system(string)
  string = paste("gzip Geno_data",c,".inp",sep="")
  system(string)
  string = paste("rm ref_data",c,".inp.gz",sep="")
  system(string)
  string = paste("gzip ref_data",c,".inp",sep="")
  system(string)
  cat("Ending Chromosome",c,"\n")
}
```

Below are a few lines of one of the files to be imputed

```
##fileformat=VCFv4.2
##FORMAT=<ID=GT,Number=1,Type=Integer,Description='Genotype'>
#CHROM  POS      ID        REF    ALT    QUAL    FILTER  INFO   FORMAT  844   854   855   856
        858      859       861    862    863     864     865    866     867   868   869   872
...
1       78635    1_78635 C        T      .       PASS    .      GT      0/0   0/0   1/0   1/0
        0/0      0/0       0/0    0/0    1/0     0/0     1/1    0/0     0/1   0/0   1/0   0/0

...
1       489702   1_489702          C     T       .       PASS   .       GT    0/0   0/1   0/0
        1/1      0/1       1/0    1/0    0/1     0/0     0/0    0/0     0/1   0/0   0/0   0/0
...
```

and of one of the reference files ('*ref_data1.inp.gz*'):

```
##fileformat=VCFv4.2
##FORMAT=<ID=GT,Number=1,Type=Integer,Description='Genotype'>
#CHROM POS    ID      REF    ALT    QUAL    FILTER  INFO    FORMAT  1008   1009   1010   1011
       1013   1014    1016   1017   1018    1019    1023    1024    1026   1027   1028   1031
...
1      78635  1_78635 C      T      .       PASS    .       GT      0|0    0|0    0|1    0|0
       0|0    0|0     0|0    0|0    0|0     0|0     1|0     0|1     0|1    0|0    1|0    0|0
...
1      153228 1_153228        C      T      .       PASS    .       GT      0|0    0|0    0|0
       0|0    0|0     0|0    0|0    0|0     0|0     0|0     0|0     0|0    0|0    0|0    0|0
...
1      243172 1_243172        C      T      .       PASS    .       GT      1|0    1|0    0|0
       1|1    0|0     0|0    1|1    0|1     0|1     1|0     0|0     1|0    0|0    1|0    1|0
...
1      334883 1_334883        C      T      .       PASS    .       GT      0|1    0|1    1|0
       1|0    0|0     0|0    1|0    0|1     1|0     0|0     0|0     0|0    0|0    0|0    0|0
...
1      417617 1_417617        C      T      .       PASS    .       GT      0|0    0|0    0|0
       0|0    0|0     0|0    0|0    0|0     0|1     1|0     0|0     1|0    0|0    0|0    0|0
...
1      489702 1_489702        C      T      .       PASS    .       GT      0|1    0|1    1|1
       1|0    1|0     0|0    1|0    0|1     1|0     0|1     0|0     0|0    1|0    0|1    0|0
...
```

Notice that the samples of individuals (the columns) differ between the two files. The reference panel has 'complete' data and is not included in the set to be imputed. Also notice that the reference panel has many more markers (rows) listed than the set for imputation; only the markers in red are in common between the files. The markers that are included in the reference data set but not in the target set will be imputed and included in the output files.

Also notice that the two files formats are essentially identical except for one difference. In the reference file (see the line highlighted in red) the | sign is used in lieu of the/to encode the diplotypes. As a consequence of the different encoding in the reference the diplotypes will be considered correctly phased and the software will not attempt to phase them. Conversely in the reduced files phase provided will be discarded and markers will be phased.

Now that we have created the correct input files for Beagle we will need to install and run the software. First, users should install the beagle program file (with extension '.jar') from this website: https://faculty.washington.edu/browning/beagle/beagle.html. The name of the file follows the format 'beagle' followed by the date of the current release, ending with '.jar'. Place the downloaded file into the folder holding the input data sets. (Some browsers may attach a '.txt' extension to the file, if so, edit the name of the file to remove the '.txt' so the filename ends in '.jar'.) A '.jar' file is an executable java program archive, this beagle program can be run in Linux by giving the command 'java -jar beagle.27Jul16.86a.jar' if 'beagle.27Jul16.86a.jar' is the name of the program file downloaded. However this command is not sufficient, we must also supply the names of the input files and set the arguments on the command line. In this case, to impute the markers on chromosome 1 using the reference panel, we need to use this command:

```
java -jar beagle.27Jul16.86a.jar gt=Geno_data1.inp.gz
ref=ref_data1.inp.gz impute=TRUE out=Phased_data1.gt
```

If we separate the call we can identify a few parts:

```
java-jar beagle.27Jul16.86a.jar
```

This is the call to the program itself.

```
gt=Geno_data1.inp.gz ref=ref_data1.inp.gz
```

This part has the two main files required (in bold): `gt`, the reduced density genotype data for the individuals to be imputed and `ref`, the phased genotype data on the reference panel used to guide the imputation.

```
impute=TRUE out=Phased_data1.gt
```

The last part, `impute=TRUE`, tells Beagle to perform imputation, and specifies the prefix name for the output file as an argument to the option `out`.

To make this easier, we have included a shell script that executes this command in a loop over both chromosomes, in the file 'BeagleLoop.sh'. The script will work for Apple OSX and Linux operating systems, but Windows users should see Box 10.1 for details on how to run this shell script and the Beagle analysis in a Windows operating system. Furthermore, the user must edit the script (using a simple text editor) so that the name of the Beagle.jar file in the script matches the one the user has downloaded.

Code example 10.2
Linux shell script to execute Beagle imputation in a loop over chromosomes (see 'BeagleLoop.sh' for details).

```
echo "=============================="
echo "BEAGLE LOOP FOR CH.10 EXAMPLE"
echo "=============================="

for i in {1..2}

#looping through chromosomes
do

#creating 2 beagles calls one for each chromosome
echo "java -jar beagle.27Jul16.86a.jar gt=Geno_data$i.inp.gz
ref=ref_data$i.inp.gz impute=TRUE out=Phased_data$i.gt" > Beagle$i.sh

#changing file permissions
chmod 755 Beagle$i.sh

#executing beagle for each chromosome
nohup sh Beagle$i.sh &

done
```

The script (which is invoked as `sh BeagleLoop.sh`) creates two calls to the Beagle software, one for each chromosome, and stores them in two files that are sequentially called. The following are a few lines from the output file ('*Phased_data_1. gt.vcf.gz*'):

```
##fileformat=VCFv4.2
##filedate=20160805
##source="beagle.27Jul16.86a.jar (version 4.1)"
##INFO=<ID=AF,Number=A,Type=Float,Description="Estimated ALT Allele
Frequencies">
##INFO=<ID=AR2,Number=1,Type=Float,Description="Allelic R-Squared:
estimated squared correlation between most probable REF dose and true
REF dose">
##INFO=<ID=DR2,Number=1,Type=Float,Description="Dosage R-Squared:
estimated squared correlation between estimated REF dose [P(RA) +
2*P(RR)] and true REF dose">
##INFO=<ID=IMP,Number=0,Type=Flag,Description="Imputed marker">
##FORMAT=<ID=GT,Number=1,Type=String,Description="Genotype">
##FORMAT=<ID=DS,Number=A,Type=Float,Description="estimated ALT dose
[P(RA) + P(AA)]">
##FORMAT=<ID=GP,Number=G,Type=Float,Description="Estimated Genotype
Probability">
#CHROM      POS     ID     REF   ALT   QUAL   FILTER      INFO   FORMAT        844
       854    855    856    858   859   861    862    863   864    865    866    867
       868    869    872
...
1      78635 1_78635     C     T     .      PASS   AR2=1.00;DR2=1.00;AF=0.22
       GT:DS 0|0:0 0|0:0 1|0:1 1|0:1 0|0:0 0|0:0 0|0:0 0|0:0 1|0:1 0|0:0 1|1:2
       0|0:0 0|1:1
...
1      153228       1_153228     C     T     .      PASS
       AR2=0.00;DR2=0.01;AF=0.062;IMP       GT:DS 0|0:0.16    0|0:0.15
       0|0:0.09     0|0:0.07     0|0:0.15     0|0:0.15     0|0:0.15     0|0:0.15
       0|0:0.09     0|0:0.16     0|0:0.03
...
1      243172       1_243172     C     T     .      PASS
       AR2=0.00;DR2=0.03;AF=0.24;IMP GT:DS 0|0:0.61     0|0:0.52     0|0:0.43
       0|0:0.25     0|0:0.52     0|0:0.52     0|0:0.52     0|0:0.52     0|0:0.43
       0|0:0.61     0|0:0.25     0|0:0.52
...
```

Again you should notice that the file format is the same (VCF). The file contains information for all non-reference samples and reports phased diplotypes. A few extra lines of meta-data have been added and some extra information is reported, but the most notable difference between this file and the previous is that next to the phased genotype the allelic dosage is reported after the: sign. For example, a genotype value of $0|0:0$ indicates that the genotype is homozygous for the major allele ($0|0$) and the estimated minor allele dosage is zero (: 0). Values of $1|0:1$ and $0|1:1$ indicate heterozygotes with minor allele dosage of 1. Values such as $0|0:0.15$ are returned for imputed scores and indicate that the most likely genotype for that SNP-individual combination is homozygous for the major allele, but its expected allele dosage is not exactly zero, rather it is 0.15. If the locus is segregating for a major allele A and a minor allele a, the expected minor allele dosage is $p(Aa) + 2p(aa)$.

Notice that the genotype calls in this set for the first marker (at position 78635) match exactly those given in the input data set, because this marker had complete genotype data in the imputed set, so nothing has changed. For the first four individuals (coded 844, 854, 855, and 856) the marker scores are 0/0, 0/0, 1/0, and 1/0 (highlighted in red above), just as they were in the original data set. For the second SNP (at position (153228), the imputed genotypes are 0/0 for the first four individuals (also highlighted in red).

To check how well imputation works we will read the phased data back into R and transform them back into the original file format.

**Code example 10.1
(Continued)**

```
#########################################################################
### To Get Output phased genotype file in original format
#########################################################################
rm(list=ls()); gc();
for(c in 1:2)
{
  cat("Starting Chromosome",c,"\n")
  string = paste("Phased_data",c,".gt.vcf.gz",sep="")
  dataframe <- read.table(file=string, sep="\t",header=FALSE,
comment.char="*", skip=9)
  ## Remove first 9 columns
  dataframe <- dataframe[ ,c(10:ncol(dataframe))]
  ## Grab Animal ID
  AnimalID <- t(dataframe[1, ])
  ## Remove Animal ID
  dataframe <- dataframe[-c(1), ]
  genotype <-
matrix(data=NA,ncol=nrow(dataframe),nrow=ncol(dataframe),byrow=TRUE)
  for(i in 1:ncol(dataframe))
  {
  # Grab animal and get genotype
    temp <- as.character(paste(dataframe[ ,c(i)]))
  ## First string split by ":"
    temp <- matrix(unlist(strsplit(temp,":")),ncol=2,byrow=TRUE)
  ## First column is maternal and paternal haplotype; second is dosage
    temp <- temp[ ,c(1)]
    temp <- matrix(unlist(strsplit(temp,"|")),ncol=3,byrow=TRUE)
    ## make genotypes 0 2 3 4
    temp[,2] <- ifelse(temp[,1] == "0" & temp[,3] == "0","0",temp[,2])
    temp[,2] <- ifelse(temp[,1] == "1" & temp[,3] == "1","2",temp[,2])
    temp[,2] <- ifelse(temp[,1] == "0" & temp[,3] == "1","3",temp[,2])
    temp[,2] <- ifelse(temp[,1] == "1" & temp[,3] == "0","4",temp[,2])
    genotype[i, ] <- temp[,2]
  }
  cat("Genotype for Chromosome ",c," Created\n")
  if(c == 1){PhasedGenotypes <- cbind(AnimalID,genotype)}
  if(c > 1){PhasedGenotypes <- cbind(PhasedGenotypes,genotype)}
  rm(genotype)
  cat("Ending Chromosome",c,"\n")
}
write.table(PhasedGenotypes,file="ImputePhasedGenotypes",sep=" ",
row.names=FALSE,col.names=FALSE, quote=FALSE)
```

We are going to write the phased result to a file *ImputePhasedGenotypes* to make it easier to perform the comparison.

Code Example 10.1
(Continued)

```
#####################################################################
###Check Accuracy of imputation since true phase was already known###
#####################################################################
rm(list=ls()) ; gc();

UnphasedGenotypes <- read.table(file="ImputeGenotype",sep=" ",
 header=FALSE)

UnphasedGenotypes <- UnphasedGenotypes[c(1:101),-c(1)];
 dim(UnphasedGenotypes)

PhasedGenotypes <- read.table(file="ImputePhasedGenotypes",sep=" ",
 header=FALSE)

PhasedGenotypes <- PhasedGenotypes[ ,-c(1)]; dim(PhasedGenotypes)

## Error Rate - compute only for the 80% subset of markers missing in
 the imputed set
f<-seq(from= 1, to=dim(PhasedGenotypes)[2],by=5)
f<-f[order(f)]
Errors <- (PhasedGenotypes[,-f] - UnphasedGenotypes[,-f])
# give all errors a value of 1
Errors[Errors != 0] <- 1
1-mean(rowSums(Errors)/ 4000)
```

To compare the actual genotypes with the imputed ones we just subtract the original values from the imputed ones and for any value different than 0 (the result if the imputation value is correct), we assign a value of 1. Then we obtain the error rate as the mean proportion of errors per row, computed across only the 4000 SNPs that were imputed.

In this example (100 individuals to be imputed 1000 markers genotyped in the reduced panel) we obtain an accuracy of approximately 0.67. The error rate includes heterozygote calls that are correct but have the wrong phase (e.g., 0|1 instead of 1|0). Some additional code is included in Code Example 10.2 to compute the accuracy ignoring phase, in which case the accuracy improves to about 0.70. The imputation accuracy will depend on the number of individuals in the reference, the number of markers to be imputed, their distribution across the genome, their allele frequency and the overall LD structure and connectedness of the sample. As stated in the previous section, in the current example data were generated with low LD and therefore, imputation accuracy is relatively low.

The example reported is intentionally minimalist and several factors influence imputation accuracy. Furthermore, in the case presented only few markers were imputed. Significant differences are to be expected when imputation happens from low density panels to sequence information.

Box 10.1: Running Linux Commands in the Windows Operating System

The examples in this chapter will not run in a Windows environment directly. Although Beagle can be run using the Windows command line, we recommend that Windows users install separate software to run the shell scripts and Linux type commands. One possibility for Windows users is to install a Unix Virtual Machine, such as VMware (http://www.vmware.com/products/player.html) or VM Virtual Box (http://www.oracle.com/technetwork/server-stor age/virtualbox/downloads/index.html). This will run a Linux environment alongside of the Windows operating system. This is a good option for users who will make extensive use of Linux programs, although it requires some effort to set up properly to share the computer's RAM and access the appropriate directories on the user's computer. A simpler approach is to install a free Linux shell program, such as Cygwin (https://www.cygwin.com/). Cygwin provides a shell program that allows users to use Linux commands and basic utilities, as well as additional Linux programs that the user can install (in our context, Beagle).

After installing Cygwin and launching the program, a command line box appears, for example:

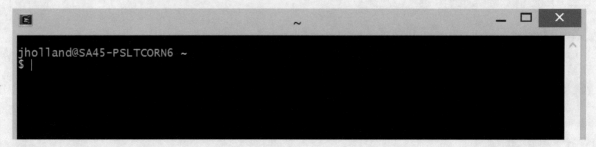

The $ symbol is the command line prompt. The default home directory is 'C:\Users\username' but a few tricks are needed to navigate the windows directories from inside Cygwin. First, Linux systems use '/' instead of '\' to indicate subdirectories. Second, folder names with spaces in them that are often used in Windows have to be provided in quotes to the Linux shell in Cygwin. Third, from inside Cygwin, a Windows path that looks like:

"C:\Users\jholland\Google Drive\Book\Book1_Examples\Ch10_impute"

has to be provided as:

"/cygdrive/c/Users/jholland/Google Drive/Book/Book1_Examples/Ch10_impute"

Notice the use of '/' instead of '\' and the use of "/cygdrive/c/Users..." instead of "C:\Users...", and finally, a subtle but important difference is that the quotation marks used inside of Cygwin are not identical to the open and close quotation marks used in some word processing programs (so copying and pasting them to the Cygwin shell will fail).

Using some basic Linux commands, such as 'cd' for change directory and 'ls' to list the files present inside a directory (see https://www.linux.com/ for tutorials and help documents), we can see the files inside the folder holding the example data sets created for this chapter, and use the gzip command to compress the file 'Geno_data1.inp' created by the R script in Code Example 10.1 to a gzip version:

Box 10.1 (continued)

```
 /cygdrive/c/Users/jholland/Google Drive/Book/Book1_Examples/Ch10...  ─  □   ✕
jholland@SA45-PSLTCORN6 ~
$ ls

jholland@SA45-PSLTCORN6 ~
$ cd "/cygdrive/c/Users/jholland/Google Drive/Book/Book1_Examples/Ch10_impute"

jholland@SA45-PSLTCORN6 /cygdrive/c/Users/jholland/Google Drive/Book/Book1_Examp
les/Ch10_impute
$ ls
beagle.09Nov15.d2a.jar          Code10-2_GenotypeSetUpBeagle4_Jim.R
Beagle1.sh                      desktop.ini
Beagle2.sh                      Geno_data1.inp
beagle2vcf.jar                  Geno_data2.inp
BeagleLoop.sh                   'OLD Original Files'
BeagleLoop2.sh                  ref_data1.inp
Code10-1_GenotypeSetUpBeagle4.R  ref_data2.inp

jholland@SA45-PSLTCORN6 /cygdrive/c/Users/jholland/Google Drive/Book/Book1_Examp
les/Ch10_impute
$ gzip Geno_data1.inp

jholland@SA45-PSLTCORN6 /cygdrive/c/Users/jholland/Google Drive/Book/Book1_Examp
les/Ch10_impute
$ |
```

You may have some different files, but you should have at least the files that end in '.inp' if you have run the R script in Code Example 10.1 Notice that the default gzip command creates a new file called 'Geno_data1.inp.gz' and deletes the original file 'Geno_data1.inp':

```
jholland@SA45-PSLTCORN6 /cygdrive/c/Users/jholland/Google Drive/Book/Book1_Examp
les/Ch10_impute
$ ls
beagle.09Nov15.d2a.jar          Code10-2_GenotypeSetUpBeagle4_Jim.R
Beagle1.sh                      desktop.ini
Beagle2.sh                      Geno_data1.inp.gz
beagle2vcf.jar                  Geno_data2.inp
BeagleLoop.sh                   'OLD Original Files'
BeagleLoop2.sh                  ref_data1.inp
Code10-1_GenotypeSetUpBeagle4.R  ref_data2.inp
```

To prepare all of the input files for the first imputation example for Beagle, we have to run gzip on the data input file for chromosome 2 and also for the reference panel data inputs for both chromosomes, so that finally we have:

```
jholland@SA45-PSLTCORN6 /cygdrive/c/Users/jholland/Google Drive/Book/Book1_Examp
les/Ch10_impute
$ ls
beagle.09Nov15.d2a.jar          Code10-2_GenotypeSetUpBeagle4_Jim.R
Beagle1.sh                      desktop.ini
Beagle2.sh                      Geno_data1.inp.gz
beagle2vcf.jar                  Geno_data2.inp.gz
BeagleLoop.sh                   'OLD Original Files'
BeagleLoop2.sh                  ref_data1.inp.gz
Code10-1_GenotypeSetUpBeagle4.R  ref_data2.inp.gz
```

Box 10.1 (continued)

Then we can run the BeagleLoop.sh shell script to execute the Beagle imputation analysis:

```
jholland@SA45-PSLTCORN6 /cygdrive/c/Users/jholland/Google Drive/Book/Book1_Examp
les/Ch10_impute
$ sh BeagleLoop.sh
============================
BEAGLE LOOP FOR CH.10 EXAMPLE
============================
nohup: appending output to 'nohup.out'

jholland@SA45-PSLTCORN6 /cygdrive/c/Users/jholland/Google Drive/Book/Book1_Examp
les/Ch10_impute
$ nohup: appending output to 'nohup.out'
```

This will run for some time, and when it finishes, it produces a series of output files:

```
jholland@SA45-PSLTCORN6 /cygdrive/c/Users/jholland/Google Drive/Book/Book1_Examp
les/Ch10_impute
$ ls
beagle.09Nov15.d2a.jar              nohup.out
Beagle1.sh                         'OLD Original Files'
Beagle2.sh                         Phased_data1.gt.log
beagle2vcf.jar                     Phased_data1.gt.vcf.gz
BeagleLoop.sh                      Phased_data1.gt.warnings
BeagleLoop2.sh                     Phased_data2.gt.log
Code10-1_GenotypeSetUpBeagle4.R    Phased_data2.gt.vcf.gz
Code10-2_GenotypeSetUpBeagle4_Jim.R Phased_data2.gt.warnings
desktop.ini                        ref_data1.inp.gz
Geno_data1.inp.gz                  ref_data2.inp.gz
Geno_data2.inp.gz
```

The imputed genotypes are found in the files 'Phased_data1.gt.vcf.gz' and 'Phased_data2.gt.vcf.gz'. These are read in and checked for accuracy against the original 'known' genotypes in the second part of the R script given in Code Example 10.1.

Imputation Without a Reference Panel

In some species, no high density reference genotypes or sequences are available to use as a means to estimate the population frequencies of haplotypes. Or, what high density genotyped samples are available may not provide reasonable estimates of the haplotype frequencies in a particular breeding population of interest. In these cases, the imputation can be carried out using information only from the samples within the population of interest. In some cases, genotype data are obtained from relatively low coverage sequencing, resulting in a data set with many SNPs called across the whole sample, but also many individual genotype calls that are missing because of the stochastic nature of some next generation sequencing platforms. So, the researcher may be faced with a case of many missing genotype calls in the sample but no reliable reference panel to use to guide the imputation. The structure of the missing data is different in this case than in the case of the high density reference panel to low density target population described previously. Instead of missing data on a consistent set of markers that were not included in the low density genotyping platform, the missing data will be closer to randomly distributed across SNPs and individuals. In this case, the imputation can proceed by inferring haplotype frequencies from the available data alone. The accuracy of this approach depends strongly on the proportion of missing data and the extent of linkage disequilibrium in the sample. More extensive linkage disequilibrium results in fewer haplotypes with higher frequencies, which helps imputation accuracy.

To simulate this scenario, Code Example 10-3 ('Code 10-3_GenotypeSetUpBeagle4NoRef.R') is an R script that takes in the same initial data as used for the reference panel imputation scenario, but instead of splitting the data into a reference panel and a target set with lower density, keeps the whole sample of 500 individuals together in a common data set and randomly introduces 20% missing data.

Code example 10.3

Generating random missing genotypes (see Code 10-3_GenotypeSetUpBeagle4NoRef.R for details).

```
######################################################################
### To create VCF File Genotype format for Beagle 4.0 or greater  ###
######################################################################
rm(list=ls()); gc();
set.seed(8022016) #set seed so that random missing data pattern is
 repeatable
## change to working directory
setwd("C:/Users/.../Ch10_impute")
## read in Marker File
input_path = "C:/Users/.../data/"
map <- read.table(file=paste0(input_path, "ImputeMarkerMap"),sep="
 ",header=FALSE)
## read in true phased genotype matrix column 1 is ID and rest of
 columns are genotypes
## 0 - 1 1; 2 - 2 2; 3 - 1 2; 4 - 2 1
geno <- read.table(file=paste0(input_path, "ImputeGenotype"),sep="
 ",header=FALSE)
## take of animal IDs
ID <- geno[,c(1)]
## remove IDs from animal
geno <- geno[ ,-c(1)]

##################RANDOM MISSING DATA
#randomly replace 80% of data with missing calls
Ngeno = nrow(geno)
Nsnps = ncol(geno)
Nmiss = round(Ngeno*Nsnps*0.8)
inds <- as.matrix(expand.grid(1:Ngeno, 1:Nsnps))
# Sample randomly a set of row column indices equal to number of
 missing data points
missing <- inds[sample(nrow(inds), Nmiss), ]
# Note that `missing` is a matrix of (row, col) indices
geno[missing] <- 5 # 5 is missing data
...
```

Details of the simulation are in the R code and will not be discussed here, but the resulting data set for the first three markers on chromosome 1 (in file 'Geno_data_miss_rand1.inp') looks like this:

```
1     78635    1_78635    C    T    .    PASS    .    GT    0/0    0/0    1/0    ./.    ./.    0/0
0/0    0/0    1/0    0/0    1/1    ./.    0/1    0/0    1/0    0/0    1/1    0/0    ./.    0/0
0/1    ./.    0/1    0/0    0/1    0/0    0/0
...
1    153228    1_153228    C    T    .    PASS    .    GT    1/1    0/0    0/0    0/0    0/0    0/1
0/0    0/0    0/0    0/0    0/0    ./.    ./.    0/0    0/0    0/0    ./.    0/0    0/0
0/0    0/0    ./.    0/0    0/0    0/0
...
```

```
1      243172   1_243172    C    T     .    PASS    .    GT    ./.   1/0   0/1   ./.   1/0   0/0
0/1    0/0      0/0      0/1     0/0   1/0   1/0   0/0   0/0   0/0   0/0   0/0   1/0   0/0
0/0    0/0      0/0      0/0     0/0   ./.
...
```

Notice that the individuals missing a genotype call at the first marker are not the same as those missing calls at the other markers. Each SNP and each individual have a unique pattern of missing data. Once the files with randomly missing data are created and compressed to .gz format, we can run the Beagle analysis using Code Example 10.4, which is a shell script named 'BeagleLoop2.sh'. This script operates very similarly to BeagleLoop.sh, but in this case Beagle is executed only with a 'gt' file input and the 'ref' input file option is not used. For example, the call to execute imputation on chromosome 1 is:

Code example 10.4
Imputation on chromosome 1 using Beagle (see shell script BeagleLoop2.sh for details)

```
java -jar beagle.27Jul16.86a.jar gt=Geno_data_miss_rand1.inp.gz
impute=TRUE out=Phased_data_miss_rand1.gt
```

Since we execute Beagle without a reference panel input, Beagle estimates the haplotype frequencies using only the data provided in the input genotype file, the same samples that we are imputing. Once the two Beagle analyses finish, they will produce two key output files:

```
Phased_data_miss_rand1.gt.vcf.gz
Phased_data_miss_rand2.gt.vcf.gz
```

These can be read back into R using the second part of Code Example 10.3 script and compared to the original complete data set. In this example, the accuracy is very poor, only about 43% (and that includes 20% of calls that were correct to begin with!). With so much missing data and low levels of LD, estimation of the haplotype frequencies is very difficult. SNPs called from low coverage sequencing often display substantial variation in the proportion of missing data among individuals, and it is advantageous to first filter out the SNPs with high missing data rates and attempt to impute only sites with a reasonable amount of data present.

Imputation with the Synbreed Package

The R package Synbreed discussed in Chap. 9 has several functions to impute missing genotypes without using a reference panel (Wimmer et al. 2012). Code Example 10.5 provides a script to read in the same set of 5000 markers on 500 individuals used previously in this chapter, randomly introduces 20% missing data, and imputes them using Synbreed. Synbreed has several options for imputing missing data; this example demonstrates the using of the 'random' imputation method, which, for each SNP, estimates the genotype frequencies from the available data, then replaces missing data with random draws from the observed genotype frequency. Obviously, this information does not use LD or haplotype information, so it will tend not to be very accurate, but it is fast and perhaps useful for situations with only a small amount of missing data and lacking marker order information.

Code example 10.5
Imputing missing genotype data with Synbreed (see code Code 10-3_ImputeWithSynbreed.R for details)

```
#####################################################################
### To Simulate Missing Geno Data and Impute with Synbreed     ###
#####################################################################
rm(list=ls()); gc();
set.seed(8062016)
## change to working directory
library(synbreed)
setwd("C:/Users/.../Ch10_impute")
input_path = "C:/Users/.../data/"

## read in Marker File
#map <- read.table(file="ImputeMarkerMap",sep=" ",header=FALSE)
map <- read.table(file=paste0(input_path,"ImputeMarkerMap"),sep=" ",
 header=F)

#for synbreed, the map object should have just chromosome and position
 and requires column names (columns 1 and 4)
#also the rownames should be the names of the markers (column 2 of the
 original map file)
rownames(map) <- map[,2]
map <- map[,c(1,4)]
colnames(map) <- c("chr", "pos") #synbreed requires these names

## read in true phased genotype matrix column 1 is ID and rest of
 columns are genotypes
## 0 - 1 1; 2 - 2 2; 3 - 1 2; 4 - 2 1
#geno <- read.table(file="ImputeGenotype",sep=" ",header=FALSE)
geno <- read.table(file=paste0(input_path,"ImputeGenotype"),sep="
```

```
",header=FALSE)
## take off animal IDs
ID <- geno[,c(1)]
## remove IDs from animal
geno <- geno[ ,-c(1)]
rownames(geno) <- ID
colnames(geno) <- rownames(map) #column names of geno are the marker
 names

#convert genotype values of 3 and 4 to '1' for heterozygotes (minor
 allele dosage code)
geno[geno == 3 | geno == 4] <- 1


####################RANDOM MISSING DATA
#randomly replace 20% of data with missing calls
Ngeno = nrow(geno)
Nsnps = ncol(geno)
Nmiss = round(Ngeno*Nsnps*0.2)
inds <- as.matrix(expand.grid(1:Ngeno, 1:Nsnps))
# Sample randomly a set of row column indices equal to number of
 missing data points
missing <- inds[sample(nrow(inds), Nmiss), ]
# Note that `missing` is a matrix of (row, col) indices
genoOrig <- geno #make a copy of original data before introducing
 missing data
geno[missing] <- NA #

#create the GP object
genoGP <- create.gpData(pheno = NULL, geno = geno, map = map, map.unit
 = 'bp')

#recode the data using codeGeno and random imputation method
genoGP2 <- codeGeno(genoGP, impute = T, impute.type = "random",
 label.heter = "1")

#compare original data, missing data set, and imputed data
Inds = rownames(genoOrig[1:10,])
genoOrig[Inds,1:5]
genoOrig[1:10,1:5] #original data
genoGP$geno[Inds, 1:5] #missing data
genoGP2$geno[Inds, 1:5] #imputed data
```

The results of this script are printouts of the first ten SNPs on the first ten individuals from the original complete data set, the data set with missing data, and the resulting imputed data set:

```
> genoOrig[Inds,1:5]
    1_78635 1_153228 1_243172 1_334883 1_417617
844        0        2        0        0        0
854        0        0        1        1        0
855        1        0        1        0        0
856        1        0        0        2        0
858        0        0        1        1        0
859        0        1        0        1        0
861        0        0        1        1        0
862        0        0        0        1        0
863        1        0        0        0        0
864        0        0        1        0        0
> genoGP$geno[Inds, 1:5] #missing data
    1_78635 1_153228 1_243172 1_334883 1_417617
844        0        2       NA        0        0
854        0        0        1        1        0
855        1       NA       NA        0        0
856        1        0       NA        2        0
858        0       NA        1       NA        0
859        0        1        0        1        0
861        0        0        1        1        0
862        0        0       NA        1        0
863        1        0        0        0        0
864        0        0        1        0        0
> genoGP2$geno[Inds, 1:5] #imputed data
    1_78635 1_153228 1_243172 1_334883 1_417617
844        0        2        0        0        0
854        0        0        1        1        0
855        1        0        1        0        0
856        1        0        0        2        0
858        0        0        1        0        0
859        0        1        0        1        0
861        0        0        1        1        0
862        0        0        1        1        0
863        1        0        0        0        0
864        0        0        1        0        0
```

The random imputation method is not generally recommended but Synbreed includes several other options for imputation. Some of the other options also do not use haplotype information, but could be useful in specific cases (e.g., the impute. type = "family" method can be useful for doubled haploids or recombinant inbred lines from biparental crosses). In addition, Synbreed has the option impute.type = "beagle", which will export the genotype data in a format that can be read by Beagle and will send a shell command to execute the beagle analysis, however, it can be tricky to set up the correct path information to make this work, and it does not always use the most current version of Beagle.

Electronic supplementary material: The online version of this chapter (doi:10.1007/978-3-319-55177-7_11) contains supplementary material, which is available to authorized users.

Abstract

The resemblance between individuals in a population has been traditionally estimated from pedigrees. High-throughput genotyping technology has enabled the use of large numbers of DNA markers to estimate the amount of genome shared by individuals. Genetic similarity estimates based on genetic markers are more precise than those based on pedigree information. Using the genomic relationships derived from markers for prediction of genetic merit of individuals has gained considerable attention in animal and plant breeding in recent years. In this chapter, we introduce the concept and provide some examples.

Realized Genomic Relationships

Traditional genetic evaluations combine phenotypic data and resemblance between relatives to predict genetic merit of individuals. The resemblance coefficients derived from pedigrees are based on probabilities that alleles are identical by descent (IBD). More recently, genetic markers distributed throughout the entire genome have been used to measure genetic similarities more precisely than pedigree information (VanRaden 2008). Genetic markers estimate the proportion of chromosome segments shared by individuals based on the identical by state (IBS) matching of marker alleles. Whereas the matrix of pairwise pedigree relationships is referred to as the **A** matrix (because the elements are pedigree-based estimates of additive genetic relationships), the matrix of pairwise realized genomic relationships estimated from marker information is referred to as the **G** matrix. In this chapter, we will describe methods to compute the **G** matrix from markers and the use of the **G** matrix to predict breeding values using genomic relationships. The methods described to compute **G** require complete data (no missing values) and numericalized marker genotypes, so raw SNP data must first be processed using the data recoding and imputation methods described in the previous two chapters.

We provide here a very small data example so the reader can follow the computations more easily. We start with a numericalized genotype data set as described in Chap. 9, reflecting the dosage of minor alleles at each locus and individual. For example, gene content values for three individuals at four loci might appear as:

	Locus1	Locus2	Locus3	Locus4
Indiv1	0	1	0	2
Indiv2	2	1	1	1
Indiv3	2	0	0	0
...				

We will refer to this matrix of raw marker scores as **M**. In the **M** matrix, the rows are individuals and the columns are loci. Dimensions of the matrix are n (number of individuals) by m (number of loci). In this example, individuals 2 and 3 are homozygous for the minor allele at locus 1. The designation of which allele is minor is based on the allele frequencies estimated over the whole sample, not just these three individuals.

We can use a simple matrix multiplication to compute the number of homozygous loci in each individual and for each pair, the number of loci with matching homozygous genotypes minus the number of loci homozygous for different alleles. This is a kind of homozygous identity in state matching coefficient. To do this, first we rescale the gene content scores in **M** by subtracting one from all elements, we get a new matrix with scores of -1, 0, and 1:

	Locus1	Locus2	Locus3	Locus4
Indiv1	−1	0	−1	1
Indiv2	1	0	0	0
Indiv3	1	−1	−1	−1

We refer here to this matrix as **M−1** (although VanRaden (2008) refers to this as **M**). If we compute the cross-product of the rows of marker scores (m_{ij} for marker j in individual i) in **M−1** for individuals 1 and 2, we get the number of loci at which they are homozygous for the same loci (matching) minus the number of loci for which they are homozygous for different loci

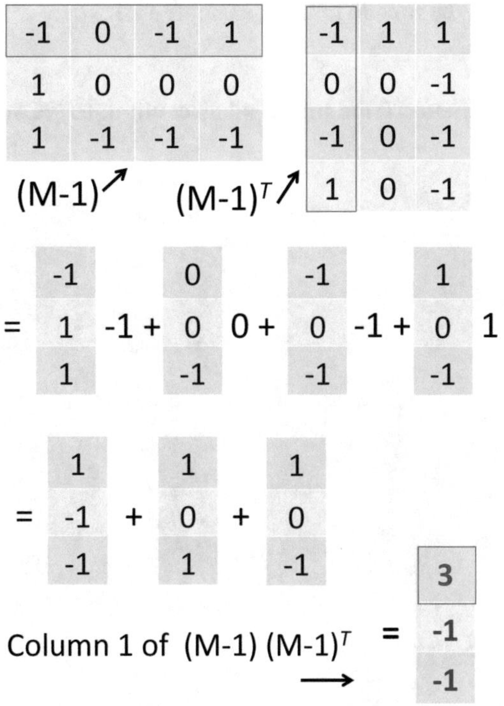

-1	0	-1	1		-1	1	1
1	0	0	0		0	0	-1
1	-1	-1	-1		-1	0	-1
$(M-1)\nearrow$		$(M-1)^T\nearrow$		1	0	-1

$$= \begin{matrix} -1 \\ 1 \\ 1 \end{matrix} \begin{matrix} -1 \\ \\ -1 \end{matrix} + \begin{matrix} 0 \\ 0 \\ \end{matrix} \begin{matrix} 0 \\ \\ \end{matrix} + \begin{matrix} -1 \\ 0 \\ -1 \end{matrix} \begin{matrix} -1 \\ \\ -1 \end{matrix} + \begin{matrix} 1 \\ 0 \\ -1 \end{matrix} \begin{matrix} 1 \\ \\ -1 \end{matrix}$$

$$= \begin{matrix} 1 \\ -1 \\ -1 \end{matrix} \begin{matrix} 1 \\ \\ \end{matrix} + \begin{matrix} 1 \\ 0 \\ 1 \end{matrix} + \begin{matrix} 1 \\ 0 \\ -1 \end{matrix}$$

Column 1 of $(M-1)(M-1)^T$ $= \begin{matrix} 3 \\ -1 \\ -1 \end{matrix}$

$\longrightarrow$

Fig. 11.1 Computing the first column of $M-1(M-1)^T$. Diagonal element at the bottom (3) counts number of homozygous loci in individual 1. Off-diagonal elements $(-1, -1)$ (row $= 2$, col $= 1$), and (row $= 3$, col $= 1$) count number of loci at which individual 1 is homozygous for same allele minus homozygous for alternate alleles as individuals 2, 3

(not matching): Cross-product of rows 1 and 2 of $\mathbf{M-1}$: $\sum_j^m m_{1j}m_{2j} = (-1^*1) + (0^*0) + (-1^*0) + (1^*0) = 1$. Similarly, the sum of squares of the first row with itself counts the number of homozygous loci in individual 1: Sum of squares of row 1 of $\mathbf{M-1}$: $\sum_j^m m_{1j}^2 = (-1)^2 + 0^2 + (-1)^2 + 1^2 = 3$.

We can scale this up to compute the cross-products and sums of squares of all the rows simultaneously by simply multiplying $\mathbf{M-1}$ by its transpose. To make this clear, consider that the first column of the result of $\mathbf{M-1}^*(\mathbf{M-1})^T$ is equal to the matrix $\mathbf{M-1}$ times the first column vector of $(\mathbf{M-1})^T$ (Fig. 11.1).

Computing the other columns of the result in the same way, we get:

$$(M-1)(M-1)^T = \begin{bmatrix} -1 & 0 & -1 & 1 \\ 1 & 0 & 0 & 0 \\ 1 & -1 & -1 & -1 \end{bmatrix} \begin{bmatrix} -1 & 1 & 1 \\ 0 & 0 & -1 \\ -1 & 0 & -1 \\ 1 & 0 & -1 \end{bmatrix} = \begin{bmatrix} 3 & -1 & -1 \\ -1 & 1 & 1 \\ -1 & 1 & 4 \end{bmatrix}$$

Consider the last column of this result, which corresponds to the cross-products of individual 3 with the others and itself: $(-1\ 1\ 4)^T$. Individuals 1 and 3 have matching homozygous genotypes at locus 3 but have non-matching homozygous genotypes at loci 1 and 4, so the element (1,3) in the result matrix is $1-2 = -1$. Individuals 2 and 3 have matching homozygous genotypes at locus 1 and no loci with non-matching homozygous genotypes, so the element (2,3) in the result is $1-0 = 1$. Individual 3 is homozygous at all four loci in this example, so the element (3,3) equals **4**.

This example can be demonstrated with the R code in Code example 11.1.

Code example 11.1
Example of similarity matrices estimated from minor allele count data ("Code 11-1_M and Z matrices for computing relationship matrix.R")

```
> M <- matrix(c(0,1,0,2,2,1,1,1,2,0,0,0), nrow=3,ncol=4, byrow=T)
> M
     [,1] [,2] [,3] [,4]
[1,]    0    1    0    2
[2,]    2    1    1    1
[3,]    2    0    0    0
> M-1
      [,1]  [,2]  [,3]  [,4]
[1,]    -1     0    -1     1
[2,]     1     0     0     0
[3,]     1    -1    -1    -1
> (M-1)%*%t(M-1)
     [,1] [,2] [,3]
[1,]    3   -1   -1
[2,]   -1    1    1
[3,]   -1    1    4
```

This cross-product matrix can be interpreted as a similarity matrix based on identity by state (IBS). However, our real interest is in estimating identity by descent (IBD) relationships among the individuals, so we need a scaling of the marker scores that will reflect covariance among the individuals in terms of IBD relationships.

As an example of why IBS measures are not directly appropriate to estimate IBD relationships, consider the contribution to $\mathbf{M}-\mathbf{1}*(\mathbf{M}-\mathbf{1})^{\mathrm{T}}$ of a locus with a low minor allele frequency of $p = 0.01$. If the population is in Hardy-Weinberg equilibrium we expect about 98% of the population to have genotype code 0 and only 0.01% of the population to have genotype code 2. The probability of sampling two unrelated individuals with matching 0 genotype codes at this locus is very high: 96%. Since a huge proportion of pairs will match for homozygous major alleles even without common ancestry, it is clear that IBS matches at this locus will provide little information about IBD relationships. In contrast to the limited information provided by common allele matches, rare allele matches at this locus will be very informative about common ancestry. The probability that an unrelated pair will both be homozygous for the minor allele at this locus is exceedingly small: $p^4 = 10^{-8}$! It should be clear that we need a method that appropriately weights the genotype matches to give more weight to matches that are less likely to occur by chance without shared ancestry.

Wright's (1922) definition of the coefficient of pedigree relationship between individuals as a correlation suggests that a realized relationship coefficient estimator should be based on the covariance or correlation among marker scores for each pair of individuals. For a single pair of individuals, i and j (members of a finite pedigree whose ancestors are unrelated), twice their single-locus coancestry coefficient can be estimated as the cross product of their centered gene content values divided by twice the allelic variance (Speed and Balding 2015).:

$$2r_{ij} = \frac{(x_i - 2p)(x_j - 2p)}{2pq} \tag{11.1}$$

In Eq. 11.1, x_i and x_j are the minor allele contents for individuals i and j, and p and q are the minor and major allele frequencies at the locus. The mean gene content value at this locus is $2p$, and this mean value is subtracted from the observed gene contents in Eq. 11.1 to get values centered on zero. To obtain the genome-wide realized relationship coefficients that can be interpreted as analogous to classical pedigree relationships coefficients, we can average these values over all markers assayed. There is sampling variance with these estimates, and the more markers used, the lower the sampling variance of the genome-wide estimate.

We will center the marker scores at each locus i by subtracting the mean score $P_i = 2\widehat{p}_i$, where $\widehat{p}_i$ is the estimated minor allele frequency based on the marker data available. As an example of this scaling, let's assume that the minor allele frequencies estimated from all the individuals in the population for the four loci in the example $\mathbf{M}$ matrix above are 0.01, 0.1,

0.25, and 0.48. We will construct a matrix **P** containing twice the minor allele frequencies for each locus as the elements in each column:

$$P = \begin{bmatrix} 0.02 & 0.2 & 0.5 & 0.96 \\ 0.02 & 0.2 & 0.5 & 0.96 \\ 0.02 & 0.2 & 0.5 & 0.96 \end{bmatrix}$$

Then we scale the original **M** matrix by subtracting **P** from it to compute a new matrix, **Z**:

$$Z = M - P = \begin{bmatrix} 0 & 1 & 0 & 2 \\ 2 & 1 & 1 & 1 \\ 2 & 0 & 0 & 0 \end{bmatrix} - \begin{bmatrix} 0.02 & 0.2 & 0.5 & 0.96 \\ 0.02 & 0.2 & 0.5 & 0.96 \\ 0.02 & 0.2 & 0.5 & 0.96 \end{bmatrix}$$

$$Z = \begin{bmatrix} -0.02 & 0.8 & -0.5 & 1.04 \\ 1.98 & 0.8 & 0.5 & 0.04 \\ 1.98 & -0.2 & -0.5 & -0.96 \end{bmatrix}$$

The difference between **M** and **Z** is a function of the allele frequencies, so it varies among columns of **Z**. For the first locus (the first column of each matrix), the minor allele frequency was very low, so **Z** is very similar to the original matrix **M**. Since homozygous common genotypes are scored as zero in **M**, they will contribute very little to the cross-products computed from **Z** for loci where the minor allele frequency is low. In contrast, the homozygous rare genotypes at this locus are near 2, so they will contribute proportionally much more to the cross-products computed from **Z**. In contrast, locus 4 had nearly balanced allele frequencies, so that heterozygous matches count for little and homozygous matches contribute about equal amounts, but still less than the homozygous matches at locus 1. The increased influence of rare allele matches in ZZ^T compared to $M-1*(M-1)^T$ is shown in Fig. 11.2, this is precisely the weighting desired to reflect IBD relationship. Notice that homozygosity of rare alleles also inflates the diagonal elements e.g., element (2,2).

We show the computations to form the ZZ^T matrix in R starting with a vector of estimated minor allele frequencies at each locus in Code example 11.1

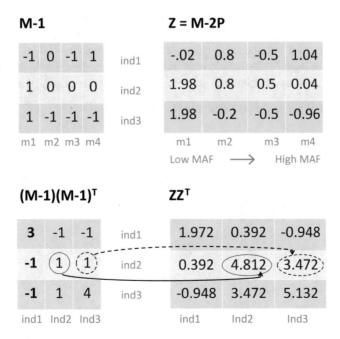

Fig. 11.2 Centering the columns of the marker data matrix M−1 to form Z matrix in a covariance matrix ZZ^T that reflects IBD relationships Individual 2 homozygosity for rare allele at locus 1 is given high weight in ZZ^T. Similarly, individuals 2 and 3 matching homozygosity for rare allele at locus 1 are given high weight in ZZ^T

Code example 11.1
(continued)

```
> maf <- c(0.01, 0.1, 0.25, 0.48)
> P <- 2*c(1,1,1)%*%t(maf)
> P
     [,1] [,2] [,3] [,4]
[1,] 0.02  0.2  0.5 0.96
[2,] 0.02  0.2  0.5 0.96
[3,] 0.02  0.2  0.5 0.96
> Z <- M-P
> Z
      [,1]   [,2]   [,3]   [,4]
[1,] -0.02   0.8   -0.5  1.04
[2,]  1.98   0.8    0.5  0.04
[3,]  1.98  -0.2   -0.5 -0.96
> Z%*%t(Z)
       [,1]    [,2]    [,3]
[1,]  1.972  0.392  -0.948
[2,]  0.392  4.812   3.472
[3,] -0.948  3.472   5.132
```

Since $\mathbf{ZZ}^T$ is a covariance matrix, its elements get larger as we include more markers in the computation. To remove the influence of marker number of the estimate, we can compute the average covariance per marker by dividing this matrix by the sum of the variances at each locus, and it is this matrix that we will use as the estimated genomic realized relationship matrix, $\mathbf{G}$:

$$\mathbf{G} = \frac{\mathbf{ZZ'}}{2 \sum \mathbf{p_i}(1 - \mathbf{p_i})} \tag{11.2}$$

Dividing by $2 \sum p_i(1 - p_i)$ (which is summed over m loci) scales $\mathbf{G}$ to be analogous to the numerator relationship matrix $\mathbf{A}$ (VanRaden 2008; Forni et al. 2011), such that the expected off-diagonal coefficient for half-sibs is 0.25, for example. Theoretical derivations of this realized relationship matrix estimator are given by Habier et al. (2007) and Endelman and Jannink (2012). This formula is the basis for the relationship matrix estimation function in the *synbreed* R package (Wimmer et al. 2012). Alternative estimators of the relationship matrix have also been proposed. These are summarized in Box 11.1.

We complete the computation of $\mathbf{G}$ for this example using **R** code:

Code example 11.1
(continued)

```
> denom = as.numeric(2*t(maf)%*%(1 - maf))
> denom
[1] 1.074
> G = (Z%*%t(Z))/denom
> G
            [,1]       [,2]       [,3]
[1,]  1.8361266  0.3649907 -0.8826816
[2,]  0.3649907  4.4804469  3.2327747
[3,] -0.8826816  3.2327747  4.7783985
```

The elements of the $\mathbf{G}$ matrix estimate the realized values of $2\theta_{ij}$, twice the probability that individuals i and j are IBD for alleles sampled at random from a common locus. The diagonal elements estimate $2\theta_{ii}$, twice the probability that two alleles sampled at random at a locus within individual i are IBD, which is related to the inbreeding coefficient (F_i) of individual i as follows:

$$E(2\theta_{ii}) = 1 + F_i \tag{11.3}$$

Thus, the diagonal elements of $\mathbf{G}$ are expected to be 1 for non-inbred individuals, and the average of the diagonal elements is the average inbreeding coefficient for the population. Even in a non-inbred population where the average of diagonal elements is about 1 as expected, there can be some variation around that number, as some individuals may have some genomic regions that are inbred due to shared recent ancestry. In the example above, the diagonal elements of $\mathbf{G}$ for individuals 2 and 3 are substantially greater than 1, suggesting that they are homozygous more often than expected based on Hardy-Weinberg expectations. Of course, a sample of four loci is not to be taken seriously, but in real world situations, one may see reliable evidence of inbreeding for some individuals based on genome-wide marker information, even if it was not predicted by pedigree.

Similarly, in a non-inbred population, the off-diagonal elements of $\mathbf{G}$ are expected to be 0.25 for half-sibs and 0.5 for full-sibs, but there will be variation around those values for specific pairs of siblings. By chance, some full-sibs will be IBD at more than 50% of their alleles and other pairs of full-sibs will be IBD at less than 50% of their alleles. Finally, unrelated individuals will have realized relationship coefficients around zero on average, but there will be variation around this value. Some pairs of individuals in unrelated families may share more alleles than expected by random chance, and these individuals can have coefficients greater than 0. These properties of the $\mathbf{G}$ matrix are what make it more effective for predicting breeding values than the pedigree relationship matrix, which consists only of the expected IBD relationships based on pedigree information. The $\mathbf{G}$ matrix more effectively weights information from siblings, based on their specific levels of relatedness to an individual whose breeding value is being predicted, rather than giving all full-sibs, for example, equal weight. Furthermore, if there are relationships between families that are not recognized by pedigree but become clear with marker data, information can be shared across families unrelated by pedigree.

One may notice that realized genomic relationship matrices can have some surprising values when viewed as estimates of coefficients of coancestry. For example, the element of $\mathbf{G}$ for individuals 1 and 3 is negative! It is not possible to reconcile a negative number as an estimate for a probability parameter (which is bounded by 0 and 1). It is perhaps better to consider the $\mathbf{G}$ matrix as estimators of two times the correlation matrix of allelic composition of individuals in the population (Powell et al. 2010). In this way, *a zero off-diagonal value* in the $\mathbf{G}$ matrix represents a pair of individuals that share alleles at a frequency equal to what is expected by random sampling of alleles at the loci tested. *A negative off-diagonal value* in the $\mathbf{G}$ matrix represents a pair of individuals that share alleles at a frequency lower than what is expected by random sampling of alleles at the loci tested, that is, a pair of individuals that is more unrelated than expected by random chance. *A positive off-diagonal value* in the $\mathbf{G}$ matrix represents a pair of individuals that share alleles at a frequency higher than what is expected by random sampling of alleles at the loci tested.

Another surprising value that can occur in realized relationship matrices are diagonal elements that are greater than 2. Having argued that the $\mathbf{G}$ matrix can be viewed as an estimator of twice the correlation matrix of alleles among individuals in a population, we might expect that the maximum value of a diagonal element is 2 since the maximum value of a correlation coefficient is 1. However, in practice, with highly inbred lines, which have an expected inbreeding coefficient of 1 and consequently an expected diagonal $\mathbf{G}$ coefficient of 2, one can observe values greater than 2. This occurs because of the heavier weight placed on rare allele matches in the computation of the $\mathbf{G}$ matrix, as mentioned previously. Across loci within an inbred individual, homozygosity for rare alleles has more weight than homozygosity for common alleles when computing the realized relationship coefficients. So, a completely inbred *individual that is homozygous for more rare alleles than expected by chance will have a diagonal elements greater than 2*. Conversely, completely inbred individuals that are homozygous for more common alleles than expected will have diagonal elements less than 2. In both cases, the lines may be completely homozygous, but their diagonal elements can vary. The average of the diagonal elements should be a reliable estimator of the average inbreeding coefficient in the population, but the individual values may not be directly interpretable as related solely to the inbreeding coefficient of the individuals.

If one observes values well out of expected bounds in a realized relationship matrix, it is likely an indication of population structure in the sample of individuals. For example, if most individuals are sampled from one common population, but a few individuals are sampled from a distinct population with different allele frequencies, the latter individuals will be much more likely to be homozygous for alleles that are rare in the first population but more common in the second population. These individuals will appear to be inbred relative to expectations based on the combined sample allele frequencies, even if they would appear non-inbred when compared to the sub-population from which they were sampled. Endelman and Jannink (2012) considered replacing the typical **G** matrix estimator with a 'shrinkage' estimator (that will 'push' extreme values of the diagonal elements toward the population mean), showing that this can improve breeding value predictions in some cases, particularly with smaller numbers of markers, lower heritability phenotypes, and for individuals with no phenotype records. In a similar vein, VanRaden (2008) and Yang et al. (2010) proposed adjusting the raw **G** matrix by regressing the pairwise realized relationship estimates onto their pedigree expectations, resulting in another form of shrinkage of the estimators with some useful properties (Yang et al. 2010). This last approach requires knowing the pedigree relationships among individuals, and can result in biased estimates of the average inbreeding coefficient (Endelman and Jannink 2012).

Because of these inconsistencies between classical theory behind describing inbreeding and genetic covariances in terms of probabilities of allelic identity by descent (IBD) on the one hand, and more recent developments in estimating genetic covariances based on identity by state (IBS) of markers, some authors have proposed new ways to conceive and measure genetic relatedness. Powell et al. (2010) suggested that IBD relationships are not really desirable for modern applications such as estimating genetic variances, predicting phenotypes (or breeding values), or association analyses. Rather, these applications rely on predicting identity in state (IBS) relationships at causal variants using IBS information at the observed markers. Thus, the interpretation of realized relationship matrices should be in terms of realized genomic correlations that reflect IBS relationships relative to the study population in hand rather than some conceptual ancestral reference population. Alternative measures of relatedness do away with the concept of non-inbred ancestral reference populations and define relatedness in terms of number of generations required to trace alleles back to a common ancestral allele, along the lines of coalescent theory. For the purposes of breeding, however, we can continue to work in the framework of IBD, since we are often able to identify reference populations, and the number of generations between a current breeding population and the reference population is usually relatively small. For applications involving more distant relationships, the newer concepts of relatedness may prove useful.

Box 11.1 Alternative estimators of the realized relationship matrix

Equation 11.2 is a generally recommended estimator for the realized relationship matrix, since the expectations of its elements equal pedigree estimators under the assumptions that the individuals are members of a known pedigree derived from unrelated ancestors. However, various other methods of estimation of the relationship matrix have been proposed. These differ in how they adjust for population allele frequencies, scale the contributions from different loci, or if they are regressed on known pedigree relationships (Habier et al. 2007; Forni et al. 2008; vanRaden 2008; Powell et al. 2010; Speed and Balding 2015).

The minor allele frequencies in the base population are unknown. In Eq. 11.2, we use the estimated minor allele frequencies based on the observed data. In some cases, the sample size may be insufficient to have reliable estimates of the allele frequencies, or the observed data may come from a sample of individuals that are not representative of the reference population. The use of estimated allele frequencies in the formula introduces some bias. Alternative forms of the relationship matrix may be useful in some circumstances. Forni et al. (2011) compared several variants of the realized relationship matrix, referring to Eq. 11.2 as the **GOF** ('**G** based on observed frequencies') estimator. They also considered alternative matrices based on substituting $p = 0.5$ for all markers (**G05**), as proposed by VanRaden (2008), or $p = $ mean of minor allele frequencies across all loci (**GMF**) in Eq. 11.2. These adjusted relationship matrices can have better predictive value than those based on estimated allele frequencies in some cases (Forni et al. 2011; VanRaden 2008).

A different approach to estimating relationships is by weighting markers by reciprocals of their expected variance as $\mathbf{GD} = \mathbf{ZDZ}'$, where **D** is diagonal with $D_{ii} = (m[2p_i(1 - p_i)])^{-1}$ (Forni et al. 2011; VanRaden 2008). (This is identical to averaging Eq. 11.1 over all loci). Both GOF and GD can be computed with the R package *synbreed*, using the arguments ret = "realized" or ret = "realizedAB" to the *kin()* function. "realizedAB" refers to the formulation

Box 11.1 (continued)

presented in Astle and Balding (2009), equivalent to **GD** in Forni et al. (2011). This scaling is commonly used in human genetics studies and has the effect of giving each marker locus equal contribution to the estimated relationship matrix, since it reduces the contribution of loci with larger minor allele frequencies relative to the unscaled version in Eq. 11.3 (Campos et al. 2013; Powell et al. 2010; Speed and Balding 2015). Another way to understand the effect of this scaling is that rare allele matches are given even greater weight than before in the computation of the relationship matrix, which is appropriate if rare alleles tend to have larger additive effects (Speed and Balding 2015).

The **G** matrix can also be normalized (**GN**) so that its average diagonal coefficients are equal to 1, as appropriate for a non-inbred population (Forni et al. 2011):

$$\mathbf{GN} = \frac{\mathbf{ZZ}'}{\{trace\,[\mathbf{ZZ}']\}/\mathbf{n}} \tag{11.4}$$

The trace of a square matrix is the sum of the diagonal elements: $\sum_{i=1}^{n} a_{ii}$. Higher levels of inbreeding can be accommodated by substituting n with $1 + F$.

Another class of modifications of the relationship matrix involves regressing the observed genomic relationships on the expected additive relationships based on pedigrees. The idea here is that the realized additive relationship estimates should be close to their pedigree expectations on average, so biases and large deviations from the expectations can be removed by regressing observed on expected relationships. VanRaden (2008) suggested first regressing the $(\mathbf{M}-\mathbf{1})(\mathbf{M}-\mathbf{1})'$ matrix on **A**, to estimate intercept (g_0) and regression coefficients (g_1):

$$(\mathbf{M} - \mathbf{1})(\mathbf{M} - \mathbf{1})' = g_0 \mathbf{1}\mathbf{1}' + g_1 \mathbf{A} + \mathbf{E} \tag{11.5}$$

Estimates of g_0 and g_1 are obtained by solving the following equation (VanRaden 2008).

$$\begin{bmatrix} n^2 & \sum_j \sum_k A_{jk} \\ \sum_j \sum_k A_{jk} & \sum_j \sum_k A_{jk}^2 \end{bmatrix} \begin{bmatrix} g_0 \\ g_1 \end{bmatrix} = \begin{bmatrix} \sum_j \sum_k (MM')_{jk} \\ \sum_j \sum_k (MM')_{jk} A_{jk} \end{bmatrix} \tag{11.6}$$

E includes real differences between true and expected fractions of DNA in common (Mendelian sampling variation), plus measurement error. Having estimated the regression coefficients, a regressed version of the **G** matrix can be computed as:

$$\mathrm{Greg} = \frac{\mathbf{MM}' - g_0 \mathbf{1}\mathbf{1}'}{g_1} \tag{11.7}$$

Similar concepts are behind alternative regression or 'shrinkage' estimators of the relationship matrix proposed by Yang et al. (2010) computed by GCTA software (Yang et al. 2011) and by Endelman and Jannink (2012) computed as an option in the R package rrBLUP (Endelman 2011).

Finally, any of these realized genomic relationship matrices may be singular (not invertible) with small numbers of markers, if some pairs of individuals have identical genotypes at all markers, or if the number of markers is smaller than the number of individuals genotyped. If a pedigree-based matrix is available for at the same set of individuals, it can be used to remove singularities in the **G** matrix by using a weighted average of the raw **G** matrix and the **A** matrix:

$$G_w = w\mathbf{G} + (1 - w)A \tag{11.8}$$

The weighting value w is typically chosen between 0.95 and 0.98, its exact value has little impact on results (Aguilar et al. 2010, 2011; VanRaden 2008).

Calculation of G Matrices

For demonstration we will use a simulated marker data set with a known pedigree to estimate a realized relationship matrix. We will demonstrate the use of Eq. 11.2 to estimate the most commonly used form of **G**, but will show how this compares to four other estimators of the relationship matrix outlined in Box 11.1. The pedigree information is contained in the file *sim_pedigree.txt*, indicating the mothers and fathers of 1246 individuals over 16 generations. The marker data are contained in the file *sim_markers.txt*, which has information on 1000 loci (in columns) genotyped on 100 individuals (in rows). The loci are coded as gene content of the minor allele. The first six observations of both files loci are printed below:

Pedigree:

```
> ped
        V1        V2        V3
1 2208166 2208112 2208052
2 2208977 2208055 2208096
3 2208220 2208166 2208977
4 2208384 2208036 2208149
5 2208282 2208220 2208384
6 2208078 2208011 2208111
```

Markers:

```
> Markers
      V1  V2  V3  V4  V5  V6  V7  V8  V9  V10  V11  V12  V13  V14  V15
1  48793   2   0   1   0   1   0   0   0   0    1    1    0    0    1
2  97315   0   1   1   0   1   0   0   0   1    1    0    0    0    1
3 110167   2   1   0   1   1   0   0   1   0    1    0    0    0    2
4 113252   0   0   0   0   2   0   1   1   1    0    1    0    0    0
5 121279   1   0   0   2   1   0   0   0   0    1    1    0    0    0
6 123530   1   0   0   0   1   1   1   0   0    0    0    0    1    0
```

An R script to estimate **G** matrices is given below and in the accompanying file Code 11-2_Gmatrix.R. In order to make the script work we need to have the *GeneticsPed* package installed. The *GeneticsPed* package is available from the Bioconductor project, and is installed using the `biocLite()` function available from the Bioconductor website; this is different than the typical R package installation from the CRAN website:

```
> source("http://bioconductor.org/biocLite.R")
> biocLite("GeneticsPed")
```

In addition, this script uses a custom function called `GenomicRel()` that we have made available in the code supplements. The `GenomicRel()` function calculates markers-based coefficients using five different methods (GOF = Eq. 11.3; GD, G05, GMF, and Greg are outlined in Box 11.2). This function is in the file *GenomicRel.R* and we make this function available in the current working environment in R by using the 'source' command and pointing to the location of this file. The 'source' and 'path' commands in this example must be modified to match the directories of the user's computer that hold the relevant files.

Code example 11.2
Genomic relationship matrices estimated for a simulated DNA marker data set. See "Code 11-2_Gmatrix.R" for more details.

```
###########################################
# Code11-2_Gmatrix.R
###########################################

# remove everything from memory in the working environment
rm(list=ls())

### load packages
library(GeneticsPed)
library(ggplot2)
library(plyr)

### source function GenomicRel() for future calls
source("/~Book/Book1_Examples/ch11_gblup/GenomicRel.R")

### location of the files. Change it for your files
path='/~Book/Book1_examples/data'
setwd(path)     # sets path as the working directory

### read Marker data
Markers=read.table("sim_markers.txt")
### read associated pedigree file
ped=read.table("sim_pedigree.txt")

# 5 different genomic relationship matrices
GOF=GenomicRel(1,Markers,ped)
GD=GenomicRel(2,Markers,ped)
G05=GenomicRel(3,Markers,ped)
GMF=GenomicRel(4,Markers,ped)
Greg=GenomicRel(5,Markers,ped)
```

After running the script, we have five estimators of the **G** matrix, each saved as a data frame with 5050 row and 4 columns. The first two variables in each data frame index the pair of individuals being compared, which correspond to the column ("col") and row indices of the **G** matrix. The last two variables are the realized marker-based ("G") and pedigree-based relationship estimators ("A"). The 5050 rows include the 100 diagonal elements and the $(100 * 99) / 2 = 4950$ off-diagonal elements from half of the symmetric 100×100 **G** matrix. A few elements of three of the matrices are shown below:

```
> head(GOF)
  col row     G     A
1   1   1  1.010  1.017
2   2   1  0.107  0.062
3   2   2  1.067  1.038
4   3   1  0.005  0.033
5   3   2 -0.014  0.017
6   3   3  1.004  1.000
> head(GD)
 col row      G     A
1  1   1   1.016  1.017
2  2   1   0.105  0.062
3  2   2   1.096  1.038
4  3   1   0.021  0.033
5  3   2  -0.020  0.017
6  3   3   1.042  1.000
> head(Greg)
 col row    G      A
1  1   1  1.044  1.017
2  2   1 .0.158  0.062
3  2   2  1.073  1.038
4  3   1  0.049  0.033
5  3   2  0.018  0.017
6  3   3  0.992  1.000
```

The cases where row and column indices are equal represent the relationship of individuals with themselves, for which the values in **G** and **A** are estimates of $1 + F$ (the inbreeding coefficient) for the individuals. Summaries of genomic relationships calculated from the markers can be compared to the **A** matrix derived from pedigree:

Code example 11.2
(continued)

```
# Create data frame called CorrOpt
CorrOpt=data.frame(row=GOF[,1],col=GOF[,2],A=GOF[,4], GOF=GOF[,3],
 GD=GD[,3],  G05=G05[,3], GMF=GMF[,3], Greg=Greg[,3])


# Correlations (first two columns are excluded)
cortable= cor(CorrOpt[,c(-1,-2)])


# Round the correlation values to 3 significant digits
round(cortable,3)


# Scatter plot and histogram of GOF as a function of A
head(CorrOpt)
plot(GOF ~ A, data=CorrOpt, col='blue')
hist(CorrOpt$GOF, col='red')
```

- We first merge five **G** matrices by pulling out the 3rd column in each and naming them GOF, GD, G05, GMF and Greg. We select the 4th column of GOF holding the pedigree estimates and name it A. The result is a data frame named CorrOpt with variables for row, col, GOF, GD, G05, GMF, Greg and A.

- A scatter plot of **GOF** and **A** matrices and a histogram of realized genomic relationships of **GOF** matrix are produced (not presented).

The first six rows of *CorrOpt* data frame appear as:

```
  row col     A     GOF     GD     G05     GMF    Greg
1   1   1  1.017   1.010  1.016   0.860   1.849  1.044
2   2   1  0.062   0.107  0.105   0.186   0.915  0.158
3   2   2  1.038   1.067  1.096   0.876   1.939  1.073
4   3   1  0.033   0.005  0.021   0.140   0.818  0.049
5   3   2  0.017  -0.014 -0.020   0.114   0.813  0.018
6   3   3  1.000   1.004  1.042   0.886   1.894  0.992
```

Correlations between the elements of the different relationship matrices are given in Table 11.1. Except **GMF**, all **G** matrices had the same high correlation ($r = 0.95$) with the **A** matrix.

The following R script calculates summary statistics of the different genetic relationships estimators compared in Table 11.2:

Code example 11.2
(continued):

```
# Summary statistics (mean, min and max) for inbreeding
# Coefficients for each G matrix and the A matrix
Inbreed=subset(CorrOpt,row-col==0)   # Take the diagonals elements
Rel=subset(CorrOpt,row-col!=0)       # Take the off diagonals
maxI=apply(Inbreed,2,max)            # Max inbreeding
maxR=apply(Rel,2,max)                # Max coefficient (off diag.)
minI=apply(Inbreed,2,min)      # Min inbreeding
minR=apply(Rel,2,min)          # Min coefficient
meanI=apply(Inbreed,2,mean)    # Mean inbreeding
meanR=apply(Rel,2,mean)        # Mean coefficient

# Combine all stats in one data frame
summary=t(rbind(meanI,minI,maxI,meanR,minR,maxR))[c(-1,-2),]
round(summary,2)
```

Notice that the range of relatedness coefficients for the pedigree-derived **A** matrix is from zero to 0.58. In contrast, some of the marker-derived realized genetic relationships have negative values. All matrices show inbreeding for some individuals.

Table 11.1 Correlations between genetic relationships derived from the pedigree and SNP markers in five different ways

	A	GOF	GD	G05	GMF	Greg
A	1	0.95	0.95	0.95	0.88	0.95
GOF	0.95	1	1.00	0.99	0.96	0.96
GD	0.95	1.00	1	0.98	0.96	0.96
G05	0.95	0.99	0.98	1	0.94	0.96
GMF	0.88	0.96	0.96	0.94	1	0.85
Greg	0.95	0.96	0.96	0.96	0.85	1

Table 11.2 Summary statistics of genetic relationships. The methods to obtain genomic relationships differed slightly, except G05 method, which assumes that the minor allele frequency in the base population is 0.5

	Diagonal elements			Off-diagonal elements		
	Mean	Min	Max	Mean	Min	Max
Chapter 1 A	1.02	1.00	1.10	0.06	0.00	0.58
Chapter 2 GOF	0.99	0.85	1.19	−0.01	−0.16	0.57
GD	0.99	0.82	1.19	−0.01	−0.15	0.60
G05	0.88	0.78	1.04	0.15	0.02	0.58
GMF	1.83	1.60	2.13	0.78	0.57	1.44
Greg	1.02	0.89	1.18	0.06	−0.14	0.62

Genomic BLUP

We have seen in previous sections how it is possible to obtain a better measure of relationships between individuals with the aid of molecular data. We can incorporate these marker-based relationship estimators into linear mixed models to predict breeding values of individuals with phenotype records and also any other individuals included in the $\mathbf{G}$ matrix even if they have no phenotype information. This is simply extending the ideas of breeding value prediction based on pedigree relationships introduced in Chaps. 4 and 5. We can use the same mixed model equations as shown for a pedigree-based prediction but substituting the $\mathbf{G}$ matrix for the $\mathbf{A}$ matrix to calculate genomic estimated breeding values (GBLUP). Here we summarize theory later we demonstrate an example of genomic prediction using GBLUP with the maritime pine data introduced in the examples in Chap. 9. The mixed model for GBLUP analysis has the usual form:

$$\mathbf{y} = \mathbf{Xb} + \mathbf{Zu} + \mathbf{e} \tag{11.9}$$

where $\mathbf{y}$ denotes the $n \times 1$ vector of observations. In the simplest case, $\mathbf{y}$ represents "raw" data on n unreplicated individuals, as is typical for animal breeding experiments, $\mathbf{b}$ is the $p \times 1$ vector of fixed effects, $\mathbf{X}$ is an $n \times p$ design matrix relating observations to the fixed effects, $\mathbf{Z}$ is an $n \times n$ identity matrix relating the n observations to the n unique individual effects, and $\mathbf{u}$ is an $n \times 1$ vector of the breeding values of each individual, and $\mathbf{e}$ is the $n \times 1$ vector of residual errors with variance $\mathbf{I}\sigma_e^2$.

More complex models can also be used. For example, often in plant breeding experiments the $\mathbf{y}$ vector represents the phenotypes of multiple-plant plots, and the g families are replicated over s environments. In such cases, the $\mathbf{Z}$ matrix will have dimensions of at least $n \times (g + s)$, relating the n observations to the g family breeding values and s environment effects. More complex structures involving family-by-environment interactions also can be fit, using the $\mathbf{G}$ structures described in Chap. 8. Furthermore, the $\mathbf{R}$ structure can be more complex than an IDV structure, allowing for environment-specific error variances (Chap. 8) and spatial correlations (Chap. 7). Using raw data from the individual experimental units along with design factors and covariates in the model has the advantage of efficiently using all of the information in the data. If the number of records is very large and the model very complex, however, model convergence may be slow and difficult to achieve. In such cases, two-step analyses can be used, usually with some loss of information (Möhring and Piepho 2009). These are described in Box 11.2.

The mixed model equations used to solve Eq. 11.9 can be extended to include realized genomic relationships as follows (assuming an IDV $\mathbf{R}$ structure):

$$\begin{bmatrix} \mathbf{X}^T\mathbf{X} & \mathbf{X}^T\mathbf{Z} \\ \mathbf{Z}^T\mathbf{X} & \mathbf{Z}^T\mathbf{Z} + \mathbf{G}^{-1}\lambda \end{bmatrix} \begin{bmatrix} \mathbf{b} \\ \mathbf{u} \end{bmatrix} = \begin{bmatrix} \mathbf{X}^T\mathbf{y} \\ \mathbf{Z}^T\mathbf{y} \end{bmatrix} \tag{11.10}$$

Equation 11.10 is exactly equal to an animal model (e.g., Eq. 4.26) but with the genomic relationship matrix, $\mathbf{G}$, substituted for the additive relationship matrix derived from the pedigree, $\mathbf{A}$. With the mixed model equations in Eq. 11.10 we can obtain predictions of $\mathbf{u}$, which represent the genomic BLUPs (GBLUPs) or genomic estimated breeding values (GEBVs) for all the individuals included in the $\mathbf{G}$ matrix whether or not they have direct phenotypic records in $\mathbf{y}$. The reliabilities of predictions for individuals with direct phenotype records will be higher than those of individuals with no phenotypes.

Box 11.2 Two-step methods for GBLUP and genomic selection

GBLUP can be conducted using data on individuals (or individual experimental units), incorporating extraneous factors or complex genotype-by-environment interactions. Such models, however, may be too computationally demanding or slow for practical use, or they may present convergence difficulties. More complex genomic selection models to be described in Chap. 12 often have even greater computational demands and are even less likely to be easily analyzed with a single analysis model based on individual data if complex interactions with the environment or complex **R** structures are desired. In these cases, two-step methods are useful. The first step is to analyze the phenotype data using an appropriate mixed model to obtain adjusted summary phenotype values (such as adjusted phenotypes or adjusted means) for the individuals or families. In the second step, the GBLUP or genomic selection model can be fit to the summary phenotype values.

A common situation in plant breeding experiments is that a series of breeding trials has been conducted to evaluate replicated families or lines across many environments. The breeding trials may be highly unbalanced, with different sets of families evaluated in different environments. In such a case, the first step of a two-step GBLUP analysis is to fit a mixed model incorporating covariates, environment and genotype-by-environment effects, and the desired **R** structure can be fit to the raw data on experimental units, using the various modeling strategies for phenotype data outlined in previous chapters. Families can be treated as fixed effects to obtain marginal predictions of family values across environments. In the second step of the analysis, the marginal predictions can be used as dependent variables in a mixed model analysis corresponding to Eq. 11.9, in which the n observations correspond to the $n = g$ families. The prediction accuracy of family effects from the first stage may vary among families when the data are unbalanced, and in such cases, a weighted **R** structure accounting for this variation can improve accuracy (Möhring and Piepho 2009). A typical weighted **R** structure for the second stage analysis is $\mathbf{wI}\sigma_e^2$, where $\mathbf{w}$ is an $n \times 1$ vector of the variances of the n family predictions from the first stage analysis.

In animal breeding situations, the "phenotypic" information available can be any of a variety of direct measurements or adjusted values: raw individual phenotypes, individual phenotypes adjusted for systematic effects, de-regressed breeding values obtained from conventional pedigree analyses weighted by corresponding reliabilities, or yield deviations of individuals' progeny weighted by either their reliability or the effective progeny contribution.

While the use of raw phenotypes is straightforward, the use of derived measures based on estimated breeding values presents some concerns. The advantage, however, especially when the genomic selection scheme is mostly applied to the male fraction of the population, is that "pseudo-phenotypes" increase the effective heritability of the trait analyzed, and therefore reduce the number of individuals needed in the training population to achieve a given accuracy. For example, if training will be performed using a trait BV as phenotype, for a population of individuals with reliabilities >0.9, the narrow-sense heritability for that trait in the analysis will effectively be ~0.9.

Garrick et al. (2009) outlined some differences among the types of phenotypic information employed in training populations for genomic selection, and how these should be weighted. Ideally, training would be performed on the true breeding values of unrelated individuals chosen from a population not under selection, however, this is rarely the case in practice. Using estimated breeding values (EBVs) as phenotypes for the training population can be less than optimal, for several reasons. First, EBVs obtained by BLUP are predictions, so they are composed of a true value plus a prediction error. The BLUP procedure has the property that prediction error reduces the variance in the EBVs, which results in underestimation of the value of superior individuals and overestimation of the value of inferior individuals.

A second issue is that BLUP, as we have seen in Chap. 2, is a shrinkage estimator, meaning that predictions are shrunk towards the mean. The amount of shrinkage for each observation will depend on how much information is available for that particular individual. The EBV for different genotypes at a particular marker will be shrunk, relative to what would be obtained using phenotypic data, but by different amounts according to the reliability of the EBV. Our interest is to obtain a contrast between the genotypes for the phenotype, but the contrast will be confounded by variation in reliability of EBV among individuals in the training population.

In most cases the best solution to account for the problem of having shrunk measures is to "re-inflate" the estimated breeding values through a de-regression process, by weighting each observation by its reliability so that $\hat{u}^*_i = \hat{u}_i / r^2_i$. De-regressed information should then be weighted in the second step of the analysis according to

Box 11.2 (continued)

$$w_i = \frac{1 - h^2}{\left[c + (1 - r_i^2)/r_i^2\right]h^2}$$

(11:11)

where r^2 is the reliability of EBV, h^2 is the heritability of the trait, and c is the proportion of variance not accounted for by genomic marker information. We don't know c in advance of doing the analysis, so usually exploratory analyses are used to establish a reasonable value of c for a particular set of data. When the mean of individuals' repeated phenotypic measures are employed for half-sib and full-sib families, a different set of weights can be applied, see Garrick et al. (2009) for details.

Below is a simple R function to obtain de-regressed breeding values along with weights to use in the training set starting from BLUP EBV and their respective reliabilities.

```
dereg<-function(gs,ps,gm,pm,gi,pi,lambda,c){

##################################################################
# calculates deregressed BVs and weights starting from EBV and
reliabilities
# gs, gm, gi are the EBVs of sire dam and individual respectively
# ps, pm and pi are the reliabilities of sire, dam and individual
respectively
#lambda is the Ve/Va ratio (obtained from the BLUP analysis)
# c is the proportion of variance (un)explained by the markers
# returns a list of de-regressed BV accuracies of de-regressed and
weights
##################################################################
        rpa<-(ps+pm)/4
        gpa<-(gs+gm)/2
        alpha<-1/(0.5-rpa)
        delta<-(0.5-rpa)/(1-pi)
        ZZpa<-lambda*(0.5*alpha-4)+0.5*lambda*sqrt(alpha^2+16/delta)
        ZZi<-delta*ZZpa+2*lambda*(2*delta-1)
        LHS<-rbind(cbind(ZZpa+4*lambda,-2*lambda),cbind(-
2*lambda,ZZi+2*lambda))
        RHS<-rbind(gpa,gi)
        so<-LHS%*%RHS
        de<-so[2]/ZZi
        rw<-1-lambda/(ZZi+lambda)
        w<-(1/(c+(1-rw)/rw))*lambda
        return(c(de,rw,w))
}
```

In addition to de-regression and weighting, Garrick et al. (2009) recommend removing the parental averages from the estimated breeding values (EBV) for two reasons. First, BLUP methods will yield an EBV for all individuals, regardless of whether phenotypic information exists for the individual itself or only for progeny or other relatives. Individuals that lack phenotypic observations for the individual itself do not contribute any genomic information that would not be contributed by their parents' genotypes and EBV. Second, if any of the parents are segregating for a QTL near a marker, approximately half of the progeny will inherit the favorable allele for the marker and half the unfavorable. Nonetheless the EBVs of both groups of progeny will be shrunk towards the parental mean by the same amount, introducing more error into the model. Different procedures can be employed to remove parental average from EBV and detailed information can be found in Garrick et al. (2009).

GBLUP with the Synbreed Package

For demonstration, we will use the maritime pine data set previously introduced in Chap. 9. In Chap. 9, the raw base pair calls on 654 individual trees at 2600 SNPs were read into R and used to create a *gpData* object using the *synbreed* package. Markers with high rates of missing data or very low minor allele frequencies were removed. The SNP genotype calls were converted to numeric minor allele content codes. The results were saved in an R data set (*maritime pine codeGeno data.rda*) that and copied to the Chap. 11 subfolder of the example data and scripts.

The *synbreed* package has a function called *kin* to estimate genetic relationships among individuals included in a *gpData* object (Wimmer et al. 2012). The '*ret =*' argument in the *kin* function determines which type of relationship is calculated. For example, '*ret = add*' will return an additive numerator coefficient relationship matrix based on the pedigree. Pedigree-based dominance relationship coefficients and other forms of additive relationship coefficients based on pedigree are also available as options. Two different marker-based realized genetic relationship matrices can also be computed. If the '*ret = realized*' argument is used, the **G** matrix is estimated using Eq. 11.2, where p is the estimated observed allele frequency at each locus (VanRaden 2008). If the '*ret = realizedAB*' argument is used, the **G** matrix is computed as '**GD**' described in Box 11.2, based on the formula in Astle and Balding (2009).

The output of the *kin* function is an object of class `relationshipMatrix` holding all of the elements of a **G** matrix for the individuals included in the *gpData* object (Wimmer et al. 2012). The `write.relationshipMatrix` of *synbreed* function converts the matrix from table format to a long format for ASReml. Code example 11.3 demonstrates the estimation of realized genetic relationships in the maritime pine data set using *synbreed*.

Code example 11.3
Calculation of genetic relationships and GBLUP using *synbreed* package (see **Code 11-3_GBLUP.R** for more details)

```
##################################################
# Statistical models of genomic selection in maritime pine
# ---> load(file='maritime pine codeGeno data.rda') R data
##################################################

library(synbreed)
library(asreml)
library(Matrix)

# Change folder according your computer!
path <- "~/.../Book/Book1_Examples/data"
setwd(path)

#Directory to save figures
gpath <- "~/.../Book /Book/Images"

### Load the R data set
load("maritime_pine_codeGeno_data.rda")
# See the list of data sets, functions in working environment
ls()
```

When we load the R data, the "gp.num" object becomes available in the environment of the R session. The `names()` function shows the names of the components of the "gp.num" object. The `head()` function is used in the script to show the first six lines of the 'geno' and 'pedigree' components. The 'geno' component is a numeric matrix with row names corresponding to individual tree IDs and column names corresponding to locus names. The elements of the matrix are the gene content values (0, 1, 2) for each combination of individual and locus. The pedigree component is a data frame with columns indicating tree ID, parent1, parent2 and generation for each individual.

Code example 11.3
(continued)

```
  ls()
"gp.num" "gpath" "path"
names(gp.num)
[1] "covar" "pheno" "geno "map" "pedigree" "phenoCovars" "info"
head(gp.num$geno[,1:3])
    F51TW9001A34NZ.1129  F51TW9001A34NZ.1367  F51TW9001A34NZ.175
0001                  1                    1                    1
0003                  1                    1                    0
0003-3                0                    0                    0
0004                  1                    1                    0
0005                  1                    1                    0
0006                  2                    2                    0
> head(gp.num$pedigree)
     ID   Par1 Par2 gener
1    0001    0    0     1
2    0003    0    0     1
172  0004    0    0     1
3    0005    0    0     1
4    0006    0    0     1
5    0008    0    0     1
```

Next we use provide the "gp.num" object to the *kin* function to estimate the pedigree-based additive relationship matrix (**A**) and the realized genomic relationship matrix (**G**):

Code example 11.3
(continued)

```
## ~~~~~~~~~ Relatedness Coefficients ~~~~~~~~~~~ ##
### ... Pedigree based ... ###
   Additive <- kin(gp.num, ret="add")    # est by 2x of kinships
   summary(Additive)

### ... Realized ... ###
   Realized <- kin(gp.num, ret="realized")
   summary(Realized)

# Plot relationship matrix with equal color keys
   plot(Realized, main='G matrix')
```

Summaries of the **A** and **G** matrices can be compared as follows:

```
> summary(Additive)
dimension                    654 x 654
rank                         654
range of off-diagonal values 0 -- 0.5
mean off-diagonal values     0.005432466
range of diagonal values     1 -- 1
mean diagonal values         1
number of unique values      3

> summary(Realized)
dimension                    654 x 654
rank                         653
range of off-diagonal values -0.146215 -- 0.6313384
mean off-diagonal values     -0.001521728
range of diagonal values     0.868 -- 1.16
mean diagonal values         0.9936881
number of unique values      208603
```

- The range of pedigree-based genetic relationship coefficients across individuals (off-diagonal components) is 0 (no pedigree relationship) to 0.5 (full-sibs) with an average of 0.0054, indicating that most pairs are unrelated by pedigree. All the diagonal elements are 1 because the pedigree indicates no inbreeding in the population.
- Off-diagonal elements of the **G** matrix range from −0.146 to 0.631, with a mean value very close to zero. As in the previous example in this chapter, we see that negative coefficients for realized genomic relationships are possible and indicate pairs of individuals that are less related than expected by random chance. The mean of the diagonal elements of **G** is very close to 1, in agreement with the pedigree estimates, but some individuals show a bit more heterozygosity than expected by random mating (diagonal elements as low as 0.87) and some have a bit more homozygosity than expected (diagonal elements as high as 1.16).
- Notice that the rank of the Additive matrix is equal to the number of individuals, 654, indicating it is a full-rank matrix that can be inverted. In this example, the rank of the Realized matrix is 653, less than the number of rows, indicating that the matrix is singular and will not be invertible.
- Another complication is that the Additive and Realized matrices are not sorted in the same order, this will cause problems if we try to compare them and may cause problems fitting them to the phenotype data.

To deal with these problems, the following code checks if the order of rows and columns is identical in the two matrices, then sorts them to match the order of the phenotype data. Next, we deal with the singularity by using the `nearPD()` function from the *Matrix* package, which returns a positive definite matrix (which will be invertible) very similar to the original matrix. The functions in the Matrix package return objects that are a special class, so we have to do a little work to extract the components and create a regular matrix class object again. Then, just to be sure we have a matrix that will be invertible by various functions (which have different tolerances for how close to singular a matrix can be), we add a very small constant value to the diagonal elements. Finally, we coerce the resulting matrix to the subclass "`relationshipMatrix`" that *synbreed* uses, and we check that its smallest eigenvalue is positive and not too small.

**Code example 11.3
(continued)**

```
### Notice that the order of rows/columns are not same
# for realized and pedigree based matrices
   identical(rownames(Realized), rownames(Additive))
   identical(sort(rownames(Realized)), sort(rownames(Additive)))

   #sort both matrices to match order of pheno object
   #this is required later in the cross-validation step
   Additive = Additive[rownames(gp.num$pheno), rownames(gp.num$pheno)]
   Realized = Realized[rownames(gp.num$pheno), rownames(gp.num$pheno)]

   ##################################################
   #~~~~~~~~~ Remove singularity from G matrix ~~~~~~~
   # use nearPD function from Matrix package to get a positive
definite matrix
   # very close to G
   RealizedPD = nearPD(Realized, keepDiag = T)
   G = matrix(RealizedPD[[1]]@x, nrow = RealizedPD[[1]]@Dim[1])
   G = G + diag(0.01, nrow(G))
   attr(G, "dimnames") = RealizedPD[[1]]@Dimnames
   class(G) = "relationshipMatrix"
   str(G)
   summary(G)
   summary(eigen(G)$values)
```

The output from the last three commands is:

```
> str(G)
relationshipMatrix [1:654, 1:654] 1.02197 -0.0617 -0.00364 0.08207 0.02484 ...
- attr(*, "dimnames")=List of 2
..$ : chr [1:654] "0001" "0003" "0003-3" "0004" ...
..$ : chr [1:654] "0001" "0003" "0003-3" "0004" ...
> summary(G)
dimension                      654 x 654
rank                           654
range of off-diagonal values   -0.146215 - 0.6313384
mean off-diagonal values       -0.001521727
range of diagonal values       0.878 - 1.17
mean diagonal values           1.003688
number of unique values        213527
> summary(eigen(G)$values)
   Min.  1st Qu.  Median  Mean   3rd Qu.  Max.
 0.0100   0.2046  0.4979  1.0040  1.2120   10.7800
```

Notice in the summary of **G** that the rank of updated matrix is equal to the row and column dimensions, and that the smallest eigenvalue is 0.01, so it should have no problem with inversion. The results of the summary also indicate that the values in the matrix have changed only slightly, which is what we want.

Fig. 11.3 Heat map of additive relationship matrix based on pedigree (**a**) and realized genomic relationship matrix (**b**) of 654 maritime pine plants based on marker information

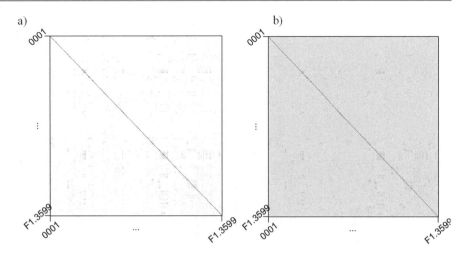

We can generate heat map representations of the two relationship matrixes very easily with the plot() function.

**Code example 11.3
(continued)**

```
plot(Additive,  main='A matrix')
plot(G, main='G matrix')
```

The pedigree-based relationship matrix is sparse with many zero elements (Fig. 11.3a), whereas the matrix of realized genetic relationships based on markers is dense (Fig. 11.3b).

We can visualize the distribution of **G** matrix coefficients according to their corresponding **A** coefficients (11.4):

**Code example 11.3
(continued)**

```
#compare G coefficients to A coefficients
boxplot(as.vector(G)~as.vector(Additive),xlab ="A", ylab="G")
```

Figure 11.4 indicates generally good agreement between the **G** and **A** coefficients, but the variation in **G** coefficients around their expected values in **A** is also obvious. In particular, there are numerous outliers with realized relationship coefficients much higher than zero for individuals with no pedigree relationship ($A = 0$). These could be cases where the pedigree is wrong or incomplete (e.g., male parent is unknown in the pedigree but is in fact one of the individuals genotyped). The converse is also true, there are cases where the pedigree indicates a half-sib ($A = 0.25$) or a full-sib or parent-offspring relationship ($A = 0.5$) but the realized relationship coefficient is near zero, suggesting that the pedigree is wrong. Users should check these outliers carefully and verify that the DNA sampling and genotyping analysis is correct or correct the pedigree information to match the observed coefficients better.

To compare specific elements of the two matrices and identify which pairs are outliers in Fig. 11.4, we can create data frames holding the elements in table form using the write.relationshipMatrix() function of *synbreed*. This function returns a data frame with observations for all the diagonal elements and all non-zero elements of the lower triangle of the matrix, a variable indicating the row index number, a variable with the column index number, and a variable for the value of the coefficient. Most of the values of the Additive matrix are zero, but none of the Realized elements are exactly zero, so the data frames have very different sizes. We can merge them together, set all of the missing row/column combinations from the Additive matrix to zero, and find the most extreme differences between the **A** and **G** coefficients:

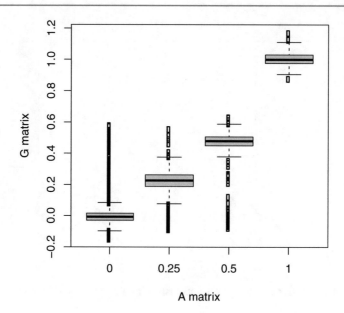

Fig. 11.4 Boxplots of the distribution of realized relationship matrix coefficients (**G**) according to their corresponding pedigree-based relationship coefficients (**A**)

Code example 11.3
(continued)

```
by(as.vector(G), as.factor(as.vector(Additive)), FUN = summary)

#convert from 'dense' matrix representation to 'table' format
   Atab = write.relationshipMatrix(Additive, sorting = "ASReml", type
= "none")
   colnames(Atab)[3] = "A"
   head(Atab)

 Gtab = write.relationshipMatrix(G, sorting = "ASReml", type = "none")
   colnames(Gtab)[3] = "G"
   head(Gtab)

   #combine A and G coefficients and compare them
   AG = merge(Atab, Gtab, all = T)
   AG[is.na(AG)] = 0

   #show the most extreme differences between A and G
   AG$diff = AG$A - AG$G
   AG = AG[order(AG$diff),]
   round(head(AG),3)

   #show the pedigrees of some pairs with extreme differences
gp.num$pedigree[gp.num$pedigree$ID %in% rownames(Additive)[c(564,
   597)],]
```

The observations with largest negative differences between **A** and **G** are shown in the output:

```
> head(AG)
         Row Column  A     G     diff
178470   597    564  0  0.579  -0.579
211404   650    479  0  0.573  -0.573
211540   650    615  0  0.573  -0.573
79004    398      1  0  0.548  -0.548
92754    431     89  0  0.545  -0.545
111310   472    154  0  0.543  -0.543
> gp.num$pedigree[gp.num$pedigree$ID %in% rownames(Additive)[c(564, 597)],]
         ID    Par1  Par2  gener
564  F1.2486  0152  1311      1
597  F1.2778  3110  3603      1
```

The last part of the output shows the pedigree information for the two individuals involved in the most extreme negative difference between **A** and **G**. The pedigree indicates no relationship, but the **G** matrix suggests they are full-sibs since they are both in the progeny generation. Their pedigree records should be checked for errors. The pairs where **A** coefficients are much bigger than their **G** coefficients are sorted to the other end of the data frame:

Code example 11.3
(continued)

```
round(tail(AG),3)
gp.num$pedigree[gp.num$pedigree$ID %in%
  rownames(Additive)[c(505, 431)],]
```

```
> tail(AG)
         Row  Column   A      G      diff
35496    266     251  0.5  -0.056  0.556
127691   505     431  0.5  -0.057  0.557
92941    431     276  0.5  -0.057  0.557
194533   624     157  0.5  -0.064  0.564
191359   619      88  0.5  -0.065  0.565
74849    387     158  0.5  -0.077  0.577
> gp.num$pedigree[gp.num$pedigree$ID %in% rownames(Additive)[c(505, 431)],]
         ID  Par1  Par2  gener
431  F1.0739  4301  3110      1
505  F1.1580  4301  3110      1
```

The individuals with the biggest difference in this group are full-sibs according to the pedigree, but their realized relationship values are below zero, so either the pedigree is wrong or there was a mixup in connecting marker data to one of the individuals. Both possibilities should be checked and errors corrected.

Now we are ready to use the realized genetic relationship created above to predict breeding values using both pedigree information ('ABLUP') and the realized relationship matrix (GBLUP). In this initial analysis we include all individuals; in this example, we have phenotype and genotype data on all individuals. This demonstrates the use of the realized relationship matrix to potentially improve predictions even where we have direct phenotype observations on all of the individuals. This is analogous to using pedigree information to estimate breeding values, which are the best predictions given the direct phenotype records as well as the information shared by relatives. Later, we will demonstrate the use of ABLUP and

GBLUP to predict breeding values for some individuals that have genotype but no phenotype records, a situation where genomic breeding value predictions may have the greatest efficiency in practice.

Code example 11.3
(continued)

```
### A-BLUP
Ablup <- gpMod(gp.num, model="BLUP", kin=Additive, trait=1)
summary(Ablup)

### G-BLUP (Direct Genetic Values)
Gblup <- gpMod(gp.num, model="BLUP", kin=G, trait=1)
summary(Gblup)

### Correlations between phenotype and predictions
cor(Ablup$y,Ablup$g)
cor(Gblup$y,Gblup$g)
cor(Gblup$y,Gblup$g,method="spearman")

# Plot the phenotype and DGV
plot(syn$g ~ syn$y, col='black', ylab='GEBV', xlab='Phenotype'
```

- The function gpMod fits genomic prediction models based on phenotypic and genotypic data in the gp.num object. Both ABLUP and GBLUP models use the option model = 'BLUP'; they differ for which relationship matrix ('Additive' or "Realized") is provided to the BLUP model.
- This is a simple mixed model. There is only one fixed effect, which is the intercept. Additional fixed effects can be included in the model using the 'fixed=' option. We can also fit more complex GBLUP models using ASReml.

The summaries of the two predictions models are:

```
> summary(Ablup)
Object of class 'gpMod'
Model used: BLUP
Nr. observations 654
Genetic performances:
  Min.    1st Qu.  Median   Mean     3rd Qu. Max
-0.85510 -0.23180 -0.02526 -0.01615 0.17680 0.84680
-
Model fit
Likelihood kernel: K = (Intercept)
Maximized log likelihood with kernel K is -125.467

Linear Coefficients:
            Estimate Std. Error
(Intercept)   10.004      0.04

Variance Coefficients:
          Estimate Std. Error
    kinTS    0.223      0.053
       In    0.348      0.045
```

```
> summary(Gblup)
Object of class 'gpMod'
Model used: BLUP
Nr. observations 654
Genetic performances:
   Min.    1st Qu.   Median Mean 3rd Qu. Max
-1.088000 -0.254300 -0.006471 0.000000 0.256000 0.973100
  -

Model fit
Likelihood kernel: K = (Intercept)
Maximized log likelihood with kernel K is -112.407

Linear Coefficients:
            Estimate Std. Error
(Intercept) 9.988             0.022

Variance Coefficients:
          Estimate Std. Error
   kinTS 0.262         0.046
      In 0.319         0.032
```

- `kinTS` identifies the additive variance components estimated with each model based on all the individuals in the 'test set'. In this example, the test set comprises all the individuals for which we have data. Later we will show how to use different samples in the estimation and test sets for cross-validation.
- `In` is the identical and independent residual variance for each model.
- The ABLUP model explained 0.223/(0.223 + 0.348) = 39% of the total variance.
- The GBLUP model explained 0.262/(0.262 + 0.319) = 45% of the total variance.

The gpMod objects returned from the `gpMod()` function are lists, from which we can extract the original phenotypic values as component 'y' and the model-based predicted values as 'g'. We can estimate the correlations between predicted and observed values for each of the models and visualize the regressions (GBLUP values in gray and ABLUP values in blue) with the following code:

Code example 11.3
(continued)

```
par(mar=c(4,4,1,1)+0.1)#sets margins of plotting area
plot( Gblup$g, Gblup$y, pch = 20, cex=1.6, col="gray",
      xlab="Observed", ylab="Predicted")
points(Ablup$g, Ablup$y, col='blue', cex=.9, pch=1)
legend("topleft",  paste("r_G = ", round(cor(Gblup$g,Gblup$y),2),
 sep=""), col="red3", bty="n")
legend("bottomright",  paste("r_A = ",round(cor(Ablup$g,Ablup$y),2),
 sep=""), col="red", bty="n")
```

The correlations between the observed values and the breeding value predictions are high in both cases (Fig. 11.5). The correlation for the ABLUP model is a bit higher than the correlation for the GBLUP model, which may be surprising given the higher proportion of variance explained by the GBLUP model in the mixed model analysis. The reason is that the correlation and regression analyses assume no covariances among observations, unlike the mixed models. Thus, in this case, the higher correlation between observed and predicted values in the ABLUP model really just indicates that the pedigree-

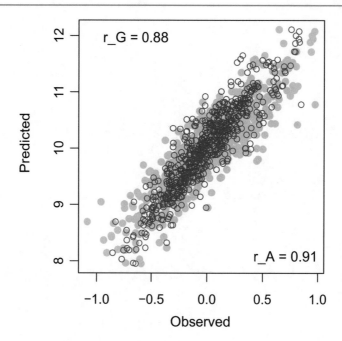

Fig. 11.5 Pedigree-base estimated breeding values (*in blue*) and genomic estimated breeding values (*in gray*) plotted against the original observations for the analysis in which all individuals are included

based EBVs are more similar to the original observations than the GBLUP estimates are. This is why this type of correlation analysis is not an adequate evaluation of the predictive ability of the models.

Cross-Validation

In the last section, we saw that we cannot evaluate the prediction accuracy of EBVs or GEBVs by comparing them against the phenotypic measurements from which the models were built, since those kinds of correlations simply tell us which model results in predictions most similar to the original phenotypic values. For example, we can get a perfect correlation between the original observations and the predicted breeding values in that case by simply using the original observations as the predictions (i.e., not fitting any model at all!). A more appropriate way to evaluate prediction accuracy of different models is to use cross-validation. Cross-validation operates by sampling phenotypic values from only part of the data set (called the 'training set' or 'estimation set'), fitting a prediction model with that subset of phenotypes, then predicting values for the validation set (or 'test set'). The estimated correlation between the observed and predicted values for the validation set (the individuals whose phenotypes were not included in the training set) is an unbiased estimate of the model's prediction accuracy (Fig. 11.6). Cross-validation is a widely used technique for evaluating prediction accuracy of many different kinds of models (Hastie et al. 2009). For example, in Chap. 12 we will use cross-validation to compare the accuracy of several different genomic selection models. *k*-fold cross-validation means the data are split into *k* folds of approximately equal size, and each fold is used once and only once as a validation set across *k* analyses.

The relative size of the prediction and validation datasets depends on the overall number of individuals for which data are available. The larger the training population used to estimate model components, the higher the predictive accuracy of the model is likely to be. The larger the validation population size, the more meaningful the correlation of predicted genetic value to known genetic values will be.

It is important to recognize that GBLUP and other genomic selection models (to be described in Chap. 12) absorb information from pedigree and more distant relationships, familial linkage, and QTL inheritance (Habier et al. 2013). Therefore, in some cases, splitting individuals between training and validation sets randomly may represent a situation where the training and validation sets are more closely related than the actual training and selection individuals are in some breeding programs. For example, a breeder may want to use models trained on a subset of the available pedigrees to select

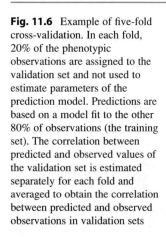

Fig. 11.6 Example of five-fold cross-validation. In each fold, 20% of the phenotypic observations are assigned to the validation set and not used to estimate parameters of the prediction model. Predictions are based on a model fit to the other 80% of observations (the training set). The correlation between predicted and observed values of the validation set is estimated separately for each fold and averaged to obtain the correlation between predicted and observed observations in validation sets

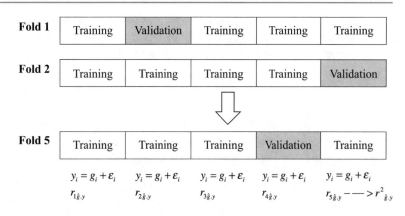

among individuals that are not well-represented in the training set. In this case, a GBLUP model may have poor accuracy because it is trained on the 'wrong' material. But a cross-validation study within the training set may provide overly optimistic estimates of the prediction accuracy because of the closer relationships within the training set. In such cases, random allocation of individuals to training and validation sets is not ideal. To better reflect a situation where the relationships are more distant between training and selection sets than within sets, allocation of individuals to training and validation sets for cross-validation should be restricted to minimize the relationships between the training and validation sets (Saatchi et al. 2011; Tiezzi et al. 2015). This issue may be particularly important for the Bayesian models described in Chap. 12 that try to prioritize significant regions of the genome.

Synbreed has a function `crossVal()` that will perform cross-validation analysis for a single type of prediction model. The function is quite flexible in allowing the user to define the assignments to training and validation sets or it will randomly assign observations to folds. BLUP models can be fit as one of the prediction model options, and we can control whether the models use ABLUP or GBLUP by specifying which relationship matrix is used. The function can take the estimated variance components from an initial analysis of the complete data set and use it for all training sets, or it can call ASReml to re-fit the model to the training data each time and re-estimate the variance components for each split of the data. The first approach is not correct, since the variance components estimates are influenced by the validation set, so our estimates of prediction accuracy can be biased upward. Re-estimating the variance components for each training set is the correct approach, although the user must have their operating system set up to understand 'ASReml' as a batch command. We show here how to use `crossVal()` based on the variance components estimated from the full data set although this is not a recommended approach. Later, we show how to perform the more appropriate approach by creating cross-validation data sets and using ASReml separately to re-fit the prediction models for each training set and evaluating the predicted and observed values of the validation sets independently. The code to perform five replications of five-fold cross-validation for each of the ABLUP and GBLUP models follows:

Code example 11.3
(continued)

```
###~~~~ (1) 5 replications of 5-fold Cross Validation
# CV A-BLUP
cv.Ablup <- crossVal(gp.num, trait=1, cov.matrix=list(Additive),
            k=5, Rep=5, Seed=1, sampling="random",
          varComp=Ablup$fit$sigma, VC.est="commit", verbose=T)
summary(cv.Ablup)

# CV GBLUP
cv.Gblup <- crossVal(gp.num, trait=1, cov.matrix=list(G),
            k=5, Rep=5, Seed=1, sampling="random",
          varComp=Gblup$fit$sigma, VC.est="commit", verbose=T)
summary(cv.Gblup)
```

This analysis takes some time because in each of five replications, the data are split into five folds, such that 25 analyses are conducted for each method. The summaries report the average correlations between predicted and observed values of each validation set, as well as the variation in this correlation among the 25 analyses for each model.

```
> summary(cv.Ablup)
Object of class 'cvData'
 5 -fold cross validation with 5 replication(s)
      Sampling:                  random
      Variance components:       committed
      Number of random effects:  654
      Number of individuals:     654 - 654
      Size of the TS:            130 - 131
Results:
                      Min            Mean +- pooled SE        Max
  Predictive ability:  0.0913         0.2518 +- 0.0066       0.3381
  Rank correlation:    0.0918         0.2338 +- 0.005098     0.3611
  Mean squared error:  0.425          0.526 +- 0.001403      0.644
  Bias (reg. slope)    0.3851         1.0347 +- 0.02564      1.5121
  10% best predicted:  10.44          10.47 +- 0.01507       10.52
> summary(cv.Gblup)
Object of class 'cvData'
 5 -fold cross validation with 5 replication(s)
      Sampling:                  random
      Variance components:       committed
      Number of random effects:  654
      Number of individuals:     654 - 654
      Size of the TS:            130 - 131
Results:
                      Min            Mean +- pooled SE        Max
  Predictive ability:  0.1905         0.3341 +- 0.006738     0.4624
  Rank correlation:    0.2094         0.3286 +- 0.005926     0.4779
  Mean squared error:  0.380          0.499 +- 0.002593      0.583
  Bias (reg. slope)    0.5265         0.9471 +- 0.02696      1.4422
  10% best predicted:  10.32          10.44 +- 0.04297       10.57
```

In our example there are 654 data points. With $k = 5$ folds we assign either 130 or 131 individuals to each fold. The statistics reported to compare models are *predictive ability*, *rank correlations*, *mean squared error* (MSE), *bias*, and *mean of the phenotypic observations for the highest 10%* of individuals within analysis (Wimmer et al. 2012). The **predictive ability** for each replication in each fold is the correlation $r(y_{VS}, GEBV)$ between phenotype and genomic estimated breeding value (GEBV) in the validation set. The **bias** is the coefficient of regression $(\widehat{\beta})$ of the observed phenotypes of the validation set on their GEBVs. A regression coefficient of 1 indicates no bias, whereas $\widehat{\beta} < 1$ indicates 'inflation' meaning more variance among the predicted (X-axis) than observed (Y-axis) values and $\widehat{\beta} > 1$ indicates 'deflation' meaning less variance among the predicted than observed values. The **mean squared error** (MSE) of each fold is calculated as the average squared deviation of GEBV from observed phenotype $\left(MSE = n^{-1} \sum_{i=1}^{n} (y_i - GEBV_i)^2 \right)$ over all individuals (Legarra et al. 2008). The **best 10%** is the mean of observed values for the individuals with the highest ranking 10% of GEBVs for each replicate.

The average prediction ability (the correlation between predicted and observed values within validation sets) averaged over 25 fold-replicate combinations was 0.25 for the ABLUP model and 0.33 for the GBLUP model. Notice that these values are much lower than the correlations estimated from the single analysis of the full data set earlier (which were around 90%), because here the predictions are made without direct phenotypic observations on the individuals. Also notice that in these independent validations, the GBLUP model has considerably higher prediction accuracy than the ABLUP model.

We can also execute cross-validation analysis with ASReml by assigning individuals randomly to folds (or, if desired, assigning them to training and validation sets based on sub-population origins or some other criterion to be used to split the data). We can then create training data sets by copying the original phenotypic data into five new variables (for five-fold cross-validation) and setting the new trait values to missing for the individuals in the validation set for each fold. The data sets can be written to the hard drive along with the relationship matrix and analyzed in ASReml standalone. We will show an example of that approach in the next section. Here we show how to submit the training sets to ASReml-R, obtain predicted values for the validation set individuals from each fold, and evaluate prediction ability from the validation sets using base R commands. To use the realized relationship matrix to model the variance-covariance structure of the individuals' genetic effects in ASReml standalone or ASReml-R, we need to write the matrix in tabular form.

Code example 11.3
(continued)

```
###~~~~ Create a 5-Fold Cross-Validation set
# Create a copy of the phenotype data
#assign fold to each individual
set.seed(110716)
new.ph = as.data.frame(gp.num$pheno)
new.ph$fold = sample(1:5,size = 654, replace = T)
new.y = as.matrix(new.ph[,1, drop = F])%*%t(rep(1, 5))
colnames(new.y) = paste0("Y", 1:5)
new.ph = cbind(new.ph, new.y)

#set one of the fold observations to NA for each line corresponding to
 its fold number
missingInFold = function(r){
  fold = r["fold"]
  index = fold + 2
  r[index] = NA
  return(r)
}
new.ph2 = t(apply(new.ph, 1, missingInFold))
new.ph.df = data.frame(new.ph2)
new.ph.df$ID = factor(rownames(new.ph.df))

########### GBLUP with ASReml ##########################
# Create inverse of G matrix in sparse format
G.giv <- write.relationshipMatrix(G,
                 file =NULL,
                 sorting=c("ASReml"),
                 type=c("ginv"), digits=10)

head(attr(G.giv, "rowNames"))
names(G.giv) <- c("row", "column", "coefficient")
head(G.giv)
```

The function `write.relationshipMatrix()` generates a data frame with three variables, corresponding to the row and column indices and the coefficient for each of the of the diagonal and lower triangular elements of the **G** matrix. In this example, we requested `type = c("ginv")`, which will return the elements of the inverse of the **G** matrix in the data frame. This is required for ASReml-R; ASreml standalone will accept either the elements of **G** itself or the elements of its inverse.

```
> round(head(G.giv),3)
     row  column  coefficient
1     1      1         4.356
2     2      1         0.046
656   2      2        19.399
3     3      1         0.308
657   3      2        -0.683
1311  3      3         2.717
```

Next, the code executes ASReml-R in a loop for each of the five new dependent variables, estimating the genetic and residual variance components separately for each fold and fitting the full relationship matrix for all individuals each time. In this way, we will get solutions for the effects for all of the individuals in every analysis, even those in the validation set whose phenotypes were set to missing in the current fold. The resulting asreml objects and predictions of random genotype effects are saved as components in two lists (result.list and pred.list). Next, we extract the variance components for each fold from the result.list object and the predictions from only the validation set for each fold from the pred.list object. We compute the reliabilities of the predictions from the validation sets and inspect the top of the resulting data frame.

Code example 11.3
(continued)

```
# execute ASReml for each fold data set and save results in a list
result.list = list()
pred.list = list()
for (trait in names(new.ph.df)[3:7])
{
  asr <- asreml(fixed = get(trait) ~ 1,
                random = ~giv(ID, var=T),
                ginverse=list(ID=G.giv),
                rcov= ~units,
                data=new.ph.df,
                na.method.X='include',
                workspace=80e6 )
  result.list[[trait]] = asr
  pred.list[[trait]] = summary(asr, all = T)$coef.rand
  rm(list = "asr")
}


########### Summarize cross-validations #########################
#extract the variance components estimates from each fold
get.varcomps = function(c){
  #c is a component of result.list
  vcs = summary(c)$varcomp$component
  names(vcs) = c("Va", "Vresid")
  return(vcs)
}
```

```
varcomps = sapply(result.list, FUN = get.varcomps)
varcomps = as.data.frame(t(varcomps))
varcomps$fold = as.numeric(gsub("Y", "", rownames(varcomps)))

#get the mean for each estimation (training) set for each fold to use
 as mean of predictions
mu = colMeans(new.ph.df[,c("Y1", "Y2", "Y3", "Y4", "Y5")],na.rm = T)
mu = as.data.frame(mu)
mu$fold = as.numeric(gsub("Y", "", rownames(mu)))

vc.mu = merge(varcomps, mu)

#for each fold, get the predictions only for the held out observations
get.pred = function(r){
  ID = r["ID"]
  ID2 = paste0("giv(ID, var = T)_", ID)
  trait = paste0("Y", r["fold"])
  return(pred.list[[trait]][ID2,])
  }

preds = t(apply(new.ph.df, 1, get.pred))

#merge predicted values from validation set only with observed values
preds.obs = merge(new.ph.df[,c("Sim.Pheno.1", "fold")],
 as.data.frame(preds), by  = 0)

# merge pred.obs with varcomps by fold
# to get the variance components for each fold
preds.obs = merge(preds.obs, vc.mu, by = "fold")
preds.obs$pred = preds.obs$mu + preds.obs$solution

# Compute reliabilities using Mrode's formula
names(preds.obs)[names(preds.obs) == "std error"] = "se"
names(preds.obs)[names(preds.obs) == "Row.names"] = "ID"
preds.obs$reliability <- with(preds.obs, 1 - se^2/Va )  # reliability
preds.obs = preds.obs[order(preds.obs$ID),]
head(preds.obs)
```

The merged validation set predictions and original observations are:

```
 > head(preds.obs)
      fold   ID    Sim.Pheno.1  solution        se          z ratio
 269  3     0001     8.747177  -0.435884423 0.4231936 -1.02998819
 270  3     0003    10.356366  -0.118972428 0.2515398 -0.47297650
 513  4     0003-3   9.597672  -0.204642434 0.4262935 -0.48005054
 241  2     0004     9.433572   0.129164649 0.4009577  0.32214038
 119  1     0005     9.799212  -0.376820347 0.4461908 -0.84452745
 274  3     0006     9.173060  -0.007441302 0.4655570 -0.01598365
      Va        Vresid       mu        pred      reliability
 269 0.2953927 0.3052332  9.986628   9.550744 0.3937128
 270 0.2953927 0.3052332  9.986628   9.867656 0.7858028
 513 0.2551090 0.3198337  9.987614   9.782971 0.2876530
 241 0.2271021 0.3340722 10.007016 10.136181 0.2920936
 119 0.3056859 0.2966048  9.970005   9.593184 0.3487230
 274 0.2953927 0.3052332  9.986628   9.979187 0.2662538
```

The very high reliability of individual with ID "0003" stands out. This individual is in the parental generation, so it is possible that it has a high reliability because it has many progenies in the data set. To verify that this value is reasonable, we count the number of progenies for individuals "0001", "0003", and "0006" to see if individual "0003" has an unusually large number of progenies:

```
# Why does 0003 have such high reliability?
# Check number of progenies it has in data compared to 0001 and 0006
print("number of progenies of 0001")
nrow(gp.num$pedigree[gp.num$pedigree$Par1 == "0001" |
 gp.num$pedigree$Par2 == "0001", ])
 print("number of progenies of 0003")
nrow(gp.num$pedigree[gp.num$pedigree$Par1 == "0003" |
 gp.num$pedigree$Par2 == "0003", ])
 print("number of progenies of 0006")
nrow(gp.num$pedigree[gp.num$pedigree$Par1 == "0006" |
 gp.num$pedigree$Par2 == "0006", ])
```

Indeed, individual "0003" has 19 progenies in the data set, and this explains its very high prediction reliability even when its value is predicted without a direct phenotypic observation. Individuals "001" and "006" had only one data point each, thus their low prediction reliability.

The validation set predictions are merged with the original phenotypic observations and we can compute the overall summary statistics for prediction ability:

Code example 11.3
(continued)

```
### summary statistics
PredAbi = with(preds.obs, cor(Sim.Pheno.1, pred))
RankCor = with(preds.obs, cor(Sim.Pheno.1, pred, method = "spearman"))
Best10 <- mean(tail(preds.obs[order(preds.obs$pred),
 "Sim.Pheno.1"],n=ceiling(nrow(preds.obs)*0.1)))
Bias   <- coef(lm(preds.obs$Sim.Pheno.1 ~ preds.obs$pred))[2]   # Bias
MSE   <- mean( (preds.obs$Sim.Pheno.1-preds.obs$pred)**2 )   # MSE
pred.sum <- rbind(PredAbi, RankCor,Best10, Bias, MSE)
round(pred.sum,2)
```

This gives the summary statistics:

	preds.obs$pred
PredAbi	0.32
RankCor	0.32
Best10	10.44
Bias	0.94
MSE	0.50

These results are congruent with the results of the `crossVal()` function reported previously.

GBLUP with Replicated Family Data in ASReml

Here will show an example of using ASReml standalone to fit a GBLUP model to raw plot-level data from a plant breeding experiment, where families were replicated across blocks and sites. Further, we demonstrate how to predict the phenotypic values of all the individuals with genotype data even if they are missing direct phenotypic observations. This can be a powerful method to predict the genetic values of untested families, which may have high economic efficiency if marker data can be collected with fewer resources or faster than phenotypes. An advantage of fitting the GBLUP model to raw plot-level data is the ability to incorporate more complex model structures for residuals and for genotype-by-environment interactions, while accounting directly for unbalance in the data and differing levels of precision among the genotype effects due to different amounts of missing data.

The example data set used consists of 508 S1 families (or "lines") from a maize population. S1 families are derived from non-inbred individuals from a randomly-mated population that were self-fertilized one generation to produce S1 generation progenies. In this example, 263 of the 508 families were measured in replicated trials in up to six environments for plant height and seed yield. The experimental design was unbalanced with three environments containing only 69 lines. This part of the data come from Horne et al. (2016). All of the phenotyped lines and an additional set of 245 untested lines were genotyped with 5677 SNPs. The genotype data are available in file "Maize_S1_genos.csv", the genetic map with marker positions in base pairs is available in "Maize_S1_map.csv" (although we don't need this information for the analysis to be described), and the plot-level trait data are available in "Maize_S1_traits.csv".

We start by using *synbreed* package to recode the marker data from base pair calls to counts of minor alleles and to compute a realized relationship matrix. The relationship matrix in tabular form appropriate for use by ASReml is written to a file "Maize_S1_G.grm".

Code example 11.4
R code to read raw genotype calls for maize S1 lines, convert to minor allele counts, and compute realized relationship matrix ("Code 11-4_Gmatrix_maizeS1.R").

```r
library(synbreed)

genos = read.csv("Maize_S1_genos.csv", stringsAsFactors = F)
rownames(genos) = genos$line
genos$line = NULL

map = read.csv("Maize_S1_map.csv", stringsAsFactors = F)
rownames(map) = map$marker
map$mark = NULL

ped = read.csv("Maize_S1_pedigree.csv", stringsAsFactors = F)
ped = create.pedigree(ID = ped$line, Par1 = ped$mother, Par2 =
 ped$father)

genos.gp = create.gpData(geno = genos, map = map, pedigree = ped,
 map.unit = "bp")

genos.num = codeGeno(genos.gp, impute = T, impute.type = "random", maf
 = 0.01, label.heter = "alleleCoding", verbose= T)

#create the realized relationship matrix
G = kin(genos.num, ret = "realized")
summary(G)

write.relationshipMatrix(G, file = "Maize_S1_G.grm", sorting =
 "ASReml", type = "none")
```

Now we can use the realized relationship matrix to model the covariance structure among lines in ASReml. This requires that we define line ID as a pedigree factor and we provide a pedigree file to ASReml as well as the file containing the elements of the relationship matrix and the phenotype data file. In this example, the lines were derived from randomly sampled plants from the population, so we have no pedigree information about the lines. The effective population size during intermating was about 20, so some close relationships (and possibly some selfing) are expected among these lines, but since pollen was bulked for fertilization and seeds were bulked at harvest, the pedigree relationships are unknown. Therefore, the parents of all individuals were set to zero in the pedigree file:

Excerpt from file "Maize_S1_pedigree.csv":

```
line          mother father
12FL0001-1         0    0
12FL0001-2         0    0
12FL0001-3         0    0
12FL0001-4         0    0
...
```

Although the pedigree file is not informative, it serves a critical function in the analysis in ASReml: the order of rows and columns of the relationship matrix must match the order of individuals (or lines, in this case) in the pedigree file. This is necessary because the elements in the relationship matrix are indexed by row and column numbers, not by line identifiers:

Excerpt from file "Maize_S1_G.grm":

```
1 1 1.0055436083
2 1 -0.0608484678
2 2 1.3310142478
3 1 -0.0287343037
3 2 -0.0495885535
3 3 0.9644275375
...
```

The first row of this file has the **G** coefficient for line 1 ("12FL0001-1") with itself, the second row has the **G** coefficient for lines 1 and 2 ("12FL0001-2"), and so forth. It is also possible to provide the elements of the inverse of **G** to ASReml, but the file extension should be .giv instead of .grm in that case. The pedigree file is needed to connect the row/column indices to the levels of the "line" factor that we will declare as a field in the trait data file and as a factor in the analysis in the ASReml file:

Code example 11.5
ASReml standalone code to fit realized relationship matrix to maize S1 line data and produce GBLUPs ("Code 11-5_Maize_S1_multivar_GBLUP.as"):

```
!NOGRAPHICS !WORKSPACE 4 !ARGS 1 2 3 !RENAME
Title: MAIZE S1 lines multivariate GBLUP model from raw data
 env   !A
 rep   !I
 plot  !I
 line  !P  #associate with pedigree
 height
 yield
Maize_S1_pedigree.csv !ALPHA !SKIP 1 #pedigree file
Maize_S1_G.grm #G matrix
Maize_S1_traits.csv  !SKIP 1  !CONTINUE !MAXITER 25 !SECTION env
!DOPART $A
```

At the beginning of the ASReml standalone file, we set some job options then declare the data fields being read in from the data file called "Maize_S1_traits.csv'). The variable line is declared as a pedigree-associated factor with qualifier "!P", and we specify a pedigree file containing all of the line names in the order that they are represented in the **G** matrix file. In this example, the **G** matrix includes all of the additional lines that were genotyped but not included in the phenotyping trials, so those lines are included in the pedigree file even though they never appear in the trait data file.

Following the specification of the data files, we have written three different parts of the analysis code to begin with simple univariate models and work up to a multivariate model that incorporates the realized relationship matrix. The first part of the analysis fits separate univariate models to the data, ignoring the relationships among the lines, but permitting the residual variances to differ among environments:

Code example 11.5
(continued)

```
!PART 1 #univariate IDV models
!CYCLE height yield
$I ~ mu !r env rep.env ,
           line line.env
   residual at(env).units
predict line !AVERAGE env rep
```

The results of these models are in result file "Code 11-5_Maize_S1_multivar_GBLUP1.asr":

```
          - - - Results from analysis of height - - -
LogL:    LogL  Residual          NEDF  NIT Cycle Text
LogL:-4407.04  1.00000           1416    8 height "LogL Converged"
Akaike Information Criterion    8834.07 (assuming 10 parameters).
Bayesian Information Criterion   8886.63


Model_Term                        Sigma    Sigma   Sigma/SE   % C
env               IDV_V    6    369.953  369.953      1.56   0 P
rep.env           IDV_V   12    9.74798  9.74798      1.51   0 P
line              NRM_V  508    233.036  233.036     10.29   0 P
line.env          IDV_V 3048    15.0444  15.0444      2.34   0 P
at(env,1).units          342 effects
Residual_1        SCA_V  342    126.730  126.730      9.38   0 P
at(env,2).units          338 effects
Residual_2        SCA_V  338    97.7683  97.7683      8.27   0 P
at(env,3).units          323 effects
Residual_3        SCA_V  323    70.2833  70.2833      7.50   0 P
at(env,4).units          138 effects
Residual_4        SCA_V  138    133.929  133.929      6.86   0 P
at(env,5).units          138 effects
Residual_5        SCA_V  138    108.312  108.312      6.60   0 P
at(env,6).units          138 effects
Residual_6        SCA_V  138    101.647  101.647      6.94   0 P
env               NRM    508


          - - - Results from analysis of yield - - -
LogL:-4788.31  1.00000           1414    8 yield "LogL Converged"
Akaike Information Criterion    9596.62 (assuming 10 parameters).
Bayesian Information Criterion   9649.16


Model_Term                        Sigma    Sigma   Sigma/SE   % C
env               IDV_V    6    102.563  102.563      1.50   0 P
rep.env           IDV_V   12    7.18482  7.18482      1.26   0 P
line              NRM_V  508    160.438  160.438      8.51   0 P
line.env          IDV_V 3048    34.8291  34.8291      2.59   0 P
at(env,1).units          342 effects
Residual_1        SCA_V  342    203.656  203.656      8.48   0 P
at(env,2).units          338 effects
Residual_2        SCA_V  338    152.418  152.418      8.19   0 P
at(env,3).units          321 effects
Residual_3        SCA_V  321    248.249  248.249      9.19   0 P
at(env,4).units          138 effects
Residual_4        SCA_V  138    216.477  216.477      6.44   0 P
at(env,5).units          138 effects
Residual_5        SCA_V  138    221.584  221.584      6.78   0 P
at(env,6).units          138 effects
Residual_6        SCA_V  138    261.549  261.549      6.85   0 P
env               NRM    508
```

In part 2 we model the covariance among lines as proportional to the realized relationship matrix for each trait:

**Code example 11.5
(continued)**

```
!PART 2 #univariate GBLUP models
!CYCLE height yield
$I ~ mu !r env rep.env ,
          grm(line) ,
          grm(line).env
     residual at(env).units
predict line !AVERAGE env rep
```

The model function `grm(line)` associates the genetic relationship matrix (which is defined previously in the code after the pedigree file and before the phenotype data file) to the factor "line" found in the trait data. Notice that we model both the main effects of lines and their interactions with environments as proportional to the realized relationship matrix. This results in the following output ("Code 11-5_Maize_S1_multivar_GBLUP2.asr"):

```
          - - - Results from analysis of height - - -
LogL:   LogL  Residual      NEDF  NIT Cycle Text
LogL:-4348.50  1.00000      1416    8 height "LogL Converged"
Akaike Information Criterion    8716.99 (assuming 10 parameters).
Bayesian Information Criterion  8769.55
```

Model_Term			Sigma	Sigma	Sigma/SE	% C
env	IDV_V	6	372.416	372.416	1.56	0 P
rep.env	IDV_V	12	9.52119	9.52119	1.51	0 P
grm(line)	GRM_V	508	304.857	304.857	9.17	0 P
at(env,1).units		342	effects			
Residual_1	SCA_V	342	127.965	127.965	10.51	0 P
at(env,2).units		338	effects			
Residual_2	SCA_V	338	89.2178	89.2178	9.55	0 P
at(env,3).units		323	effects			
Residual_3	SCA_V	323	80.8619	80.8619	9.12	0 P
at(env,4).units		138	effects			
Residual_4	SCA_V	138	135.262	135.262	7.09	0 P
at(env,5).units		138	effects			
Residual_5	SCA_V	138	105.792	105.792	6.84	0 P
at(env,6).units		138	effects			
Residual_6	SCA_V	138	100.651	100.651	7.07	0 P
grm(line)	Maiz	508				
grm(line).env		3048	effects			
env	ID_V	1	15.6113	15.6113	3.36	0 P

```
          - - - Results from analysis of yield - - -
LogL:-4777.75  1.00000      1414    8 yield "LogL Converged"
Akaike Information Criterion    9575.50 (assuming 10 parameters).
Bayesian Information Criterion  9628.04
```

Model_Term			Sigma	Sigma	Sigma/SE	% C
env	IDV_V	6	103.692	103.692	1.51	0 P
rep.env	IDV_V	12	6.40997	6.40997	1.19	0 P
grm(line)	GRM_V	508	258.379	258.379	7.44	0 P
at(env,1).units		342	effects			
Residual_1	SCA_V	342	211.005	211.005	10.31	0 P

```
at(env,2).units              338  effects
Residual_2        SCA_V      338    174.462    174.462      9.99    0 P
at(env,3).units              321  effects
Residual_3        SCA_V      321    275.683    275.683     10.71    0 P
at(env,4).units              138  effects
Residual_4        SCA_V      138    231.603    231.603      6.96    0 P
at(env,5).units              138  effects
Residual_5        SCA_V      138    242.984    242.984      7.18    0 P
at(env,6).units              138  effects
Residual_6        SCA_V      138    279.659    279.659      7.28    0 P
grm(line)         Maiz       508
grm(line).env               3048  effects
env               ID_V         1   19.2736    19.2736      2.30    0 P
```

The models fit in part 2 do not differ in the number of parameters compared to the models fit in part 1, they differ only in changing the covariance structure of lines from IDV in part 1 to proportional to the realized relationship matrix in part 2. Therefore, we can directly compare the likelihoods of the two models for each trait. For height, the log likelihood improved from -4407.04 (IDV model) to -4348.50 (GBLUP model); for yield the difference was -4788.31 (IDV model) versus -4777.75 (GBLUP model). The estimated genetic variance components changed from 233.036 (IDV) to 304.857 (GBLUP) for height and from 160.438 (IDV) to 258.379 (GBLUP) for yield. The line by environment interaction term changed little for height but decreased from about 35 (IDV) to about 19 (GBLUP) for yield.

Finally, we extend the GBLUP model to a multivariate model for height and yield in model 3. The key term in this model is "$\texttt{glm(line).us(Trait)}$", which fits the variance covariance for line effects on height and yield as:

$$
\mathbf{Var}\left(\begin{bmatrix}\mathbf{G}_{1h}\\\mathbf{G}_{1y}\\\mathbf{G}_{2h}\\\mathbf{G}_{2y}\\\vdots\end{bmatrix}\right) = \begin{bmatrix}\mathbf{G}_{11}\sigma_{gh}^2 & \mathbf{G}_{11}\sigma_{ghy} & \mathbf{G}_{12}\sigma_{gh}^2 & \mathbf{G}_{12}\sigma_{ghy} & \cdots \\ \mathbf{G}_{11}\sigma_{ghy} & \mathbf{G}_{11}\sigma_{gy}^2 & \mathbf{G}_{12}\sigma_{ghy} & \mathbf{G}_{12}\sigma_{gy}^2 & \\ \mathbf{G}_{12}\sigma_{gh}^2 & \mathbf{G}_{12}\sigma_{ghy} & \mathbf{G}_{22}\sigma_{gh}^2 & \mathbf{G}_{22}\sigma_{ghy} & \\ \mathbf{G}_{12}\sigma_{ghy} & \mathbf{G}_{12}\sigma_{gy}^2 & \mathbf{G}_{22}\sigma_{ghy} & \mathbf{G}_{22}\sigma_{gy}^2 & \\ \vdots & & & & \ddots\end{bmatrix}
\tag{11.12}
$$

$$
= \mathbf{G} \otimes \begin{bmatrix}\sigma_{gh}^2 & \sigma_{ghy} \\ \sigma_{ghy} & \sigma_{gy}^2\end{bmatrix}
$$

Similarly, the line-by-environment interactions are modeled proportional to the direct product of a 6×6 identity matrix (for six environments) by the $\mathbf{G}$ matrix by the unstructured trait variance-covariance matrix with the term "$\texttt{env.grm(line).us(Trait)}$".

Code example 11.5
(continued)

```
!PART 3 #multivariate with grm(line)*id(env)
!ASUV
height yield ~ Trait !f mv,
        !r us(Trait).env ,
          us(Trait).rep.env ,
          grm(line).us(Trait) ,
          env.grm(line).us(Trait)

          residual sat(env).id(units).us(Trait)
predict line !AVERAGE env rep
```

The results of this analysis include 2×2 trait covariance components for the residuals at each of six environments as well as for the line and line-by-environment effects ("Code 11-5_Maize_S1_multivar_GBLUP3.asr"):

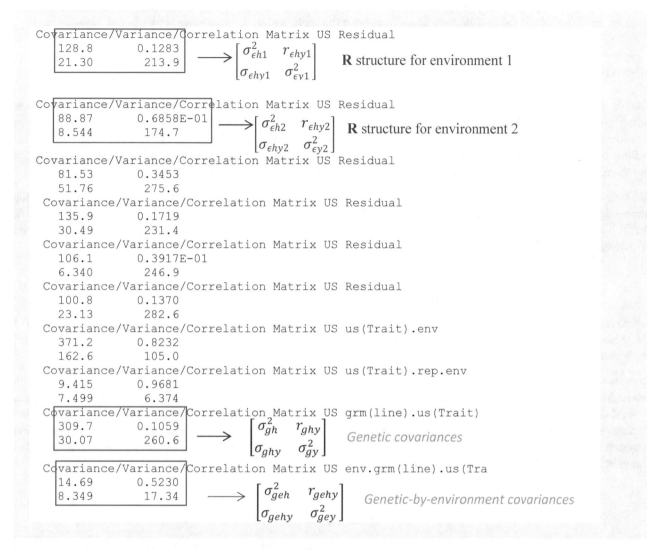

In Table 11.3 we compare the predictions and their standard errors for three lines with phenotypic data (their line identifiers start with "12FL") and three untested lines (their line identifiers start with "15CL") for yield for each of the three models. The most notable differences among the models are the predictions of the untested lines from Model 1, which all have a constant prediction at the population mean and zero reliability. Models 2 and 3 use the realized relationship matrix to predict the values of those lines based on their covariances with the tested lines. The Model 2 and 3 predictions of the untested lines

Table 11.3 Predicted yields of maize S1 lines from three models. Tested lines were evaluated in up to six environments; untested lines have no phenotypic records. All lines were genotyped and included in a realized relationship matrix

Line	Model 1 (IDV) Prediction (SE)	Rel.	Model 2 (GBLUP) Prediction (SE)	Rel.	Model 3 (Multivariate GBLUP) Prediction (SE)	Rel.
Tested lines						
12FL0001-1	91.05 (4.40)	0.88	93.97 (4.44)	0.92	93.91 (4.45)	0.92
12FL0001-2	84.68 (4.40)	0.88	86.04 (4.40)	0.93	86.03 (4.40)	0.93
12FL0001-3	70.11 (4.37)	0.88	70.71 (4.30)	0.93	70.37 (4.30)	0.93
Untested lines						
15CL1006-11	83.73 (12.9)	0	80.47 (8.69)	0.71	80.20 (8.72)	0.71
15CL1006-16	83.73 (12.9)	0	88.84 (9.72)	0.63	89.05 (9.73)	0.64
15CL1006-17	83.73 (12.9)	0	86.16 (8.84)	0.70	85.71 (8.85)	0.70

have much higher standard errors than the tested lines, but nevertheless they are reasonably good predictions with reliabilities around 0.7. The predictions from Model 3 are only slightly different than from Model 2 with very small increases in reliability. As discussed in Chap. 6, the multivariate models will have greatest effect when the missing data patterns are different and the genetic correlation is high for the traits analyzed, and neither is true in this example.

Blended Genetic Relationships

Genotyping the entire breeding population may not be reasonable due to high cost and logistical limitations. In order to incorporate information from individuals that were not genotyped but have pedigree information, a blended genetic relationship matrix (**H** matrix) was proposed (Misztal et al. 2009; Legarra, et al. 2009; Christensen and Lund 2010). This is done by adding the genomic relationships (**G** matrix) of a subset of the population to the numerator relationship matrix **A** derived from pedigree for the all the population (Misztal et al. 2009; Legarra, et al. 2009).

Let **u** be a vector of genetic effects. Starting with a model of expected additive genetic relationships based on pedigree, the variance of **u** is a product of the **A** matrix and the variance (σ_u^2) explained by polygenic effects: var(**u**) = $\mathbf{A}\sigma_u^2$ (Legarra et al. 2009). Consider, however, that there are two types of individuals in **u**: non-genotyped individuals (**u**$_1$) and genotyped individuals (**u**$_2$). Then **A** can be partitioned as:

$$\mathbf{A} = \begin{vmatrix} \mathbf{A}_{11} & \mathbf{A}_{12} \\ \mathbf{A}_{21} & \mathbf{A}_{22} \end{vmatrix} \tag{11.13}$$

where **A**$_{11}$ is the pedigree-based relationship matrix of non-genotyped individuals, **A**$_{22}$ is the pedigree-based relationship of the genotyped individuals, and **A**$_{12}$ and its transpose **A**$_{21}$ are the relationships between genotyped and non-genotyped individuals.

Since we have genotype information on the individuals in **u**$_2$, we can replace the pedigree expected relationships in **A**$_{22}$ with the realized genomic relationships among those individuals, which would be a **G** matrix as described earlier in this chapter. The relationships between the genotyped individuals and non-genotyped individuals in **A**$_{12}$ then need to be adjusted to take into account the realized relationships in **G** as well as their known pedigree relationships. The unified relationship matrix that accounts for both the realized relationships among the genotyped individuals and the pedigree relationships of all other individuals is referred to as an **H** matrix (Legarra et al. 2009):

$$\mathbf{H} = \begin{vmatrix} \mathbf{A}_{11} + \mathbf{A}_{12}\mathbf{A}_{22}^{-1}(\mathbf{G} - \mathbf{A}_{22})\mathbf{A}_{22}^{-1}\mathbf{A}_{21} & \mathbf{A}_{12}\mathbf{A}_{22}^{-1}\mathbf{G} \\ \mathbf{G}\,\mathbf{A}_{22}^{-1}\mathbf{A}_{21} & \mathbf{G} \end{vmatrix} \tag{11.14}$$

The upper left corner in the **H** matrix is the variance of ungenotyped individuals, notice that it is no longer simply a function of their pedigree relationships because now it incorporates additional information from the **G** matrix that is shared via the pedigree relationships with the genotyped individuals. The **H** matrix is a semi-positive or positive definite matrix by construction (Legarra et al. 2009). The inverse of **H** matrix has a simple form that is easier to understand:

$$\mathbf{H}^{-1} = \mathbf{A}^{-1} + \begin{vmatrix} 0 & 0 \\ 0 & \mathbf{G}^{-1} - \mathbf{A}_{22}^{-1} \end{vmatrix} \tag{11.15}$$

In this equation $\mathbf{A}^{-1}$ is the inverse of numerator relationship matrix for all individuals, $\mathbf{G}^{-1}$ is the inverse of realized genomic relationship matrix and $\mathbf{A}_{22}^{-1}$ is the inverse of **A** matrix for the genotyped individuals (Aguilar et al. 2010; Legarra and Ducrocq 2012; Legarra et al. 2009).

$$\mathrm{Var}(\mathbf{u}_1) = \left[\mathbf{A}_{11} + \mathbf{A}_{12}\mathbf{A}_{22}^{-1}(\mathbf{G} - \mathbf{A}_{22})\mathbf{A}_{22}^{-1}\mathbf{A}_{21}\right]\sigma_A^2 \tag{11.16}$$

$\mathbf{G}\sigma_A^2$ is the variance of genotyped individuals and $\mathbf{A}_{12}\mathbf{A}_{22}^{-1}\mathbf{G}\sigma_A^2$ is the covariance between **u**$_1$ and **u**$_2$ (Legarra et al. 2009).

Example Calculation of H Matrix

To demonstrate computation of a blended relationship matrix, **H**, we use data from a tree breeding study in which nine crosses were generated using 13 individuals as females and males (Ogut et al. 2012; Zapata-Valenzuela et al. 2013). A single-pair mating design was used. A total 354 progeny trees were evaluated. Out of 354 trees, 166 were genotyped using 3461 SNP markers. There are two pedigree files. The first pedigree file is called H_complete_ped.txt, with 367×3 dimension (354 individuals + 13 parents = 367). The second pedigree file is called H_geno_ped.txt with dimension of 178×3. This second file includes parents at the top of the file (the first 13 individuals) followed by 166 genotyped trees. The genotyped trees are numbered from 202 to 367. A subset of the H_geno_ped.txt is given below.

```
> H_geno_ped
    tree par1 par2
1     1    0    0
2     2    0    0
3     3    0    0
...
11   11    0    0
12   12    0    0
13   13    4   12
14  202    1    2
15  203    1    2
16  204    1    2
...
177 365    9    8
178 366    9    8
179 367    9    8
```

An R script to obtain blended genomic relationship matrix **H** is given below. This part loads the **GeneticPed** package from CRAN website, reads the pedigree file and creates the **A** matrix.

Code example 11.6
R code to calculate blended genetic relationship matrix (see Code 11-5_Hmatrix.R for details).

```
###########################################
# Code11-6_HMatrix.R
# Contributed by Funda Ogut
###########################################

# remove everything in working environment
rm(list=ls())

### load packages
library(ggplot2)
library(GeneticsPed)

# ~~~~ Calculation of A matrix ~~~~#
path='~/Book/Book1_Examples/data'
setwd(path)     # sets path as the working directory
ped <- read.table('H_complete_ped.txt', header=F)

s <- ped[,2]    # sire (male) vector
d <- ped[,3]    # dam (female) vector
```

```
# ~~~~~~~ Create function called 'createA' ~~~~~~~~~~~~~~#
## Given the vectors of sire and dam,
# return additive relationship matrix 'A' (TABULAR METHOD).

`createA` <-
function(s, d){
 if (nargs()==1){
     stop("sire vector and dam vector are required")
 }

 if (length(s) != length(d)){
     stop("size of the sire vector and dam vector are different!")
 }

n <- length(s)
N <- n + 1
A <- matrix(0, ncol=N, nrow=N)   # 368 x 368 matrix

s <- (s == 0)*(N) + s
d <- (d == 0)*N + d

for(i in 1:n){

     A[i,i] <- 1 + A[s[i], d[i]]/2

     for(j in (i+1):n){
             if (j > n) break
             A[i,j] <- ( A[i, s[j]] + A[i,d[j]] )/2
             A[j,i] <- A[i,j]
     }
}
   return(A[1:n, 1:n])
}
```

The following **R** script partitioned the **A** matrix into four sub matrices as **A_un** (A_{11}) is the matrix of trees with *no genotype* information, **A_gen** (A_{22}) is the matrix of trees with genotypes, **A_covup** (A_{12}) upper triangle and **A_covlow** (A_{21}) lower triangle covariances between non-genotyped and genotyped individuals.

Code example 11.6
(continued)

```
# ~~~~~~~~~~ Create A and sub matrices from A ~~~~~~~~~~#
# A, full pedigree for non-genotyped and genotyped individual
A_full <- createA(s,d)     # Dim 367 x 367

# A11, subset of A for non-genotyped individuals Dim=201 x 201
A_un <- A_full[1:201,1:201]

# A22, subset of the A for genotyped individuals Dim=166 x 166
A_gen <- A_full[202:367,202:367]

# A12, covariance matrix between non-genotyped & genotyped individuals
A_covup <- A_full[1:201,202:367]     # Dim = 201 x 166
# A21, covariance matrix between genotyped & non-genotyped individuals
A_covlow <- A_full[202:367,1:201]    # Dim = 166 x 201
```

The following script combines the **A** matrix derived from pedigree for the whole population (N = 367) and genomic relationships matrix **G** of n = 166 individuals.

Code example 11.6
(continued)

```
#~~~~~~~~~~~~~~~ calculation of G matrix ~~~~~~~~~~~~~~~~~~#
# Given markers and pedigree, return G matrix
# read pedigree for genotyped individuals
geno_ped <-read.table("H_geno_ped.txt",header=T,sep="\t")
# read markers
Hmarkers <- read.table("H_PineMarkers.txt",header=F)

# Load GeneticRel function
load("~/Book/.../GenomicRel1.rda")
G=GenomicRel(1,Markers,pedigree)   # G based on obs allele freq

#~~~~~~ Calculation of H matrix ~~~~~~~~~~~~~~~~~~~#
Z=solve(A_gen)        # Dim 166 x 166, A^22
T=(G-A_gen)*1         # Dim 166 x 166, G - A22

A= A_covup %*% Z %*% T %*% Z %*% A_covlow     # Dim 367 x 367
B= A_covup %*% Z %*% T                        # Dim 201 x 166
C= T %*% Z %*% A_covlow          # Dim 166 x 201
D= T                             # G-A22   Dim 166 x 166
E= G %*% Z %*% A_covlow          # Dim 166 x 201

p <- cbind(A,B)   # 201 x 367
o <- cbind(C,D)   # 166 x 367
J <- rbind(p,o)   # 367 x 367

H= A_full + J    # 367 x 367
round(H[1:10,1:8],2)
```

$$\begin{bmatrix} A_{un} + A & B \\ C & T \end{bmatrix}$$

This example involved a small data set. For large data sets, computing the inverses of **A** and **G** matrices can be difficult. Readers should consider optimized software that can handle large data sets. For example, Ignacio Aguilar and Ignacy Misztal at University of Georgia developed the *PreGSF90* module to calculate genetic relationships for their BLUPF90 software (http://nce.ads.uga.edu/wiki/doku.php?id=application_programs).

A partial output from **H** matrix in table format (367 × 367) is given below.

```
> round(H[1:10,1:8],2)
  1.05   0.01   0.01  -0.06   0.02   0.02   0.02  -0.06
  0.01   0.80  -0.06   0.11  -0.02  -0.03  -0.03  -0.04
  0.01  -0.06   0.95  -0.16   0.02  -0.01  -0.01  -0.10
 -0.06   0.11  -0.16   0.77  -0.01  -0.06  -0.06  -0.08
  0.02  -0.02   0.02  -0.01   0.98   0.08   0.08  -0.01
  0.02  -0.03  -0.01  -0.06   0.08   0.99  -0.01   0.02
  0.02  -0.03  -0.01  -0.06   0.08  -0.01   0.99   0.02
 -0.06  -0.04  -0.10  -0.08  -0.01   0.02   0.02   0.73
  0.04  -0.10   0.01  -0.16   0.08   0.14   0.14  -0.19
 -0.02   0.01  -0.06  -0.05   0.03   0.09   0.09   0.15
```

The following section of Code 11-6_Hmatrix.R code converts the **H** inverse to a sparse format for ASReml to use in prediction of genomic estimated breeding values.

Code example 11.6
(continued)

```
###~~~~~~~~ Inverse of H ~~~~~~~~~~~~~~~~~~~~~~~#
Hinv <-solve(H)    # inverse of H, Dim 367 x 367
round(Hinv[1:10,1:8],2)

# Convert to sparse format, appropriate for the ASReml
 col1=NA
 col2=NA
 col3=NA
 print("Still working, this takes a while")
 for (i in 1:nrow(Hinv)){
     for (j in 1:ncol(Hinv)){
         if (i >= j){
             col1=cbind(col1,i)
             col2=cbind(col2,j)
             col3=cbind(col3,Hinv[i,j])
         }
     }
 }

Hinv.sparse=cbind(t(col1),t(col2),t(col3))
row.names(Hinv.sparse)=c(0:(nrow(Hinv.sparse)-1))

#~~~ Visualize H matrix ~~~#
Hinv.sp.df <- data.frame(Hinv.sparse)
# Delete the first row 0 NA NA NA
Hinv.sp.df <- Hinv.sp.df[-c(1),]

# Rename the columns
names(Hinv.sp.df)[names(Hinv.sp.df)=="X1"] <-"row"
names(Hinv.sp.df)[names(Hinv.sp.df)=="X2"] <-"col"
names(Hinv.sp.df)[names(Hinv.sp.df)=="X3"] <-"coefficient"
round(head(Hinv.sp.df), 2)

### Save the inverse of H matrix for ASReml to use
 write.table(Hinv.sp.df,"HinvPine.giv",row.names=F,col.names=F,sep=" ")
```

The output file `HinvPine.giv` has three columns named 'row', 'col' and 'coefficient'. A subset of the file is given below.

```
  round(head(Hinv.sp.df), 2)
    row col coefficient
1   1   1    35.5
2   2   1     8.0
3   2   2    31.0
4   3   1     0.0
5   3   2     0.0
6   3   3    36.5
```

The coefficients are inversed genetic covariances between individuals and the inbreeding coefficients of individuals.

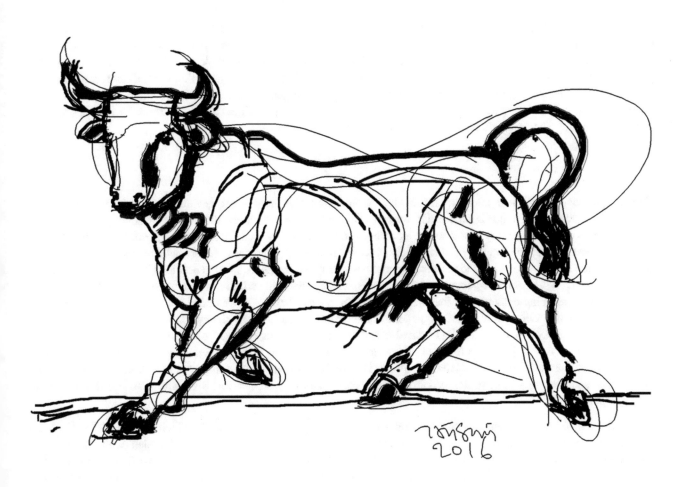

Electronic supplementary material: The online version of this chapter (doi:10.1007/978-3-319-55177-7_12) contains supplementary material, which is available to authorized users.

Abstract

The whole genome regression approach to genomic selection is based on using large numbers of DNA markers to predict breeding values of individuals in animal and plant breeding programs. It is related to GBLUP introduced in Chapter 11, but it is distinguished from GBLUP by simultaneously modeling the effects of many DNA markers, and using the sum of marker allele effect estimates to predict breeding values of individuals. Since the introduction of the concept in 2001 (Meuwissen et al. 2001), it has revolutionized crop and livestock breeding. Genomic selection is a very active area of research so that new algorithms, software and methods are constantly being developed. Within this context we will briefly introduce a few key ideas and provide some examples for demonstration.

Regression Models for Genomic Prediction

Meuwissen et al. (2001) suggested the use of all available molecular markers in regression models to predict the overall genomic merit of an individual. Since then the application of that concept, which is known as genomic selection (GS) has been widely adopted in livestock (VanRaden et al. 2009; Hayes and Goddard 2010; Wolc et al. 2011; Duchemin et al. 2012) and as well as crop (Rutkowski et al. 2011; Zhao et al. 2012) species. Inclusion of genomic information in many livestock and some plant breeding schemes has become routine, permitting the generation of fast and accurate individual breeding value predictions. Considering the breeder's equation for genetic gain (Δ_g) provides some insight into how GS has impacted breeding programs. The breeder's equation can be written in this form:

$$\Delta_g = \frac{r i \sigma_a}{L} \tag{12.1}$$

where r is the accuracy of the prediction; i is the selection intensity; σ_a is the standard deviation of genetic variance and L is the generation interval. A major impact of GS on genetic gain has been to reduce the generation interval (L), because it enables identification of selection candidates at an earlier age than was previously possible. Furthermore, GS has enabled a significant boost in selection intensity i, through an increased number of candidates genotyped and available for selection. Yet, GS has not substantially changed the fundamental basis of breeding gains, since genomic information is used to rank individuals based on their additive genetic breeding values, which is the same criterion for selection that breeders have used for most of the preceding century, and that we have used in earlier chapters of this book. In Chap. 11 we described the prediction of breeding values from molecular genetic marker data using a **G** matrix that describes the genomic relationships among genotyped individuals. This method is flexible and easy to implement, and works well since replacing the numerator relationship matrix **A** with **G** allows to better characterize the relationship among individuals and improve within-family selection by better accounting for Mendelian sampling. The GBLUP approach relies on the 'infinitesimal' model of classical quantitative genetics, in which genetic control of complex traits is assumed to be equally distributed across many (infinite) loci.

A second class of models is equally popular, and arguably more powerful for traits in which individual DNA variants account for a significant fraction of phenotypic variance. We will refer to this large class of models with the loose (and somewhat misleading) term of Bayesian alphabet models. We have seen in previous chapters how BLUPs are shrinkage estimators. The idea behind whole genome regression via Bayesian alphabet models is similar. The advantage of shrinkage estimators (at the marker level) in the context of whole genome regression is that shrinkage can improve prediction based on markers by reducing the mean squared error of the estimate, especially with a large number of markers, with a relatively small trade-off of introducing some bias because it forces some of the estimated regression coefficients towards zero.

We will provide a brief introduction of this class of models here and provide a general overview of some differences between few of the most popular options. For a formal treatment, readers are referred to the original Meuwissen et al. (2001) paper, and to reviews papers on the subject (Gianola et al. 2009; Habier et al. 2011).

A Brief Tour of Bayesian Concepts

It is important to remark that while historically the vast majority of the models used in genomic prediction have been developed in a Bayesian context, this is not a requirement. It is nonetheless a convenient and relatively computationally parsimonious way to obtain shrinkage estimators for all markers simultaneously. It is also important to recognize that a Bayesian implementation is not prerogative of whole genome regression models, and even most of the BLUP machinery developed in previous chapters can be implemented efficiently in a Bayesian framework. A key concept of all Bayesian approaches resides on the fact that all the parameters in the model are treated as random variables. In a Bayesian context, loosely speaking, we can quantify our knowledge (which can be very vague) of a particular parameter before observing the data by assigning a prior probability to that parameter. This is then combined with the evidence arising from the data (the likelihood) through the Bayes theorem. Bayes theorem gives the conditional probability of a set of parameter values given some observed data (also called the "prior probability") as:

$$f(\boldsymbol{\theta}|\mathbf{y}) = \frac{f(\mathbf{y}|\boldsymbol{\theta})f(\boldsymbol{\theta})}{f(\mathbf{y})} \propto f(\mathbf{y}|\boldsymbol{\theta})f(\boldsymbol{\theta}) \tag{12.2}$$

$f(\mathbf{y}|\boldsymbol{\theta})$ = The data likelihood

$f(\boldsymbol{\theta})$ = The prior

$\boldsymbol{\theta}$ = The parameter vector

$\mathbf{y}$ = The data vector

Effectively this equation "weighs" the evidence on the parameter we are interested in estimating arising from the data (my current experiment), with the previous knowledge we might have on the parameter (which comes from previous experiments). As you might see intuitively, the stronger is the assumption we make about the prior probability the higher the influence it will have on the posterior probability. For example, consider an experiment in which you want to determine the yield of a particular crop. You could measure the yield on a number of plants (say 100) and then use the average (the natural estimator and also the maximum likelihood estimator) as your estimate of the yield for that crop. Let's say that the average yield measured in your experiment is 17.8 kg with a standard error of 5 kg. Now let's assume that a scientist before you carried out 3 separate experiments similar to yours and found that the average yield was 17.4 kg with a standard error of 5 kg. If we apply the Bayes theorem with $\mathbf{y}$ being the data and theta ($\boldsymbol{\theta}$) being the average crop yield we want to estimate (and we assume a normal prior distribution), we find that using the $\propto f(\mathbf{y}|\boldsymbol{\theta})f(\boldsymbol{\theta})$ of Bayes theorem will produce a posterior mean of 17.6 kg, so that effectively the information of the previous experiment has been incorporated. The posterior mean is closer to the prior mean than the mean based only on the current measurements. Now, let's say that after a 1000 experiments the crop yield measured by the previous scientist was still 17.4 kg but with a standard error of only 1 kg. In this case, the posterior mean will be 17.5 kg, so that the stronger assumption regarding the prior will be reflected in the posterior. If no experiment was previously run, we might still want use a prior but we might want to choose a relatively vague one. So we might still assume that the average yield was 17.4 kg but the standard error can be very large, say 20 kg. In this case, we would be effectively saying that *a priori* we know very little about the yield of the crop and we will strongly rely on the data collected in our experiment (the likelihood), so that the posterior estimate for our crop yield will be 17.8 kg.

The overall application of this simple principle becomes complex relatively quickly, because $f(\boldsymbol{\theta}|\mathbf{y})$ often does not have closed form, or its calculation may involve multiple integrations. For this reason, Bayesian analyses are often carried out with the use of Markov Chain Monte Carlo (MCMC) methods, in which inferences on the parameters are obtained from statistics of samples obtained empirically from $f(\boldsymbol{\theta}|\mathbf{y})$. Among the most popular approaches, and the one used in most Bayesian models for GS, is the Gibbs sampler. This method obtains samples for all the parameters sequentially for each iteration from the distribution of a particular parameter conditional on all others. Further details and a rigorous treatment of the topic can be found in Gelman and Rubin (1992), and specifically for breeding applications in Sorensen and Gianola (2007).

The general model employed in most of whole genome regression analyses, following notation provided by Gianola et al. (2009) is:

$$\mathbf{y} = \mathbf{Xb} + \mathbf{Wa}_m + \mathrm{e} \tag{12.3}$$

where $\mathbf{y}$ is a vector of phenotypic values. Note that the choice of the phenotypes employed has implications on the application of GS; see Garrick et al. (2009) for an in-depth discussion. $\mathbf{X}$ is an incidence matrix for the fixed effects (which in the simplest case reduces to a vector of 1s for the overall mean), $\mathbf{b}$ is a vector of fixed effects, including the overall mean, μ. $\mathbf{W}$ is a known matrix of numeric genotype scores for each marker. This could be a matrix of minor allele frequency dosage (e.g., 0, 1, or 2 copies of the minor allele at each marker) as was used in Chaps. 10 and 11 or it could be the deviation of minor allele counts from the heterozygous genotype (-1, 0, and 1 codes). The $\mathbf{a_m}$ is a vector of molecular marker additive effects and $\mathbf{e}$ is a vector of residuals with distribution $\mathbf{e} \sim N(0, \mathbf{I}\sigma_e^2)$. The vectors of coefficients has a multivariate normal distribution, and in most implementation marker-specific prior variances of different form are implemented (de los Campos et al. 2009, 2010; Pérez et al. 2010). The τ_j^2 are exponential priors, $\tau_j^2 \sim Exp(\lambda^2)$. The λ^2 can have a Gamma prior distribution $\lambda^2 \sim \text{Gamma}(r, \delta)$ where r (shape) and δ (rate) are hyper parameters. The residual variance is commonly a scaled inverse chi-square prior distribution with $\sigma_e^2 \sim \chi^{-2}(df_e, S_e)$ (Park and Casella 2008). Several authors have discussed advantages and disadvantages of this implementation (de los Campos et al. 2009; Yi and Xu 2008).

In the following section we briefly introduce a general framework of whole genome regression methods with a Bayesian interpretation.

Ridge Regression

When the number of predictors (p) is larger than the number of observations (n), as in analysis of genomic data for breeding, regularization (shrinkage) is commonly used to overcome singularity in estimation of regression coefficients. Ridge regression introduced by Hoerl and Kennard (1970) constrains regression coefficients by adding a small constant (λ) to the diagonal matrix of ($\mathbf{W}^T\mathbf{W}$) to estimate the coefficients.

$$\widehat{\mathbf{a}}_{\text{Ridge}} = \left(\mathbf{W}^T\mathbf{W} + \lambda\mathbf{I}_p\right)^{-1}\mathbf{W}^T\mathbf{y} \qquad (12.4)$$

The solutions for ridge regression also minimize the following equation

$$\widehat{\mathbf{a}}_{\text{Ridge}} = \sum_{i}^{n}\left(y_i - W_i^T a\right)^2 + \lambda\sum_{j=1}^{p} a_j^2 \qquad (12.5)$$

The first part of the model $\sum_{i}^{n}\left(y_i - W_i^T a\right)^2$ minimizes the residual sum of squares, the second part $\lambda\sum_{j=1}^{p} a_j^2$ puts constrains on the coefficients $\widehat{a}$. The penalty term is a product of a constant (λ) and the squared vector $\left(a_j^2\right)$ of coefficients. As λ approaches 0, the $\widehat{a}$ coefficients approach ordinary least-squares solutions. As λ approaches infinity, the coefficients approach 0. The ridge regression penalty introduces some bias by shrinking the estimates toward zero. At the same time though, it reduces the variance of estimates and thus, usually, results in a lower mean square error $MSE = \text{Bias}^2 + \sigma_a^2$.

The same concept can be given a Bayesian interpretation by assigning marker effects to a multivariate normal prior distribution with a common variance (σ_a^2). The common variance of marker effects can be modelled hierarchically usually through a scaled inverted χ^2 distribution, $\sigma_a^2 \sim \chi^{-2}(df_a, S_a)$ where the degrees of freedom df_a and scale S_a are normally treated as hyperpriors (they are assigned fixed values). Similarly a scaled inverted χ^2 prior distribution with degrees of freedom df_e, and scale parameter S_e, $\sigma_e^2 \sim \chi^{-2}(df_e, S_e)$ is normally employed for the residual variance (de los Campos et al. 2009; Gianola 2013; Pérez et al. 2010).

It is important to point out that ridge regression obtains coefficients predictions for all the loci so that even for large λ values the coefficients are not shrunk to exactly zero. It is also worth noting that ridge regression uncannily resembles the GBLUP approach described in previous chapters. In reality the two approaches are equivalent (see for example VanRaden 2008) as an equivalent model can be produced utilizing the $\mathbf{W'W}$ product (a m markers by m markers matrix) or the $\mathbf{WW'}$ product (an n individuals by n individuals matrix that is proportional to the $\mathbf{G}$ matrix estimators introduced in Chap. 11). In the first case solving the model will produce predictions of each individual marker effect, while in the second predictions of each individual's breeding value will be obtained. Nonetheless, with minimal algebraic manipulation it is possible to go from one solution to the other (Ruppert et al. 2003).

BayesA and BayesB

In Meuwissen et al. (2001) two models were proposed for GS with Bayesian methods.

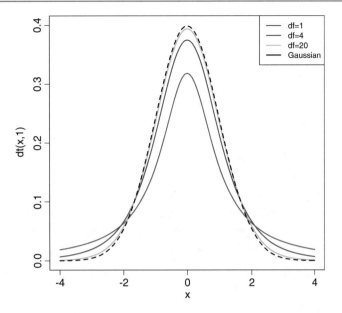

Fig. 12.1 Student's t distributions for various degrees of freedom (v = 1, 4 and 20) and comparison with Gaussian distribution. The tails for t-distribution are fatter compared to the Gaussian distribution. When used as priors they allow QTLs to have a larger effect

In **BayesA** the prior distribution (again our prior belief about the parameters) of each marker (a_x) is assumed normal with mean 0 and variance $\sigma^2_{a_x}$. The value of a_x depends on the variance parameter, so a further prior distribution is placed on each $\sigma^2_{a_x}$, namely a scaled inverted χ^2 with 2 hyperpriors: ν (degrees of freedom) and S^2 (scale). Thus (following again Gianola's notation)

$$a_x \mid \sigma^2_{a_x} \sim N\left(0, \sigma^2_{a_x}\right)$$
$$\sigma^2_{a_x} \mid \nu, S^2 \sim \nu S^2 \chi_{\nu}^{-2}$$

(12.6)

so that the prior for each a_x (the marker effect) is a t-distribution $[a_k \mid 0, \nu, S^2]$.

The result is that this t-distribution (with fatter tails than a normal distribution) will allow some markers to have a large effect, "escaping" the shrinkage toward their expected value (0), effectively capturing QTL (individual loci with large effects on trait variation) (Fig. 12.1).

It should be noted that in this case each marker is effectively assigned its own variance. Furthermore, in most implementations the residuals are assigned the following prior distribution:

$$e_n \mid \sigma^2_e \sim N\left(0, \sigma^2_e\right)$$
$$\sigma^2_e \mid \nu_e, S^2_e \sim \nu_e S^2_e \chi_{\nu_e}^{-2}$$

(12.7)

Additional prior distributions are also assigned to other effects (including the mean). We will not present these and will just point out that the sampling scheme remains the same.

An additional condition is introduced in **BayesB** where

$$a_k \mid \sigma^2_{a_x} \sim \begin{cases} 0 & \text{for } \sigma^2_{a_x} = 0 \\ N\left(0, \sigma^2_{a_x}\right) & \sigma^2_{a_x} > 0 \end{cases}$$

(12.8)

and

$$\sigma^2_{a_x} \mid \pi \sim \begin{cases} 0 & prob\ \pi \\ \nu S^2 \chi_{\nu}^{-2} & prob\ (1 - \pi) \end{cases}$$

(12.9)

So that effectively:

$$p\left(a_x, \sigma^2_{a_x} | \pi\right) \sim \begin{cases} a_x, \sigma^2_{a_x} = 0 & prob\ \pi \\ N\left(0, \sigma^2_{a_x}\right), p\left(\nu S^2 \chi_\nu^{-2}\right) & prob\ (1-\pi) \end{cases} \tag{12.10}$$

Finally, we have:

$$p(a_x | \pi) \sim \begin{cases} a_x = 0 & prob\ \pi \\ t[0, \nu, s^2] & prob\ (1-\pi) \end{cases} \tag{12.11}$$

Again, in this model each marker possesses its own variance. The difference is introduced by the parameter π that represents the probability that a marker's effect is exactly 0. Two things should be noted. The first is that by setting π (again a hyper-parameter in this implementation) at different levels we specify how many markers (at each iteration) will escape this strong shrinkage. Does this seem arbitrary? It is, but it can be generalized (see below). The second is that by setting $\pi = 0$ we revert back to **BayesA**.

One last important note: Bayes-B is **not** an explicit model selection approach, at least not in the traditional sense. During the Gibbs sampling process, at each iteration each marker can be assigned zero effect (with probability π), so that no marker is permanently discarded from the analysis.

Although these two models are extremely popular they are not free of drawbacks. Because of their structure, both Bayes A and B models are sensitive to the choice of priors and hyper-parameters employed and are dependent on the prior information used. An extensive review of the properties of these models is found in Gianola et al. (2009). Attempts to generalize the choice of the degrees of freedom and scale parameters (assigning them a prior distribution) can be found in literature (Cleveland et al. 2012; Maltecca et al. 2012; Yi and Xu 2008). It should be noted, though, that the overall impact of these efforts on the actual predictive ability of the models with real data is somewhat limited.

BayesC and BayesCpi

A different implementation of Bayesian regression analyses was developed by Habier et al. (2011). In this paper, the problems arising in BayesA and B are tackled by using a single marker variance rather than a locus specific variance, so that the influence of the scale parameter is lessened. In addition, the authors introduced methods to treat the π parameter as an extra unknown and to estimate it from the data. The model in this case is similar to the previous ones, with the difference that in BayesC:

$$\sigma^2_{a_x} = \sigma^2_a \tag{12.12}$$

meaning that all the markers have *a priori* a common variance, which in its most common implementation is:

$$\begin{aligned} a_x \mid \sigma^2_a &\sim N\left(0, \sigma^2_a\right) \\ \sigma^2_a \mid \nu, S^2 &\sim \nu S^2 \chi_\nu^{-2} \end{aligned} \tag{12.13}$$

with $\nu = 4.2$ and S^2 obtained similarly to BayesA and B. The result of this change is that SNPs belonging to the $(1-\pi)$ bin (those loci with effects) are from a mixture of multivariate t distributions $t[0, \nu, \mathbf{I}S^2]$. The remaining structure of the model is similar to BayesA and BayesB although the sampling scheme of the models is slightly different.

Bayesian Lasso

Another alternative model for marker effects involves double exponential priors for the marker effects (with the distribution of residuals typically remaining as an inverted chi-square distribution). This leads to the so-called Bayesian Lasso implementation with a different kind of shrinkage on the markers (Park and Casella 2008; Tibshirani 1996). The distribution is defined by the location μ and a scale parameter β (Andrews and Mallows 1974) and has fatter tails than the normal distribution (Fig. 12.2), so that it gives more weight to some markers while shrinking the effects of a large number of markers, including the possibility of shrinkage to exactly zero effect.

There is no explicit variable selection in the Bayesian version of Lasso, but the effects of some markers become zero. The only difference from ridge regression is that the absolute values of coefficients are used in the penalty part instead of squares. In the frequentist version of Lasso the marker effects are estimated as

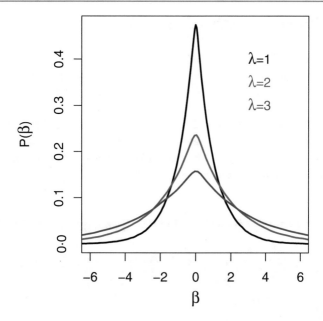

Fig. 12.2 Double exponential (Laplace) distribution for different regularization parameter (lambda) values used as prior density in Bayesian regression

$$\widehat{\mathbf{a}}_{\text{Lasso}} = \underset{\mathbf{a}}{\operatorname{argmin}} \left\{ |\mathbf{y} - \mathbf{Wa}|^2 + \lambda \sum_{i=1}^{p} |a_i| \right\} \tag{12.14}$$

Where $\mathbf{y}$ is the vector of phenotypes, $\mathbf{W}$ is the genotypes matrix and $\mathbf{a}$ is the vector of markers effects, The λ is again a regularization parameter. The argument (*argmin*) refers to estimation of regression coefficients (*a*) while minimizing the expression inside the curly brackets. The expression $|\mathbf{y} - \mathbf{Wa}|^2$ refers to the sum of squares penalty (minimizing the residual sum of squares), while $\lambda \sum_{i=1}^{p} |a_i|$ puts a stronger shrinkage on markers with small effect, pushing some to zero effect, and making the model sparse (Gianola 2013; Heslot et al. 2012). Shrinkage of markers is dependent on the regularization parameter λ and shrinkage is not uniform. Compared to the ridge regression, the change is subtle but can improve the model performance substantially in some cases.

Choice of Statistical Models

The true genetic architecture of the trait of interest should dictate whether the Bayesian models provide greater predictive power than GBLUP (de los Campos et al. 2013). The models described in the previous section are just a fraction of the possible marker regression models. Many models that appear very different are in reality fundamentally related. Often the change of a few model assumptions leads from one model to the other. A good review of the relationships between several of the most popular models is provided in de los Campos et al. (2013). It is important to note that with simulated data, differences between models become more marked, often reflecting the assumptions made in simulating the data. However, in real life, model performance will depend more on the trait genetic architecture (which we do not really know) and the population structure than on the choice of priors.

Furthermore, the bulk of these models have been built with a "QTL mapping" attitude, meaning that they attempt to effectively capture the LD between markers and QTL and weed out uninformative markers. Nevertheless, a surprising number of complex traits seem to be (or at least behave) as truly polygenic with a very large number of QTL with small effects (Buckler et al. 2009; Hill et al. 2008). Under these conditions, differences between models are typically small. Lastly GS models trace not only the LD between markers and QTL but also the familial linkage (Habier et al. 2007; Saatchi et al. 2011), so that the population structure becomes as important as the choice of the model for the analysis. There is no one model that is best for all circumstances and a careful consideration of models, training/prediction splits, and nature of the trait is needed when analyzing real data (de los Campos et al. 2010).

Bayesian Regression Examples with *BGLR* Package

In the following section we will use the *BGLR* R package (Pérez and de los Campos 2014) to demonstrate the use of Bayesian whole genome regression prediction models. The *BGLR* package implements a variety of Bayesian and semi-parametric methods, such as Bayesian reproducing kernel Hilbert spaces regression, Bayesian Ridge, Bayesian Lasso, BayesA, BayesB and BayesC etc. We will start by running a script that demonstrates the use of Bayesian Lasso genomic prediction, which will introduce the basic commands. For demonstration, the maritime pine data set introduced in Chap. 9 and also used for GBLUP in Chap. 11 will be used. Briefly, 654 pine trees were genotyped using 2600 SNP markers to establish associations with simulated phenotypes. Several different traits with different degrees of polygenicity were simulated. We will start with some data manipulation.

Code example 12.1
**Bayesian Lasso regression for genomic prediction using *BGLR* package. See R code *Code12-1_BayesianModels.
R* for more details.**

```
##################################################
# Code12-1_BayesianModels.R
# Isik, Holland, Maltecca
##################################################

# Remove everything in the working environment
rm(list=ls())
set.seed(12345)

# install BGLR package
require("BGLR")
library(coda)
library(MCMCpack)


#################################################
# Change folder according your computer!
path <- "/Users/.../Book/Book1_Examples/data"
setwd(path)
# Save temp files here
outpath <- getwd()

### Load the R data set
load('maritime pine gpData.rda')

### * GENOTYPES
genotype <- as.matrix(gp.num$geno)
# genotype[1:15, 1:5]

# BGLR requires matrix of markers.
# marker matrix is centered on zero for ease of calculation
geno_centered <- genotype - 1


### * PEHNOTYPE
phenotypes <- data.frame(gp.num$pheno)
names(phenotypes) <- c('Trait1','Trait2','Trait3','Trait4')
y <- as.array(phenotypes$Trait1)
head(y) <- gp.num$pheno
```

In the next section of the code we randomly split the data into $k = 2$ folds, and we use one set for training (TS contains 554 individuals) and the smaller set for validation (VS contains 100 individuals). In the training set, individuals have both phenotype and genotypes. In the validation set the trait is set to missing. Individual marker effects in the training set are obtained by fitting the Bayesian Lasso regression model. The sum of all marker effects for an individual in the validation set is the genomic estimated breeding value (GEBV) of the individual. The correlation between GEBV and the phenotype of individuals in the validation set are used to evaluate the predictive ability of markers. This process is normally repeated many times to get a distribution of the accuracy and other model fit statistics.

Code example 12.1
(continued)

```
# * sampling of population for cross validation
samplesize <- 100
fold <- 2

# * sample vector y from observation 1, using samplesize
whichNa <- sample(1:length(y), size = samplesize, replace = FALSE)
yNa <- y                 # make a copy of y. yNa and y are numeric
yNa[whichNa] <- NA       # set the values of sampled trees to missing


#############################################
## 0   BGLR
#############################################

fmR <- BGLR(y  = yNa, response_type = "gaussian", a = NULL, b = NULL,
            ETA = list(MRK = list(X = geno_centered, model = "BL")),
            nIter = 20000, burnIn = 1000, thin = 10,
            saveAt = outpath)

summary(fmR)
```

The *BGLR* function fits the Bayesian marker regression model (Bayesian Lasso in this case). The selection of the Bayesian Lasso model is determined by the 'model = BL' option in the code. For this demonstration we ran 20,000 iterations, out of which 1000 iterations were used as burn-in and only one of every 100 posterior samples was kept to thin the chain. In an MCMC run, the **"burn-in"** is a number of iterations discarded at the beginning of the chain to make sure that the samples taken come from the posterior distribution desired. Some authors describe this process as an unnecessary step and mention some undesirable effects (Toft et al. 2007). In sampling, **"thinning"** is used so that only a certain proportion of the iteration results are kept to reduce the autocorrelation between samples. In this example, every 10th draw was kept from the posterior sequence and the rest were discarded.

In this example, we ran only 20,000 iterations, but for large numbers of markers (or more complex models) much longer chains should be run. The R package *coda* provides a set of functions to help the user to monitor the status of Markov Chain convergence (Plummer et al. 2006).

While the BGLR package accepts categorical traits, we are analyzing a continuous trait in this case so that the type = "gaussian" argument was employed. The 'a=' and 'b=' parameters are vectors for upper and lower limits of the censored response variables, respectively. Since there is no censoring in this data, they were set to null. The argument ETA = refers to a two tier list that specifies the linear predictors, such as intercept and other fixed or random effects. This is an important part of the model and allows selecting the type of regression chosen.

For this first example we used a Bayesian Lasso with the default priors provided by the package. The "BL" option allows one to set the λ as a fixed parameter. In alternative implementations λ can have two possible prior distributions: a gamma or a beta distribution. The program default is type = 'gamma', which fixes the shape parameter of a gamma density to 1.1 and then solves for the rate parameter to match the expected proportion of variance accounted for by the corresponding element of the linear predictor (for details see http://genomics.cimmyt.org/BGLR-extdoc.pdf). The proportion of variance explained by the linear element has a general interpretation but in the case of marker effects it can be (at least loosely) be interpreted as the proportion of variance accounted for by the markers. This parameter is governed by the argument R2. The default value of R2 is set to 0.5.

We have assigned the output to an R object called fmR. The object contains a wealth of information on the structure, fit and outputs of the model. The structure of the object can be seen by typing str(fmR) in the R console. It is a list object and each element of the list can be accessed by the use of the dollar sign. For example, the R2 employed in the model can be seen by typing fmRETAMRK$R2. A subsection of the output is given below.

```
> str(fmR)

List of 21
$ y           : num [1:654] NA 10.4 9.6 NA NA ...
...

$ df0         : num 5
$ S0          : num 1.62
$ yHat        : num [1:654] 9.57 9.91 9.67 10.07 10.07 ...
$ SD.yHat     : num [1:654] 0.385 0.225 0.326 0.401 0.411 ...
$ mu          : num 10.1
$ SD.mu       : num 0.309
$ varE        : num 0.344
...

$ ETA :List of 1
  ..$ MRK:List of 16
  .. ..$ model     : chr "BL"
...

  .. ..$ R2        : num 0.5
  .. ..$ lambda    : num 57
  .. ..$ type      : chr "gamma"
  .. ..$ shape     : num 1.1
  .. ..$ rate      : num 4.97e-05
...

  .. ..$ tau2      : num [1:2567] 0.000689 0.000675 0.000694 ...
...
```

The R2 is reported as 0.5, because we allowed the program to set this parameter at its default value. The lambda value (regularization parameter) was set to $\lambda = 57$. Larger values of lambda would shrink the marker effects more strongly. The prior density of λ^2 is set to a gamma distribution with shape r and rate δ parameters, with expectations $\lambda^2 \sim gamma(r, \delta)$. The shape parameter was set as previously mentioned at 1.1 and the rate was $\delta = 4.97e - 05$. The package assigns a flat prior (meaning conveying vague almost uniform prior knowledge) to the vector of fixed effects (intercept in this example), thus the expectation of the intercept is $p(\mu) \propto 1$. 'tau2' is the vector of marker-specific variances. The tau values for the first three markers are shown in the str(fmR) output above. The complete vector of values can be accessed as fmRETAMRK $tau2. In a Bayesian Lasso model, markers are assigned identical and independent distributed exponential priors with expectations $\tau_j^2 \sim Exp(\lambda^2)$. At the beginning of the output we see two parameters (df0 and S0). These two values represent the degrees of freedom and scale parameter for the prior distribution of the residual variance. The residual variance is assigned a scaled inverse χ^2 prior distribution with degrees of freedom $df_e = 5$, and scale parameter $S_e = 1.62$. That is, the expectations of residuals are $\sigma_e^2 \sim \chi^{-2} (df_e, S_e)$ (de los Campos et al. 2013; de los Campos and Perez Rodriguez 2014).

The summary of the model can be obtained by using the function 'summary()'

```
> summary(fmR)
-------------------> Summary of data & model <--------------------

 Number of phenotypes= 554
```

```
Min (TRN)=  7.949793
Max (TRN)= 12.11167
Variance of phenotypes (TRN)= 0.5719
Residual variance= 0.3681
N-TRN= 554  N-TST= 100
Correlation TRN= 0.8696

-- Linear Predictor --

Intercept included by default
Coefficientes in ETA[ 1 ] ( MRK ) modeled as in  BL
```

N-TRN =554 is the size of the training set, N-TST = 100 is the size of the validation set. The correlation of direct genetic values and the phenotype in the training set is 0.8696. Markers explained about 61% of the variance in phenotype in the training set ~0.5719/(0.5719 + 0.3681).

In order to understand the predictive power of markers we can estimate model fit statistics; the predicted ability of markers, rank correlations, model bias and the phenotypic mean of the top 10% of individuals ranked on their GEBV (Wimmer et al. 2012). See Chap. 11 for the explanation of model fit statistics.

Code example 12-1
(continued)

```
#### *  Calculate model statistics
df1 <- data.frame(fmR$yHat[whichNa], y[whichNa])
PredAbi = round( cor(df1[, 2], df1[, 1]), 2)
Rank = round( cor(df1[, 2], df1[, 1], method = 'spearman'), 2)
MSE =   round( mean((df1[, 2] - df1[,1])^2), 2)
Bias  = round( coef(lm(y[whichNa] ~ fmR$yHat[whichNa]))[2], 2)
Best10 =    round( mean(tail(sort(df1[, 2]), n =
 ceiling(nrow(df1)*0.1))), 2)
# organize in a table
table <- cbind(PredAbi, Rank, MSE, Bias, Best10)
rownames(table) <- c('Height')
table
```

	PredAbi	Rank	MSE	Bias	Best10
Height	0.41	0.39	0.41	1.03	11.1

For this one-time cross validation, the prediction ability of the markers is 0.41, with little bias on the predictions (1.03). The rank correlation between phenotype and GEBV in the validation set is 0.39. When individuals are ranked based on their GEBV, the top 10 had height mean of 11.1. See the R code "Code example 12.1" for visualization of the relationships between phenotype and GEBV.

Comparisons of some Bayesian models

In the following code, we compare Bayesian Ridge Regression, Bayesian Lasso, BayesA, BayesB and BayesC models, following an example in de los Campos and Perez Rodriguez (2014). We leave the priors at their default settings and do not try to optimize any of the models, as the objective is to compare a large variety of models. A particular note in this case is necessary for the BayesB model. While in the traditional implementation of the model, as described in the previous section of the chapter, the π is treated as a hyper prior, in BGLR this is treated as another random parameter with its own prior distribution and assigned a beta distribution $\pi \sim Beta(p_0, \pi_0)$. That is actually one of the nice properties of Bayesian hierarcical models. Choosing values $p_0 = 2$, $\pi_0 = 0.5$ will produce a flat uniform [0,1] prior for π. Conversely, increasing the value of p_0 will increase the mass of the distribution towards the value of π_0. So for example, if we want a strong prior on π_0 for a oligogenic architecture (a trait controlled by a few major genes) we could choose a large p_0 and a $\pi_0 = 0.05$. The π governs the proportion of markers at any iteration that have a non 0 effect. The default values of BGLR are $p_0 = 10$, $\pi_0 = 0.5$. You might have noticed that while in the previous sections we have presented π as the prior probability of a marker of being 0 (the usual way in which is presented in the majority of implementations) in BGLR the interpretation of the parameter π_0 is reversed representing the prior probability of a maker to be $\neq 0$. So from this point on we will refer to π_0 as a prior probability of a marker of having non null value.

Code example 12.1
(continued)

```
###############################################
## * 4 Comparing models
###############################################

nIter = 20000;  burnIn = 5000

## Bayesian LASSO
ETA<-list(MRK=list(X=geno_centered, model="BL"))
fmBL <- BGLR(y = yNa,  ETA = ETA,  nIter=nIter,  burnIn=burnIn,
 saveAt = "BL_")

## Bayesian Ridge Regression (Gaussian prior),
ETA<-list(MRK=list(X=geno_centered,model="BRR"))
 fmBRR<-BGLR(y=yNa,ETA=ETA, nIter=nIter,burnIn=burnIn, saveAt="BRR_")

## Bayes A (Scaled-t prior)
ETA$MRK$model<-"BayesA"
 fmBA<-BGLR(y=yNa,ETA=ETA,nIter=nIter,burnIn=burnIn,saveAt="BA_")

## Bayes B (point of mass at zero + scaled-t slab)
 ETA$MRK$model<-"BayesB"
 fmBB<-BGLR(y=yNa,ETA=ETA,nIter=nIter,burnIn=burnIn,saveAt="BB_")

## Bayes C (point of mass at zero + scaled-t slab)
 ETA$MRK$model<-"BayesC"
 fmBC<-BGLR(y=yNa,ETA=ETA,nIter=nIter,burnIn=burnIn,saveAt="BC_")

#### * Calculate correlations
 r_BL  <- cor(y[whichNa],  fmBL$yHat[whichNa] )
 r_BRR <- cor(y[whichNa],fmBRR$yHat[whichNa] )
 r_BA  <- cor(y[whichNa], fmBA$yHat[whichNa] )
 r_BB  <- cor(y[whichNa], fmBB$yHat[whichNa] )
 r_BC  <- cor(y[whichNa], fmBC$yHat[whichNa] )

 # organize in a table
table <- data.frame(rbind( r_BL, r_BRR, r_BA, r_BB,  r_BC) ,
rbind(fmBL$varE,fmBRR$varE,fmBA$varE,  fmBB$varE,fmBC$varE),
rbind(fmBL$fit$pD,fmBRR$fit$pD,fmBA$fit$pD,fmBB$fit$pD,fmBC$fit$pD),
rbind( fmBL$fit$DIC,fmBRR$fit$DIC,fmBA$fit$DIC,fmBB$fit$DIC,
 fmBC$fit$DIC)  )

colnames(table) <- c('PredAbi', "varE", "pD", "DIC")
rownames(table) <- c('BLasso',"BRidge","BayesA","BayesB","BayesC")
round(table, 3)
```

The 'pD' is the effective number of predictors, showing the complexity of the model (Spiegelhalter et al. 2002). The 'DIC' is deviance information criteria. Small DIC values are preferred. Model fit statistics pD, DIC, residual variance and predicted ability of markers for statistical models are summarized below.

```
       PredAbi  varE      pD      DIC
BLasso   0.410 0.361 166.310 1172.716
BRidge   0.421 0.348 179.770 1163.091
BayesA   0.416 0.353 174.289 1165.808
BayesB   0.399 0.352 176.262 1166.047
BayesC   0.415 0.348 179.192 1162.593
```

For this small example, the models did not differ for prediction ability of markers. Bayesian Lasso had the highest residual error variance, lowest effective number of predictors (pD) and highest deviance information criteria (DIC). The estimated residual variances were similar for all the models.

The predicted marker coefficients from Bayesian Ridge regression and BayesB are plotted in Fig. 12.3 for demonstration using code *Code 12-1_BayesianModels.R*. It can be seen from the plot how BayesB puts less shrinkage on markers with large effect (red dots) compared to Ridge Regression (black dots).

Model Fit Statistics and Model Convergence

In maximum likelihood-based methods, convergence is reached when the change in likelihood between the current and previous iterations is less than some chosen threshold. The same approach is not possible with MCMC methods because in this case, parameter estimates are constructed as statistics of the respective posterior distributions. Therefore, it is necessary to check model convergence in other ways. The simplest way is to inspect the *trace plots* of the residual variance and other parameter estimates that indicate how their values change over iterations. After the burn-in phase, the trace plots should appear as random samples from the target posterior without large "jumps" or erratic behaviors. In the example below we plot the convergence of residual variance and the regularization parameter (lambda) over iterations.

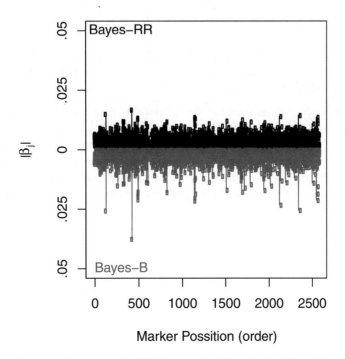

Fig. 12.3 Markers effects (absolute beta values) from Bayesian Ridge Regression (*black dots*) and BayesB models (*red dots*). BayesB favors loci with large effect as shown by *red dots*

Code example 12.1
(continued)

```
##########################################################
### *  5  TRACE PLOTS FOR LASSO
##########################################################

### plot of residuals convergence
varE <-scan(paste(outpath,'/BL_varE.dat',sep=''))

par(mfrow=c(1,2))
par(mar=c(4,4.5,0,0)+0.2)#sets margins of plotting area

plot(varE, type="o", col = "gray60", pch = 20, cex = 1, ylim =
c(0,0.6), ylab = expression(paste(sigma[epsilon]^2)))

abline(h=fmBL$varE, col=1, lwd=2, lty=1)
abline(v=fmBL$burnIn/fmBL$thin,col=1, lwd=2, lty=1)

### lambda (regularization parameter of BL)
lambda <- scan(paste(outpath,'/BL_ETA_MRK_lambda.dat',sep=''))

plot(lambda, type='o', col="gray60",cex=.5, ylim = c(0,100),
     ylab=expression(lambda))

abline(v=fmBL$burnIn/fmBL$thin, col=1, lwd=2, lty=1) # vertical line
abline(h=fmBL$ETA$MRK$lambda, col=1, lwd=2, lty=1) # horizontal line
```

In the residual plot (Fig. 12.4), the residual variance (h = fmBL$varE) estimated from the model is the horizontal line, the X-axis labelled "index" represents the sequential iteration number that was maintained after thinning. The ratio of "burn-in" to "thinning" is used as the vertical line (v = fmBL$burnIn/fmBL$thin).

More formal methods to check for convergence are available, but we will not discuss them here in detail. R packages offer a wealth of possibilities in this regard. Below is a simple example using our data and the R package *coda*, one of the most used and intuitive packages. There is unfortunately no silver bullet to check convergence in MCMC analyses. For this example, we will use a commonly employed measure, the Gelman-Rubin statistic (in practice, multiple tests of convergence are usually recommended). The Gelman-Rubin statistic checks the convergence of MCMC by comparing differences between multiple chains. The test aims at comparing the estimated between-chain and within-chain variances for a particular

Fig. 12.4 Trace plots of the Bayesian Lasso regression model for 20,000 iterations. The horizontal axes represent the sequential iteration retained after thinning. Vertical axes are the model parameters residual variance (*left*) and regularization parameter lambda (*right*). The vertical lines represent the end of the burn-in phase. The horizontal lines are the posterior means of the residual variance and the shrinkage factor lambda

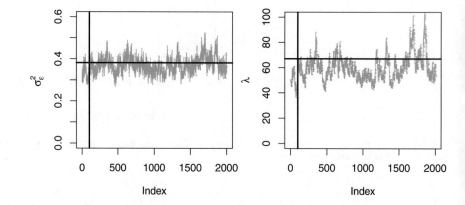

parameter. The difference is measured through the *potential scale reduction factor* (Brooks and Roberts 1998). Large values of potential scale reduction factor indicate lack of convergence. A potential scale reduction factor of 1 means that between- and within-chain variances are equal and therefore convergence has been reached. As a rule of thumb, values below 1.1 are considered acceptable. When the test fails, a longer chain is required or a different model specification is needed. The Gelman plot shows potential scale reduction factor over time, and can be used to determine whether a chain reduction is also stable as large variation over-time of the potential scale reduction factor might be indicative of poor convergence.

As the Gelman-Rubin statistic requires multiple chains we will start by running five chains using the model specifications outlined in the previous sections.

**Code example 12.1
(continued)**

```
###################################
# 8 MODEL DIAGNOSTICS
###################################


# Bayesian LASSO Regression
# Number of samples
nIter = 20000
# Burn-in period for the Gibbs sampler
burnIn = 0
# Number of chains
n.chains=5

# Proportion of phenotypic variance attributed to model residuals
r=0.5

# Prior hyperparameter values
# sigmaE2 (residual variance)
mode.sigE=r*var(y)
dfe=3
Se=mode.sigE*(dfe + 2)

# lambda
mode.sigL=(1-r)*var(y) # proportion attributed to genetics
lambda.hat=sqrt(2*mode.sigE/mode.sigL*sum(colMeans(W)^2))
delta.lambda=0.05                # rate
r.lambda=lambda.hat*delta.lambda # shape

# Set priors
prior=list( varE=list(S0=Se,df0=dfe,value=runif(1,min=0,max=100)),
            lambda=list(type='random',
                        value=runif(1,min=0,max=100),
                        shape=shape,
                        rate=rate) )

ETA<-list(MRK=list(X=geno_centered, model="BL", prior))

for (k in 1:n.chains){
  # Fit Bayesian LASSO Regression
  fmR<-BGLR(y=y,ETA=ETA,
          nIter=nIter,
          burnIn=burnIn,
          thin=1,
          saveAt=paste(outpath,'/Gelman-Rubin_Chain_L_',k,sep=''))
}
```

Next, we load the saved MCMC samples for σ_e^2 into an R object.

Code example 12.1
(continued)

```
# Load MCMC draws of varE in matrix varE_L
 varE_L=matrix(NA,ncol=n.chains,nrow=nIter)
 for (k in 1:n.chains){
   varE_L[,k]=read.table( file=paste(outpath, '/Gelman-
Rubin_Chain_L_',k,'varE.dat',sep='') )$V1
 }
```

We will then create an MCMC object with all the chains (note that the *coda* package requires objects of class "MCMC").

Code example 12.1
(continued)

```
# MCMC object with all chains
 idx=500:20000
 THETA=mcmc.list(mcmc(draws[idx,1]),
                 mcmc(draws[idx,2]),
                 mcmc(draws[idx,3]),
                 mcmc(draws[idx,4]),
                 mcmc(draws[idx,5]))

# Gelman-Plot
# Create output file
par(mfrow=c(1,1))
par(mar = c(5, 5, 5, 5), mgp=c(3,1,0) )
gelman.plot(THETA,
            main='',
            xlab='Iteration',
            cex.lab=1.2,
            cex=1.5,
            lwd=2)
 dev.off()
```

Finally, we compute the Gelman-Rubin statistic.

Code example 12.1
(continued)

```
# Gelman-Rubin statistic
 gelman.rubin=gelman.diag(THETA)
 gelman.rubin
```

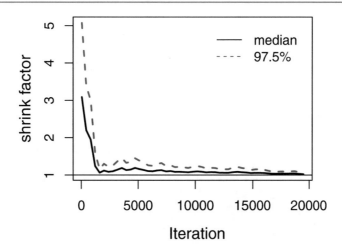

Fig. 12.5 Gelman plot of the shrinkage factor for σ_e^2. The X axis is the iteration number. The Y axis is a measure of convergence, the potential scale reduction factor

The output of the call is reported below:

```
Potential scale reduction factors:

       Point est.   Upper C.I.
[1,]       1.03        1.06
```

As you can see both the point estimate and the upper confidence interval for the potential scale reduction factor of σ_e^2 are below 1.1 indicating convergence. Additionally, we can also produce a Gelman plot to examine the stability of the potential scale reduction factor over samples and verify that the decrease in potential scale reduction factor is stable (Fig. 12.5).

Choice of Priors

BGLR provides a convenient way to choose the shape of prior distributions shape via the R2 argument. By default the R2 is set to 0.5. That is, the effect modeled explains 50% of the total variance. If there is more than one linear predictor, each will have its own R2 values.

In the following example, we will set the residual variance priors as well as lambda mimicking the default values of the "BL" option (Bayesian Lasso). As before, residuals are assigned a scaled inverse χ^2 distribution. By default, the degrees of freedom are set to 5 and that the relevant scale parameter is obtained from R2 by solving $S_e = Var(\mathbf{y})(1 - R2)(df_e + 2)$. This guarantees that the prior mode of the residual variance is $Var(\mathbf{y})(1 - R2)$. The same process can be applied for other variances too. The following code will reproduce default values for the residual variance:

**Code example 12.1
(continued)**

```
#########################################
### 6 PRIOR SPECIFICATION
#########################################

  # Proportion of phenotypic variance attributed to model residuals
  R2=0.5

  # Prior hyperparameter values
  # sigmaE2 (residual variance)
  mode.sigE=R2*var(y)
  dfe=5
  Se=mode.sigE*(dfe + 2)
```

Similarly, using the BL specification, the default lambda (if not fixed a *priori*) can also be specified with the option: $\texttt{type} = $ 'gamma'. This fixes the shape parameter of a gamma density to 1.1 and solve for the rate parameter to reflect the R2. The rate is calculated as $(s-1)/\left(\frac{2(1-R2)}{R2\ MSx}\right)$, where *MSx* is the sum of sample variances of the columns of genotypes and *s* is the shape parameter.

**Code example 12.1
(continued)**

```
# lambda
  mode.sigL=(1-R2)*var(y) # variance explained by markers
  rate=2*Se/mode.sigL*sum(colMeans(geno_centered)^2)
  shape=1.1

  # Set priors
  prior=list( varE=list(S0=Se,df0=dfe),
              lambda=list(type='random',
                          shape=shape,
                          rate=rate) )
```

We can now fit the same example with the updated prior list:

**Code example 12.1
(continued)**

```
ETA<-list(MRK=list(X=geno_centered, model="BL", prior))
  fmBL <- BGLR(y = y,  ETA = ETA,  nIter=nIter,  burnIn=burnIn,
  saveAt=outpath )
```

In the example dataset, varying the R2 parameter does not dramatically affect the posterior distributions (solid lines in Fig. 12.6). This is not uncommon and in many cases, most of the information in a run will come from the likelihood, rather than from the prior. A sensitivity analysis is always a sensible choice to empirically evaluated how much the prior and hyper-parameter settings influence the results. Further details on rules for choosing prior specifications and hyper-parameters can be found at http://genomics.cimmyt.org/BGLR-extdoc.pdf.

Here we demonstrate a small sensitivity analysis with the example data set:

Code example 12.1
(continued)

```
#######################################
### 6 PRIOR INFLUENCE
#######################################

# Bayesian LASSO
# propE, proportion of phenotypic variance attributed to model
residuals
# Number of samples
nIter = 20000
# Burn-in period for the Gibbs sampler
burnIn = 5000
# propE, proportion of phenotypic variance attributed to model
residuals
propE=c(0.8,0.6,0.4)
for (r in propE ){
  # Prior hyperparameter values
  # sigmaE2
  mode.sigE=r*var(y)
  dfe=3
  Se=mode.sigE*(dfe + 2)

  # lambda
  mode.sigL=(1-r)*var(y)

lambda.hat=sqrt(2*mode.sigE/mode.sigL*sum(colMeans(geno_centered)^2))
  delta.lambda=0.05                    # rate
  r.lambda=lambda.hat*delta.lambda     # shape

  # Set priors
  prior=list( varE=list(S=Se,df=dfe),
              lambda=list(type='random',
                          value=lambda.hat,
                          shape=r.lambda,
                          rate=delta.lambda) )
```

We begin by setting the different values for σ_e^2 R2 using propE = c(0.8, 0.6, 0.4). Then we specify the different prior values as outlined in the previous section. Finally we run the sequential calls of BGLR using the different priors and save the outputs.

Code example 12.1
(continued)

```
  ETA<-list(MRK=list(X=geno_centered, model="BL", prior))
  # Fit Bayesian LASSO Regression
  fmR<-BGLR(y=y,ETA=ETA,
            nIter=nIter,
            burnIn=burnIn,
            thin=1,
            saveAt=paste(outpath,'/LASSO_r_',r,'_',sep=''))
}
```

After the analysis we can load the MCMC runs (for σ_e^2) and put them all in an R object for further manipulation.

**Code example 12.1
(continued)**

```
# Load MCMC draws of varE in matrix varE_L
 k=0
 varE_L=matrix(NA,ncol=length(propE),nrow=nIter)
 for (r in propE){
    k=k+1

 varE_L[,k]=read.table(file=paste(outpath,'/LASSO_r_',r,'_varE.dat',sep
='')),sep=' ')$V1
    }
```

We can now plot prior and posterior densities. Note that we are going to plot the prior using the function `dinvgamma()` in R, so that we are not using the MCMC samples but a plot generated using the density function and parameters specified to match the priors used.

**Code example 12.1
(continued)**

```
# Prior densities for varE
 k=0
 N=2^9
 priorE=matrix(NA,ncol=length(propE),nrow=N)
 for(r in propE){
    k=k+1
    mode.sigE=r*var(y)
    dfe=3
    Se=mode.sigE*(dfe + 2)
    xE=seq(from=0.01,to=0.8,length.out=N)
    priorE[,k]=dinvgamma(x=xE,shape=dfe/2, scale=Se/2)
 }
```

Then we produce the posterior density, making use of the generated MCMC samples.

**Code example 12.1
(continued)**

```
# Posterior densities for varE
 for (k in 1:length(propE)){
    namx=paste('dLx_',k,sep='')
    namy=paste('dLy_',k,sep='')
    dL=density(varE_L[(burnIn+1):nIter,k],n=N)
    assign(namx,dL$x)
    assign(namy,dL$y)
 }
```

Finally, we plot the results.

**Code example 12.1
(continued)**

```
# Plot
 par(mfrow=c(1,1))
 par(mar = c(4.3, 4.3, 0.9, 0.9), mgp=c(3,1,0) )

 plot(xE,priorE[,1]/max(priorE[,1]),
      xlab=expression(sigma[e]^2),ylab='Scaled density',
      main='', xlim=c(0,1),
      type='l',lty=2,lwd=2,cex.lab=1.2,col=2)

 for (k in 2:length(propE)){
   lines(xE,priorE[,k]/max(priorE[,k]),type='l',lty=2,lwd=2,col=k+1) }

 for (k in 1:length(propE)){
   xx=get(paste('dLx_',k,sep=''))
   yy=get(paste('dLy_',k,sep=''))
   lines(xx,yy/max(yy),type='l',lwd=2,col=k+1)
 }

 legend('bottomright',lty=c(1,1,1),lwd=2,col=c(2,3,4),
        legend=c('Posterior (Prior: 20% Genetics)',
                 'Posterior (Prior: 40% Genetics)',
                 'Posterior (Prior: 60% Genetics)'),
        bty='n',cex=0.9)
```

The figure suggests that priors' effect on the posterior distribution of residual variance is limited. Despite choosing three different priors (residual variance is assumed 80%, 60% and 40%), the posterior means of the residual variance were close to each other (0.34, 0.33, 0.33). The exercise could be repeated with all other parameters for which a prior distribution is specified.

Fig. 12.6 Prior (*dashed lines*) and estimated scaled posterior densities (*solid lines*) for the residual variance assuming three informative priors in Bayesian Lasso. The prior densities reflect the proportion of phenotypic variance attributed to the model residuals: 80% (*red*), 60% (*green*), or 40% (*blue*). The solid lines are posterior distributions of genetics effects

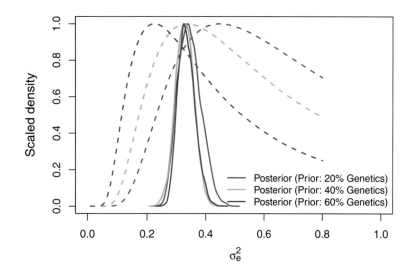

Genetic Architecture

We have seen in the previous example how model choice did not significantly affect the result of our analysis. In the following section we will explore the influence of model choice when the trait analyzed has different genomic architectures. The example data set includes four simulated traits with different genetic architectures so we can demonstrate the influence of genetic architecture on Bayesian analysis results. The first simulated trait has a completely additive polygenic architecture (no QTL) while for traits 2 to 4, we simulated an increasing number of causative loci underlying the trait (2, 10 and 100 QTL, respectively) in addition to additive polygenic effects. For a given trait, all QTL have identical true effects. Although this is unrealistic, it simplifies interpretation of differences between the models employed. For each of the four traits we will run Bayesian Ridge Regression, BayesA, BayesB, BayesC, and Bayesian LASSO models (using default values for priors) and compare the predictive ability of the models. The code example below loops through the four traits, fitting each of the models to each trait and evaluating their prediction abilities.

Code example 12.1
(continued)

```
########################################
# 9   GENETIC ARCHITECTURE
########################################

Sim_tr<-read.table("maritime pine simul pheno.csv", sep=",",
header=T) #reading simulated data

### checking distribution of traits
par(mfrow=c(2,2))
for(i in 2:(ncol(Sim_tr))){
 plot(density(Sim_tr[,i]),main=names(Sim_tr)[i])
}

nIter = 30000;  burnIn = 10000
# storing results
results<-list(TRAIT_1=NULL,TRAIT_2=NULL,TRAIT_3=NULL,TRAIT_4=NULL)

for(trait  in 2:5){
y<- Sim_tr[,trait]
# * sample vector y from observation 1, using samplesize
whichNa <- sample(1:length(y), size = samplesize, replace = FALSE)
yNa <- y                 # make a copy of y. yNa and y are numeric
yNa[whichNa] <- NA    # set the values of sampled trees to NA

## Bayesian Ridge Regression (Gaussian prior),
ETA<-list(MRK=list(X=geno_centered,model="BRR"))
fmBRR<-BGLR(y=yNa,ETA=ETA, nIter=nIter,burnIn=burnIn, saveAt="BRR_")
## Bayes A (Scaled-t prior)
ETA$MRK$model<-"BayesA"
fmBA<-BGLR(y=yNa,ETA=ETA,nIter=nIter,burnIn=burnIn,saveAt="BA_")
## Bayes B (point of mass at zero + scaled-t slab)
ETA$MRK$model<-"BayesB"
fmBB<-BGLR(y=yNa,ETA=ETA,nIter=nIter,burnIn=burnIn,saveAt="BB_")
# Bayes C (point of mass at zero + scaled-t slab)
ETA$MRK$model<-"BayesC"
fmBC<-BGLR(y=yNa,ETA=ETA,nIter=nIter,burnIn=burnIn,saveAt="BC_")
# Bayesian LASSO
ETA<-list(MRK=list(X=geno_centered, model="BL"))
fmBL <- BGLR(y = yNa,  ETA = ETA,  nIter=nIter,  burnIn=burnIn,
saveAt = "BL_")
```

```
#### *  Calculate correlations
r_BRR <- cor(y[whichNa], fmBRR$yHat[whichNa] )
r_BA  <- cor(y[whichNa], fmBA$yHat[whichNa] )
r_BB  <- cor(y[whichNa], fmBB$yHat[whichNa] )
r_BC  <- cor(y[whichNa], fmBC$yHat[whichNa] )
r_BL  <- cor(y[whichNa], fmBL$yHat[whichNa] )

# organize in a table
table <- data.frame(rbind( r_BRR,r_BA,r_BB,  r_BC,r_BL) ,
rbind( fmBRR$varE, fmBA$varE,fmBB$varE,fmBC$varE,fmBL$varE) ,
rbind(fmBRR$fit$pD,fmBA$fit$pD,fmBB$fit$pD,fmBC$fit$pD,fmBL$fit$pD),
rbind(fmBRR$fit$DIC,fmBA$fit$DIC,fmBB$fit$DIC,fmBC$fit$DIC,
fmBL$fit$DIC)   )
colnames(table) <- c( 'PredAbi', "varE", "pD", "DIC")
rownames(table) <- c( "BRidge", "BayesA", "BayesB",
"BayesC","BLasso")
table<-round(table, 3)
results[[trait-1]]<-table
}
```

The resulting output is reported below.

```
$TRAIT_1

          PredAbi    varE       pD      DIC
BRidge     0.455   0.341  185.324  1227.982
BayesA     0.456   0.344  183.208  1231.820
BayesB     0.452   0.353  175.644  1237.907
BayesC     0.453   0.343  185.707  1231.653
BLasso     0.448   0.353  173.912  1237.622

$TRAIT_2

          PredAbi    varE       pD      DIC
BRidge     0.502   2.144  250.393  2383.088
BayesA     0.735   1.750   90.324  2103.156
BayesB     0.734   1.730   96.520  2101.898
BayesC     0.727   1.959   31.998  2113.096
BLasso     0.694   1.313  275.636  2111.773

$TRAIT_3

          PredAbi    varE       pD      DIC
BRidge     0.408   1.331  200.816  2052.399
BayesA     0.563   1.203  208.085  1998.584
BayesB     0.700   1.301  107.687  1947.973
BayesC     0.694   1.296  111.528  1949.099
BLasso     0.491   1.295  197.169  2028.970
```

```
$TRAIT_4

          PredAbi    varE        pD         DIC
BRidge     0.356    0.508    204.830    1483.607
BayesA     0.368    0.498    204.444    1471.021
BayesB     0.407    0.477    216.179    1457.777
BayesC     0.391    0.490    211.429    1467.967
BLasso     0.371    0.518    195.375    1483.115
```

For the first trait (controlled by a purely polygenic architecture), there is almost no difference among models; ridge regression performed as well as the "QTL-oriented" Bayesian models. For trait 2 (2 QTL plus polygenic background), however, Bayes A, B, C and LASSO all perform well, and are substantially better than Bayesian Ridge Regression. As the number of QTL increases from trait 2 to 100 (trait 2 to trait 4), the differences among models decreases. Even in the presence of a finite number of QTL, the use of a Bayesian alphabet model does not automatically guarantee good performance and for QTL of moderate size a large number of observations would be needed to truly take advantage of these models.

Next, we will attempt to tune model priors to increase predictive ability. In the following example we show the impact of varying the π_0 parameter (roughly speaking, the prior proportion of markers we expect to have a significant contribution for a trait) on a BayesC analysis. We created a function called PiInf to make running the comparison easier.

Code example 12.1
(continued)

```
##########################################
 ## 10 Pi INFLUENCE
##########################################

PiInf<-function(nIter=30000,
 burnIn=15000,mod=c("BayesC","BayesC","BayesC","BayesC","BayesC"),prob=
 c(0.01,0.1,0.2,0.5,0.9),count=c(200,200,200,200,200)){
...
```

The function takes few parameters as input and passes them sequentially to different BGLR calls. Specifically, the number of iterations and burn-in, the models to be run and the two parameters that define π in BGLR: probIn (not shown in above piece of code), defining the probability π_0 and counts, which regulates the degree of belief in the probability of inclusion (p_0). Large values of p_0 correspond to stronger belief in the prior. For each simulated trait we initialized a series of BayesC models with different value of π_0 (0.99, 0.9, 0.8, 0.5, and 0.10). Remember BGLR models π_0 as the probability of inclusion so a *probIn* of 0.01 will means that we assume a prior probability of not being 0 of 1%). For all models we use a large prior credibility with counts values of 200 to maximize the influence of the priors on the results. See 'Code 12-1_BayesianModels. R' and the next page for details.

Code example 12.1
(continued)

```
  # storing results
results<-list(TRAIT_1=NULL,TRAIT_2=NULL,TRAIT_3=NULL,TRAIT_4=NULL)
  for(trait  in 2:5){
  y<- Sim_tr[,trait]
  # * sample vector y from observation 1, using samplesize
  whichNa <- sample(1:length(y), size = samplesize, replace = FALSE)
  yNa <- y                # make a copy of y. yNa and y are numeric
  yNa[whichNa] <- NA     # set the values of sampled trees to NA
  ## Bayes C pi .99
  ETA<-list(MRK=list(X=geno_centered, model=mod[1],
 probIn=prob[1],counts=count[1]))
  fmBB1<-BGLR(y=yNa,ETA=ETA,nIter=nIter,burnIn=burnIn)
  ## Bayes B pi .90
  ETA<-list(MRK=list(X=geno_centered, model=mod[2],
 probIn=prob[2],counts=count[2]))
  fmBB2<-BGLR(y=yNa,ETA=ETA,nIter=nIter,burnIn=burnIn)
  ## Bayes B pi .80
  ETA<-list(MRK=list(X=geno_centered, model=mod[3],
 probIn=prob[3],counts=count[3]))
  fmBB3<-BGLR(y=yNa,ETA=ETA,nIter=nIter,burnIn=burnIn)
  ## Bayes B pi .50
  ETA<-list(MRK=list(X=geno_centered, model=mod[4],
 probIn=prob[4],counts=count[4]))
  fmBB4<-BGLR(y=yNa,ETA=ETA,nIter=nIter,burnIn=burnIn)
  ## Bayes B pi .10
  ETA<-list(MRK=list(X=geno_centered, model=mod[5],
 probIn=prob[5],counts=count[5]))
  fmBB5<-BGLR(y=yNa,ETA=ETA,nIter=nIter,burnIn=burnIn)
  #### *  Calculate correlations
  r_BB99  <- cor(y[whichNa], fmBB1$yHat[whichNa] )
  r_BB90 <- cor(y[whichNa],fmBB2$yHat[whichNa] )
  r_BB80  <- cor(y[whichNa], fmBB3$yHat[whichNa] )
  r_BB50  <- cor(y[whichNa], fmBB4$yHat[whichNa] )
  r_BB10  <- cor(y[whichNa], fmBB5$yHat[whichNa] )
  # organize in a table
  table <- data.frame(rbind(r_BB99, r_BB90,r_BB80,r_BB50, r_BB10) ,
rbind(fmBB1$varE, fmBB2$varE, fmBB3$varE, fmBB5$varE, fmBB5$varE) ,
rbind(fmBB1$fit$pD,fmBB2$fit$pD,fmBB3$fit$pD,fmBB5$fit$pD,fmBB5$fit$p)
,
 rbind(fmBB1$fit$DIC,fmBB2$fit$DIC,fmBB3$fit$DIC,fmBB4$fit$DIC,
 fmBB5$fit$DIC)   )

colnames(table) <- c('PredAbi',"varE", "pD", "DIC")
rownames(table) <- c('BayesC_01',"BayesC_10","BayesC_20", "BayesC_50",
 "BayesC_90")
  table<-round(table, 3)
  results[[trait-1]]<-table
 }
return(results)
}
 rs<-PiInf()
 rs<-PiInf(count=c(0.02,0.02,0.02,0.02,0.02))
```

A call to the function with default values and results assigned to an object called `rs<-PiInf()` will produce the following results.

```
$TRAIT_1

          PredAbi    varE        pD        DIC
BayesC_01   0.481   0.450   106.764   1317.615
BayesC_10   0.456   0.367   170.406   1259.039
BayesC_20   0.445   0.350   183.514   1242.488
BayesC_50   0.428   0.332   197.749   1225.903
BayesC_90   0.426   0.332   197.749   1224.992

$TRAIT_2

          PredAbi    varE        pD        DIC
BayesC_01   0.817   1.971    15.438   2101.023
BayesC_10   0.820   1.838    66.254   2109.304
BayesC_20   0.807   1.604   149.464   2110.596
BayesC_50   0.584   1.996   253.924   2239.406
BayesC_90   0.463   1.996   253.924   2344.012

$TRAIT_3

          PredAbi    varE        pD        DIC
BayesC_01   0.618   1.354    66.554   1929.406
BayesC_10   0.553   1.189   151.510   1936.810
BayesC_20   0.498   1.129   211.090   1963.667
BayesC_50   0.421   1.313   206.500   2023.452
BayesC_90   0.391   1.313   206.500   2049.896

$TRAIT_4

          PredAbi    varE        pD        DIC
BayesC_01   0.555   0.594   146.064   1519.501
BayesC_10   0.545   0.501   199.313   1471.533
BayesC_20   0.508   0.487   210.120   1464.278
BayesC_50   0.461   0.506   205.252   1474.328
BayesC_90   0.439   0.506   205.252   1482.828
```

Once again for the first trait (completely polygenic) no significant differences are apparent for the different choices of the π_0 parameter. Conversely for the second trait (with two large-effect QTL), two things are evident. The first is that the tuned version of BayesC performs better with respect to the previous non tuned example (`BayesC 0.727`). Second, it is possible to see that the choice of π_0 does influence the results. Increasing π_0 from 0.01 to 0.90 reduces the predictive ability almost in half for this trait. At the lower values of π_0, the model will include the effects of fewer markers and allow their effects to be larger, which better matches the genetic architecture of this trait. Results for a larger number of QTL per traits are more nuanced but show the same pattern. Once again such striking results are due to the oversimplified simulated data employed here. Yet results show that different genetic architectures can be effectively accounted for by choosing and tuning the right model. Although we used only BayesC for this example, tuning the parameters for the other models could also improve their performance.

We can check the effect of the counts parameter (which models our "degree of belief" in π_0) by running a similar analysis with a much smaller value of the counts.

Code example 12.1
(continued)

```
rs<-PiInf(count=c(0.02,0.02,0.02,0.02,0.02))
```

The results are reported below.

$TRAIT_1

	PredAbi	varE	pD	DIC
BayesC_01	0.317	0.420	127.547	1295.687
BayesC_10	0.424	0.353	185.147	1248.054
BayesC_20	0.444	0.346	190.197	1241.106
BayesC_50	0.446	0.350	183.732	1237.405
BayesC_90	0.443	0.350	183.732	1242.333

$TRAIT_2

	PredAbi	varE	pD	DIC
BayesC_01	0.763	2.018	11.427	2110.254
BayesC_10	0.763	2.017	12.071	2110.518
BayesC_20	0.762	2.017	13.353	2111.469
BayesC_50	0.762	2.014	12.914	2110.567
BayesC_90	0.763	2.014	12.914	2110.674

$TRAIT_3

	PredAbi	varE	pD	DIC
BayesC_01	0.641	1.460	53.416	1962.122
BayesC_10	0.642	1.423	72.817	1966.113
BayesC_20	0.645	1.427	71.235	1965.750
BayesC_50	0.641	1.428	69.472	1964.425
BayesC_90	0.644	1.428	69.472	1965.036

$TRAIT_4

	PredAbi	varE	pD	DIC
BayesC_01	0.282	0.561	168.985	1508.213
BayesC_10	0.334	0.509	197.754	1478.732
BayesC_20	0.346	0.512	196.119	1478.667
BayesC_50	0.353	0.511	203.157	1481.997
BayesC_90	0.366	0.511	203.157	1485.769

In this case, focusing on the second trait, we see that by putting less "certainty" on π, the model results rely more heavily on the evidence provided by the current experiment (the likelihood) than the prior, so the results of models with very different values of π are more similar.

Cross-Validation

In the previous sections we have used a simple strategy to assess accuracy of prediction by splitting data into a single training and prediction set. To help avoid inflation of estimated prediction accuracy by hidden data structure, we should split the data into many subsets (k-folds) and evaluate prediction accuracy for many combinations of training and validation sets. Because the estimate of prediction ability is subject to sampling variation, more so when the training set is not large. In the code below we demonstrate how to randomly sample 60 individuals as the validation set for $k = 5$ folds for multiple traits, run a BayesC model on each training set, evaluate its fit on the appropriate validation set, and summarize model fit statistics over replicated folds.

**Code example 12.1
(continued)**

```r
###########################################
#11 CROSS VALIDATION
###########################################

model <- "BayesC"

# sampling of population for cross validation
samplesize <- 60
nfolds <- 5

## Matrix to store the summary of on each trait
tabRes  <- data.frame(row.names=c('PredAbi', 'RankCor', 'MSE',
'Bias', 'Best10'))

## data frame for sampled and predicted data
rawdf <- data.frame()

# data frame for summary statistics
tabdf <- data.frame()

tt<-Sim_tr[, -1]
trait<-names(tt)

# 1st Loop --> TRAITS
for(trait in trait)
{
  # Phenotype of a single trait
  y <- tt[, trait]

  # statistics for all folds of a trait are saved in ftab
  ftab <- data.frame()

  # 2nd Loop --> FOLDS
  for(fold in 1:nfolds)
  {
    # Sample from y starting from observation 1, using samplesize
    whichNa <- sample(1:length(y), size = samplesize, replace = T)
    yNa <- y               # make a copy of y. yNa and y are numeric
    yNa[whichNa] <- NA     # set the values of sampled trees to NA

# Run the model
    fmR <- BGLR(y = yNa, response_type ="gaussian", a=NULL, b=NULL,
          ETA = list(MRK = list(X = geno_centered, model = model)),
            nIter = 10000, burnIn = 1000, thin = 10, saveAt = "")

    # Data frame, combine predicted (yHat and observed y)
    df1 <- data.frame(rep(fold, samplesize), rep(trait, samplesize),
                    fmR$yHat[whichNa], y[whichNa])

    colnames(df1) <- c('fold', 'trait', 'GEBV', 'Phenotype')
```

```r
# Calculate statistics, put them in matrix 'tab' created above
  tab = list()
  tab$fold <- fold
  tab$trait <- trait
  tab$PredAbi <- round(cor(df1$Phenotype, df1$GEBV), 2) # PredAbi
  tab$RankCor <-round(cor(df1$Phenotype,df1$GEBV,method='spearman'), 2)
  tab$MSE <- round(mean((df1$Phenotype - df1$GEBV)^2), 2) # MSE
  tab$Bias <- round(coef(lm(y[whichNa] ~ fmR$yHat[whichNa]))[2], 2)
  tab$Best10 <- round(mean(tail(sort(df1$Phenotype), n =
ceiling(nrow(df1)*0.1))), 2) # Best 10%

  # Save df of GEBV and phenotype
      rawdf = rbind(rawdf, df1)
      # Save statistics
      ftab <- rbind(ftab, tab)
    }

    # Save tab results in a df
    tabdf <- rbind(tabdf, ftab)

    # Function tp create a summary for the folds
    lambda   <- function(x) paste(round(mean(x), 2), " (",
                                  round( min(x), 2), "-",
                                  round( max(x), 2), ")", sep = "")
    # Make a summary of tabs for each folds of the trait
    tabs <- data.frame()
    tabs['PredAbi', trait] <- lambda(ftab$PredAbi)
    tabs['RankCor', trait] <- lambda(ftab$RankCor)
    tabs['MSE', trait] <- lambda(ftab$MSE)
    tabs['Bias', trait] <- lambda(ftab$Bias)
    tabs['Best10',trait] <- lambda(ftab$Best10)

    # See results
    tabRes[, trait] <- tabs[, trait]
}

# Sample ebv and GEBV
colnames(rawdf) <- c('fold', 'trait', 'GEBV', 'Phenotype')
head(rawdf)

# Print final SUMMARY stats
tabRes
```

The summary statistics of the cross-validation analysis are provided below.

	Sim.Pheno	Sim.Pheno.2	Sim.Pheno.3
PredAbi	0.41 (0.27-0.54)	0.77 (0.72-0.83)	0.61 (0.58-0.67)
RankCor	0.39 (0.20-0.52)	0.72 (0.65-0.80)	0.58 (0.52-0.62)
MSE	0.50 (0.42-0.57)	2.12 (1.95-2.35)	1.57 (1.16-1.90)
Bias	1.15 (0.74-1.53)	1.07 (0.92-1.19)	1.04 (0.87-1.23)
Best10	11.39 (11.09-11.57)	55.03 (54.42-55.68)	87.99 (87.43-88.28)

	Sim.Pheno.4
PredAbi	0.44 (0.35-0.52)
RankCor	0.41 (0.32-0.49)
MSE	0.81 (0.64-1.03)
Bias	1.15 (0.93-1.53)
Best10	77.01 (76.67-77.43)

In this simple cross validation example the prediction ability for trait 'Sim.Pheno' ranged from 0.27 to 0.54 with a mean of 0.41. This wide variation in results suggests that in practice, we should conduct additional replications of the cross-validation to get a better estimate of the mean prediction ability.

Index of Figures

© Springer International Publishing AG 2017
F. Isik et al., *Genetic Data Analysis for Plant and Animal Breeding*, DOI 10.1007/978-3-319-55177-7

Literature Cited

Aguilar, I., Misztal, I., Johnson, D. L., Legarra, A., Tsuruta, S., & Lawlor, T. J. (2010). Hot topic: A unified approach to utilize phenotypic, full pedigree, and genomic information for genetic evaluation of Holstein final score. *Journal of Dairy Science, 93*, 743–752. doi:10.3168/jds.2009-2730.

Aguilar, I., Misztal, I., Legarra, A., & Tsuruta, S. (2011). Efficient computation of the genomic relationship matrix and other matrices used in single-step evaluation. *Journal of Animal Breeding and Genetics, 128*, 422–428. doi:10.1111/j.1439-0388.2010.00912.x.

Andrews, D. F., & Mallows, C. L. (1974). Scale mixtures of normal distributions. *Journal of the Royal Statistical Society: Series B: Methodological, 36*, 99–102.

Astle, W., & Balding, D. J. (2009). Population structure and cryptic relatedness in genetic association studies. *Statistical Science, 24*, 451–471. doi:10.1214/09-STS307.

Baker, R. J. (1978). Issues in diallel analysis. *Crop Science, 18*, 533. doi:10.2135/cropsci1978.0011183X001800040001x.

Bates, D., Maechler, M., Bolker, B., Walker, S. (2013). Fitting linear mixed-effects models using lme4. *Journal of Statistical Software, 67*(1), 1–48.

Botstein, D., White, R. L., Skolnick, M., & Davis, R. W. (1980). Construction of a genetic linkage map in man using restriction fragment length polymorphisms. *American Journal of Human Genetics, 32*, 314–331.

Brooks, S. P., & Roberts, G. O. (1998). Convergence assessment techniques for Markov chain Monte Carlo. *Statistics and Computing, 8*, 319–335. doi:10.1023/A:1008820505350.

Brownie, C., & Gumpertz, M. L. (1997). Validity of spatial analyses for large field trials. *Journal of Agricultural, Biological, and Environmental Statistics, 2*, 1–23.

Brownie, C., Bowman, D. T., & Burton, J. W. (1993). Estimating spatial variation in analysis of data from yield trials: A comparison of methods. *Agronomy Journal, 85*, 1244–1253.

Browning, S. R. (2006). Multilocus association mapping using variable-length markov chains. *The American Journal of Human Genetics, 78*, 903–913. doi:10.1086/503876.

Browning, S. R., & Browning, B. L. (2007). Rapid and accurate haplotype phasing and missing-data inference for whole-genome association studies by use of localized haplotype clustering. *The American Journal of Human Genetics, 81*, 1084–1097. doi:10.1086/521987.

Browning, S. R., & Browning, B. L. (2011). Haplotype phasing: Existing methods and new developments. *Nature Reviews. Genetics, 12*, 703–714. doi:10.1038/nrg3054.

Browning, B. L., & Browning, S. R. (2013). Improving the accuracy and efficiency of identity-by-descent detection in population data. *Genetics, 194*, 459–471. doi:10.1534/genetics.113.150029.

Browning, B. L., & Yu, Z. (2009). Simultaneous genotype calling and haplotype phasing improves genotype accuracy and reduces false-positive associations for genome-wide association studies. *The American Journal of Human Genetics, 85*, 847–861. doi:10.1016/j.ajhg.2009.11.004.

Buckler, E. S., Holland, J. B., Bradbury, P. J., Acharya, C. B., Brown, P. J., Browne, C., Ersoz, E., Flint-Garcia, S., Garcia, A., Glaubitz, J. C., Goodman, M. M., Harjes, C., Guill, K., Kroon, D. E., Larsson, S., Lepak, N. K., Li, H., Mitchell, S. E., Pressoir, G., Peiffer, J. A., Rosas, M. O., Rocheford, T. R., Romay, M. C., Romero, S., Salvo, S., Villeda, H. S., da Silva, H. S., Sun, Q., Tian, F., Upadyayula, N., Ware, D., Yates, H., Yu, J., Zhang, Z., Kresovich, S., & McMullen, M. D. (2009). The genetic architecture of maize flowering time. *Science, 325*, 714–718. doi:10.1126/science.1174276.

Burdon, R. D. (1977). Genetic correlation as a concept for studying genotype-environment interaction in forest tree breeding. *Silvae Genetica, 26*, 5–6.

Butler, D. G., Cullis, B. R., Gilmour, A. R., & Gogel, B. J. (2009). *Mixed models for S language environments ASReml-R reference manual*. Brisbane: The State of Queensland, Department of Primary Industries and Fisheries.

Chancerel, E., Lepoittevin, C., Provost, G. L., Lin, Y.-C., Jaramillo-Correa, J. P., Eckert, A. J., Wegrzyn, J. L., Zelenika, D., Boland, A., Frigerio, J.-M., Chaumeil, P., Garnier-Géré, P., Boury, C., Grivet, D., González-Martínez, S. C., Rouzé, P., de Peer, Y. V., Neale, D. B., Cervera, M. T., Kremer, A., & Plomion, C. (2011). Development and implementation of a highly-multiplexed SNP array for genetic mapping in maritime pine and comparative mapping with loblolly pine. *BMC Genomics, 12*, 368. doi:10.1186/1471-2164-12-368.

Chancerel, E., Lamy, J.-B., Lesur, I., Noirot, C., Klopp, C., Ehrenmann, F., Boury, C., Le Provost, G., Label, P., & Lalanne, C. (2013). High-density linkage mapping in a pine tree reveals a genomic region associated with inbreeding depression and provides clues to the extent and distribution of meiotic recombination. *BMC Biology, 11*, 50.

Christensen, O. F., & Lund, M. S. (2010). Genomic prediction when some animals are not genotyped. *Genetics Selection Evolution, 42*(1), 2.

Cleveland, M., Hickey, J., & Forni, S. (2012). A common dataset for genomic analysis of livestock populations. *G3 (Bethesda), 2*, 429–435. doi:10.1534/g3.111.001453.

Cochran, W. G., & Cox, G. M. (1957). *Experimental designs* (2nd ed.). New York: Wiley.

Cockerham, C. C. (1963). Estimation of genetic variances. In W. D. Hanson & H. F. Robinson (Eds.), *Statistical genetics and plant breeding* (pp. 53–94). Washington, DC: National Research Council.

© Springer International Publishing AG 2017

F. Isik et al., *Genetic Data Analysis for Plant and Animal Breeding*, DOI 10.1007/978-3-319-55177-7

Cooper, M., & DeLacy, I. H. (1994). Relationships among analytical methods used to study genotypic variation and genotype-by-environment interaction in plant breeding multi-environment experiments. *Theoretical and Applied Genetics, 88*, 561–572. doi:10.1007/BF01240919.

Crossa, J., Burgueño, J., Cornelius, P. L., McLaren, G., Trethowan, R., & Krishnamachari, A. (2006). Modeling genotype × environment interaction using additive genetic covariances of relatives for predicting breeding values of wheat genotypes. *Crop Science, 46*, 1722. doi:10.2135/cropsci2005.11-0427.

Cullis, B., Gogel, B., Verbyla, A., & Thompson, R. (1998). Spatial analysis of multi-environment early generation variety trials. *Biometrics, 54*, 1–18.

Cullis, B. R., Smith, A. B., & Coombes, N. E. (2006). On the design of early generation variety trials with correlated data. *JABES, 11*, 381–393. doi:10.1198/108571106X154443.

Cullis, B. R., Smith, A. B., Beeck, C. P., & Cowling, W. A. (2010). Analysis of yield and oil from a series of canola breeding trials. Part II. Exploring variety by environment interaction using factor analysis. This article is one of a selection of papers from the conference "exploiting genome-wide association in oilseed Brassicas: a model for genetic improvement of major OECD crops for sustainable farming". *Genome, 53*, 1002–1016. doi:10.1139/G10-080.

Cullis, B. R., Jefferson, P., Thompson, R., & Smith, A. B. (2014). Factor analytic and reduced animal models for the investigation of additive genotype-by-environment interaction in outcrossing plant species with application to a *Pinus radiata* breeding programme. *Theoretical and Applied Genetics, 127*, 2193–2210. doi:10.1007/s00122-014-2373-0.

de los Campos, G., Naya, H., Gianola, D., Crossa, J., Legarra, A., Manfredi, E., Weigel, K., & Cotes, J. M. (2009). Predicting quantitative traits with regression models for dense molecular markers and pedigree. *Genetics, 182*, 375–385. doi:10.1534/genetics.109.101501.

de los Campos, G., Gianola, D., Rosa, G. J., Weigel, K. A., & Crossa, J. (2010). Semi-parametric genomic-enabled prediction of genetic values using reproducing kernel Hilbert spaces methods. *Genetics Research, 92*, 295–308.

de los Campos, G., Hickey, J. M., Pong-Wong, R., Daetwyler, H. D., & Calus, M. P. L. (2013). Whole-genome regression and prediction methods applied to plant and animal breeding. *Genetics, 193*, 327–345. doi:10.1534/genetics.112.143313.

Dickerson, G. E. (1969). Techniques for research in quantitative animal genetics. In *Techniques and procedures in animal science research* (pp. 36–79).

Duchemin, S. I., Colombani, C., Legarra, A., Baloche, G., Larroque, H., Astruc, J. M., ... & Manfredi, E. (2012). Genomic selection in the French Lacaune dairy sheep breed. *Journal of dairy science, 95*(5), 2723–2733.

Dudley, J. W., & Moll, R. H. (1969). Interpretation and use of estimates of heritability and genetic variances in plant breeding1. *Crop Science, 9*, 257. doi:10.2135/cropsci1969.0011183X000900030001x.

Endelman, J. B. (2011). Ridge regression and other kernels for genomic selection with R package rrBLUP. *The Plant Genome Journal, 4*, 250. doi:10.3835/plantgenome2011.08.0024.

Endelman, J. B., & Jannink, J.-L. (2012). Shrinkage estimation of the realized relationship matrix. *G3, 2*, 1405–1413. doi:10.1534/g3.112.004259.

Falconer, D. S, & Mackay, T. F. (1996). Introduction to Quantitative Genetics (4th edition). 480 p. Longman Group Ltd.

Falconer, D. S., Mackay, T. F., & Frankham, R. (1996). Introduction to quantitative genetics. (4th edn). *Trends in Genetics, 12*, 280.

Federer, W. T., & Raghavarao, D. (1975). On augmented designs. *Biometrics, 31*, 29–35. doi:10.2307/2529707.

Fisher, R. A. (1918). 009: The Correlation Between Relatives on the Supposition of Mendelian Inheritance.

Flint-Garcia, S. A., Thornsberry, J. M., & Buckler IV, E. S. (2003). Structure of linkage disequilibrium in plants. *Annual Review of Plant Biology, 54*, 357–374. doi:10.1146/annurev.arplant.54.031902.134907.

Forni, S., Aguilar, I., & Misztal, I. (2011). Different genomic relationship matrices for single-step analysis using phenotypic, pedigree and genomic information. *Genetics Selection Evolution, 43*, 1.

Garrick, D. J., Taylor, J. F., Fernando, R. L., et al. (2009). Deregressing estimated breeding values and weighting information for genomic regression analyses. *Genetics, Selection, Evolution, 41*, 10–1186.

Gelman, A., & Rubin, D. B. (1992). Inference from iterative simulation using multiplesequences. *Statistical Science, 7*, 457–472.

Gianola, D. (2013). Priors in whole-genome regression: The Bayesian alphabet returns. *Genetics, 194*, 573–596.

Gianola, D., de los Campos, G., Hill, W. G., Manfredi, E., & Fernando, R. (2009). Additive genetic variability and the Bayesian alphabet. *Genetics, 183*, 347–363. doi:10.1534/genetics.109.103952.

Gilmour, A. R., Cullis, B. R., & Verbyla, A. P. (1997). Accounting for natural and extraneous variation in the analysis of field experiments. *Journal of Agricultural, Biological, and Environmental Statistics, 2*, 269–293.

Gilmour, A. R., Gogel, B., Cullis, B., & Thompson, R. (2009). *ASReml user's guide. Release 3.0*. Hemel Hempstead: VSN International, Ltd.

Gilmour, A. R., Gogel, B. J., Cullis, B. R., Welham, C. J., & Thompson, R. (2014). *ASReml user guide release 4.1, functional specification*. Hemel Hempstead: VSN International Ltd. www.vsni.co.uk.

Grattapaglia, D., & Resende, M. D. (2011). Genomic selection in forest tree breeding. *Tree Genetics & Genomes, 7*, 241–255.

Grattapaglia, D., Ribeiro, V. J., & Rezende, G. D. S. P. (2004). Retrospective selection of elite parent trees using paternity testing with microsatellite markers: An alternative short term breeding tactic for Eucalyptus. *Theoretical and Applied Genetics, 109*, 192–199. doi:10.1007/s00122-004-1617-9.

Habier, D., Fernando, R. L., & Dekkers, J. C. M. (2007). The impact of genetic relationship information on genome-assisted breeding values. *Genetics, 177*, 2389–2397. doi:10.1534/genetics.107.081190.

Habier, D., Fernando, R. L., & Dekkers, J. C. M. (2009). Genomic selection using low-density marker panels. *Genetics, 182*, 343–353. doi:10.1534/genetics.108.100289.

Habier, D., Fernando, R. L., Kizilkaya, K., & Garrick, D. J. (2011). Extension of the bayesian alphabet for genomic selection. *BMC Bioinformatics, 12*, 186. doi:10.1186/1471-2105-12-186.

Habier, D., Fernando, R. L., & Garrick, D. J. (2013). Genomic BLUP decoded: A look into the black box of genomic prediction. *Genetics, 194*, 597–607. doi:10.1534/genetics.113.152207.

Hallauer, A. R., & Miranda, J. B. (1988). *Quantitative genetics in maize breeding* (2nd ed.). Ames: Iowa State Univ. Press.

Hallauer, A.R., Carena, M.J., Filho, J.B.M. (2010). Testers and combining ability. In *Quantitative genetics in maize breeding, handbook of plant breeding* (pp. 383–423). New York: Springer. doi:10.1007/978-1-4419-0766-0_8.

Hastie, T., Tibshirani, R., & Friedman, J. (2009). *The elements of statistical learning data mining, inference, and prediction* (2nd ed.). New York: Springer.

Hayes, B. J., Bowman, P. J., Chamberlain, A. J., & Goddard, M. E. (2009). Invited review: Genomic selection in dairy cattle: Progress and challenges. *Journal of Dairy Science, 92*, 433–443. doi:10.3168/jds.2008-1646.

Hayes, B., & Goddard, M. (2010). Genome-wide association and genomic selection in animal breeding This article is one of a selection of papers from the conference "Exploiting Genome-wide Association in Oilseed Brassicas: a model for genetic improvement of major OECD crops for sustainable farming". *Genome, 53*(11), 876–883.

Henderson, C. R. (1949). Estimation of changes in herd environment. *Journal of Dairy Science, 32*, 706.

Henderson, C. R. (1990). Statistical methods in animal improvement: Historical overview. In P. D. D. Gianola & D. K. Hammond (Eds.), *Advances in statistical methods for genetic improvement of livestock, advanced series in agricultural sciences* (pp. 2–14). Berlin: Springer. doi:10.1007/978-3-642-74487-7_1.

Heslot, N., Yang, H.-P., Sorrells, M. E., & Jannink, J.-L. (2012). Genomic selection in plant breeding: A comparison of models. *Crop Science, 52*, 146. doi:10.2135/cropsci2011.06.0297.

Hickey, J. M., Kinghorn, B. P., Tier, B., van der Werf, J. H., & Cleveland, M. A. (2012). A phasing and imputation method for pedigreed populations that results in a single-stage genomic evaluation. *Genetics, Selection, Evolution, 44*, 10–1186.

Hill, W. G. (1984). On selection among groups with heterogeneous variance. *Animal Science, 39*, 473–477. doi:10.1017/S0003356100032220.

Hill, W. G., Goddard, M. E., & Visscher, P. M. (2008). Data and theory point to mainly additive genetic variance for complex traits. *PLoS Genetics, 4*, e1000008. doi:10.1371/journal.pgen.1000008.

Hoerl, A. E., & Kennard, R. W. (1970). Ridge regression: Biased estimation for nonorthogonal problems. *Technometrics, 12*, 55–67. doi:10.1080/00401706.1970.10488634.

Holland, J. B. (2006). Estimating genotypic correlations and their standard errors using multivariate restricted maximum likelihood estimation with SAS Proc MIXED. *Crop Science, 46*, 642–654.

Holland, J. B., Nyquist, W. E., & Cervantes-Martinez, C. T. (2003). Estimating and interpreting heritability for plant breeding: An update. *Plant Breeding Reviews, 22*, 9–111.

Horne, D. W., Eller, M. S., & Holland, J. B. (2016). Responses to recurrent index selection for reduced fusarium ear rot and lodging and for increased yield in maize. *Crop Science, 56*, 85–94. doi:10.2135/cropsci2015.06.0333.

Howie, B. N., Donnelly, P., & Marchini, J. (2009). A flexible and accurate genotype imputation method for the next generation of genome-wide association studies. *PLoS Genetics, 5*, e1000529. doi:10.1371/journal.pgen.1000529.

Huang, Y., Hickey, J. M., Cleveland, M. A., & Maltecca, C. (2012). Assessment of alternative genotyping strategies to maximize imputation accuracy at minimal cost. *Genetics, Selection, Evolution, 44*, 25.

Isik, F., Isik, K., Lee, S. (1999). Genetic variation in *Pinus brutia* ten. In *Turkey*: I. *Growth, biomass and stem quality trats. International Journal of Forest Genetics, 6*, 89–99.

Isik, F., Li, B., Frampton, J., & Goldfarb, B. (2004). Efficiency of seedlings and rooted cuttings for testing and selection in *Pinus taeda. Forest Science, 50*, 44–53.

Isik, F., Boos, D. D., & Li, B. (2005). The distribution of genetic parameter estimates and confidence intervals from small disconnected diallels. *Theoretical and Applied Genetics, 110*, 1236–1243. doi:10.1007/s00122-005-1957-0.

Isik, F., Bartholomé, J., Farjat, A., Chancerel, E., Raffin, A., Sanchez, L., Plomion, C., & Bouffier, L. (2016). Genomic selection in maritime pine. *Plant Science, 242*, 108–119.

John, J. A., & Eccleston, J. A. (1986). Row-column α-designs. *Biometrika, 73*, 301–306. doi:10.1093/biomet/73.2.301.

Kehel, Z., Habash, D. Z., Gezan, S. A., Welham, S. J., & Nachit, M. M. (2010). Estimation of spatial trend and automatic model selection in augmented designs. *Agronomy Journal, 102*, 1542. doi:10.2134/agronj2010.0175.

Kenward, M. G., & Roger, J. H. (1997). Small sample inference for fixed effects from restricted maximum likelihood. *Biometrics, 53*, 983–997. doi:10.2307/2533558.

Kenward, M. G., & Roger, J. H. (2009). An improved approximation to the precision of fixed effects from restricted maximum likelihood. *Computational Statistics and Data Analysis, 53*, 2583–2595. doi:10.1016/j.csda.2008.12.013.

Legarra, A., & Ducrocq, V. (2012). Computational strategies for national integration of phenotypic, genomic, and pedigree data in a single-step best linear unbiased prediction. *Journal of Dairy Science, 95*, 4629–4645. doi:10.3168/jds.2011-4982.

Legarra, A., Robert-Granié, C., Manfredi, E., & Elsen, J.-M. (2008). Performance of genomic selection in mice. *Genetics, 180*, 611–618. doi:10.1534/genetics.108.088575.

Legarra, A., Aguilar, I., & Misztal, I. (2009). A relationship matrix including full pedigree and genomic information. *Journal of Dairy Science, 92*, 4656–4663. doi:10.3168/jds.2009-2061.

Lewontin, R. C. (1964). The interaction of selection and linkage. I. General considerations; heterotic models. *Genetics, 49*, 49–67.

Li, Y., Willer, C. J., Ding, J., Scheet, P., & Abecasis, G. R. (2010). MaCH: Using sequence and genotype data to estimate haplotypes and unobserved genotypes. *Genetic Epidemiology, 34*, 816–834. doi:10.1002/gepi.20533.

Lynch, M., & Walsh, B. (1998). *Genetics and analysis of quantitative traits* (1st ed.). Sunderland: Sinauer Associates.

Maltecca, C., Parker, K. L., & Cassady, J. P. (2012). Application of multiple shrinkage methods to genomic predictions. *Journal of Animal Science, 90*, 1777–1787. doi:10.2527/jas.2011-4350.

Mangin, B., Siberchicot, A., Nicolas, S., Doligez, A., This, P., & Cierco-Ayrolles, C. (2012). Novel measures of linkage disequilibrium that correct the bias due to population structure and relatedness. *Heredity, 108*, 285–291. doi:10.1038/hdy.2011.73.

Marchini, J., & Howie, B. (2010). Genotype imputation for genome-wide association studies. *Nature Reviews. Genetics, 11*, 499–511. doi:10.1038/nrg2796.

McKeand, S. E., & Bridgwater, F. E. (1998). A strategy for the third breeding cycle of loblolly pine in the southeastern US. *Silvae Genetica, 47*, 223–234.

Meuwissen, T. H. E., Hayes, B. J., & Goddard, M. E. (2001). Prediction of total genetic value using genome-wide dense marker maps. *Genetics, 157*, 1819–1829.

Meyer, K. (2009). Factor-analytic models for genotype x environment type problems and structured covariance matrices. *Genetics Selection Evolution, 41*, 21.

Milliken, G. A., & Johnson, D. E. (2004). *Analysis of messy data, Designed Experiments* (Vol. 1). Boca Raton: Chapman & Hall/CRC.

Misztal, I., Legarra, A., & Aguilar, I. (2009). Computing procedures for genetic evaluation including phenotypic, full pedigree, and genomic information. *Journal of Dairy Science, 92*, 4648–4655. doi:10.3168/jds.2009-2064.

Möhring, J., & Piepho, H.-P. (2009). Comparison of weighting in two-stage analysis of plant breeding trials. *Crop Science, 49*, 1977. doi:10.2135/cropsci2009.02.0083.

Möhring, J., Melchinger, A. E., & Piepho, H. P. (2011). REML-based diallel analysis. *Crop Science, 51*, 470. doi:10.2135/cropsci2010.05.0272.

Money, D., Gardner, K., Migicovsky, Z., Schwaninger, H., Zhong, G.-Y., & Myles, S. (2015). LinkImpute: Fast and accurate genotype imputation for nonmodel organisms. *G3: Genes|Genomes|Genetics, 5*, 2383–2390.

Mrode, R. A. (2014). *Linear models for the prediction of animal breeding values* (3rd ed.). Boston: CABI.

Müller, B. U., Schützenmeister, A., & Piepho, H.-P. (2010). Arrangement of check plots in augmented block designs when spatial analysis is used. *Plant Breeding, 129*, 581–589. doi:10.1111/j.1439-0523.2010.01803.x.

Mullin, T., Andersson, B., Bastien, J. -C., Beaulieu, J., Burdon, R. D., Dvorak, W. S., King, J. N., Kondo, T., Krakowski, J., Lee, S. J., McKeand, S. E., Pâques, L., Raffin, A., Russell, J. H., Skr?ppa, T., Stoehr, M., Yanchuk, A. (2011). Economic importance, breeding objectives and achievements. In *Genetics, genomics and breeding of conifers* (pp. 40–127). Science Publishers.

Nyquist, W. E. (1991). Estimation of heritability and prediction of selection response in plant populations. *Critical Reviews in Plant Sciences, 10*, 235–322.

Ogut, F., Maltecca, C., Whetten, R., McKeand, S., & Isik, F. (2014). Genetic analysis of diallel progeny test data using factor analytic linear mixed models. *Forest Science, 60*(1), 119–127.

Park, T., & Casella, G. (2008). The Bayesian Lasso. *Journal of the American Statistical Association, 103*, 681–686. doi:10.1198/016214508000000337.

Patterson, H. D., & Williams, E. R. (1976). A new class of resolvable incomplete block designs. *Biometrika, 63*, 83–92. doi:10.1093/biomet/63.1.83.

Patterson, B., Vaillancourt, R. E., Pilbeam, D. J., & Potts, B. M. (2004). Factors affecting variation in outcrossing rate in *Eucalyptus globulus*. *Australian Journal of Botany, 52*, 773–780.

Pérez, P., & de Los Campos, G. (2014). Genome-wide regression & prediction with the BGLR statistical package. *Genetics, 198*(2), 483–495.

Pérez, P., de los Campos, G., Crossa, J., & Gianola, D. (2010). Genomic-enabled prediction based on molecular markers and pedigree using the Bayesian linear regression package in R. *The Plant Genome Journal, 3*, 106. doi:10.3835/plantgenome2010.04.0005.

Piepho, H.-P., & Möhring, J. (2007). Computing heritability and selection response from unbalanced plant breeding trials. *Genetics, 177*, 1881–1888. doi:10.1534/genetics.107.074229.

Piepho, H.-P., Mohring, J., Melchinger, A. E., & Buscher, A. (2007). BLUP for phenotypic selection in plant breeding and variety testing. *Euphytica, 161*, 209–228.

Plomion, C., Chancerel, E., Endelman, J., Lamy, J.-B., Mandrou, E., Lesur, I., Ehrenmann, F., Isik, F., Bink, M. C., Bouffier, L., et al. (2014). Genome-wide distribution of genetic diversity and linkage disequilibrium in a mass-selected population of maritime pine. *BMC Genomics, 15*, 171.

Plummer, M., Best, N., Cowles, K., & Vines, K. (2006). CODA: Convergence diagnosis and output analysis for MCMC. *R News, 6*, 7–11.

Powell, J. E., Visscher, P. M., & Goddard, M. E. (2010). Reconciling the analysis of IBD and IBS in complex trait studies. *Nature Reviews Genetics, 11*, 800–805. doi:10.1038/nrg2865.

Qiao, C. G., Basford, K. E., DeLacy, I. H., & Cooper, M. (2000). Evaluation of experimental designs and spatial analyses in wheat breeding trials. *TAG Theoretical and Applied Genetics, 100*, 9–16. doi:10.1007/s001220050002.

Qiao, C. G., Basford, K. E., DeLacy, I. H., & Cooper, M. (2004). Advantage of single-trial models for response to selection in wheat breeding multi-environment trials. *Theoretical and Applied Genetics, 108*, 1256–1264.

Quaas, R. L., & Pollak, E. J. (1981). Modified equations for sire models with groups. *Journal of Dairy Science, 64*, 1868–1872. doi:10.3168/jds.S0022-0302(81)82778-6.

Rawlings, J. O., Pantula, S., & Dickey, D. A. (2001). *Applied regression analysis: A research tool*. New York: Springer.

Reich, D. E., Cargill, M., Bolk, S., Ireland, J., Sabeti, P. C., Richter, D. J., Lavery, T., Kouyoumjian, R., Farhadian, S. F., Ward, R., & Lander, E. S. (2001). Linkage disequilibrium in the human genome. *Nature, 411*, 199–204. doi:10.1038/35075590.

Ripke, S., Isaacs, A., & van Duijn, C. M. (2007). GenABEL: An R library for genome-wide association analysis. *Bioinformatics, 23*, 1294–1296. doi:10.1093/bioinformatics/btm108.

Robinson, G. K. (1991). That BLUP is a good thing: The estimation of random effects. *Statistical Science, 6*, 15–32.

Rodríguez-Álvarez, M. X., Boer, M. P., van Eeuwijk, F. A., Eilers, P. H. C. (2016). Spatial models for field trials. arXiv:1607.08255 [stat].

Ron, D., Singer, Y., Tishby, N. (1995). On the learnability and usage of acyclic probabilistic finite automata. In *Journal of computer and system sciences* (pp. 31–40). ACM Press.

Ruppert, D., Wand, M. P., & Carroll, R. J. (2003). *Semiparametric regression*. Cambridge, UK: Cambridge University Press.

Rutkoski, J. E., Poland, J., Jannink, J.-L., & Sorrells, M. E. (2013). Imputation of unordered markers and the impact on genomic selection accuracy. *G3, 3*, 427–439. doi:10.1534/g3.112.005363.

Saatchi, M., McClure, M. C., McKay, S. D., Rolf, M. M., Kim, J., Decker, J. E., Taxis, T. M., Chapple, R. H., Ramey, H. R., Northcutt, S. L., Bauck, S., Woodward, B., Dekkers, J. C., Fernando, R. L., Schnabel, R. D., Garrick, D. J., & Taylor, J. F. (2011). Accuracies of genomic breeding values in American Angus beef cattle using K-means clustering for cross-validation. *Genetics Selection Evolution, 43*, 40. doi:10.1186/1297-9686-43-40.

Sargolzaei, M., Chesnais, J. P., & Schenkel, F. S. (2014). A new approach for efficient genotype imputation using information from relatives. *BMC Genomics, 15*, 478. doi:10.1186/1471-2164-15-478.

SAS Institute, Inc. (2011a). The MIXED procedure. In *SAS/STAT(R) 9.3 user's guide*.

SAS Institute, Inc. (2011b). The HPMIXED procedure. In *SAS/STAT(R) 9.3 user's guide*.

Scheet, P., & Stephens, M. (2006). A fast and flexible statistical model for large-scale population genotype data: Applications to inferring missing genotypes and haplotypic phase. *American Journal of Human Genetics, 78*, 629–644. doi:10.1086/502802.

Self, S. G., & Liang, K.-Y. (1987). Asymptotic properties of maximum likelihood estimators and likelihood ratio tests under nonstandard conditions. *Journal of the American Statistical Association, 82*, 605–610. doi:10.2307/2289471.

Shin, J.-H., Blay, S., McNeney, B., & Graham, J. (2006). LDheatmap: An R function for graphical display of pairwise linkage disequilibria between single nucleotide polymorphisms. *Journal of Statistical Software, 16*, 1–10.

Smith, A., Cullis, B., & Gilmour, A. (2001). The analysis of crop variety evaluation data in Australia. *Australian & New Zealand Journal of Statistics, 43*, 129–145.

Smith, A., Cullis, B., & Thompson, R. (2002). Exploring variety-environment data using random effects AMMI models with adjustments for spatial field trend: Part 1: Theory. In M. S. Kang (Ed.), *Quantitative genetics, genomics, and plant breeding* (pp. 323–335). Wallingford: CABI.

Smith, A. B., Cullis, B. R., & Thompson, R. (2005). The analysis of crop cultivar breeding and evaluation trials: An overview of current mixed model approaches. *The Journal of Agricultural Science, 143*, 449–462. doi:10.1017/S0021859605005587.

Smith, A. B., Ganesalingam, A., Kuchel, H., & Cullis, B. R. (2014). Factor analytic mixed models for the provision of grower information from national crop variety testing programs. *Theoretical and Applied Genetics, 128*, 55–72. doi:10.1007/s00122-014-2412-x.

Sorensen, D., & Gianola, D. (2007). *Likelihood, Bayesian, and MCMC methods in quantitative genetics*. New York: Springer Science & Business Media.

Speed, D., & Balding, D. J. (2015). Relatedness in the post-genomic era: Is it still useful? *Nature Reviews Genetics, 16*, 33–44. doi:10.1038/nrg3821.

Spiegelhalter, D. J., Best, N. G., Carlin, B. P., & Van Der Linde, A. (2002). Bayesian measures of model complexity and fit. *Journal of the Royal Statistical Society, Series B: Statistical Methodology, 64*, 583–639. doi:10.1111/1467-9868.00353.

Steel, R. G. D., Torrie, J. H., & Dickey, D. A. (1997). *Principles and procedures of statistics. A biometrical approach*. New York: McGraw-Hill.

Thompson, R., Cullis, B., Smith, A., & Gilmour, A. (2003). A sparse implementation of the average information algorithm for factor analytic and reduced rank variance models. *Australian & New Zealand Journal of Statistics, 45*, 445–459. doi:10.1111/1467-842X.00297.

Tibshirani, R. (1996). Regression shrinkage and selection via the Lasso. *Journal of the Royal Statistical Society: Series B: Methodological, 58*, 267–288.

Tiezzi, F., Maltecca, C., Cecchinato, A., & Bittante, G. (2015). Comparison between different statistical models for the prediction of direct genetic component on embryo establishment and survival in Italian Brown Swiss dairy cattle. *Livestock Science, 180*, 6–13.

Toft, N., Innocent, G. T., Gettinby, G., & Reid, S. W. J. (2007). Assessing the convergence of Markov Chain Monte Carlo methods: An example from evaluation of diagnostic tests in absence of a gold standard. *Preventive Veterinary Medicine, 79*, 244–256. doi:10.1016/j.prevetmed.2007.01.003.

VanRaden, P. M. (2008). Efficient methods to compute genomic predictions. *Journal of Dairy Science, 91*, 4414–4423.

VanRaden, P. M., Van Tassell, C. P., Wiggans, G. R., Sonstegard, T. S., Schnabel, R. D., Taylor, J. F., & Schenkel, F. S. (2009). Invited review: Reliability of genomic predictions for North American Holstein bulls. *Journal of dairy science, 92*(1), 16–24.

VanRaden, P. M., Null, D. J., Sargolzaei, M., Wiggans, G. R., Tooker, M. E., Cole, J. B., Sonstegard, T. S., Connor, E. E., Winters, M., van Kaam, J. B. C. H. M., Valentini, A., Van Doormaal, B. J., Faust, M. A., & Doak, G. A. (2013). Genomic imputation and evaluation using high-density Holstein genotypes. *Journal of Dairy Science, 96*, 668–678. doi:10.3168/jds.2012-5702.

Warnes, G., Gorjanc, G., Leisch, F., & Man, M. (2013). Genetics: a package for population genetics. R Package (Version 1.3. 8.1). R Foundation. Accessed 2016 Oct 1.

Welham, S. J., Gogel, B. J., Smith, A. B., Thompson, R., & Cullis, B. R. (2010). A comparison of analysis methods for late-stage variety evaluation trials. *Australian & New Zealand Journal of Statistics, 52*, 125–149. doi:10.1111/j.1467-842X.2010.00570.x.

White, T. L., Adams, T. W., & Neale, D. B. (2007). *Forest genetics*. Cambridge, MA: CABI.

Wickham, H. (Ed.). (2010). *ggplot2: Elegant graphics for data analysis* (1. 2009. Corr. 3rd printing 2010 edition. ed.). New York: Springer.

Wimmer, V., Albrecht, T., Auinger, H.-J., & Schön, C.-C. (2012). Synbreed: A framework for the analysis of genomic prediction data using R. *Bioinformatics, 28*, 2086–2087. doi:10.1093/bioinformatics/bts335.

Wolc, A., Stricker, C., Arango, J., Settar, P., Fulton, J. E., O'Sullivan, N. P., ... & Lamont, S. J. (2011). Breeding value prediction for production traits in layer chickens using pedigree or genomic relationships in a reduced animal model. *Genetics Selection Evolution, 43*(1), 5.

Wright, S. (1922). Coefficients of inbreeding and relationship. *The American Naturalist, 56*, 330–338.

Yamada, Y. (1962). Genotype by environment interaction and genetic correlation of the same trait under different environments. *The Japanese Journal of Genetics, 37*, 498–509. doi:10.1266/jjg.37.498.

Yang, R.-C., Crossa, J., Cornelius, P. L., & Burgueño, J. (2009). Biplot analysis of genotype × environment interaction: Proceed with caution. *Crop Science, 49*, 1564–1576. doi:10.2135/cropsci2008.11.0665.

Yang, J., Benyamin, B., McEvoy, B. P., Gordon, S., Henders, A. K., Nyholt, D. R., Madden, P. A., Heath, A. C., Martin, N. G., Montgomery, G. W., Goddard, M. E., & Visscher, P. M. (2010). Common SNPs explain a large proportion of the heritability for human height. *Nature Genetics, 42*, 565–569. doi:10.1038/ng.608.

Yang, J., Lee, S. H., Goddard, M. E., & Visscher, P. M. (2011). GCTA: A tool for genome-wide complex trait analysis. *The American Journal of Human Genetics, 88*, 76–82. doi:10.1016/j.ajhg.2010.11.011.

Yi, N., & Xu, S. (2008). Bayesian LASSO for quantitative trait loci mapping. *Genetics, 179*, 1045–1055.

Yu, J., Holland, J. B., McMullen, M. D., & Buckler, E. S. (2008). Genetic design and statistical power of nested association mapping in maize. *Genetics, 178*, 539–551. doi:10.1534/genetics.107.074245.

Zapata-Valenzuela, J., Whetten, R. W., Neale, D., McKeand, S., & Isik, F. (2013). Genomic estimated breeding values using genomic relationship matrices in a cloned population of loblolly pine. *G3, 3*, 909–916. doi:10.1534/g3.113.005975.

Index

Printed in the United States
By Bookmasters